Methods in Enzymology

Volume 237
HETEROTRIMERIC G PROTEINS

METHODS IN ENZYMOLOGY

EDITORS-IN-CHIEF

John N. Abelson Melvin I. Simon

DIVISION OF BIOLOGY
CALIFORNIA INSTITUTE OF TECHNOLOGY
PASADENA, CALIFORNIA

FOUNDING EDITORS

Sidney P. Colowick and Nathan O. Kaplan

Methods in Enzymology

Volume 237

Heterotrimeric G Proteins

EDITED BY

Ravi Iyengar

DEPARTMENT OF PHARMACOLOGY
MOUNT SINAI SCHOOL OF MEDICINE
NEW YORK, NEW YORK

ACADEMIC PRESS
A Division of Harcourt Brace & Company
San Diego New York Boston London Sydney Tokyo Toronto

This book is printed on acid-free paper.

Copyright © 1994 by ACADEMIC PRESS, INC.

All Rights Reserved.

No part of this publication may be reproduced or transmitted in any form or by any means, electronic or mechanical, including photocopy, recording, or any information storage and retrieval system, without permission in writing from the publisher.

Academic Press, Inc.
525 B Street, Suite 1900, San Diego, California 92101-4495

United Kingdom Edition published by
Academic Press Limited
24–28 Oval Road, London NW1 7DX

International Standard Serial Number: 0076-6879

International Standard Book Number: 0-12-182138-2

PRINTED IN THE UNITED STATES OF AMERICA
94 95 96 97 98 99   MM   9 8 7 6 5 4 3 2 1

Table of Contents

CONTRIBUTORS TO VOLUME 237 ix

PREFACE xv

VOLUMES IN SERIES xvii

Section I. Gα Subunits

1. Measurement of Receptor-Stimulated Guanosine 5'-O-(γ-Thio)triphosphate Binding by G Proteins	THOMAS WIELAND AND KARL H. JAKOBS	1
2. Receptor-Stimulated Hydrolysis of Guanosine 5'-Triphosphate in Membrane Preparations	PETER GIERSCHIK, THOMAS BOUILLON, AND KARL H. JAKOBS	13
3. Regulation of G-Protein Activation by Mastoparans and Other Cationic Peptides	ELLIOTT M. ROSS AND TSUTOMU HIGASHIJIMA	26
4. Guanosine 5'-O-(γ-Thio)triphosphate Binding Assay for Solubilized G Proteins	DONNA J. CARTY AND RAVI IYENGAR	38
5. Activation of Cholera Toxin by ADP-Ribosylation Factors: 20-kDa Guanine Nucleotide-Binding Proteins	JOEL MOSS, RANDY S. HAUN, SU-CHEN TSAI, CATHERINE F. WELSH, FANG-JEN SCOTT LEE, S. RUSS PRICE, AND MARTHA VAUGHAN	44
6. Pertussis Toxin-Catalyzed ADP-Ribosylation of G Proteins	DONNA J. CARTY	63
7. Synthesis and Use of Radioactive Photoactivatable NAD$^+$ Derivatives as Probes for G-Protein Structure	RICHARD R. VAILLANCOURT, N. DHANASEKARAN, AND ARNOLD E. RUOHO	70
8. Photoaffinity Guanosine 5'-Triphosphate Analogs as a Tool for the Study of GTP-Binding Proteins	MARK M. RASENICK, MADHAVI TALLURI, AND WILLIAM J. DUNN III	100
9. Preparation of Activated α Subunits of G_s and G_is: From Erythrocyte to Activated Subunit	LUTZ BIRNBAUMER, DAGOBERTO GRENET, FERNANDO RIBEIRO-NETO, AND JUAN CODINA	110

10. Purification and Separation of Closely Related Members of Pertussis Toxin–Substrate G Proteins	TOSHIAKI KATADA, KENJI KONTANI, ATSUSHI INANOBE, ICHIRO KOBAYASHI, YOSHIHARU OHOKA, HIROSHI NISHINA, AND KATSUNOBU TAKAHASHI	131
11. Purification of Transducin	JOËLLE BIGAY AND MARC CHABRE	139
12. Expression of G-Protein α Subunits in *Escherichia coli*	ETHAN LEE, MAURINE E. LINDER, AND ALFRED G. GILMAN	146
13. Synthesis and Applications of Affinity Matrix Containing Immobilized $\beta\gamma$ Subunits of G Proteins	IOK-HOU PANG, ALAN V. SMRCKA, AND PAUL C. STERNWEIS	164
14. Purification of Activated and Heterotrimeric Forms of G_q Proteins	JONATHAN L. BLANK AND JOHN H. EXTON	174
15. Purification of Phospholipase C-Activating G Protein, G_{11}, from Turkey Erythrocytes	GARY L. WALDO, JOSÉ L. BOYER, AND T. KENDALL HARDEN	182
16. Purification of Recombinant $G_q\alpha$, $G_{11}\alpha$, and $G_{16}\alpha$ from Sf9 Cells	JOHN R. HEPLER, TOHRU KOZASA, AND ALFRED G. GILMAN	191
17. Expression and Purification of G-Protein α Subunits Using Baculovirus Expression System	STEPHEN G. GRABER, ROBERT A. FIGLER, AND JAMES C. GARRISON	212
18. Analysis of G-Protein α and $\beta\gamma$ Subunits by *in Vitro* Translation	EVA J. NEER, BRADLEY M. DENKER, THOMAS C. THOMAS, AND CARL J. SCHMIDT	226
19. Assays for Studying Functional Properties of *in Vitro* Translated $G_s\alpha$ Subunit	YVES AUDIGIER	239
20. Myristoylation of G-Protein α Subunits	SUSANNE M. MUMBY AND MAURINE E. LINDER	254
21. Specificity and Functional Applications of Antipeptide Antisera Which Identify G-Protein α Subunits	GRAEME MILLIGAN	268
22. Identification of Receptor-Activated G Proteins: Selective Immunoprecipitation of Photolabeled G-Protein α Subunits	KARL-LUDWIG LAUGWITZ, KARSTEN SPICHER, GÜNTER SCHULTZ, AND STEFAN OFFERMANNS	283

23. Identification of Mutant Forms of G-Protein α Subunits in Human Neoplasia by Polymerase Chain Reaction-Based Techniques	JOHN LYONS	295
24. Detection of Mutations and Polymorphisms of $G_s\alpha$ Subunit Gene by Denaturing Gradient Gel Electrophoresis	PABLO V. GEJMAN AND LEE S. WEINSTEIN	308
25. Construction of Mutant and Chimeric G-Protein α Subunits	SIM WINITZ, MARIJANE RUSSELL, AND GARY L. JOHNSON	321
26. Design of Degenerate Oligonucleotide Primers for Cloning of G-Protein α Subunits	THOMAS M. WILKIE, ANNA M. ARAGAY, A. JOHN WATSON, AND MELVIN I. SIMON	327
27. Microinjection of Antisense Oligonucleotides to Assess G-Protein Subunit Function	CHRISTIANE KLEUSS, GÜNTER SCHULTZ, AND BURGHARDT WITTIG	345
28. Inactivation of G-Protein Genes: Double Knockout in Cell Lines	RICHARD M. MORTENSEN AND J. G. SEIDMAN	356
29. Targeted Inactivation of the $G_{i2}\alpha$ Gene with Replacement and Insertion Vectors: Analysis in a 96-Well Plate Format	UWE RUDOLPH, ALLAN BRADLEY, AND LUTZ BIRNBAUMER	366
30. G-Protein Assays in *Dictyostelium*	B. EWA SNAAR-JAGALSKA AND PETER J. M. VAN HAASTERT	387
31. Fluorescence Assays for G-Protein Interactions	RICHARD A. CERIONE	409
32. Specific Peptide Probes for G-Protein Interactions with Receptors	HEIDI E. HAMM AND HELEN M. RARICK	423
33. Vaccinia Virus Systems for Expression of Gα Genes in S49 Cells	FRANKLIN QUAN AND MICHAEL FORTE	436

Section II. Gβγ Subunits

34. Purification of Tβγ Subunit of Transducin	JOËLLE BIGAY AND MARC CHABRE	449
35. Adenylyl Cyclase Assay for βγ Subunits of G Proteins	JIANQIANG CHEN, DONNA J. CARTY, AND RAVI IYENGAR	451
36. Synthesis and Use of Biotinylated βγ Complexes Prepared from Bovine Brain G Proteins	JANE DINGUS, MICHAEL D. WILCOX, RUSSELL KOHNKEN, AND JOHN D. HILDEBRANDT	457

37. Design of Oligonucleotide Probes for Molecular Cloning of β and γ Subunits	CHRISTINE GALLAGHER AND NARASIMHAN GAUTAM	471
38. Characterization of Antibodies for Various G-Protein β and γ Subunits	ALEXEY N. PRONIN AND NARASIMHAN GAUTAM	482
39. Preparation, Characterization, and Use of Antibodies with Specificity for G-Protein γ Subunits	JANET D. ROBISHAW AND ERIC A. BALCUEVA	498
40. Isoprenylation of γ Subunits and G-Protein Effectors	BERNARD K. -K. FUNG, JANMEET S. ANANT, WUN-CHEN LIN, OLIVIA C. ONG, AND HARVEY K. YAMANE	509

AUTHOR INDEX 521

SUBJECT INDEX 541

Contributors to Volume 237

Article numbers are in parentheses following the names of contributors.
Affiliations listed are current.

JANMEET S. ANANT (40), *Jules Stein Eye Institute, University of California at Los Angeles School of Medicine, Los Angeles, California 90024*

ANNA M. ARAGAY (26), *Department of Biology, California Institute of Technology, Pasadena, California 91125*

YVES AUDIGIER (19), *UMR 9925, Université Paul Sabatier, 31062 Toulouse Cedex, France*

ERIC A. BALCUEVA (39), *Weis Center for Research, Geisinger Clinic, Danville, Pennsylvania 17822*

JOËLLE BIGAY (11, 34), *Institut de Pharmacologie moléculaire et cellulaire, Unité propre 411 du Centre National de la Recherche Scientifique, 06560 Valbonne, France*

LUTZ BIRNBAUMER (9, 29), *Departments of Cell Biology, Medicine, Molecular Physiology, and Biophysics, and Division of Neurosciences, Baylor College of Medicine, Houston, Texas 77030*

JONATHAN L. BLANK (14), *Division of Basic Sciences, National Jewish Center for Immunology and Respiratory Medicine, Denver, Colorado 80206*

THOMAS BOUILLON (2), *Department of Clinical Pharmacology, University of Göttingen, 37075 Göttingen, Germany*

JOSÉ L. BOYER (15), *Department of Pharmacology, University of North Carolina School of Medicine, Chapel Hill, North Carolina 27599*

ALLAN BRADLEY (29), *Institute for Molecular Genetics and Howard Hughes Medical Institute, Baylor College of Medicine, Houston, Texas 77030*

DONNA J. CARTY (4, 6, 35), *Department of Pharmacology, Mount Sinai School of Medicine, New York, New York 10029*

RICHARD A. CERIONE (31), *Department of Pharmacology, Cornell University, Ithaca, New York 14853*

MARC CHABRE (11, 34), *Institut de Pharmacologie moléculaire et cellulaire, Unité propre 411 du Centre National de la Recherche Scientifique, 06560 Valbonne, France*

JIANQIANG CHEN (35), *Department of Pharmacology, Mount Sinai School of Medicine, New York, New York 10029*

JUAN CODINA (9), *Department of Cell Biology, Baylor College of Medicine, Houston, Texas 77030*

BRADLEY M. DENKER (18), *Department of Medicine, Brigham and Women's Hospital and Harvard Medical School, Boston, Massachusetts 02115*

N. DHANASEKARAN (7), *Department of Biochemistry and Fels Institute for Cancer and Molecular Biology, Temple University School of Medicine, Philadelphia, Pennsylvania 19140*

JANE DINGUS (36), *Department of Cell and Molecular Pharmacology and Experimental Therapeutics, Medical University of South Carolina, Charleston, South Carolina 29425*

WILLIAM J. DUNN III (8), *Departments of Medicinal Chemistry and Pharmacognosy, University of Illinois College of Pharmacy, Chicago, Illinois 60612*

JOHN H. EXTON (14), *Department of Molecular Physiology and Biophysics, Vanderbilt University School of Medicine, Nashville, Tennessee 37232*

ROBERT A. FIGLER (17), *Departments of Molecular Physiology and Biological Physics, University of Virginia Health Sciences Center, Charlottesville, Virginia 22908*

MICHAEL FORTE (33), *Vollum Institute, Oregon Health Sciences University, Portland, Oregon 97201*

BERNARD K. -K. FUNG (40), *Jules Stein Eye Institute, University of California at Los Angeles School of Medicine, Los Angeles, California 90024*

CHRISTINE GALLAGHER (37), *Departments of Anesthesiology and Genetics, Washington University School of Medicine, St. Louis, Missouri 63110*

JAMES C. GARRISON (17), *Department of Pharmacology, University of Virginia Health Sciences Center, Charlottesville, Virginia 22908*

NARASIMHAN GAUTAM (37, 38), *Departments of Anesthesiology and Genetics, Washington University School of Medicine, St. Louis, Missouri 63110*

PABLO V. GEJMAN (24), *Unit on Molecular Clinical Investigation, Clinical Neurogenetics Branch, National Institute of Mental Health, National Institutes of Health, Bethesda, Maryland 20892*

PETER GIERSCHIK (2), *Department of Pharmacology and Toxicology, University of Ulm, 89069 Ulm, Germany*

ALFRED G. GILMAN (12, 16), *Department of Pharmacology, University of Texas Southwestern Medical Center, Dallas, Texas 75235*

STEPHEN G. GRABER (17), *Department of Pharmacology and Toxicology, West Virginia University, Morgantown, West Virginia 26505*

DAGOBERTO GRENET (9), *Department of Cell Biology, Baylor College of Medicine, Houston, Texas 77030*

HEIDI E. HAMM (32), *Department of Physiology and Biophysics, University of Illinois at Chicago, Chicago, Illinois 60612*

T. KENDALL HARDEN (15), *Department of Pharmacology, University of North Carolina School of Medicine, Chapel Hill, North Carolina 27599*

RANDY S. HAUN (5), *Laboratory of Cellular Metabolism, National Heart, Lung, and Blood Institute, National Institutes of Health, Bethesda, Maryland 20892*

JOHN R. HEPLER (16), *Department of Pharmacology, University of Texas Southwestern Medical Center, Dallas, Texas 75235*

TSUTOMU HIGASHIJIMA[1] (3), *Department of Pharmacology, University of Texas Southwestern Medical Center, Dallas, Texas 75235*

JOHN D. HILDEBRANDT (36), *Department of Cell and Molecular Pharmacology and Experimental Therapeutics, Medical University of South Carolina, Charleston, South Carolina 29425*

ATSUSHI INANOBE (10), *Department of Life Science, Tokyo Institute of Technology, Yokohama, Kanagawa 227, Japan*

RAVI IYENGAR (4, 35), *Department of Pharmacology, Mount Sinai School of Medicine, New York, New York 10029*

KARL H. JAKOBS (1, 2), *Institut für Pharmakologie, Universitätsklinikum Essen, 45122 Essen, Germany*

GARY L. JOHNSON (25), *Division of Basic Sciences, National Jewish Center for Immunology and Respiratory Medicine, Denver, Colorado 80206*

TOSHIAKI KATADA (10), *Department of Physiological Chemistry, Faculty of Pharmaceutical Sciences, University of Tokyo, Bunkyo-ku, Tokyo 113, Japan*

CHRISTIANE KLEUSS (27), *Bereich Molekularbiologie und BioInformatik, Institut für Molekularbiologie und Biochemie, Freie Universität Berlin, D-14195 Berlin, Germany*

ICHIRO KOBAYASHI (10), *Department of Life Science, Tokyo Institute of Technology, Yokohama, Kanagawa 227, Japan*

RUSSELL KOHNKEN (36), *Molecular Geriatrics, Libertyville, Illinois 60048*

KENJI KONTANI (10), *Department of Life Science, Tokyo Institute of Technology, Yokohama, Kanagawa 227, Japan*

---

[1] Deceased.

TOHRU KOZASA (16), *Department of Pharmacology, University of Texas Southwestern Medical Center, Dallas, Texas 75235*

KARL-LUDWIG LAUGWITZ (22), *Institut für Pharmakologie, Freie Universität Berlin, 14195 Berlin, Germany*

ETHAN LEE (12), *Department of Pharmacology, University of Texas Southwestern Medical Center, Dallas, Texas 75235*

FANG-JEN SCOTT LEE (5), *Laboratory of Cellular Metabolism, National Heart, Lung, and Blood Institute, National Institutes of Health, Bethesda, Maryland 20892*

WUN-CHEN LIN (40), *Jules Stein Eye Institute, University of California at Los Angeles School of Medicine, Los Angeles, California 90024*

MAURINE E. LINDER (12, 20), *Department of Cell Biology and Physiology, Washington University School of Medicine, St. Louis, Missouri 63110*

JOHN LYONS (23), *Department of Drug Discovery, Onyx Pharmaceuticals, Richmond, California 94806*

GRAEME MILLIGAN (21), *Molecular Pharmacology Group, Departments of Biochemistry and Pharmacology, University of Glasgow, G12 8QQ Glasgow, Scotland, United Kingdom*

RICHARD M. MORTENSEN (28), *Department of Medicine, Brigham and Women's Hospital and Harvard Medical School, Boston, Massachusetts 02115*

JOEL MOSS (5), *Laboratory of Cellular Metabolism, National Heart, Lung, and Blood Institute, National Institutes of Health, Bethesda, Maryland 20892*

SUSANNE M. MUMBY (20), *Department of Pharmacology, University of Texas Southwestern Medical Center, Dallas, Texas 75235*

EVA J. NEER (18), *Department of Medicine, Brigham and Women's Hospital and Harvard Medical School, Boston, Massachusetts 02115*

HIROSHI NISHINA (10), *Department of Life Science, Tokyo Institute of Technology, Yokohama, Kanagawa 227, Japan*

STEFAN OFFERMANNS (22), *Department of Biology, California Institute of Technology, Pasadena, California 91125*

YOSHIHARU OHOKA (10), *Department of Life Science, Tokyo Institute of Technology, Yokohama, Kanagawa 227, Japan*

OLIVIA C. ONG (40), *Jules Stein Eye Institute, University of California at Los Angeles School of Medicine, Los Angeles, California 90024*

IOK-HOU PANG (13), *Department of Glaucoma Research, Alcon Laboratories, Fort Worth, Texas 76134*

S. RUSS PRICE (5), *Renal Division, Emory University Hospital, Atlanta, Georgia 30322*

ALEXEY N. PRONIN (38), *Departments of Anesthesiology and Genetics, Washington University School of Medicine, St. Louis, Missouri 63110*

FRANKLIN QUAN (33), *Vollum Institute, Oregon Health Sciences University, Portland, Oregon 97201*

HELEN M. RARICK (32), *Department of Physiology and Biophysics, University of Illinois at Chicago, Chicago, Illinois 60612*

MARK M. RASENICK (8), *Departments of Physiology and Biophysics, University of Illinois College of Medicine, Chicago, Illinois 60612*

FERNANDO RIBEIRO-NETO (9), *Department of Pharmacology, Duke University Medical Center, Durham, North Carolina 27710*

JANET D. ROBISHAW (39), *Weis Center for Research, Geisinger Clinic, Danville, Pennsylvania 17822*

ELLIOTT M. ROSS (3), *Department of Pharmacology, University of Texas Southwestern Medical Center, Dallas, Texas 75235*

UWE RUDOLPH (29), *Institute of Pharmacology, University of Zurich, CH-8057 Zurich, Switzerland*

ARNOLD E. RUOHO (7), *Department of Pharmacology, University of Wisconsin Medical School, Madison, Wisconsin 53706*

MARIJANE RUSSELL (25), *Division of Basic Sciences, National Jewish Center for Immunology and Respiratory Medicine, Denver, Colorado 80206*

CARL J. SCHMIDT (18), *Department of Medicine, Brigham and Women's Hospital and Harvard Medical School, Boston, Massachusetts 02115*

GÜNTER SCHULTZ (22, 27), *Institut für Pharmakologie, Freie Universität Berlin, D-14195 Berlin, Germany*

J. G. SEIDMAN (28), *Department of Genetics, Harvard Medical School, Boston, Massachusetts 02115*

MELVIN I. SIMON (26), *Department of Biology, California Institute of Technology, Pasadena, California 91125*

ALAN V. SMRCKA (13), *Department of Pharmacology, University of Texas Southwestern Medical Center, Dallas, Texas 75235*

B. EWA SNAAR-JAGALSKA (30), *Cell Biology and Genetics Unit, Clusius Laboratory, Leiden University, 2333 AL Leiden, The Netherlands*

KARSTEN SPICHER (22), *Institut für Pharmakologie, Freie Universität Berlin, 14195 Berlin, Germany*

PAUL C. STERNWEIS (13), *Department of Pharmacology, University of Texas Southwestern Medical Center, Dallas, Texas 75235*

KATSUNOBU TAKAHASHI (10), *Department of Life Science, Tokyo Institute of Technology, Yokohama, Kanagawa 227, Japan*

MADHAVI TALLURI (8), *Departments of Physiology and Biophysics, University of Illinois College of Medicine, Chicago, Illinois 60612*

THOMAS C. THOMAS (18), *Alexion Pharmaceuticals, Inc., New Haven, Connecticut 06511*

SU-CHEN TSAI (5), *Laboratory of Cellular Metabolism, National Heart, Lung, and Blood Institute, National Institutes of Health, Bethesda, Maryland 20892*

RICHARD R. VAILLANCOURT (7), *Division of Basic Sciences, Department of Pediatrics, National Jewish Center for Immunology and Respiratory Medicine, Denver, Colorado 80206*

PETER J. M. VAN HAASTERT (30), *Department of Biochemistry, University of Gröningen, 9747 AG Gröningen, The Netherlands*

MARTHA VAUGHAN (5), *Laboratory of Cellular Metabolism, National Heart, Lung, and Blood Institute, National Institutes of Health, Bethesda, Maryland 20892*

GARY L. WALDO (15), *Department of Pharmacology, University of North Carolina School of Medicine, Chapel Hill, North Carolina 27599*

A. JOHN WATSON (26), *Department of Biology, California Institute of Technology, Pasadena, California 91125*

LEE S. WEINSTEIN (24), *Molecular Pathophysiology Branch, National Institute of Diabetes and Digestive and Kidney Diseases, National Institutes of Health, Bethesda, Maryland 20892*

CATHERINE F. WELSH (5), *Department of Cell Biology and Anatomy, University of Miami, Miami, Florida 33136*

THOMAS WIELAND (1), *Institut für Pharmakologie, Universitätsklinikum Essen, 45122 Essen, Germany*

MICHAEL D. WILCOX (36), *Department of Cell and Molecular Pharmacology and Experimental Therapeutics, Medical University of South Carolina, Charleston, South Carolina 29425*

THOMAS M. WILKIE (26), *Department of Pharmacology, University of Texas Southwestern Medical Center, Dallas, Texas 75235*

SIM WINITZ (25), *Division of Basic Sciences, National Jewish Center for Immunology and Respiratory Medicine, Denver, Colorado 80206*

BURGHARDT WITTIG (27), *Bereich Molekularbiologie und BioInformatik, Institut für Molekularbiologie und Biochemie, Freie Universität Berlin, D-14195 Berlin, Germany*

HARVEY K. YAMANE (40), *Jules Stein Eye Institute, University of California at Los Angeles School of Medicine, Los Angeles, California 90024*

Preface

Heterotrimeric ($\alpha\beta\gamma$) G proteins function as cell surface signal transducers for a large number of hormones, neurotransmitters, and for autocrine and paracrine factors. It is now known that a hundred or so (not counting the olfactory) receptors are coupled to various effectors through G proteins. This large number of receptors couple to members of one of the four families of G proteins to transmit their signals. The specificity of the receptor–G protein interactions determines the transmission of the signal to different downstream pathways. As with all real life situations, the specificity of interactions between receptors and G proteins is varied and complex. Consequently, signals from a single receptor may be transmitted through several pathways simultaneously. Each of the four families of G proteins has many members. Cloning studies have indicated that there are twenty α, four β, and six γ subunits. It has been suggested that very large numbers of heterotrimeric G proteins with defined subunit compositions can be generated from these individual subunits. "Knockout" experiments indicate that G proteins of defined subunit composition communicate signals from different receptors. However, at this time there is not sufficient general information to indicate that the functional identity of a G protein is defined by the molecular identity of all of its subunits. Consequently, in spite of the molecular heterogeneity of the β and γ subunits, the different G proteins are still classified by the identity of their α subunits. This classification is useful because it indicates which intracellular messenger pathway is used. In this volume the nomenclature of G proteins is based on the molecular identity of the α subunit. Hence it should be noted that native purified G proteins may have a single molecular species of α subunits, but multiple forms of β and γ subunits.

In the past decade there has been a substantial increase in our understanding of signal-transducing G proteins. This advance has been brought about by a combination of molecular biological and biochemical techniques. Both approaches are represented in this book. Two types of chapters are included. The first presents techniques unique to the field of signal-transducing G proteins, the second general techniques that have been applied to the study of G-protein systems. Chapters of the latter type may be useful even to researchers who do not work on signal-transducing G proteins.

The field of heterotrimeric G proteins has been covered in part in Volumes 109 and 195 of *Methods in Enzymology*. However, the subject has not been covered in a systematic fashion. Consequently, in planning this volume, an attempt was made to include all or at least most of the techniques in one volume. Thus techniques such as cholera and pertussis toxin labeling as well as some techniques for the purification of native G proteins are covered but by different authors. It should be noted that in G protein research many laboratories use different protocols for the same overall experiments. Since a single experimental protocol does not always work for the same G protein in different systems it can be useful to have more than one experimental procedure available to tackle the same question. Several, but not all subjects, are thus addressed by multiple laboratories in this volume.

I thank Lutz Birnbaumer, Henry Bourne, and Buzz Brown for their useful suggestions during the initial planning of this volume. I also thank Ms. Lina Mazzella for her valuable assistance. Last, but not least, I thank the authors for their contributions and for complying with my suggestions for change in a timely fashion.

RAVI IYENGAR

METHODS IN ENZYMOLOGY

VOLUME I. Preparation and Assay of Enzymes
Edited by SIDNEY P. COLOWICK AND NATHAN O. KAPLAN

VOLUME II. Preparation and Assay of Enzymes
Edited by SIDNEY P. COLOWICK AND NATHAN O. KAPLAN

VOLUME III. Preparation and Assay of Substrates
Edited by SIDNEY P. COLOWICK AND NATHAN O. KAPLAN

VOLUME IV. Special Techniques for the Enzymologist
Edited by SIDNEY P. COLOWICK AND NATHAN O. KAPLAN

VOLUME V. Preparation and Assay of Enzymes
Edited by SIDNEY P. COLOWICK AND NATHAN O. KAPLAN

VOLUME VI. Preparation and Assay of Enzymes (*Continued*)
Preparation and Assay of Substrates
Special Techniques
Edited by SIDNEY P. COLOWICK AND NATHAN O. KAPLAN

VOLUME VII. Cumulative Subject Index
Edited by SIDNEY P. COLOWICK AND NATHAN O. KAPLAN

VOLUME VIII. Complex Carbohydrates
Edited by ELIZABETH F. NEUFELD AND VICTOR GINSBURG

VOLUME IX. Carbohydrate Metabolism
Edited by WILLIS A. WOOD

VOLUME X. Oxidation and Phosphorylation
Edited by RONALD W. ESTABROOK AND MAYNARD E. PULLMAN

VOLUME XI. Enzyme Structure
Edited by C. H. W. HIRS

VOLUME XII. Nucleic Acids (Parts A and B)
Edited by LAWRENCE GROSSMAN AND KIVIE MOLDAVE

VOLUME XIII. Citric Acid Cycle
Edited by J. M. LOWENSTEIN

VOLUME XIV. Lipids
Edited by J. M. LOWENSTEIN

VOLUME XV. Steroids and Terpenoids
Edited by RAYMOND B. CLAYTON

VOLUME XVI. Fast Reactions
Edited by KENNETH KUSTIN

VOLUME XVII. Metabolism of Amino Acids and Amines (Parts A and B)
Edited by HERBERT TABOR AND CELIA WHITE TABOR

VOLUME XVIII. Vitamins and Coenzymes (Parts A, B, and C)
Edited by DONALD B. MCCORMICK AND LEMUEL D. WRIGHT

VOLUME XIX. Proteolytic Enzymes
Edited by GERTRUDE E. PERLMANN AND LASZLO LORAND

VOLUME XX. Nucleic Acids and Protein Synthesis (Part C)
Edited by KIVIE MOLDAVE AND LAWRENCE GROSSMAN

VOLUME XXI. Nucleic Acids (Part D)
Edited by LAWRENCE GROSSMAN AND KIVIE MOLDAVE

VOLUME XXII. Enzyme Purification and Related Techniques
Edited by WILLIAM B. JAKOBY

VOLUME XXIII. Photosynthesis (Part A)
Edited by ANTHONY SAN PIETRO

VOLUME XXIV. Photosynthesis and Nitrogen Fixation (Part B)
Edited by ANTHONY SAN PIETRO

VOLUME XXV. Enzyme Structure (Part B)
Edited by C. H. W. HIRS AND SERGE N. TIMASHEFF

VOLUME XXVI. Enzyme Structure (Part C)
Edited by C. H. W. HIRS AND SERGE N. TIMASHEFF

VOLUME XXVII. Enzyme Structure (Part D)
Edited by C. H. W. HIRS AND SERGE N. TIMASHEFF

VOLUME XXVIII. Complex Carbohydrates (Part B)
Edited by VICTOR GINSBURG

VOLUME XXIX. Nucleic Acids and Protein Synthesis (Part E)
Edited by LAWRENCE GROSSMAN AND KIVIE MOLDAVE

VOLUME XXX. Nucleic Acids and Protein Synthesis (Part F)
Edited by KIVIE MOLDAVE AND LAWRENCE GROSSMAN

VOLUME XXXI. Biomembranes (Part A)
Edited by SIDNEY FLEISCHER AND LESTER PACKER

VOLUME XXXII. Biomembranes (Part B)
Edited by SIDNEY FLEISCHER AND LESTER PACKER

VOLUME XXXIII. Cumulative Subject Index Volumes I–XXX
Edited by MARTHA G. DENNIS AND EDWARD A. DENNIS

VOLUME XXXIV. Affinity Techniques (Enzyme Purification: Part B)
Edited by WILLIAM B. JAKOBY AND MEIR WILCHEK

VOLUME XXXV. Lipids (Part B)
Edited by JOHN M. LOWENSTEIN

VOLUME XXXVI. Hormone Action (Part A: Steroid Hormones)
Edited by BERT W. O'MALLEY AND JOEL G. HARDMAN

VOLUME XXXVII. Hormone Action (Part B: Peptide Hormones)
Edited by BERT W. O'MALLEY AND JOEL G. HARDMAN

VOLUME XXXVIII. Hormone Action (Part C: Cyclic Nucleotides)
Edited by JOEL G. HARDMAN AND BERT W. O'MALLEY

VOLUME XXXIX. Hormone Action (Part D: Isolated Cells, Tissues, and Organ Systems)
Edited by JOEL G. HARDMAN AND BERT W. O'MALLEY

VOLUME XL. Hormone Action (Part E: Nuclear Structure and Function)
Edited by BERT W. O'MALLEY AND JOEL G. HARDMAN

VOLUME XLI. Carbohydrate Metabolism (Part B)
Edited by W. A. WOOD

VOLUME XLII. Carbohydrate Metabolism (Part C)
Edited by W. A. WOOD

VOLUME XLIII. Antibiotics
Edited by JOHN H. HASH

VOLUME XLIV. Immobilized Enzymes
Edited by KLAUS MOSBACH

VOLUME XLV. Proteolytic Enzymes (Part B)
Edited by LASZLO LORAND

VOLUME XLVI. Affinity Labeling
Edited by WILLIAM B. JAKOBY AND MEIR WILCHEK

VOLUME XLVII. Enzyme Structure (Part E)
Edited by C. H. W. HIRS AND SERGE N. TIMASHEFF

VOLUME XLVIII. Enzyme Structure (Part F)
Edited by C. H. W. HIRS AND SERGE N. TIMASHEFF

VOLUME XLIX. Enzyme Structure (Part G)
Edited by C. H. W. HIRS AND SERGE N. TIMASHEFF

VOLUME L. Complex Carbohydrates (Part C)
Edited by VICTOR GINSBURG

VOLUME LI. Purine and Pyrimidine Nucleotide Metabolism
Edited by PATRICIA A. HOFFEE AND MARY ELLEN JONES

VOLUME LII. Biomembranes (Part C: Biological Oxidations)
Edited by SIDNEY FLEISCHER AND LESTER PACKER

VOLUME LIII. Biomembranes (Part D: Biological Oxidations)
Edited by SIDNEY FLEISCHER AND LESTER PACKER

VOLUME LIV. Biomembranes (Part E: Biological Oxidations)
Edited by SIDNEY FLEISCHER AND LESTER PACKER

VOLUME LV. Biomembranes (Part F: Bioenergetics)
Edited by SIDNEY FLEISCHER AND LESTER PACKER

VOLUME LVI. Biomembranes (Part G: Bioenergetics)
Edited by SIDNEY FLEISCHER AND LESTER PACKER

VOLUME LVII. Bioluminescence and Chemiluminescence
Edited by MARLENE A. DELUCA

VOLUME LVIII. Cell Culture
Edited by WILLIAM B. JAKOBY AND IRA PASTAN

VOLUME LIX. Nucleic Acids and Protein Synthesis (Part G)
Edited by KIVIE MOLDAVE AND LAWRENCE GROSSMAN

VOLUME LX. Nucleic Acids and Protein Synthesis (Part H)
Edited by KIVIE MOLDAVE AND LAWRENCE GROSSMAN

VOLUME 61. Enzyme Structure (Part H)
Edited by C. H. W. HIRS AND SERGE N. TIMASHEFF

VOLUME 62. Vitamins and Coenzymes (Part D)
Edited by DONALD B. MCCORMICK AND LEMUEL D. WRIGHT

VOLUME 63. Enzyme Kinetics and Mechanism (Part A: Initial Rate and Inhibitor Methods)
Edited by DANIEL L. PURICH

VOLUME 64. Enzyme Kinetics and Mechanism (Part B: Isotopic Probes and Complex Enzyme Systems)
Edited by DANIEL L. PURICH

VOLUME 65. Nucleic Acids (Part I)
Edited by LAWRENCE GROSSMAN AND KIVIE MOLDAVE

VOLUME 66. Vitamins and Coenzymes (Part E)
Edited by DONALD B. MCCORMICK AND LEMUEL D. WRIGHT

VOLUME 67. Vitamins and Coenzymes (Part F)
Edited by DONALD B. MCCORMICK AND LEMUEL D. WRIGHT

VOLUME 68. Recombinant DNA
Edited by RAY WU

VOLUME 69. Photosynthesis and Nitrogen Fixation (Part C)
Edited by ANTHONY SAN PIETRO

VOLUME 70. Immunochemical Techniques (Part A)
Edited by HELEN VAN VUNAKIS AND JOHN J. LANGONE

VOLUME 71. Lipids (Part C)
Edited by JOHN M. LOWENSTEIN

VOLUME 72. Lipids (Part D)
Edited by JOHN M. LOWENSTEIN

VOLUME 73. Immunochemical Techniques (Part B)
Edited by JOHN J. LANGONE AND HELEN VAN VUNAKIS

VOLUME 74. Immunochemical Techniques (Part C)
Edited by JOHN J. LANGONE AND HELEN VAN VUNAKIS

VOLUME 75. Cumulative Subject Index Volumes XXXI, XXXII, XXXIV–LX
Edited by EDWARD A. DENNIS AND MARTHA G. DENNIS

VOLUME 76. Hemoglobins
Edited by ERALDO ANTONINI, LUIGI ROSSI-BERNARDI, AND EMILIA CHIANCONE

VOLUME 77. Detoxication and Drug Metabolism
Edited by WILLIAM B. JAKOBY

VOLUME 78. Interferons (Part A)
Edited by SIDNEY PESTKA

VOLUME 79. Interferons (Part B)
Edited by SIDNEY PESTKA

VOLUME 80. Proteolytic Enzymes (Part C)
Edited by LASZLO LORAND

VOLUME 81. Biomembranes (Part H: Visual Pigments and Purple Membranes, I)
Edited by LESTER PACKER

VOLUME 82. Structural and Contractile Proteins (Part A: Extracellular Matrix)
Edited by LEON W. CUNNINGHAM AND DIXIE W. FREDERIKSEN

VOLUME 83. Complex Carbohydrates (Part D)
Edited by VICTOR GINSBURG

VOLUME 84. Immunochemical Techniques (Part D: Selected Immunoassays)
Edited by JOHN J. LANGONE AND HELEN VAN VUNAKIS

VOLUME 85. Structural and Contractile Proteins (Part B: The Contractile Apparatus and the Cytoskeleton)
Edited by DIXIE W. FREDERIKSEN AND LEON W. CUNNINGHAM

VOLUME 86. Prostaglandins and Arachidonate Metabolites
Edited by WILLIAM E. M. LANDS AND WILLIAM L. SMITH

VOLUME 87. Enzyme Kinetics and Mechanism (Part C: Intermediates, Stereochemistry, and Rate Studies)
Edited by DANIEL L. PURICH

VOLUME 88. Biomembranes (Part I: Visual Pigments and Purple Membranes, II)
Edited by LESTER PACKER

VOLUME 89. Carbohydrate Metabolism (Part D)
Edited by WILLIS A. WOOD

VOLUME 90. Carbohydrate Metabolism (Part E)
Edited by WILLIS A. WOOD

VOLUME 91. Enzyme Structure (Part I)
Edited by C. H. W. HIRS AND SERGE N. TIMASHEFF

VOLUME 92. Immunochemical Techniques (Part E: Monoclonal Antibodies and General Immunoassay Methods)
Edited by JOHN J. LANGONE AND HELEN VAN VUNAKIS

VOLUME 93. Immunochemical Techniques (Part F: Conventional Antibodies, Fc Receptors, and Cytotoxicity)
Edited by JOHN J. LANGONE AND HELEN VAN VUNAKIS

VOLUME 94. Polyamines
Edited by HERBERT TABOR AND CELIA WHITE TABOR

VOLUME 95. Cumulative Subject Index Volumes 61–74, 76–80
Edited by EDWARD A. DENNIS AND MARTHA G. DENNIS

VOLUME 96. Biomembranes [Part J: Membrane Biogenesis: Assembly and Targeting (General Methods; Eukaryotes)]
Edited by SIDNEY FLEISCHER AND BECCA FLEISCHER

VOLUME 97. Biomembranes [Part K: Membrane Biogenesis: Assembly and Targeting (Prokaryotes, Mitochondria, and Chloroplasts)]
Edited by SIDNEY FLEISCHER AND BECCA FLEISCHER

VOLUME 98. Biomembranes (Part L: Membrane Biogenesis: Processing and Recycling)
Edited by SIDNEY FLEISCHER AND BECCA FLEISCHER

VOLUME 99. Hormone Action (Part F: Protein Kinases)
Edited by JACKIE D. CORBIN AND JOEL G. HARDMAN

VOLUME 100. Recombinant DNA (Part B)
Edited by RAY WU, LAWRENCE GROSSMAN, AND KIVIE MOLDAVE

VOLUME 101. Recombinant DNA (Part C)
Edited by RAY WU, LAWRENCE GROSSMAN, AND KIVIE MOLDAVE

VOLUME 102. Hormone Action (Part G: Calmodulin and Calcium-Binding Proteins)
Edited by ANTHONY R. MEANS AND BERT W. O'MALLEY

VOLUME 103. Hormone Action (Part H: Neuroendocrine Peptides)
Edited by P. MICHAEL CONN

VOLUME 104. Enzyme Purification and Related Techniques (Part C)
Edited by WILLIAM B. JAKOBY

VOLUME 105. Oxygen Radicals in Biological Systems
Edited by LESTER PACKER

VOLUME 106. Posttranslational Modifications (Part A)
Edited by FINN WOLD AND KIVIE MOLDAVE

VOLUME 107. Posttranslational Modifications (Part B)
Edited by FINN WOLD AND KIVIE MOLDAVE

VOLUME 108. Immunochemical Techniques (Part G: Separation and Characterization of Lymphoid Cells)
Edited by GIOVANNI DI SABATO, JOHN J. LANGONE, AND HELEN VAN VUNAKIS

VOLUME 109. Hormone Action (Part I: Peptide Hormones)
Edited by LUTZ BIRNBAUMER AND BERT W. O'MALLEY

VOLUME 110. Steroids and Isoprenoids (Part A)
Edited by JOHN H. LAW AND HANS C. RILLING

VOLUME 111. Steroids and Isoprenoids (Part B)
Edited by JOHN H. LAW AND HANS C. RILLING

VOLUME 112. Drug and Enzyme Targeting (Part A)
Edited by KENNETH J. WIDDER AND RALPH GREEN

VOLUME 113. Glutamate, Glutamine, Glutathione, and Related Compounds
Edited by ALTON MEISTER

VOLUME 114. Diffraction Methods for Biological Macromolecules (Part A)
Edited by HAROLD W. WYCKOFF, C. H. W. HIRS, AND SERGE N. TIMASHEFF

VOLUME 115. Diffraction Methods for Biological Macromolecules (Part B)
Edited by HAROLD W. WYCKOFF, C. H. W. HIRS, AND SERGE N. TIMASHEFF

VOLUME 116. Immunochemical Techniques (Part H: Effectors and Mediators of Lymphoid Cell Functions)
Edited by GIOVANNI DI SABATO, JOHN J. LANGONE, AND HELEN VAN VUNAKIS

VOLUME 117. Enzyme Structure (Part J)
Edited by C. H. W. HIRS AND SERGE N. TIMASHEFF

VOLUME 118. Plant Molecular Biology
Edited by ARTHUR WEISSBACH AND HERBERT WEISSBACH

VOLUME 119. Interferons (Part C)
Edited by SIDNEY PESTKA

VOLUME 120. Cumulative Subject Index Volumes 81–94, 96–101

VOLUME 121. Immunochemical Techniques (Part I: Hybridoma Technology and Monoclonal Antibodies)
Edited by JOHN J. LANGONE AND HELEN VAN VUNAKIS

VOLUME 122. Vitamins and Coenzymes (Part G)
Edited by FRANK CHYTIL AND DONALD B. MCCORMICK

VOLUME 123. Vitamins and Coenzymes (Part H)
Edited by FRANK CHYTIL AND DONALD B. MCCORMICK

VOLUME 124. Hormone Action (Part J: Neuroendocrine Peptides)
Edited by P. MICHAEL CONN

VOLUME 125. Biomembranes (Part M: Transport in Bacteria, Mitochondria, and Chloroplasts: General Approaches and Transport Systems)
Edited by SIDNEY FLEISCHER AND BECCA FLEISCHER

VOLUME 126. Biomembranes (Part N: Transport in Bacteria, Mitochondria, and Chloroplasts: Protonmotive Force)
Edited by SIDNEY FLEISCHER AND BECCA FLEISCHER

VOLUME 127. Biomembranes (Part O: Protons and Water: Structure and Translocation)
Edited by LESTER PACKER

VOLUME 128. Plasma Lipoproteins (Part A: Preparation, Structure, and Molecular Biology)
Edited by JERE P. SEGREST AND JOHN J. ALBERS

VOLUME 129. Plasma Lipoproteins (Part B: Characterization, Cell Biology, and Metabolism)
Edited by JOHN J. ALBERS AND JERE P. SEGREST

VOLUME 130. Enzyme Structure (Part K)
Edited by C. H. W. HIRS AND SERGE N. TIMASHEFF

VOLUME 131. Enzyme Structure (Part L)
Edited by C. H. W. HIRS AND SERGE N. TIMASHEFF

VOLUME 132. Immunochemical Techniques (Part J: Phagocytosis and Cell-Mediated Cytotoxicity)
Edited by GIOVANNI DI SABATO AND JOHANNES EVERSE

VOLUME 133. Bioluminescence and Chemiluminescence (Part B)
Edited by MARLENE DELUCA AND WILLIAM D. MCELROY

VOLUME 134. Structural and Contractile Proteins (Part C: The Contractile Apparatus and the Cytoskeleton)
Edited by RICHARD B. VALLEE

VOLUME 135. Immobilized Enzymes and Cells (Part B)
Edited by KLAUS MOSBACH

VOLUME 136. Immobilized Enzymes and Cells (Part C)
Edited by KLAUS MOSBACH

VOLUME 137. Immobilized Enzymes and Cells (Part D)
Edited by KLAUS MOSBACH

VOLUME 138. Complex Carbohydrates (Part E)
Edited by VICTOR GINSBURG

VOLUME 139. Cellular Regulators (Part A: Calcium- and Calmodulin-Binding Proteins)
Edited by ANTHONY R. MEANS AND P. MICHAEL CONN

VOLUME 140. Cumulative Subject Index Volumes 102–119, 121–134

VOLUME 141. Cellular Regulators (Part B: Calcium and Lipids)
Edited by P. MICHAEL CONN AND ANTHONY R. MEANS

VOLUME 142. Metabolism of Aromatic Amino Acids and Amines
Edited by SEYMOUR KAUFMAN

VOLUME 143. Sulfur and Sulfur Amino Acids
Edited by WILLIAM B. JAKOBY AND OWEN GRIFFITH

VOLUME 144. Structural and Contractile Proteins (Part D: Extracellular Matrix)
Edited by LEON W. CUNNINGHAM

VOLUME 145. Structural and Contractile Proteins (Part E: Extracellular Matrix)
Edited by LEON W. CUNNINGHAM

VOLUME 146. Peptide Growth Factors (Part A)
Edited by DAVID BARNES AND DAVID A. SIRBASKU

VOLUME 147. Peptide Growth Factors (Part B)
Edited by DAVID BARNES AND DAVID A. SIRBASKU

VOLUME 148. Plant Cell Membranes
Edited by LESTER PACKER AND ROLAND DOUCE

VOLUME 149. Drug and Enzyme Targeting (Part B)
Edited by RALPH GREEN AND KENNETH J. WIDDER

VOLUME 150. Immunochemical Techniques (Part K: *In Vitro* Models of B and T Cell Functions and Lymphoid Cell Receptors)
Edited by GIOVANNI DI SABATO

VOLUME 151. Molecular Genetics of Mammalian Cells
Edited by MICHAEL M. GOTTESMAN

VOLUME 152. Guide to Molecular Cloning Techniques
Edited by SHELBY L. BERGER AND ALAN R. KIMMEL

VOLUME 153. Recombinant DNA (Part D)
Edited by RAY WU AND LAWRENCE GROSSMAN

VOLUME 154. Recombinant DNA (Part E)
Edited by RAY WU AND LAWRENCE GROSSMAN

VOLUME 155. Recombinant DNA (Part F)
Edited by RAY WU

VOLUME 156. Biomembranes (Part P: ATP-Driven Pumps and Related Transport: The Na,K-Pump)
Edited by SIDNEY FLEISCHER AND BECCA FLEISCHER

VOLUME 157. Biomembranes (Part Q: ATP-Driven Pumps and Related Transport: Calcium, Proton, and Potassium Pumps)
Edited by SIDNEY FLEISCHER AND BECCA FLEISCHER

VOLUME 158. Metalloproteins (Part A)
Edited by JAMES F. RIORDAN AND BERT L. VALLEE

VOLUME 159. Initiation and Termination of Cyclic Nucleotide Action
Edited by JACKIE D. CORBIN AND ROGER A. JOHNSON

VOLUME 160. Biomass (Part A: Cellulose and Hemicellulose)
Edited by WILLIS A. WOOD AND SCOTT T. KELLOGG

VOLUME 161. Biomass (Part B: Lignin, Pectin, and Chitin)
Edited by WILLIS A. WOOD AND SCOTT T. KELLOGG

VOLUME 162. Immunochemical Techniques (Part L: Chemotaxis and Inflammation)
Edited by GIOVANNI DI SABATO

VOLUME 163. Immunochemical Techniques (Part M: Chemotaxis and Inflammation)
Edited by GIOVANNI DI SABATO

VOLUME 164. Ribosomes
Edited by HARRY F. NOLLER, JR., AND KIVIE MOLDAVE

VOLUME 165. Microbial Toxins: Tools for Enzymology
Edited by SIDNEY HARSHMAN

VOLUME 166. Branched-Chain Amino Acids
Edited by ROBERT HARRIS AND JOHN R. SOKATCH

VOLUME 167. Cyanobacteria
Edited by LESTER PACKER AND ALEXANDER N. GLAZER

VOLUME 168. Hormone Action (Part K: Neuroendocrine Peptides)
Edited by P. MICHAEL CONN

VOLUME 169. Platelets: Receptors, Adhesion, Secretion (Part A)
Edited by JACEK HAWIGER

VOLUME 170. Nucleosomes
Edited by PAUL M. WASSARMAN AND ROGER D. KORNBERG

VOLUME 171. Biomembranes (Part R: Transport Theory: Cells and Model Membranes)
Edited by SIDNEY FLEISCHER AND BECCA FLEISCHER

VOLUME 172. Biomembranes (Part S: Transport: Membrane Isolation and Characterization)
Edited by SIDNEY FLEISCHER AND BECCA FLEISCHER

VOLUME 173. Biomembranes [Part T: Cellular and Subcellular Transport: Eukaryotic (Nonepithelial) Cells]
Edited by SIDNEY FLEISCHER AND BECCA FLEISCHER

VOLUME 174. Biomembranes [Part U: Cellular and Subcellular Transport: Eukaryotic (Nonepithelial) Cells]
Edited by SIDNEY FLEISCHER AND BECCA FLEISCHER

VOLUME 175. Cumulative Subject Index Volumes 135–139, 141–167

VOLUME 176. Nuclear Magnetic Resonance (Part A: Spectral Techniques and Dynamics)
Edited by NORMAN J. OPPENHEIMER AND THOMAS L. JAMES

VOLUME 177. Nuclear Magnetic Resonance (Part B: Structure and Mechanism)
Edited by NORMAN J. OPPENHEIMER AND THOMAS L. JAMES

VOLUME 178. Antibodies, Antigens, and Molecular Mimicry
Edited by JOHN J. LANGONE

VOLUME 179. Complex Carbohydrates (Part F)
Edited by VICTOR GINSBURG

VOLUME 180. RNA Processing (Part A: General Methods)
Edited by JAMES E. DAHLBERG AND JOHN N. ABELSON

VOLUME 181. RNA Processing (Part B: Specific Methods)
Edited by JAMES E. DAHLBERG AND JOHN N. ABELSON

VOLUME 182. Guide to Protein Purification
Edited by MURRAY P. DEUTSCHER

VOLUME 183. Molecular Evolution: Computer Analysis of Protein and Nucleic Acid Sequences
Edited by RUSSELL F. DOOLITTLE

VOLUME 184. Avidin–Biotin Technology
Edited by MEIR WILCHEK AND EDWARD A. BAYER

VOLUME 185. Gene Expression Technology
Edited by DAVID V. GOEDDEL

VOLUME 186. Oxygen Radicals in Biological Systems (Part B: Oxygen Radicals and Antioxidants)
Edited by LESTER PACKER AND ALEXANDER N. GLAZER

VOLUME 187. Arachidonate Related Lipid Mediators
Edited by ROBERT C. MURPHY AND FRANK A. FITZPATRICK

VOLUME 188. Hydrocarbons and Methylotrophy
Edited by MARY E. LIDSTROM

VOLUME 189. Retinoids (Part A: Molecular and Metabolic Aspects)
Edited by LESTER PACKER

VOLUME 190. Retinoids (Part B: Cell Differentiation and Clinical Applications)
Edited by LESTER PACKER

VOLUME 191. Biomembranes (Part V: Cellular and Subcellular Transport: Epithelial Cells)
Edited by SIDNEY FLEISCHER AND BECCA FLEISCHER

VOLUME 192. Biomembranes (Part W: Cellular and Subcellular Transport: Epithelial Cells)
Edited by SIDNEY FLEISCHER AND BECCA FLEISCHER

VOLUME 193. Mass Spectrometry
Edited by JAMES A. MCCLOSKEY

VOLUME 194. Guide to Yeast Genetics and Molecular Biology
Edited by CHRISTINE GUTHRIE AND GERALD R. FINK

VOLUME 195. Adenylyl Cyclase, G Proteins, and Guanylyl Cyclase
Edited by ROGER A. JOHNSON AND JACKIE D. CORBIN

VOLUME 196. Molecular Motors and the Cytoskeleton
Edited by RICHARD B. VALLEE

VOLUME 197. Phospholipases
Edited by EDWARD A. DENNIS

VOLUME 198. Peptide Growth Factors (Part C)
Edited by DAVID BARNES, J. P. MATHER, AND GORDON H. SATO

VOLUME 199. Cumulative Subject Index Volumes 168–174, 176–194

VOLUME 200. Protein Phosphorylation (Part A: Protein Kinases: Assays, Purification, Antibodies, Functional Analysis, Cloning, and Expression)
Edited by TONY HUNTER AND BARTHOLOMEW M. SEFTON

VOLUME 201. Protein Phosphorylation (Part B: Analysis of Protein Phosphorylation, Protein Kinase Inhibitors, and Protein Phosphatases)
Edited by TONY HUNTER AND BARTHOLOMEW M. SEFTON

VOLUME 202. Molecular Design and Modeling: Concepts and Applications (Part A: Proteins, Peptides, and Enzymes)
Edited by JOHN J. LANGONE

VOLUME 203. Molecular Design and Modeling: Concepts and Applications (Part B: Antibodies and Antigens, Nucleic Acids, Polysaccharides, and Drugs)
Edited by JOHN J. LANGONE

VOLUME 204. Bacterial Genetic Systems
Edited by JEFFREY H. MILLER

VOLUME 205. Metallobiochemistry (Part B: Metallothionein and Related Molecules)
Edited by JAMES F. RIORDAN AND BERT L. VALLEE

VOLUME 206. Cytochrome P450
Edited by MICHAEL R. WATERMAN AND ERIC F. JOHNSON

VOLUME 207. Ion Channels
Edited by BERNARDO RUDY AND LINDA E. IVERSON

VOLUME 208. Protein–DNA Interactions
Edited by ROBERT T. SAUER

VOLUME 209. Phospholipid Biosynthesis
Edited by EDWARD A. DENNIS AND DENNIS E. VANCE

VOLUME 210. Numerical Computer Methods
Edited by LUDWIG BRAND AND MICHAEL L. JOHNSON

VOLUME 211. DNA Structures (Part A: Synthesis and Physical Analysis of DNA)
Edited by DAVID M. J. LILLEY AND JAMES E. DAHLBERG

VOLUME 212. DNA Structures (Part B: Chemical and Electrophoretic Analysis of DNA)
Edited by DAVID M. J. LILLEY AND JAMES E. DAHLBERG

VOLUME 213. Carotenoids (Part A: Chemistry, Separation, Quantitation, and Antioxidation)
Edited by LESTER PACKER

VOLUME 214. Carotenoids (Part B: Metabolism, Genetics, and Biosynthesis)
Edited by LESTER PACKER

VOLUME 215. Platelets: Receptors, Adhesion, Secretion (Part B)
Edited by JACEK J. HAWIGER

VOLUME 216. Recombinant DNA (Part G)
Edited by RAY WU

VOLUME 217. Recombinant DNA (Part H)
Edited by RAY WU

VOLUME 218. Recombinant DNA (Part I)
Edited by RAY WU

VOLUME 219. Reconstitution of Intracellular Transport
Edited by JAMES E. ROTHMAN

VOLUME 220. Membrane Fusion Techniques (Part A)
Edited by NEJAT DÜZGÜNEŞ

VOLUME 221. Membrane Fusion Techniques (Part B)
Edited by NEJAT DÜZGÜNEŞ

VOLUME 222. Proteolytic Enzymes in Coagulation, Fibrinolysis, and Complement Activation (Part A: Mammalian Blood Coagulation Factors and Inhibitors)
Edited by LASZLO LORAND AND KENNETH G. MANN

VOLUME 223. Proteolytic Enzymes in Coagulation, Fibrinolysis, and Complement Activation (Part B: Complement Activation, Fibrinolysis, and Nonmammalian Blood Coagulation Factors)
Edited by LASZLO LORAND AND KENNETH G. MANN

VOLUME 224. Molecular Evolution: Producing the Biochemical Data
Edited by ELIZABETH ANNE ZIMMER, THOMAS J. WHITE, REBECCA L. CANN, AND ALLAN C. WILSON

VOLUME 225. Guide to Techniques in Mouse Development
Edited by PAUL M. WASSARMAN AND MELVIN L. DEPAMPHILIS

VOLUME 226. Metallobiochemistry (Part C: Spectroscopic and Physical Methods for Probing Metal Ion Environments in Metalloenzymes and Metalloproteins)
Edited by JAMES F. RIORDAN AND BERT L. VALLEE

VOLUME 227. Metallobiochemistry (Part D: Physical and Spectroscopic Methods for Probing Metal Ion Environments in Metalloproteins)
Edited by JAMES F. RIORDAN AND BERT L. VALLEE

VOLUME 228. Aqueous Two-Phase Systems
Edited by HARRY WALTER AND GÖTE JOHANSSON

VOLUME 229. Cumulative Subject Index Volumes 195–198, 200–227 (in preparation)

VOLUME 230. Guide to Techniques in Glycobiology
Edited by WILLIAM J. LENNARZ AND GERALD W. HART

VOLUME 231. Hemoglobins (Part B: Biochemical and Analytical Methods)
Edited by JOHANNES EVERSE, KIM D. VANDEGRIFF, AND ROBERT M. WINSLOW

VOLUME 232. Hemoglobins (Part C: Biophysical Methods)
Edited by JOHANNES EVERSE, KIM D. VANDEGRIFF, AND ROBERT M. WINSLOW

VOLUME 233. Oxygen Radicals in Biological Systems (Part C)
Edited by LESTER PACKER

VOLUME 234. Oxygen Radicals in Biological Systems (Part D) (in preparation)
Edited by LESTER PACKER

VOLUME 235. Bacterial Pathogenesis (Part A: Identification and Regulation of Virulence Factors)
Edited by VIRGINIA L. CLARK AND PATRIK M. BAVOIL

VOLUME 236. Bacterial Pathogenesis (Part B: Integration of Pathogenic Bacteria with Host Cells) (in preparation)
Edited by VIRGINIA L. CLARK AND PATRIK M. BAVOIL

VOLUME 237. Heterotrimeric G Proteins
Edited by RAVI IYENGAR

VOLUME 238. Heterotrimeric G Protein Effectors (in preparation)
Edited by RAVI IYENGAR

VOLUME 239. Nuclear Magnetic Resonance (Part C) (in preparation)
Edited by THOMAS L. JAMES AND NORMAN J. OPPENHEIMER

VOLUME 240. Numerical Computer Methods (Part B) (in preparation)
Edited by MICHAEL L. JOHNSON AND LUDWIG BRAND

VOLUME 241. Retroviral Proteases (in preparation)
Edited by LAWRENCE C. KUO AND JULES A. SHAFER

VOLUME 242. Neoglycoconjugates (in preparation)
Edited by Y. C. LEE AND REIKO T. LEE

VOLUME 243. Inorganic Microbial Sulfur Metabolism (in preparation)
Edited by HARRY D. PECK, JR., AND JEAN LEGALL

Section I

Gα Subunits

[1] Measurement of Receptor-Stimulated Guanosine 5'-O-(γ-Thio)triphosphate Binding by G Proteins

By THOMAS WIELAND *and* KARL H. JAKOBS

Introduction

Many transmembrane signaling processes caused by extracellular hormones and neurotransmitters are mediated by receptors interacting with heterotrimeric ($\alpha\beta\gamma$) guanine nucleotide-binding proteins (G proteins) attached to the inner face of the plasma membrane. Agonist-liganded receptors apparently initiate activation of G proteins by catalyzing the exchange of guanosine 5'-diphosphate (GDP) by guanosine 5'-triphosphate (GTP) bound to the α subunits.[1,2] In membrane preparations and reconstituted systems, this activation process is frequently monitored by studying agonist stimulation of high-affinity GTPase, an enzymatic activity of G-protein α subunits.[3] However, the measurement of G-protein GTPase activity reflects steady-state kinetics of the overall G-protein activity cycle and not only the first step in the signal transduction cascade (i.e., the GDP/GTP exchange reaction). Furthermore, with regard to the molecular stoichiometry of receptor–G-protein interactions, only qualitative but not quantitative data can be obtained.

To study the initial steps of G-protein activation by agonist-liganded receptors in a quantitative manner, the binding of radiolabeled GTP analogs, which are not hydrolyzed by the GTPase activity of G-protein α subunits, to G proteins is determined. Of these GTP analogs, guanosine 5'-O-(γ-[^{35}S]thio)triphosphate ([^{35}S]GTPγS) is most frequently used. This nucleotide has a high affinity for all types of G proteins and is available with a relatively high specific radioactivity (1000–1400 Ci/mmol; physical half-life 87.4 days). Here we describe the measurement of receptor-induced binding of [^{35}S]GTPγS to membranous and detergent-solubilized G proteins and how this method can be adapted to different G proteins for an optimal response to receptor stimulation.

Materials

The [^{35}S]GTPγS (1000–1400 Ci/mmol) is obtained from Du Pont New England Nuclear (Bad Homburg, Germany). The substance is delivered

---

[1] A. G. Gilman, *Annu. Rev. Biochem.* **56**, 615 (1987).
[2] L. Birnbaumer, J. Abramovitz, and A. M. Brown, *Biochim. Biophys. Acta* **1031**, 163 (1990).
[3] D. Cassel and Z. Selinger, *Biochim. Biophys. Acta* **452**, 538 (1976).

in a buffer containing 10 mM N-tris(hydroxymethyl)methylglycine–NaOH, pH 7.6, and 10 mM dithiothreitol (DTT). To minimize decomposition, the solution is diluted 100-fold in this buffer and stored in aliquots at or below −70° before use. If the reagent is not stored at these recommended conditions, and after repeated freezing and thawing, chemical decomposition is rather high.

Unlabeled nucleotides and 3-[(3-cholamidopropyl)dimethylammonio]-1-propane sulfonate (CHAPS) are from Boehringer Mannheim (Mannheim, Germany). N-Ethylmaleimide, N-formylmethionylleucylphenylalanine (fMet-Leu-Phe), isoproterenol, and carbachol are from Sigma (St. Louis, MO). Glass fiber filters (GF/C) are from Whatman (Clifton, NJ), and nitrocellulose filters (pore size 0.45 µm) are from Schleicher and Schuell (Keene, NH).

Membranes of various cells and tissues are prepared as previously described[4-7] and stored in aliquots at −70°. Before use in the binding assay, the membranes are thawed, diluted with 10 mM triethanolamine hydrochloride, pH 7.4, containing 5 mM EDTA, centrifuged for 10–30 min at 30,000 g, and resuspended in 10 mM triethanolamine hydrochloride, pH 7.4, at the appropriate membrane protein concentration.

Equipment

    Incubator or water bath
    Filtration funnel with vacuum pump
    Cooled centrifuge (4°, up to 30,000 g) for membrane preparation
    Ultracentrifuge with fixed-angle and swing-out rotors for preparation of membranes, solubilized proteins, and sucrose density gradient centrifugation
    Shaker to equilibrate the filters with the scintillation cocktail
    Liquid scintillation spectrometer
    Freezer (preferably −70° or lower) for storage of membranes and [^{35}S]GTPγS

Measurement of Agonist-Induced [^{35}S]GTPγS Binding to G Proteins in Membranes

The assay is performed in 3-ml plastic reaction tubes. The assay volume is 100 µl. The final concentrations of the reaction mixture constituents

---

[4] P. Gierschik, M. Steisslinger, D. Sidiropoulos, E. Herrmann, and K. H. Jakobs, *Eur. J. Biochem.* **283**, 97 (1989).
[5] G. Hilf and K. H. Jakobs, *Eur. J. Pharmacol.* **172**, 155 (1989).
[6] D. S. Papermaster and W. J. Dreyer, *Biochemistry* **13**, 2438 (1974).
[7] G. Puchwein, T. Pfeuffer, and E. J. M. Helmreich, *J. Biol. Chem.* **249**, 3232 (1974).

are as follows: triethanolamine hydrochloride (pH 7.4), 50 mM; MgCl$_2$, 5 mM; EDTA, 1 mM; DTT, 1 mM; NaCl, 0–150 mM; GDP, 0–100 μM; [^{35}S]GTPγS, 0.3–0.5 nM (~50 nCi). The incubation temperature and membrane concentration as well as the concentrations of NaCl and GDP have to be adjusted to the individual cell type and the G-protein subtype activated by the receptor under study.

1. The reaction mixture (40 μl) together with the receptor agonist or its diluent (10 μl) are thermally preequilibrated for 5 min at the desired reaction temperature.
2. The binding reaction is started by addition of the membrane suspension (50 μl) and vortexing.
3. Samples are incubated for the appropriate incubation time, for example, 60 min at 30°.
4. The incubation is terminated by the addition of 2.5 ml of an ice-cold washing buffer (50 mM Tris-HCl, pH 7.5, 5 mM MgCl$_2$).
5. This mixture is passed through the filtration funnel. For systems containing only membrane-bound G proteins, Whatman GF/C glass fiber filters are used. In systems containing soluble G proteins (e.g., transducin), nitrocellulose filters are required.
6. The reaction tube is washed two times with 2.5 ml of the washing buffer, and this solution is also passed through the same filter.
7. The filter is additionally washed two times with 2.5 ml of the washing buffer and then dried at room temperature.
8. The dried filters are put into 5-ml counting vials and equilibrated with 4 ml of a scintillation cocktail for 20 min at room temperature by moderate shaking. Any commercially available scintillation cocktail suitable for counting of ^{35}S can be used. Also a self-made cocktail consisting of 2 liters toluene, 1 liter Triton X-100, 15 g 2,5-diphenyloxazole, and 3 g 2,2'-p-phenylenebis(4-methyl-5-phenyloxazole) can be used.

Application to Various Cell Types and Different G Proteins

In membranes of various cell types, including human neutrophils,[8] human platelets,[9] human leukemia cells (HL-60),[10,11] rat myometrium,[12]

---

[8] R. Kupper, B. Dewald, K. H. Jakobs, M. Baggiolini, and P. Gierschik, *Biochem. J.* **282**, 429 (1992).

[9] C. Gachet, J.-P. Cazenave, P. Ohlmann, G. Hilf, T. Wieland, and K. H. Jakobs, *Eur. J. Biochem.* **207**, 259 (1992).

[10] P. Gierschik, R. Moghtader, C. Straub, K. Dieterich, and K. H. Jakobs, *Eur. J. Biochem.* **197**, 725 (1991).

[11] T. M. Schepers, M. E. Brier, and K. R. McLeish, *J. Biol. Chem.* **267**, 159 (1992).

[12] C. Liebmann, M. Schnittler, M. Nawrath, and K. H. Jakobs, *Eur. J. Pharmacol.* **207**, 67 (1991).

and pig atrium,[13] as well as in reconstituted systems,[14-16] an agonist-induced increase in [^{35}S]GTPγS binding to G proteins is optimal at millimolar concentrations (1–5 mM) of free Mg^{2+}. In general, the membrane concentration (and the incubation time) should be kept at a level at which not more than 20–30% of the total added [^{35}S]GTPγS is bound. For most tissues, 1–10 μg of membrane protein/tube is appropriate. Even lower amounts are required in systems with a high content of G proteins (e.g., rod outer segment membranes). To optimize the response to agonist-activated receptors, the assay system has to be adapted to the respective G protein interacting with these receptors. The major strategy is to minimize the agonist-independent binding to G proteins without lowering the agonist-induced binding.

Regulation by Guanosine 5'-Diphosphate, Sodium Chloride, and Temperature

As reported for native HL-60 plasma membranes, the G proteins are initially in a GDP-liganded form, and, thus, the agonist-independent binding of [^{35}S]GTPγS to G proteins is limited by the dissociation of GDP from the binding sites.[17] Two components may contribute to this agonist-independent reaction: (1) spontaneous, receptor-independent dissociation of G-protein-bound GDP and (2) GDP release induced by agonist-unliganded receptors.[18] The extent of the spontaneous agonist-independent release of GDP from G proteins differs between various cell types and apparently even more between different G-protein subtypes.

Therefore, several approaches can be used to adapt the assay to the respective system. In Fig. 1, receptor-stimulated [^{35}S]GTPγS binding to three different G proteins is shown. In HL-60 membranes, G proteins of the G$_i$ subtype interacting with chemotactic receptors are predominantly seen.[17] When these membranes are incubated with [^{35}S]GTPγS at 30° (Fig. 1A), which condition leads to a rapid spontaneous dissociation of GDP from the G proteins, a rather high basal binding rate is obtained, and addition of a receptor agonist (e.g., fMet-Leu-Phe) does not cause a further increase in binding. GDP has to be added to keep the G$_i$ proteins in the GDP-liganded form required for optimal receptor action. The additional presence of NaCl at an optimal concentration of 100–150 mM increases

---

[13] G. Hilf, P. Gierschik, and K. H. Jakobs, *Eur. J. Biochem.* **186**, 725 (1989).
[14] A. B. Fawzi and J. K. Northup, *Biochemistry* **29**, 3804 (1990).
[15] M. R. Tota, K. R. Kahler, and M. I. Schimerlik, *Biochemistry* **26**, 8175 (1987).
[16] V. A. Florio and P. C. Sternweis, *J. Biol. Chem.* **264**, 3909 (1989).
[17] T. Wieland, J. Kreiss, P. Gierschik, and K. H. Jakobs, *Eur. J. Biochem.* **205**, 1201 (1992).
[18] G. Hilf and K. H. Jakobs, *Eur. J. Pharmacol.* **225**, 245 (1992).

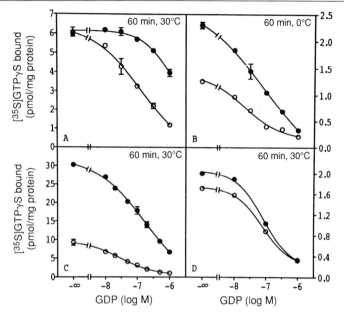

FIG. 1. Regulation by GDP of receptor-stimulated [^{35}S]GTPγS binding to different G proteins. HL-60 membranes (A, B), bovine rod outer segment membranes (C), and turkey erythrocyte membranes (D) were incubated with approximately 0.5 nM [^{35}S]GTPγS in a reaction mixture containing 50 mM triethanolamine hydrochloride, pH 7.4, 5 mM MgCl$_2$, 1 mM EDTA, 1 mM DTT, 150 mM NaCl, and GDP, at the indicated concentrations, for 60 min at either 30° (A, C, D) or 0° (B) in the absence (○) and presence of receptor stimuli (●) (10 μM fMet-Leu-Phe in A and B, bright white light in C, 10 μM isoproterenol in D). For further details, see text. Means plus or minus the standard deviation of triplicate determinations are given.

further the inhibitory influence of GDP on basal binding, probably by uncoupling G proteins from unoccupied receptors.[10,19] In membranes from other cell types, addition of KCl may give similar or even better results.

In contrast, when binding is performed at 0° with a low spontaneous rate of GDP release, basal binding of [^{35}S]GTPγS is much lower, and, most importantly, fMet-Leu-Phe-induced binding is observed without addition of GDP (Fig. 1B). At 0°, addition of either GDP or NaCl exerted only negative effects on receptor-induced binding.[17] Thus, for measuring receptor-stimulated binding of [^{35}S]GTPγS to G$_i$ or G$_o$[16] proteins either incubation at low temperature (e.g., 0°) or, at higher temperatures

[19] P. Gierschik, D. Sidiropoulos, M. Steisslinger, and K. H. Jakobs, *Eur. J. Pharmacol.* **172**, 481 (1989).

($25°-37°$), addition of GDP ($0.1-10 \mu M$) and NaCl ($100-150$ mM) is recommended.

The retinal G protein transducin is known to have a very low spontaneous dissociation rate for GDP, even when incubated at higher temperatures.[14] Thus, as shown in Fig. 1C, addition of GDP is not required for stimulation of GTPγS binding by light-activated rhodopsin (by illumination of bovine rod outer segment membranes adapted to dim red light). Because the binding of GTPγS observed in dim red light is more sensitive to inhibition by GDP and ionic strength than that observed in bright white light, addition of GDP and NaCl (or KCl) will result in a marked increase in the level (-fold) stimulation by illumination.

The membranous G_s protein seems to have a low spontaneous rate of GDP dissociation. As shown in Fig. 1D, agonist (isoproterenol) activation of β-adrenoceptors interacting with G_s proteins induces a slight but significant increase in binding of [^{35}S]GTPγS when turkey erythrocyte membranes are incubated for 60 min at $30°$. Similarly, as described for HL-60 membranes at $0°$, addition of GDP and NaCl only decreases receptor-induced binding and, therefore, should be avoided in this system. Moreover, in turkey erythrocyte membranes, the isoproterenol-independent binding of [^{35}S]GTPγS is probably due to binding to GDP-free G proteins other than G_s. Thus, additional strategies have to be used to reduce the agonist-independent binding of the radioligand.

Treatment of Membranes with N-Ethylmaleimide and/or Unlabeled GTPγS to Reduce Agonist-Independent Binding of [^{35}S]GTPγS

1. Turkey erythrocyte membranes (3–5 mg protein) are incubated for 30 min on ice in 50-ml centrifugation tubes in a reaction mixture (total volume 9 ml) containing 50 mM triethanolamine hydrochloride, pH 7.4, 10 mM N-ethylmaleimide (NEM), and 1 mM EDTA.

2. One milliliter of a 150 mM 2-mercaptoethanol solution is added to stop protein alkylation by NEM, followed by the addition of 30 ml of ice-cold 10 mM triethanolamine hydrochloride, pH 7.4.

3. Membranes are pelleted by centrifugation (10 min at 30,000 g), resuspended in 1 ml of 10 mM triethanolamine hydrochloride, pH 7.4, and further treated with unlabeled GTPγS or stored in aliquots at $-70°$.

4. The NEM-treated or untreated membranes are incubated for 30 min at $30°$ in a reaction mixture (final volume 2 ml) containing 50 mM triethanolamine hydrochloride, pH 7.4, 5 mM $MgCl_2$, 1 mM EDTA, and 1 μM unlabeled GTPγS.

5. After addition of 8 ml of ice-cold 10 mM triethanolamine hydrochloride, pH 7.4, membranes are pelleted by centrifugation (10 min at

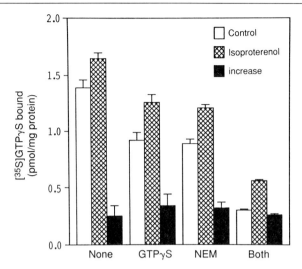

FIG. 2. Optimization of β-adrenoceptor-induced [^{35}S]GTPγS binding to turkey erythrocyte membrane G_s proteins. Turkey erythrocyte membranes were preincubated with either 10 mM NEM, 1 μM unlabeled GTPγS, or both agents as described in the text. Thereafter, binding of 0.45 nM [^{35}S]GTPγS to untreated (none), GTPγS-treated, NEM-treated, and membranes treated with both agents (both) (10 μg protein each) was determined in the absence (open bars) and presence of 10 μM isoproterenol (cross-hatched bars). The β-adrenoceptor-induced increase in [^{35}S]GTPγS binding is given by the filled bars. Means of assay triplicates plus or minus the standard deviation are shown.

30,000 g), resuspended in 1 ml of 10 mM triethanolamine hydrochloride, pH 7.4, and stored in aliquots at $-70°$.

The thiol group-alkylating agent NEM has been shown to modify cysteine residues on pertussis toxin-sensitive G proteins. Similar to $G_{i/o}$ proteins ADP-ribosylated by pertussis toxin, NEM-treated $G_{i/o}$ proteins cannot be activated by receptors.[20] When turkey erythrocyte membranes are preincubated with NEM, the isoproterenol-independent binding of GTPγS is lowered by about 35%, whereas the β-adrenergic agonist-induced binding is not altered (Fig. 2). A similar reduction in agonist-independent binding is obtained by treatment of the membranes with a saturating concentration of unlabeled GTPγS in the absence of agonists. Most likely, unlabeled GTPγS binds preferentially to G proteins with a high spontaneous GDP release and, therefore, prevents binding of radiolabeled GTPγS to these G proteins. In turkey erythrocyte membranes, a combination of

[20] M. Ui, in "ADP-Ribosylating Toxins and G Proteins" (J. Moss and M. Vaughan, eds.), p. 45. American Society for Microbiology, Washington, D.C., 1990.

both treatments gives the best results in lowering isoprotereno1-independent binding (~75% reduction) without affecting the β-adrenoceptor-mediated increase in [^{35}S]GTPγS binding. In human platelet membranes (not shown), NEM treatment alone was sufficient for improving relative prostaglandin E_1-induced binding of [^{35}S]GTPγS to G_s. Thus, to study agonist-induced GTPγS binding to G_s in membranes of other tissues, a stepwise treatment and testing after each treatment should be performed.

Stoichiometry of Receptor–G-Protein Interactions

For quantitative analysis of the stoichiometry of receptor–G-protein interactions, the amount of receptors under study as well as the amount of G proteins present in the given membrane preparation have to be measured. The concentration of receptors is determined by binding saturation analysis with receptor-specific radioligands. The concentration of G proteins activated by the receptors is analyzed by measuring agonist-stimulated binding of [^{35}S]GTPγS under conditions optimal for the given receptor and G protein under study (see above). Furthermore, the incubation time has to be adjusted to give a maximal absolute increase in agonist stimulation of [^{35}S]GTPγS binding. In addition, the assay contains increasing concentrations of GTPγS (up to 0.5–1 μM) in the absence and presence of the receptor agonist at a maximally effective concentration. Under ideal conditions, the resulting binding data can be analyzed by Scatchard transformation. Such an analysis performed for formyl peptide receptor-stimulated binding of GTPγS to G proteins in HL-60 membranes is exemplified in Fig. 3. In these membranes, a stoichiometry of up to 20 G proteins activated by a one agonist-liganded formyl peptide receptor has been determined by this technique.[10]

Measurement of Agonist-Induced [^{35}S]GTPγS Binding in Solubilized Systems

Solubilization of Membrane Proteins and Sucrose Density Gradient Centrifugation

1. Membrane aliquots are thawed, pelleted by centrifugation (30 min at 30,000 g) and resuspended in solubilization buffer [Tris-HCl, pH 7.4, 50 mM; EDTA, 1 mM; DTT, 1 mM; CHAPS, 1% (by mass)] at a protein concentration of 1% (by mass).

2. Proteins are solubilized by mild agitation for 1 hr at 4°.

3. The mixture is centrifuged for 1 hr at 200,000 g, and the supernatant is passed through a nitrocellulose membrane (pore size 0.45 μm).

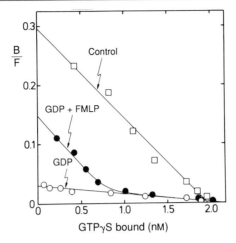

FIG. 3. Quantification of G proteins activated by an agonist-activated receptor in HL-60 membranes. Binding of GTPγS to HL-60 membranes (3 μg protein) was determined in a total volume of 100 μl in the absence (control, □) and presence of 1 μM GDP (○) or 1 μM GDP plus 10 μM fMet-Leu-Phe (GDP + FMLP, ●) at increasing concentrations of GTPγS (0.3–500 nM). A Scatchard analysis of the binding data is shown. For further experimental details, see text.

4. Beckman 5-ml Ultraclear tubes are stepwise filled by sublayering with 1.2 ml of each gradient buffer containing 50 mM Tris-HCl, pH 7.4, 1 mM EDTA, 1 mM DTT, 0.3% CHAPS, and 5, 10, 15, or 20% sucrose. Gradients are kept for 3 hr at room temperature for linearizing.

5. Next, 150 μl solubilizate and 50 μl solubilization buffer including molecular weight marker proteins (e.g., catalase and cytochrome c) are layered on each gradient.

6. Gradients are centrifuged for 12 hr at 50,000 rpm in a Beckman SW 50 rotor. Thereafter, tubes are punctured and fractions (~250 μl) are collected by trapping drops into 1.5-ml Eppendorf plastic tubes.

7. The distribution of the marker proteins is measured by spectrometric methods.

Receptor-Stimulated [^{35}S]GTPγS Binding in Fractions Containing Both Receptor and G Proteins

The distribution of the receptor in the sucrose density gradient fractions is assessed by measuring binding of radiolabeled agonists or antagonists for the respective receptor. For identification of G proteins, 10 μl of each fraction is incubated for 60 min at 30° in 3-ml plastic reaction tubes with 40 μl of a reaction mixture containing 25 mM HEPES, pH 8.0, 20 mM

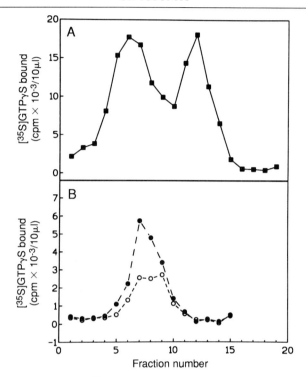

FIG. 4. Receptor-stimulated [^{35}S]GTPγS binding in sucrose density gradient fractions of solubilized pig atrial membranes. Pig atrial membranes were solubilized and sucrose gradient fractions prepared as described in the text. (A) The distribution of G proteins was determined by measuring binding of 10 nM [^{35}S]GTPγS in 10-μl aliquots of each fraction. (B) Muscarinic acetylcholine receptor-stimulated binding of [^{35}S]GTPγS (0.6 nM) was determined in 10-μl aliquots of the fractions in the absence (○) and presence of 100 μM carbachol (●). For experimental details of [^{35}S]GTPγS binding, see text.

$MgCl_2$, 1 mM EDTA, 1 mM DTT, 100 mM NaCl, 0.124% Lubrol PX, and 10 nM GTPγS, including 0.6–0.8 nM [^{35}S]GTPγS. The incubation is terminated by the addition of 2.5 ml of ice-cold washing buffer (50 mM Tris-HCl, pH 7.5, 5 mM $MgCl_2$). The mixture is rapidly passed through nitrocellulose filters, and filters are washed and treated as described above. As shown for solubilized G proteins from pig atrial membranes,[21] two distinct peaks of G proteins are obtained by sucrose density gradient centrifugation (Fig. 4A). The first peak with a higher sedimentation coefficient (about 6.4 S) contains essentially heterotrimeric G proteins, whereas

[21] G. Hilf and K. H. Jakobs, *Cellular Signalling* **4,** 787 (1992).

the second peak (sedimentation coefficient about 3.3 S) is mainly due to small molecular mass G proteins.

In fractions containing both heterotrimeric G proteins and receptor(s), agonist-induced binding of [^{35}S]GTPγS can be measured using assay conditions similarly as described for membrane preparations. For example, 10 μl of each sucrose density gradient fraction obtained from solubilized pig atrial membranes are incubated in a total volume of 60 μl for 90 min at 25° with a reaction mixture containing 50 mM Tris-HCl, pH 7.4, 5 mM MgCl$_2$, 1 mM EDTA, 1 mM DTT, 100 mM NaCl, 5 μM GDP, and 0.6 nM [^{35}S]GTPγS, with or without the muscarinic agonist carbachol at 100 μM. The final concentration of the detergent CHAPS should be maintained at 0.1%. At lower concentrations, precipitation of solubilized proteins may occur, whereas higher CHAPS concentrations will lower the agonist-induced binding. The reaction is stopped, and binding of [^{35}S]GTPγS to solubilized G proteins is analyzed by filtration through nitrocellulose filters as described above. In fractions containing both muscarinic acetylcholine receptors and heterotrimeric G proteins, carbachol stimulates binding of [^{35}S]GTPγS to solubilized G proteins by about 2-fold (Fig. 4B). Similar data have been obtained for fMet-Leu-Phe-stimulated binding of [^{35}S]GTPγS in sucrose density gradient fractions of HL-60 membranes containing both solubilized formyl peptide receptors and G$_i$ proteins (not shown). These fractions may then be pooled and used for identification and/or purification of the G protein(s) interacting with the respective receptors.

[2] Receptor-Stimulated Hydrolysis of Guanosine 5'-Triphosphate in Membrane Preparations

By PETER GIERSCHIK, THOMAS BOUILLON, and KARL H. JAKOBS

Introduction

The most prominent functional characteristic of signal-transducing heterotrimeric guanine nucleotide-binding proteins (G proteins) is their cyclic movement through a series of activation and deactivation steps, which is characterized by changes in the nature of the bound guanine nucleotide and in the status of the subunit association. Activation of G proteins is initiated by the release of guanosine 5'-diphosphate (GDP) from the heterotrimeric G protein, followed by the binding of guanosine 5'-triphosphate (GTP) and the dissociation of the αβγ heterotrimer into GTP-liganded α subunit and free βγ dimer. Both free α subunits and free βγ dimers are capable of effector regulation.[1,2] G-protein deactivation results from hydrolysis of GTP bound to the α subunit by its intrinsic GTPase activity, followed by the reassociation of the now GDP-bound α subunit with the βγ dimer. Agonist-occupied receptors accelerate the movement of the G protein through the cycle by facilitating both the release of GDP from the GTP-mediated activation of the G protein (see Ref. 3 for review). The actual hydrolysis of GTP by the α subunit does not appear to be influenced by the receptor,[3] but it may be accelerated by effectors, at least in certain cases.[4-7]

The activation/deactivation cycle just described was first proposed by Cassel and Selinger on the basis of studies on the effects of β-adrenoceptor stimuli and cholera toxin on the hydrolysis of [γ-^{32}P]GTP by turkey erythrocyte membranes.[8-10] Aside from laying the foundation of our current understanding of the molecular mechanisms of G-protein action, the pioneering work by Cassel and Selinger established the methodology for measuring receptor-stimulated hydrolysis of GTP in membrane preparations. This methodology has since been used by many investigators to study the interaction of receptors with G proteins in native and reconstituted systems, and, in conjunction with the use of bacterial toxins, to characterize the nature of the G protein(s) coupled to a given receptor in a given membrane system.

The purpose of this chapter is to describe the methods used in our laboratories to assay receptor-stimulated GTP hydrolysis in native membrane preparations. Experiments done with membranes from myeloid differentiated HL-60 cells and human peripheral blood neutrophils are shown to exemplify important features of this reaction. The stimuli used are the formyl peptide receptor agonist N-formylmethionylleucylphenylalanine (fMet-Leu-Phe) and the chemokine interleukin 8. The receptors for the two stimuli are coupled to pertussis toxin-sensitive G proteins, most likely to G_{i2} and G_{i3}.[11,12]

[1] J. R. Hepler and A. G. Gilman, *Trends Biochem. Sci.* **17**, 383 (1992).
[2] L. Birnbaumer, *Cell (Cambridge, Mass.)* **71**, 1069 (1992).
[3] A. G. Gilman, *Annu. Rev. Biochem.* **56**, 615 (1987).
[4] V. Y. Arshavsky and M. D. Bownds, *Nature (London)* **357**, 416 (1992).
[5] F. Pagès, P. Deterre, and C. Pfister, *J. Biol. Chem.* **267**, 22018 (1992).
[6] G. Berstein, J. L. Blank, D.-Y. Jhon, J. H. Exton, S. G. Rhee, and E. M. Ross, *Cell (Cambridge, Mass.)* **70**, 411 (1992).
[7] H. R. Bourne and L. Stryer, *Nature (London)* **358**, 541 (1992).
[8] D. Cassel and Z. Selinger, *Biochim. Biophys. Acta* **452**, 538 (1976).
[9] D. Cassel and Z. Selinger, *Proc. Natl. Acad. Sci. U.S.A.* **74**, 3307 (1977).
[10] D. Cassel, H. Levkovitz, and Z. Selinger, *J. Cyclic Nucleotide Res.* **3**, 393 (1977).
[11] P. Gierschik, D. Sidiropulos, and K. H. Jakobs, *J. Biol. Chem.* **264**, 21470 (1989).
[12] R. W. Kupper, B. Dewald, K. H. Jakobs, M. Baggiolini, and P. Gierschik, *Biochem. J.* **282**, 429 (1992).

The first of the two methods we describe is a modification of the protocol originally developed by Cassel and Selinger.[8] This method has been used to assay GTP hydrolysis stimulated by many agonists in membranes prepared from a great number of cell types and tissues. Examples are the stimulation of GTP hydrolysis in rat pancreatic plasma membranes by cholecystokinin and secretin,[13] in frog erythrocyte membranes by β-adrenoceptor agonists and prostaglandin (PG) E_1,[14] in rat liver plasma membranes by glucagon,[15] in membranes of human mononuclear cells by prostaglandins (PGE_1, PGE_2, PGA_1, PGB_1, $PGF_{1\alpha}$),[16] in membranes of human platelets by α_2-adrenoceptor agonists,[17–19] prostaglandin E_1,[17,18–20] thrombin,[21,22] platelet-activating factor,[23] and thromboxane (TX) A_2 receptor agonists,[24] in membranes of hamster adipocytes by α_2-adrenoceptor agonists,[25] adenosine receptor agonists,[26,27] and prostaglandins ($PGE_1 \geq PGE_2 > PGF_{2\alpha} > PGD_2$),[25,26] in S49 mouse lymphoma cell membranes by somatostatin,[28–30] in membranes of NG108-15 neuroblastoma × glioma hybrid cells by opioid receptor agonists,[31–33] α_2-adrenoceptor agonists,[33] and bradykinin,[34] in rat striatal synaptic plasma membranes by muscarinic cholinoceptor agonists,[35] in rat brain membranes by opioid receptor agonists,[36] in cardiac membranes by muscarinic cholinoceptor agonists,[37,38]

[13] M. Lambert, M. Svoboda, and J. Christophe, *FEBS Lett.* **99**, 303 (1979).
[14] L. J. Pike and R. J. Lefkowitz, *J. Biol. Chem.* **255**, 6860 (1980).
[15] N. Kimura and N. Shimada, *FEBS Lett.* **117**, 172 (1980).
[16] A. J. Bitonti, J. Moss, N. N. Tandon, and M. Vaughan, *J. Biol. Chem.* **255**, 2026 (1980).
[17] K. Aktories and K. H. Jakobs, *FEBS Lett.* **130**, 235 (1981).
[18] K. Aktories, G. Schultz, and K. H. Jakobs, *FEBS Lett.* **146**, 65 (1982).
[19] K. Aktories, G. Schultz, and K. H. Jakobs, *Naunyn-Schmiedeberg's Arch. Pharmacol.* **324**, 196 (1983).
[20] H. A. Lester, M. L. Steer, and A. Levitzki, *Proc. Natl. Acad. Sci. U.S.A.* **79**, 719 (1982).
[21] K. Aktories and K. H. Jakobs, *Eur. J. Biochem.* **145**, 333 (1984).
[22] R. Grandt, K. Aktories, and K. H. Jakobs, *Biochem. J.* **237**, 669 (1986).
[23] S.-B. Hwang, *J. Biol. Chem.* **263**, 3225 (1988).
[24] A. Shenker, P. K. Goldsmith, C. G. Unson, and A. M. Spiegel, *J. Biol. Chem.* **266**, 9309 (1991).
[25] K. Aktories, G. Schultz, and K. H. Jakobs, *Mol. Pharmacol.* **21**, 336 (1982).
[26] K. Aktories, G. Schultz, and K. H. Jakobs, *Life Sci.* **30**, 269 (1982).
[27] K. Aktories, G. Schultz, and K. H. Jakobs, *Biochim. Biophys. Acta* **719**, 58 (1982).
[28] K. H. Jakobs, K. Aktories, and G. Schultz, *Nature (London)* **303**, 177 (1983).
[29] K. Aktories, G. Schultz, and K. H. Jakobs, *Mol. Pharmacol.* **24**, 183 (1983).
[30] K. Aktories, G. Schultz, and K. H. Jakobs, *FEBS Lett.* **158**, 169 (1983).
[31] G. Koski and W. A. Klee, *Proc. Natl. Acad. Sci. U.S.A.* **78**, 4185 (1981).
[32] G. Koski, R. A. Streaty, and W. A. Klee, *J. Biol. Chem.* **257**, 14035 (1982).
[33] D. L. Burns, E. L. Hewlett, J. Moss, and M. Vaughan, *J. Biol. Chem.* **258**, 1435 (1983).
[34] R. Grandt, C. Greiner, P. Zubin, and K. H. Jakobs, *FEBS Lett.* **196**, 279 (1986).
[35] P. Onali, M. C. Olianas, J. P. Schwartz, and E. Costa, *Mol. Pharmacol.* **24**, 380 (1983).
[36] P. H. Franklin and W. Hoss, *J. Neurochem.* **43**, 1132 (1984).
[37] J. W. Fleming and A. M. Watanabe, *Circ. Res.* **64**, 340 (1988).
[38] G. Hilf and K. H. Jakobs, *Eur. J. Pharmacol.* **172**, 155 (1989).

in *Dictyostelium discoideum* membranes by cAMP receptor agonists,[39] in membranes of peripheral blood neutrophils membranes by formyl peptide receptor agonists,[40-43] complement C5a,[41,42] leukotriene B_4,[41,43] platelet-activating factor,[23] and interleukin 8,[12] and in membranes of myeloid differentiated HL-60 cells by fMet-Leu-Phe[44,45] and leukotriene B_4.[46]

The second method is a simplified version of the standard protocol, which we have developed for membranes of myeloid differentiated HL-60 and peripheral blood neutrophils. We expect that this methodology will be applicable to other membrane systems and offer the advantages discussed below in these systems as well.

Preparation of Plasma Membranes

Preparation of Membranes from Myeloid Differentiated HL-60 Cells

Human promyelocytic HL-60 cells[47] (ATCC, Rockville, MD, CCL 240) are grown in suspension culture as described elsewhere in this series.[48] Cells are induced to differentiate into mature myeloid forms by cultivation in the presence of 1.25% (v/v) dimethyl sulfoxide for 5 days.[49] The preparation of membranes from differentiated cells is described in Ref. 48.

Preparation of Plasma Membranes from Human Peripheral Neutrophils

Human neutrophils are prepared from donor blood and homogenized as described.[12] The protocol we use to prepare plasma membranes from

[39] B. E. Snaar-Jagalska, K. H. Jakobs, and P. J. M. van Haastert, *FEBS Lett.* **236**, 139 (1988).
[40] F. Okajima, T. Katada, and M. Ui, *J. Biol. Chem.* **260**, 6761 (1985).
[41] D. E. Feltner, R. S. Smith, and W. A. Marasco, *J. Immunol.* **137**, 1961 (1986).
[42] M. W. Wilde, K. E. Carlson, D. R. Manning, and S. H. Zigmond, *J. Biol. Chem.* **264**, 190 (1989).
[43] C. Pelz, T. Matsumoto, T. F. P. Molski, E. L. Becker, and R. I. Sha'afi, *J. Cell. Biochem.* **39**, 197 (1989).
[44] P. Gierschik, D. Sidiropoulos, M. Steissinger, and K. H. Jakobs, *Eur. J. Pharmacol.* **172**, 481 (1989).
[45] P. Gierschik, M. Steissinger, D. Sidiropoulos, E. Herrmann, and K. H. Jakobs, *Eur. J. Biochem.* **183**, 97 (1989).
[46] K. R. McLeish, P. Gierschik, T. Schepers, D. Sidiropoulos, and K. H. Jakobs, *Biochem. J.* **260**, 427 (1989).
[47] S. J. Collins, R. C. Gallo, and R. E. Gallagher, *Nature (London)* **270**, 347 (1977).
[48] P. Gierschik and M. Camps, this series, Vol. 238 [14].
[49] S. J. Collins, F. W. Ruscetti, R. E. Gallagher, and R. C. Gallo, *Proc. Natl. Acad. Sci. U.S.A.* **75**, 2458 (1978).

the homogenate is a modification of the procedure developed by Borregard *et al.*[50] Details are given in Ref. 12.

Determination of GTP Hydrolysis in Membrane Preparations

Method I

Hydrolysis of [γ-^{32}P]GTP (NEN, Boston, MA, Cat. No. NEG-040; 30 Ci/mmol) is measured in a reaction mixture (100 μl) containing 50 mM triethanolamine hydrochloride, pH 7.4 at 30°, 2 mM MgCl$_2$, 1 mM dithiothreitol (DTT), 0.1 mM EGTA, 2 mg/ml bovine serum albumin (BSA, Boehringer Mannheim, Mannheim, Germany, Cat. No. 735 086), 0.8 mM adenosine 5'-(β,γ-imino)triphosphate [App(NH)p], 0.1 mM ATP, 0.4 mg/ml creatine kinase (Boehringer Mannheim, Cat. NO. 120 969), 5 mM creatine phosphate (di-Tris salt, Sigma, St. Louis, MO, Cat. No. P 4635), and 0.1 μM [γ-^{32}P]GTP (10 Ci/mmol; mixed from [γ-^{32}P]GTP and unlabeled GTP). All unlabeled nucleotides are obtained from Boehringer Mannheim [App(NH)p, Cat. No. 102 547; ATP, Cat. No. 126 888; GTP, Cat. No. 106 356].

The reactions are started by the addition of membranes (5–15 μg of protein in 50 μl of 10 mM triethanolamine hydrochloride, pH 7.4) to the prewarmed (5 min, 30°) reaction mixture (50 μl) and are then allowed to proceed for 15 min at 30°. The reactions are terminated by the addition of 700 μl of ice-cold sodium phosphate (10 mM, pH 2.0) containing 5% (w/v) activated charcoal (Riedel de Haën, Seelze, Germany, Cat. No. 18001). The reaction tubes are then centrifuged for 30 min at 12,000 g, and 500 μl of the supernatant is used for determination of Cerenkov radiation. Low K_m GTPase activity is calculated by subtracting the high K_m GTPase activity (determined at 0.1 μM [γ-^{32}P]GTP plus 50 μM unlabeled GTP) from the total GTPase activity (determined at 0.1 μM [γ-^{32}P]GTP).[8,25] This is conveniently done by assaying one membrane sample without stimulus in the presence of 0.1 μM [γ-^{32}P]GTP plus 50 μM unlabeled GTP and subtracting the Cerenkov value of this sample as a blank from all other values. This procedure also compensates for the [^{32}P]P$_i$ present as a contaminant in the [γ-^{32}P]GTP preparation. We routinely adjust the assay conditions (e.g., amount of membrane protein, incubation time, temperature) so that the blank value is equal to or smaller than 20% of the value obtained for unstimulated low K_m GTPase activity.

Comments. The charcoal suspension is stirred on ice in a beaker with a magnetic stirring bar. The App(NH)p is included in the incubation me-

---

[50] N. Borregard, J. M. Heiple, E. R. Simons, and R. A. Clark, *J. Cell Biol.* **97**, 52 (1983).

dium to suppress nonspecific nucleoside triphosphatase activity present in many membranes.[8] ATP and the ATP regeneration system creatine kinase plus creatine phosphate are present to prevent the nucleoside diphosphokinase-mediated transfer of the [γ-^{32}P]phosphate group of [γ-^{32}P]GTP to endogenously present ADP, followed by the degradation of newly generated [γ-^{32}P]ATP by specific ATPases.[8] Additional assay components, for example, ouabain[31] or protease inhibitors,[13] may be necessary in specific membranes to inhibit Na$^+$,K$^+$-ATPase or proteolytic degradation of GTPase stimuli and/or membrane components involved in receptor-stimulated GTP hydrolysis, respectively. Adenosine 5'-O-(3-thio)triphosphate (ATPγS) was used as a concentration of 0.1 mM in addition to ATP and an ATP regenerating system in *Dictyostelium discoideum* membranes to maximally reduce redistribution of radioactivity among guanine and adenine nucleotides.[39]

Although stimulation of GTPase by receptors coupled to the stimulatory G protein of adenylyl cyclase (adenylate cyclase), namely, G$_s$, has been reported in the literature,[8,13–17,20,25,27] the degree of this stimulation is frequently relatively modest, and problems have been encountered in obtaining this response in many cases.[8,13,15,16,25,31,32,51] A potential solution is treatment of the membranes with the sulfhydryl group alkylating agent *N*-ethylmaleimide, which has been shown to suppress basal low K_m GTPase activity and thereby improve the detection of GTP hydrolysis stimulated by receptors coupled to G$_s$.[8,18,20] Note, however, that *N*-ethylmaleimide is a potent inhibitor of receptor-stimulated GTP hydrolysis by pertussis toxin-sensitive G proteins.[29,52] An alternative approach is to determine GTP hydrolysis by G$_s$ indirectly by measuring the decay of receptor-stimulated adenylate cyclase activity.[10,53–55] The maximal hydrolysis of [γ-^{32}P]GTP should not exceed 10–20% of the total [γ-^{32}P]GTP present in the assay to avoid problems associated with large changes in the substrate concentration and/or isotope dilution (i.e., generation of unlabeled GTP from GDP) by the regenerating system.

The stimulatory effect of agonists on GTP hydrolysis typically ensues without delay. Under the conditions specified above, both basal and agonist-stimulated hydrolysis of GTP are linear with time for at least 15 min

---

[51] H. R. Kaslow, Z. Farfel, G. L. Johnson, and H. R. Bourne, *Mol. Pharmacol.* **15**, 472 (1979).

[52] K. H. Jakobs, P. Lasch, M. Minuth, K. Aktories, and G. Schultz, *J. Biol. Chem.* **257**, 2829 (1982).

[53] D. Cassel, F. Eckstein, M. Lowe, and Z. Selinger, *J. Biol. Chem.* **254**, 9835 (1979).

[54] S. B. Masters, R. T. Miller, M.-H. Chi, F.-H. Chang, B. Beiderman, N. G. Lopez, and H. R. Bourne, *J. Biol. Chem.* **264**, 15467 (1989).

[55] C. A. Landis, S. B. Masters, A. Spada, A. M. Pace, H. R. Bourne, and L. Vallar, *Nature (London)* **340**, 692 (1989).

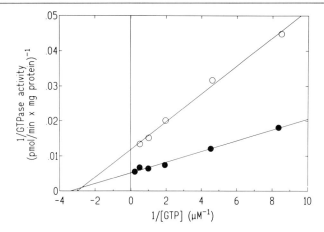

FIG. 1. Effect of interleukin 8 on the kinetic parameters of high-affinity GTPase in human neutrophil plasma membranes. Low K_m GTPase activity was determined at GTP concentrations ranging from 0.12 to 5 μM in the absence (○) or presence (●) of 0.3 μM interleukin 8. Double-reciprocal plots of low K_m GTPase activity versus GTP concentration are shown. Basal low K_m GTPase exhibited an apparent K_m value for GTP of approximately 0.3 μM and a V_{max} of around 85 pmol of GTP hydrolyzed/min × mg protein. Interleukin 8 at 0.3 μM had no effect on the apparent affinity of the enzyme for GTP, but it increased the V_{max} value by about 2.2-fold.

(see, however, Ref. 39). Finally, agonists usually enhance steady-state GTPase activity by increasing the V_{max} of the reaction, rather than changing the apparent K_m for GTP. This is shown for the stimulatory effect of interleukin 8 on the GTP hydrolysis by human neutrophil membranes in Fig. 1.

Method II

Hydrolysis of GTP is assayed essentially as described above, except that the incubation medium is composed of 50 mM triethanolamine hydrochloride, pH 7.4 at 30°, 5 mM MgCl$_2$, 1 mM DTT, 1 mM EDTA, 150 mM NaCl, 20 nM [γ-^{32}P]GTP (NEN, NEG-004, 30 Ci/mmol), and 10 μM GDP. The incubation is performed for 15 min at 30°.

Comments. The second protocol offers several potential advantages in comparison to method I. First, the incubation medium is composed of fewer reagents. The assay is thus more conveniently set up. Second, the protocol allows one to investigate the effect of GDP on receptor-stimulated hydrolysis of [γ-^{32}P]GTP and to directly compare this function to the receptor-stimulated binding of [^{35}S]GTPγS, which can be assayed in mem-

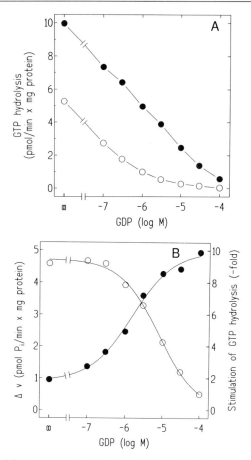

FIG. 2. Effect of GDP on agonist-stimulated GTP hydrolysis in HL-60 membranes. (A) Low K_m GTPase activity was determined in HL-60 membranes using method II at increasing concentrations of GDP in the absence (○) or presence (●) of 10 μM fMet-Leu-Phe. See text for experimental details. (B) Replot of the absolute (○) and relative (●) increase in GTPase activity caused by the addition of fMet-Leu-Phe.

brane preparations under very similar conditions.[56,57] Third, the relative stimulation of GTP hydrolysis caused by receptor activation is much higher. For example, the chemotactic peptide fMet-Leu-Phe is capable of stimulating the hydrolysis of GTP by membranes of HL-60 granulocytes up to 10-fold when assayed under the conditions of method II (Fig. 2),

[56] P. Gierschik, R. Moghtader, C. Straub, K. Dieterich, and K. H. Jakobs, *Eur. J. Biochem.* **197**, 725 (1991).
[57] T. Wieland and K. H. Jakobs, this volume [1].

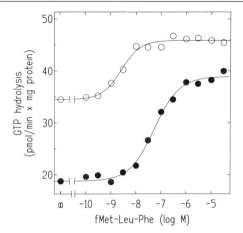

FIG. 3. Sodium ions decrease the potency of agonists to stimulate GTP hydrolysis but enhance the magnitude of the response. Low K_m GTPase activity was determined in HL-60 membranes using method I in the absence (○) or presence (●) of 100 mM NaCl and in the presence of the indicated concentrations of fMet-Leu-Phe. See text for experimental details. The EC_{50} of fMet-Leu-Phe was 2.6 and 54 nM in the absence and presence of NaCl, respectively. The net increase in GTPase activity caused by the addition of 10 μM fMet-Leu-Phe was 11.3 and 20.2 pmol/min × mg protein in the absence and presence of NaCl, respectively.

whereas method I allows for only an approximately 2-fold stimulation (see Fig. 3). On the other hand, the second method is probably not well suited for membranes containing high levels of nonspecific nucleoside triphosphatases and/or nucleoside diphosphokinases (see above). In addition, analyzing the kinetic parameters of GTP hydrolysis is more complicated when this protocol is used because both GDP and GTP are present in the incubation medium.

Parameters Affecting Receptor-Mediated Stimulation of GTP Hydrolysis in Membrane Preparations

Monovalent Cations

Monovalent cations, in particular sodium ions, have been shown to decrease the potency of agonists to stimulate GTP hydrolysis and to enhance the magnitude of the response.[44] The second effect is frequently due to a marked inhibition of basal GTP hydrolysis. This is shown in Fig. 3 for the stimulation of GTP hydrolysis by fMet-Leu-Phe in HL-60 membranes. Similar findings were reported for opiate agonists[32] and bradykinin[34] in membranes of NG108-15 cells, and for somatostatin in membranes of S49 lymphoma cells.[58] Sodium ion is the most effective cation in this regard, followed by Li$^+$, K$^+$, and choline cation. Half-maximal and maximal effects are observed at 20–40 and 100–150 mM Na$^+$, respectively (see Refs. 44 and 59 for more detailed discussions on the regulation of G-protein-mediated signal transfer by sodium ions).

Magnesium Ions

Magnesium ions have profound effects on agonist stimulation of GTP hydrolysis in membrane preparations. Two main effects have been observed. First, Mg^{2+} is absolutely required and markedly enhances agonist stimulation of GTPase. In membranes of HL-60 cells, stimulation of GTP hydrolysis by fMet-Leu-Phe is half-maximal and maximal at approximately 2 and 20 μM free Mg^{2+}, respectively (Fig. 4A). Second, magnesium ions markedly enhance the potency of agonists to stimulate GTP hydrolysis in this membrane system (Fig. 4B; see Refs. 45 and 59 for detailed discussions of the effect of Mg^{2+} on receptor–G-protein coupling in membrane preparations).

Guanine Nucleotides

The potency of agonists to stimulate GTP hydrolysis is influenced by the concentration of guanine nucleotides present in the assay. Thus, when GTP hydrolysis is assayed in HL-60 membranes according to method II in the presence of 10 nM [γ-^{32}P]GTP, but in the absence of GDP, the EC$_{50}$ of the stimulus fMet-Leu-Phe is approximately 60 nM (Fig. 5). The Hill coefficient of this stimulation is around 0.4, suggesting an involvement of more than one fMet-Leu-Phe receptor affinity state in stimulating GTPase under these conditions. In contrast, when GDP is present in the assay at a concentration of 10 μM together with the substrate [γ-^{32}P]GTP, the concentration–response curve is not only shifted to the right (EC$_{50}$ ≈ 580 nM) but also exhibits a Hill coefficient (0.9) much closer to unity, suggesting the involvement of essentially one low-affinity binding state of the fMet-Leu-Phe receptor in mediating GTPase stimulation. Similar effects are seen when GTP hydrolysis is assayed in HL-60 membranes according to method I at low versus high concentrations of GTP (see Ref. 58 for results and a more detailed discussion of the influence of the guanine nucleotide level on the regulation of receptor-stimulated GTPase in membrane preparations).

[58] P. Gierschik and K. H. Jakobs, in "G-Proteins as Mediators of Cellular Signalling Processes" (M. D. Houslay and G. Milligan, eds.), p. 67. Wiley, Chichester, 1990.

[59] P. Gierschik, K. McLeish, and K. H. Jakobs, J. Cardiovasc. Pharmacol. 12(Suppl. 5), S20 (1988).

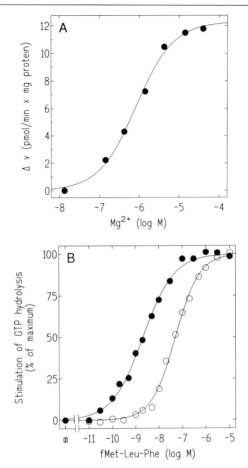

FIG. 4. Effect of magnesium ions on agonist-stimulated GTP hydrolysis in HL-60 membranes. (A) Low K_m GTPase activity was determined in HL-60 membranes at increasing concentrations of free Mg^{2+} according to method I, except that the incubation was performed in the presence of 3 mM EDTA and NaCl was absent from the incubation medium. The concentrations of free Mg^{2+} were calculated according to Sullivan [W. J. O'Sullivan, in "Data for Biochemical Research" (R. M. C. Dawson, D. C. Elliot, W. H. Elliot, and K. M. Jones, eds.), 2nd Ed., p. 423. Oxford Univ. Press, New York and Oxford, 1969]. The increase in GTP hydrolysis caused by the addition of 10 μM fMet-Leu-Phe is shown. (B) Low K_m GTPase activity was determined at 0.45 μM (○) and 4 mM (●) free Mg^{2+} in the presence of the indicated concentrations of fMet-Leu-Phe. The assay was performed according to method I, except that NaCl was absent from the incubation medium. Basal GTPase activities in the absence of fMet-Leu-Phe were 11.0 ± 0.9 and 40.0 ± 1.2 pmol/min × mg protein at 0.45 μM and 4 mM free Mg^{2+}, respectively. Maximal GTPase activities were 18.4 ± 1.0 and 51.0 ± 2.0 pmol/min × mg protein at 0.45 μM and 4 mM free Mg^{2+}, respectively. The EC_{50} of fMet-Leu-Phe was 2.1 and 50 nM at 0.45 μM and 4 mM free Mg^{2+}, respectively.

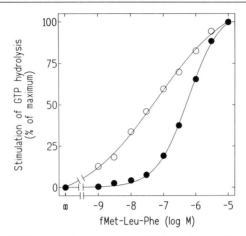

FIG. 5. Effect of GDP on agonist-stimulated GTP hydrolysis in HL-60 membranes. Low K_m GTPase activity was determined in HL-60 membranes according to method II in the absence (○) or presence (●) of 10 μM GDP. Basal GTPase activities in the absence of fMet-Leu-Phe were 9.5 and 1.0 pmol/min × mg protein in the absence and presence of GDP, respectively. Maximal GTPase activities were 18.6 and 4.4 pmol/min × mg protein in the absence and presence of GDP, respectively.

ADP-Ribosylation of G Proteins by Cholera or Pertussis Toxin

Cholera toxin ADP-ribosylates the α subunits of G_s and, at least under certain conditions, the α subunits of G_t, G_i, and G_o (see Ref. 60 for review). Cholera toxin-mediated ADP-ribosylation of α_s has been shown to inhibit agonist-stimulated but not basal GTPase activity.[9,17,20,61] The mechanism of inhibition is likely to be a specific inhibition of the actual GTP hydrolysis by α_s.[9,10] Owing to the difficulties in observing receptor-stimulated G_s GTPase activity in membrane preparations (see above), it is frequently not possible to demonstrate an inhibitory effect of cholera toxin-mediated ADP-ribosylation of G_s on this function (e.g., see Refs. 16 and 27). ADP-ribosylation of retinal G_t by cholera toxin has been shown to reduce markedly light-stimulated GTP hydrolysis by G_t.[62,63] Treatment of membranes of differentiated HL-60 cells with cholera toxin and NAD^+ reduced GTPase activity in response to various chemotactic stimuli, which

[60] P. Gierschik and K. H. Jakobs, in "Handbook of Experimental Pharmacology, Vol. 102: Selective Neurotoxicity" (H. Herken and F. Hucho, eds.), p. 807. Springer-Verlag, Berlin, Heidelberg, and New York, 1992.
[61] M. Svoboda, M. Lambert, and J. Christophe, Biochim. Biophys. Acta 675, 46 (1981).
[62] M. E. Abood, J. B. Hurley, M.-C. Pappone, H. R. Bourne, and L. Stryer, J. Biol. Chem. 257, 10540 (1982).
[63] S. E. Navon and B. K.-K. Fung, J. Biol. Chem. 259, 6686 (1984).

bind to receptors coupled via G_{i2} and/or G_{i3} to stimulation of phospholipase C.[11,46,64] Similar results have been reported for membranes from peripheral blood neutrophils[23,43] (see Ref. 60 for a detailed discussion of the effect of cholera toxin-mediated ADP-ribosylation on G-protein function).

Pertussis toxin ADP-ribosylates the α subunits of the α_i, α_o, and α_t subfamilies of the G-protein α subunits (see Ref. 65 for review). ADP-ribosylation by pertussis toxin specifically inhibits the interaction of G proteins with receptors and has thus been shown to block receptor-stimulated GTP hydrolysis in many cases (e.g., see Refs. 19, 23, 30, 33, 37, 40, and 66–68). Interestingly, pertussis toxin-mediated G-protein ADP-ribosylation frequently also reduces basal GTPase.[19,33,40,44,69] The latter effect is particularly obvious when GTP hydrolysis is assayed in the absence of sodium ions.[44] As GTP hydrolysis by purified G proteins is not affected by pertussis toxin-mediated ADP-ribosylation,[70,71] we have previously suggested that G proteins interact with and are activated by receptors even in the absence of agonists, and that G-protein ADP-ribosylation by pertussis toxin uncouples unoccupied receptors from G-protein interaction and activation.[44,65] This view is supported by observations suggesting that opioid receptors present in membranes of NG108-15 cells spontaneously associate with G proteins and give rise to increased GTPase activity even in the absence of agonists[72] and by findings demonstrating that this increased basal GTPase activity is inhibited by certain antagonists ("negative antagonists") in a stereospecific fashion[72,73] (see Refs. 74 and 75 for more detailed discussions on the negative intrinsic activity of antagonists on G-protein-coupled receptors). Members of the phospholipase C-stimulating α_q subfamily of the G-protein α subunits lack the consensus sequence for pertussis toxin-mediated ADP-ribosylation.[76] Their ability

[64] P. Gierschik and K. H. Jakobs, *FEBS Lett.* **224,** 219 (1987).
[65] P. Gierschik, *Curr. Top. Microbiol. Immunol.* **175,** 69 (1992).
[66] K. Aktories, G. Schultz, and K. H. Jakobs, *FEBS Lett.* **156,** 88 (1983).
[67] C. Van Dop, G. Yamanaka, F. Steinberg, R. D. Sekura, C. R. Manclark, L. Stryer, and H. R. Bourne, *J. Biol. Chem.* **259,** 23 (1984).
[68] P. A. Watkins, J. Moss, D. L. Burns, E. L. Hewlett, and M. Vaughan, *J. Biol. Chem.* **259,** 1378 (1984).
[69] T. Katada, T. Amano, and M. Ui, *J. Biol. Chem.* **257,** 3739 (1982).
[70] K. Haga, T. Haga, A. Ichiyama, T. Katada, H. Kurose, and M. Ui, *Nature (London)* **316,** 731 (1985).
[71] K. Enomoto and T. Asakawa, *FEBS Lett.* **202,** 63 (1986).
[72] T. Costa, J. Lang, C. Gless, and A. Herz, *Mol. Pharmacol.* **37,** 383 (1990).
[73] T. Costa and A. Herz, *Proc. Natl. Acad. Sci. U.S.A.* **86,** 7321 (1989).
[74] T. Costa, Y. Ogino, P. J. Munson, H. O. Onaran, and D. Rodbard, *Mol. Pharmacol.* **41,** 549 (1992).
[75] W. Schütz and M. Freissmuth, *Trends Pharmacol. Sci.* **13,** 376 (1992).
[76] M. I. Simon, M. Strathmann, and N. Gautam, *Science* **252,** 802 (1991).

to hydrolyze GTP is thus not affected by treatment of cells or membranes with this toxin.[22,34] Although α_q subunits carry the consensus sequence for ADP-ribosylation by cholera toxin and can be activated by mutations in this sequence,[77,78] neither cholera toxin-mediated ADP-ribosylation of these α subunits nor effects of cholera toxin treatment on their GTPase activity have been reported in the literature (cf. Refs. 22 and 23).

Acknowledgments

The expert technical assistance of Christina Stannek and Susanne Gierschik is greatly appreciated. Studies performed in the authors' laboratory reported herein were supported by grants from the Deutsche Forschungsgemeinschaft and the Fritz Thyssen Stiftung.

[77] B. R. Conklin, O. Chabre, Y. H. Wong, A. D. Federman, and H. R. Bourne, *J. Biol. Chem.* **267,** 31 (1992).
[78] P. Schnabel, R. Schreck, D. L. Schiller, M. Camps, and P. Gierschik, *Biochem. Biophys. Res. Commun.* **188,** 1018 (1992).

[3] Regulation of G-Protein Activation by Mastoparans and Other Cationic Peptides

By ELLIOTT M. ROSS and TSUTOMU HIGASHIJIMA†

Introduction

Mastoparan, Ile-Asn-Leu-Lys-Ala-Leu-Ala-Ala-Leu-Ala-Lys-Lys-Ile-Leu-amide (MP), activates G proteins by catalyzing GTP/GDP exchange, the mechanism of action of G-protein-coupled receptors. Like receptors, MP accelerates guanine nucleotide exchange at micromolar Mg^{2+}, it does not alter hydrolysis of bound GTP, its action is markedly potentiated by G-protein $\beta\gamma$ subunits, and it is blocked by pertussis toxin-catalyzed ADP-ribosylation of the target α subunit. MP binds to G proteins as an amphipathic, cationic α helix, the predicted structure of the G-protein-binding domains on receptors, further suggesting that MPs are true receptor mimics that bind to the receptor-recognition site on G-protein α subunits.[1,2]

The MPs are prototypical of a wide variety of amphiphilic, cationic peptides that activate G proteins. MP is a natural component of wasp venom, and at least seven MP analogs are produced by different species.

† Deceased.
[1] T. Higashijima, J. Burnier, and E. M. Ross, *J. Biol. Chem.* **265,** 14176 (1990).
[2] T. Higashijima, S. Uzu, T. Nakajima, and E. M. Ross, *J. Biol. Chem.* **263,** 6491 (1988).

Peptide fragments of G-protein-coupled receptors, several natural cationic peptides, and some nonpeptide cationic amphiphiles display similar activities. Maximum activity, potency (concentration dependence), and selectivity for specific G proteins vary by orders of magnitude among these different compounds.

Mastoparans are selective in their activation of different G proteins. Under optimal conditions, MP accelerates nucleotide exchange by G_o more than 30-fold. It is almost as active on G_i, but it is less potent and produces far lower maximal rates with G_s, G_t, G_z, G_q, and $p21^{ras}$. In this context, it is important to note that almost any cationic compound, even NH_4^+,[3] can accelerate nucleotide exchange somewhat. Many cationic peptides can stimulate exchange by 50–100%,[1] a level that we consider insignificant in comparison to the large and specific effects observed for the most active peptides.

Synthetic congeners of MP display diverse maximal activities, patterns of selectivity among different G proteins, and cytolytic effects. For example, [Lys^{10},Ala^{12},Leu^{13}]MP is far less cytolytic than its Ala^{10} homolog. Both [Leu^2]MP and several peptides based on the sequence of G_s-coupled receptors stimulate the activation of G_s over 20-fold. Other peptides display their own spectrum of selectivities among G proteins.

Regulatory activity requires that a G-protein-stimulating peptide be both amphiphilic and cationic. Helix-breaking residues or charged residues on what should be the hydrophobic face of the helix both diminish activity dramatically. Mas17, with a lysyl residue on the hydrophobic side of the helix, is a commercially available negative control for experiments where MP is used to alter G-protein activation (Fig. 1). [Gln^4,Leu^{13}]MP is an alternative negative control that does not activate G proteins but is similar to MP both in overall hydropathy and in hydrophobic moment. As the conformations of mastoparans bound to G-protein residues become better understood, the design of G-protein-selective analogs should improve rapidly.

General Considerations for Using G-Protein-Activating Peptides

The greatest potential utility of MP and related G-protein regulators is as specific probes of cellular signaling pathways. Because of their large effects and independence of receptors, they provide a unique means for manipulating G-protein systems. Wise use requires understanding of their biochemical actions, which have been defined in studies of purified G

---

[3] K. M. Ferguson, T. Higashijima, M. D. Smigel, and A. G. Gilman, *J. Biol. Chem.* **261,** 7393 (1986).

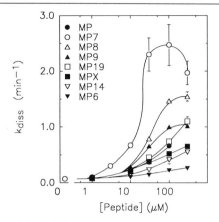

FIG. 1. Structure–activity relationships for some MP analogs. The ability of MP and several congeners to stimulate equilibrium GDP exchange (see Fig. 4) was measured using G_o that had been reconstituted into phospholipid vesicles (see Fig. 2). The most active analog, Mas7 ([Ala^{12},Leu^{13}]MP), is both more potent and more effective at its optimal concentration than is MP itself. The least active analog, Mas6 ([Lys^{13}]MP), has a positive charge on what should be the hydrophobic face of the α helix. MPX was discovered in the venom of a wasp that does not make MP, and the other peptides are designed MP congeners. (From Higashijima et al.,[1] where structure–function relationships of these and other peptides are discussed.)

proteins. These principles are just as validly applied to studies of cells and cell lysates.

The effects of MP on nucleotide exchange by G proteins can be assayed according to any of the protocols used to measure exchange. These include the equilibrium exchange of bound GDP, the binding of GTPγS or GTP, or steady-state GTPase activity under conditions where exchange is rate limiting. Below, we first discuss factors that influence the utility, applicability, or interpretability of all the assays. Each assay is then discussed individually.

Effects of Detergents and Lipids

The G proteins, especially αβγ trimers, bind to lipid bilayers and detergent micelles. With few exceptions, however, only membrane-associated G proteins are sensitive to regulation by receptors. Although MPs are also most effective in regulating membrane-bound G proteins, they demonstrate appreciable activity in the presence of many, but not all, detergents.[1] The effect of MPs on a G protein reflects both the identity

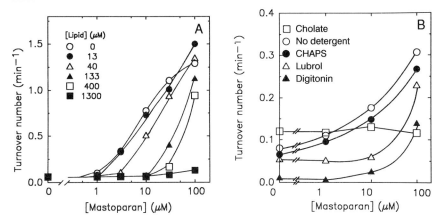

FIG. 2. Effects of lipids and detergents on MP stimulation of the GTPase activity of G_o. (A) Increasing the phospholipid concentrations decreases the potency of MP. G_o was reconstituted into vesicles composed of phosphatidylcholine : phosphatidylethanolamine : phosphatidylserine (3:4:3) and assayed for GTPase activity in the presence of increasing concentrations of MP. The concentration of G_o was 0.25 nM (determined by [^{35}S]GTPγS binding), and the total lipid concentration was 1.3 μM. To demonstrate the effects of lipid on the potency of MP, other assays included excess added lipid at the concentrations shown. Data are expressed as molar turnover numbers. (B) MP stimulation depends on what detergent or lipid is present. MP-stimulated GTPase activity was assayed in the presence of 1% cholate, 0.1% digitonin, 0.1% Lubrol PX, 0.5% 3-[(3-cholamidopropyl)dimethylammonio]-1-propane sulfonate (CHAPS), or 10 ppm Lubrol. Note that activities in (A) were assayed for different periods of time depending on the MP concentration, from 2.5 min at 100 μM MP to 30 min in the absence of MP, to maintain linear assay time courses. (From Higashijima et al.[1])

and the concentration of lipid or detergent (Fig. 2). There is no striking dependence on the identity of the lipid, and either pure phosphatidylcholine or mixtures of neutral and anionic lipids support regulation by MP.

Although lipid-associated G proteins are more sensitive to MP than are G proteins in detergent solution, increasing concentrations of lipid increase the concentrations of MP needed to obtain a given degree of stimulation (Fig. 2A). This phenomenon probably reflects the fact that MPs and other regulatory peptides partition into lipid membranes, where they fold into an α helix.[4] However, increasing the concentration of total lipid increases the total surface area to which peptide can bind and thus decreases its surface concentration. Thus, maximum effects of MP are observed under conditions where G proteins are membrane-associated

[4] K. Wakamatsu, A. Okada, T. Miyazawa, M. Ohya, and T. Higashijima, *Biochemistry* **31**, 5654 (1992).

but the total lipid concentration is minimized. This appears true both in preparations of reconstituted purified G proteins and in suspensions of biological membranes, where the concentration of membrane should be decreased as much as is consistent with the sensitivity of the assay.

Last, we have noticed that at least one MP congener, Mas7, causes p21^{ras} to stick to polypropylene test tubes. We do not know how many small GTP-binding proteins are affected in this way nor how many MP analogs share this annoying property. Sticking is prevented by 0.1% Lubrol, but Lubrol inhibits stimulation.

Effects of βγ Subunits

The G-protein α subunits are most sensitive to regulation by peptides in the presence of stoichiometric amounts of βγ subunits (Fig. 3). The effects of βγ subunits are complex, in that they include the well-documented suppression of the basal exchange rate in addition to an enhancement of the maximally peptide-stimulated rate. We have not noted a significant effect of βγ subunits on the concentration dependence on MP. The effect of βγ subunits on nucleotide exchange is also dependent on the concentration of Mg^{2+} and on the presence of detergent or lipid.

Effects of Magnesium Ion

Mastoparan stimulates nucleotide exchange by G proteins in the absence of Mg^{2+} (Fig. 3A). The only direct effect of Mg^{2+} on MP-stimulated exchange is inhibition, which is observed above 1 mM Mg^{2+}. In some assay formats, however, interactions between MP and Mg^{2+} are apparently complex (Fig. 3). First, Mg^{2+} (≥ 0.1 μM) is required both for the activation of G proteins by bound GTP (or GTP analogs) and for subsequent GTP hydrolysis. Thus, steady-state GTPase activity or the rate of G-protein activation will display broad Mg^{2+} optima in the 0.1–100 μM range, where activation and hydrolysis proceed but where inhibition of exchange is negligible. Second, Mg^{2+} in the 1–100 mM range markedly stimulates GDP release and may contribute to GTP binding. These effects decrease the increments in nucleotide exchange rates caused by MP even though absolute rates may remain high. Third, high millimolar concentrations of Mg^{2+} can act synergistically with G-protein βγ subunits to stimulate nucleotide exchange, also independently of MP. Thus, Mg^{2+} can reverse the effect of G-protein βγ subunits and thus further complicate the effects of MP.

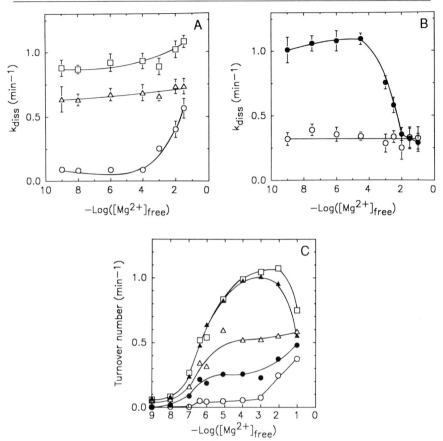

FIG. 3. Real and apparent effects of Mg^{2+} on the stimulation of nucleotide exchange by MP. Data shown are for G_o. (A) In the presence of the $\beta\gamma$ subunits, MP (□, 100 μM; △, 10 μM) stimulates equilibrium GDP exchange in the presence of less than 1 nM Mg^{2+}. Exchange is minimal in the absence of MP (○). (B) In the absence of $\beta\gamma$ subunits, 100 μM MP (●) stimulates exchange to similar levels, but basal exchange (○) is much faster and millimolar Mg^{2+} inhibits stimulation. (C) Effect of Mg^{2+} on steady-state GTPase activity in absence (lower curve) or presence of 3 μM, 10 μM, 30 μM, and 100 μM MP. Because the hydrolysis of bound GTP only proceeds in the presence of Mg^{2+} in the 0.1–1 μM range, MP-stimulated steady-state GTPase activity requires Mg^{2+}, but not for the MP-stimulated exchange step. The MP-stimulated GTPase activity is inhibited above 10 mM. In the absence of MP, Mg^{2+} stimulates GTPase by promoting GDP release. (From Higashijima et al.[1])

Peptide Purification

Modern methods of solid-phase peptide synthesis frequently produce the desired peptide with good yields and few side products. However, single substitutions or deletions in the MP sequence can have profound

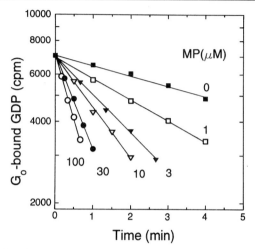

FIG. 4. Equilibrium GDP exchange stimulated by MP. About 100 pmol of Lubrol-solubilized G_o was allowed to bind and hydrolyze [α-^{32}P]GTP (~100 cpm/fmol) at 20° for 20 min and reconstituted into phospholipid vesicles by a standard gel filtration protocol that also removes unbound nucleotide. The exchange of bound [α-^{32}P]GDP with added 1 μM unlabeled GDP was monitored as the loss of bound ^{32}P at 20° in medium that contained 0.1 mM free Mg^{2+} and the concentrations of MP shown. Nonlinear least-squares fits of the original duplicate data points yield first-order exchange rate constants that range from 0.07 min^{-1} in the absence of MP to 1.1 min^{-1} at 100 μM MP. Note that the entire reaction takes place at equilibrium over a relatively short time.

effects on activity and selectivity, and it is therefore important to purify all G-protein regulatory peptides. High-performance liquid chromatography (HPLC) on a large-pore C_4 column, Vydac 214TP510 (The Separations Group, Hesperia, CA) or equivalent, using an acetonitrile gradient in 0.1% trifluoroacetic acid provides efficient purification with good yield. Depending on their hydrophobicities, peptides may elute between 25 and 50% acetonitrile; MP elutes at about 40% acetonitrile.

In Vitro Assays of Nucleotide Exchange

Equilibrium Exchange of GDP or GTPγS

Measurement of the rate of dissociation of radiolabeled GDP at equilibrium (i.e., in the presence of a saturating concentration of unlabeled GDP) is conceptually the simplest way to assay the rate of ligand exchange and the stimulation of exchange by MPs (Fig. 4).[5] The only slow step

[5] R. C. Rubenstein, M. E. Linder, and E. M. Ross, *Biochemistry* **30**, 10769 (1991).

is dissociation; neither hydrolysis, activation, nor alteration in subunit dissociation are involved. The dissociation rate constant is determined directly as the slope of a plot of ln[bound GDP] versus time. Data are usually obtained over a period of less than 5 min, often in the first 20 sec.

The G protein bound to radiolabeled GDP is prepared by incubating either free α subunit or $\alpha\beta\gamma$ trimer, in detergent solution or after reconstitution, with [^{3}H]GDP, [α-^{32}P]GDP, or [α-^{32}P]GTP (which hydrolyzes rapidly) under conditions appropriate for the subsequent assay.[1,5] The G protein may also be labeled in detergent solution prior to reconstitution, thus removing excess unbound nucleotide, but there is usually no need to remove unbound nucleotide by gel filtration unless background is a particular problem. Binding is usually allowed to proceed for 30–60 min at 30° in the presence of 10 mM Mg^{2+} in suitable buffer (generally 50 mM sodium HEPES, pH 8.0). Labeled nucleotide is added at 1–5 μM and at a specific activity appropriate for the amount of G protein to be used in the assay. GDP binding is quenched by the addition of EDTA to chelate Mg^{2+}. ([^{35}S]GTPγS binding is stabilized at this step by 0.1 mM Mg^{2+}.) The extent of binding, which should approach completion, is estimated by standard methods of nitrocellulose filter binding (see below).

Equilibrium nucleotide exchange is initiated by adding the nucleotide–G-protein complex to a reaction mixture of 30–100 μl that contains 50 mM sodium HEPES (pH 8.0), 1 mM EDTA, MgSO$_4$ to yield the desired concentration of free Mg^{2+}, 1–10 μM free unlabeled nucleotide, and MP or other regulatory peptide (and 0.1% Lubrol PX if desired). The concentration of nucleotide-liganded G protein is chosen according to availability, manipulation of the α–$\beta\gamma$ subunit dissociation equilibrium, and its radiochemical specific activity. Release of GDP is usually measured at 30°, but 20° is more convenient for G$_o$. Release of bound radiolabeled nucleotide is quenched by dilution with 2 volumes of buffer that contains 0.1% Lubrol and 10 mM free Mg^{2+}, 10 mM NaF, and 10 μM AlCl$_3$ to stabilize bound GDP. Remaining bound nucleotide is determined by filtration over nitrocellulose filters (Schleicher and Schuell, Keene, NH, BA85 or equivalent), which are washed with 15 ml of cold 20 mM Tris-Cl (pH 8.0), 0.1 M NaCl, 10 mM MgCl$_2$.

Mastoparan is unique in that it promotes exchange of GTPγS, which in most conditions binds nearly irreversibly.[1] Equilibrium exchange of [^{35}S]GTPγS is measured according to the same protocol used for GDP. Excess free radiolabeled ligand should be removed prior to initiation of the assay because of its tendency to bind to the nitrocellulose filter (see below). It should also be noted that GTPγS promotes dissociation of α and $\beta\gamma$ subunits. Equilibrium exchange of GDP and GTPγS therefore measures mechanistically different reactions, in one case binding to a nonactivated trimer and in the other binding to largely dissociated, activated α subunit.

GTPγS Binding

The first and most common nucleotide binding assay for G proteins uses [^{35}S]GTPγS,[6] and acceleration of the rate of binding provides a measure of MP action. GTPγS binding is apparently first-order at reasonable concentrations. The rate constant, proportional to the initial rate, is stimulated up to 20-fold by MPs. Assays for [^{35}S]GTPγS binding are conducted under the same conditions as described above for equilibrium exchange: HEPES/EDTA buffer with 100 μM Mg^{2+} to minimize Mg^{2+}-stimulated exchange, with or without 0.1% Lubrol, at 20° or 30°. App(NH)p or ATP (0.1 mM) may be included to block nonspecific nucleoside triphosphate binding activity contributed by contaminants or by other proteins added to the assay (kinases, effectors). For measurement of MP-stimulated [^{35}S]GTPγS binding, 100–300 nM [^{35}S]GTPγS is usually adequate to minimize denaturation (see below) while maintaining low nonspecific binding. For measurements of total G protein using Mg^{2+}-stimulated exchange in the presence of detergent, at least 1 μM [^{35}S]GTPγS is required to obtain maximum binding to G_s and 0.3 μM is needed for G_i or G_o. Bound [^{35}S]GTPγS is trapped on nitrocellulose filters as described above.

The [^{35}S]GTPγS binding assays are convenient and popular, but they present two problems. First, some cationic peptides cause [^{35}S]GTPγS to bind to the nitrocellulose filters. This problem can be severe, particularly for N-terminally truncated MP analogs, and can completely invalidate the assay. Other peptides may raise the background only slightly. It is therefore crucial to test for G-protein-independent binding whenever a new peptide or altered assay condition is used. Second, the binding of [^{35}S]GTPγS converts a nonactivated, trimeric G protein to an active, dissociated α subunit. Thus, the assay of [^{35}S]GTPγS binding includes at least two reactions subsequent to nucleotide exchange, one of which is Mg^{2+}-dependent. It should also be noted that prolonged incubation of G proteins with MPs can cause significant denaturation, particularly in detergent solution and at low concentrations of guanine nucleotides. This effect, which presumably reflects deceased affinity for nucleotides and the well-known instability of unliganded G proteins, causes binding to reach a maximum and then decline. Consequently, it is important to establish times and temperatures that maintain the activity of the G protein to be assayed.

[6] J. K. Northup, M. D. Smigel, and A. G. Gilman, *J. Biol. Chem.* **257,** 11416 (1982).

If GTPγS binding appears to be the assay of choice but background binding is high, an alternative to measuring the binding of [^{35}S]GTPγS is to measure functional activation of G protein by unlabeled GTPγS. This assay is particularly valuable in impure systems, and it was initially used to identify and purify both G_s and transducin. Functional assays are available for G_s, using adenylyl cyclase,[7,8] transducin, using cyclic GMP phosphodiesterase, and G_q, using phospholipase C-β.[9,10] In either case, the GTPγS binding reaction is quenched with excess GDP, 1 mM Mg^{2+}, and detergent (Lubrol for cyclase, cholate for phospholipase), the diluted reaction mixture is added to the effector enzyme, and the G-protein-stimulated activity is measured by standard assays (described elsewhere in this series).

Steady-State GTPase

The easiest high-throughput assay for approximating MP-stimulated exchange is the measurement of steady-state GTPase activity.[1] Because the hydrolysis of bound GTP is relatively slow ($t_{1/2}$ approximately 10–60 sec, longer for G_z), steady-state hydrolysis is usually dependent only on the rates of GDP release and GTP binding. Standard assay protocols for measuring steady-state GTPase are available elsewhere.[11] The reaction utilizes the buffers described above for the binding assays and [γ-^{32}P]GTP, usually at 100–300 nM, as substrate. Tris-Cl is substituted for sodium HEPES (pH 8.0) because it gives a lower assay background. App(NH)p (0.1 mM) may be added to suppress any contaminating nucleoside triphosphatase activity. Reactions are allowed to proceed for 1–30 min (see below) and quenched with 800 μl of a 5% slurry of activated charcoal (Norit A or equivalent) in 50 mM NaH$_2$PO$_4$, followed by centrifugation. The [^{32}P]P$_i$ in the supernatant is measured by Cerenkov counting.

The absolute sensitivity of GTPase assays is often limited by the purity of [γ-^{32}P]GTP. Note that the difference between 99% purity and 99.8% purity is a 5-fold difference in assay background. Synthesis of [γ-^{32}P]GTP, which is easy and inexpensive,[12] is usually only 98% complete, and commercial [γ-^{32}P]GTP is often only 98–99% pure after shipping. ^{32}P-Labeled

---

[7] E. M. Ross and A. G. Gilman, *J. Biol. Chem.* **252**, 6966 (1977).

[8] J. K. Northup, P. C. Sternweis, M. D. Smigel, L. S. Schleifer, E. M. Ross, and A. G. Gilman, *Proc. Natl. Acad. Sci. U.S.A.* **77**, 6516 (1980).

[9] A. V. Smrcka, J. R. Hepler, K. O. Brown, and P. C. Sternweis, *Science* **251**, 804 (1991).

[10] G. Berstein, J. L. Blank, A. V. Smrcka, T. Higashijima, P. C. Sternweis, J. H. Exton, and E. M. Ross, *J. Biol. Chem.* **267**, 8081 (1992).

[11] T. Higashijima, K. M. Ferguson, M. D. Smigel, and A. G. Gilman, *J. Biol. Chem.* **262**, 757 (1987).

[12] R. A. Johnson and T. F. Walseth, *Adv. Cyclic Nucleotide Res.* **10**, 135 (1979).

impurities may be orthophosphate, pyrophosphate, or a silicophosphate. [γ-^{32}P]GTP can be purified easily to 99.9 + % as follows. [γ-^{32}P]GTP is applied to a small (0.45 × 2.5 cm) anion-exchange column (Vydac 303NT or equivalent) and eluted with 0.3–0.5 M neutral sodium phosphate buffer. Elution conditions vary with the column used; pure [γ-^{32}P]GTP can usually be obtained in 0.5 ml. Initially pure [γ-^{32}P]GTP can be stored for weeks at $-80°$ in sodium phosphate supplemented with 10 mM Tricine and 50% (v/v) methanol. A syringe, inexpensive injector, and column should be designated for ^{32}P use and kept separate to avoid contaminating the HPLC apparatus.

Two special considerations are demanded in using steady-state GTPase to approximate the rate of GDP/GTP exchange. First, because MPs are such effective exchange catalysts, the MP-stimulated exchange rate can easily exceed the rate of hydrolysis of bound GTP. If this occurs, the maximal steady-state GTPase activity will not correspond to the maximal exchange rate, and maximal stimulation by MP will be underestimated, thus altering assessments of efficacy of the regulatory peptide. Concentration–response curves will also be shifted artifactually to lower concentration ranges, thus invalidating estimates of potency. It is therefore essential to modulate assay conditions to maintain maximally stimulated exchange rates below the rate of hydrolysis. In general, GTPase rates with a molar turnover number greater than or equal to 1 min^{-1} may not be reliable.

A second caveat in measuring MP-stimulated GTPase is that the GTPase rate at high MP concentrations is likely to decline with time because of G-protein denaturation.[1] Such denaturation, which probably reflects the instability of nucleotide-free G protein, is faster in detergent solution than in membranes. It can cause underestimation of maximal stimulation and overestimates of potency, the latter resulting from concentration curves that reach artifactual maxima when increasing denaturation overshadows increasing GTPase rates. Fixed-time assays must be held to within the linear range, sometimes 2 min or less.

G-Protein-Mediated Effects of Mastoparans on Cells

Given the ability of MPs and related peptides to stimulate G-protein activation, it is tempting to try to use MPs to manipulate cellular signaling pathways, either in intact cells per se or in cell lysates or fractions. Indeed, MPs are potential prototypes for drugs that would be designed to manipulate G proteins for therapeutic purposes. To this end, novel peptide and peptide-mimetic G-protein regulators are being designed with appropriate activity, selectivity, and pharmacokinetic properties to allow such use.

An intermediate goal is to develop a panel of reagents that can selectively stimulate individual G proteins in cells to allow more direct study of their signal transducing activities.

Mastoparan was discovered as a secretagogue for mast cells, where its predominant effect in the 1–30 μM range is to cause pertussis toxin-sensitive secretion that is apparently G_i-mediated. Other cellular effects of MP have also been shown to be G_i-mediated. However, it should not be assumed that an effect of MP on cells necessarily indicates the involvement of a G protein. Cationic, amphiphilic peptides interact with diverse proteins with poorly predictable specificity, and the site of action of any cellular effect of MPs must be verified before it can be exploited. Mastoparan can bind to calmodulin with a K_d of about 1 nM and block its regulatory activity.[13] Our guess is that MP is relatively concentrated on or near the inner face of the plasma membrane, thus favoring activity of G proteins with micromolar EC_{50} values rather than higher affinity interactions with calmodulin. MP also activates phospholipase A_2[14] and can inhibit phospholipase C,[15,16] and most amphiphilic peptides cause cell lysis at high concentrations. The contribution of these non-G-protein-related activities can vary markedly from cell to cell. For example, we recently found that HL-60 cells are quite sensitive to the lytic effects of [Ala^{12},Ile^{13}]MP, but mast cells are much less so.

The prudent conclusion is that MPs may be used with cells or subcellular fractions to promote known or probable G-protein-mediated events, but a cellular response to MPs cannot yet be taken as firm indication that the response is controlled by a G-protein signaling pathway. Corroborating information, such as appropriate second messenger responses, concentration dependence, pertussis toxin sensitivity when appropriate, and peptide antagonists that inhibit the actions of both MPs and receptors,[17] is still required. With such controls, however, MPs and related peptides can be used to manipulate G-protein signaling without depending on the fortuitous presence of receptors.

Acknowledgments

Work from the authors' laboratory was supported by National Institutes of Health Grants GM30355 and GM40676 and by R. A. Welch Foundation Grant I-0982.

---

[13] D. A. Malencik and S. R. Anderson, *Biochem. Biophys. Res. Commun.* **114**, 50 (1983).
[14] A. Argiolas and J. J. Pisano, *J. Biol. Chem.* **258**, 13697 (1983).
[15] R. J. H. Wojcikiewicz and S. R. Nahorski, *FEBS Lett.* **247**, 341 (1989).
[16] M. A. Wallace and H. Carter, *Biochim. Biophys. Acta* **1006**, 311 (1989).
[17] H. Mukai, E. Munekata, and T. Higashijima, *J. Biol. Chem.* **267**, 16237 (1992).

[4] Guanosine 5'-O-(γ-Thio)triphosphate Binding Assay for Solubilized G Proteins

By DONNA J. CARTY and RAVI IYENGAR

Introduction

Receptor activation of trimeric ($\alpha\beta\gamma$) G proteins involves the release of bound GDP and the binding of GTP to the α subunit. When a receptor coupled to a G protein binds agonist, the receptor changes conformation in such a way as to promote this exchange. The nucleotide exchange results in activation of the α subunit and its dissociation from the $\beta\gamma$ subunits. The α subunit has GTPase activity and will eventually hydrolyze bound GTP to GDP and thus return to its inactive state. It is only able to affect its target enzyme or channel during that time period when GTP is bound but not yet hydrolyzed. If a nonhydrolyzable analog of GTP is bound, the α subunit becomes persistently activated. Guanosine 5'-O-(γ-thio)triphosphate (GTPγS) is such an analog. GTPγS was originally shown to be capable of persistently activating adenylyl cyclase (adenylate cyclase).[1] With the availability of ^{35}S-labeled GTPγS, a high specific activity radioligand capable of binding to G-protein α subunits, binding assays for G-protein α subunits have become feasible.

Even in the detergent solutions used for isolation of G proteins, α subunits retain the capability to bind guanine nucleotides. Under certain conditions, [^{35}S]GTPγS binding can be used as an assay to quantify functional G proteins during purification from tissue. The major advantage of the binding assay is that it is very sensitive and requires very little protein. It is also quite fast. There are, however, several disadvantages.

The G proteins are by no means the only proteins that bind GTP (or GTPγS). In membranes from many tissues, such as liver, G proteins represent only a small proportion of the proteins capable of binding GTP. Hence, in these tissues, GTPγS binding is not useful for G-protein quantification during early stages of G-protein purification. However, the assay described in this chapter can be used during all stages of G-protein purification from brain tissue, because in brain G proteins are very abundant and are responsible for almost all the GTP binding activity.

The nature of the G proteins also determines if GTPγS binding is a suitable assay. The major brain G proteins are members of the G_o and G_i

[1] L. Birnbaumer, T. L. Swartz, J. Abramowitz, P. W. Mintz, and R. Iyengar, *J. Biol. Chem.* **255**, 3542 (1980).

subfamilies. These G proteins bind GTPγS rapidly and irreversibly even in the absence of receptors. The conditions required for optimal binding of GTPγS to these G proteins have been well defined. This is not the case for many other G proteins, which may vary in their requirements for the presence of an agonist, optimal Mg^{2+} requirements, or the time required to bind GTPγS maximally. Generally G_s α subunits require much higher (25–50 mM) Mg^{2+} than G_i or G_o.[2-4] G_q and G_{11} bind GDP tightly and do not release the bound nucleotide except in the presence of an activated receptor.[5]

It should be noted the ability to bind GTPγS is not indicative of the ability of a G-protein α subunit to be functional in an effector-stimulation assay. This has been shown for both recombinant $G_s\alpha$ and $G_i\alpha$.[6,7] Notwithstanding these reservations, the GTPγS binding assay is useful for purification of bovine brain G proteins, and we routinely use it in our laboratories.

Measurement of GTPγS Binding

Materials

[^{35}S]GTPγS (~1300 Ci/mmol) is purchased from NEN–Du Pont (Boston, MA). Nitrocellulose filters (Cat. No. HAWP 025 00) are purchased from Millipore (Bedford, MA). All other reagents are the best grade available. Lubrol PX and cholate are purified as previously described.[4]

Assay Conditions

The assay described below is a modification of that described originally by Northrup et al.[8] and has been optimized as described for purification of brain G_i and G_o or for preparation of active α subunits derived from these G proteins. Several experiments were performed to determine optimal conditions for GTPγS binding to bovine brain G_i and G_o. These are individually described after a generic assay protocol.

[2] R. Iyengar and L. Birnbaumer, *J. Biol. Chem.* **256**, 11036 (1981).
[3] D. J. Carty, E. Padrell, J. Codina, L. Birnbaumer, J. D. Hildebrandt, and R. Iyengar, *J. Biol. Chem.* **265**, 6268 (1990).
[4] E. Padrell, D. J. Carty, T. M. Moriarty, J. D. Hildebrandt, E. M. Landau, and R. Iyengar, *J. Biol. Chem.* **266**, 9771 (1991).
[5] I.-H. Pang and P. Sternweis, *J. Biol. Chem.* **265**, 18707 (1991).
[6] M. P. Graziano, M. Freissmuth, and A. G. Gilman, this series, Vol. 195, p. 192.
[7] M. E. Linder, D. A. Ewald, R. J. Miller, and A. G. Gilman, *J. Biol. Chem.* **265**, 8243 (1990).
[8] J. K. Northrup, M. D. Smigel, and A. G. Gilman, *J. Biol. Chem.* **257**, 11416 (1982).

A typical binding reaction is carried out in a final volume of 50 μl. The incubation mixture contains 25 mM sodium HEPES, 1 mM EDTA, 100 mM NaCl, 50–100 nM [^{35}S]GTPγS [~200,000 counts/min (cpm)/assay], and the required amount of Mg^{2+} ion, generally added as MgCl$_2$. In general, when mixtures of G proteins are assayed, 25 mM MgCl$_2$ is a reasonable concentration to use. The reaction mixture is incubated for 60 min at 32°.

It is necessary to include a sulfhydryl group reducing agent such as dithiothreitol (DTT) or 2-mercaptoethanol. We typically use 20 mM 2-mercaptoethanol, which is added with the protein. DTT in the assay at 1 mM is also adequate. If proteins are incubated without sulfhydryl group reducing agents, then little or no specific binding is observed.

Although we routinely use 100 mM NaCl, the binding reaction can tolerate higher NaCl concentrations. At 5 mM MgCl$_2$ and 0.1% Lubrol PX, 100 mM NaCl does not inhibit binding, and even at 1 M NaCl binding was reduced only 20%. Therefore, there is no compelling need to normalize NaCl concentrations when assaying and comparing fractions eluting from a salt gradient for G proteins. However, G-protein activity is not stable in high salt, and G proteins should not be stored for more than 2 days at more than 100 mM NaCl.

The concentration of detergent in the assay is very crucial. Generally the concentration of Lubrol should be 0.05% or less. Up to 0.5% cholate can be tolerated in the assay (see below). The assay requires 5–10 μg of brain membrane detergent extracts or 5–10 ng of purified G proteins. If purified proteins are used, the assay should be performed in plastic tubes. We use polystyrene tubes (Sarstedt, Princeton, NJ, Cat. No. 55.478). Nonspecific binding can be measured by the addition of 100 μM unlabeled GTPγS or by omission of the proteins. In either case, nonspecific binding is generally 2–5% of total binding.

At the end of the incubation, the samples are rapidly diluted with 2 ml of ice-cold stopping solution (25 mM Tris-HCl, 100 mM NaCl, and 25 mM MgCl$_2$, pH 7.5) and filtered through nitrocellulose filters under vacuum. The filters are washed twice with 10 ml stopping solution, dried, and counted in a liquid scintillation counter. Because the persistent binding of GTPγS requires Mg^{2+}, it is crucial that there be at least 1–2 mM free Mg^{2+} in the wash buffer.

Tolerable Detergent Concentrations

The experiment in Fig. 1 shows the effect of various concentrations of Lubrol PX or cholate on GTPγS binding. In this experiment, the concentration of cholate was varied from 0.09% (as contributed by the sample)

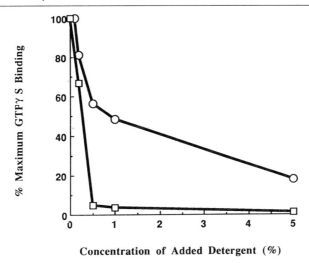

FIG. 1. [^{35}S]GTPγS binding to solubilized bovine brain G proteins at varying concentrations of added sodium cholate (○) or Lubrol PX (□). Binding was measured in the presence of 5 mM Mg^{2+} and 0.09% cholate (included in the sample). Concentrations of the other assay components and assay conditions are described in the text. Data are plotted as a percentage of maximum bound after subtraction of nonspecific binding measured in the presence of 10 μM GTPγS.

to 5.09% in the absence of Lubrol and NaCl, and the concentration of Lubrol was varied from 0 to 5% in the presence of 0.09% cholate. The Mg^{2+} concentration was 5 mM. The GTPγS binding was reduced 19% when the concentration of cholate was approximately doubled by adding 0.1% additional cholate, and it was reduced even further by greater concentrations. A total cholate concentration of 5.09% resulted in 82% reduction in binding. Lubrol inhibited GTPγS binding even more than cholate. Binding was reduced 33% by 0.1% added Lubrol and was inhibited nearly completely (95%) by as little as 0.5%. From such studies it is recommended that Lubrol concentrations in the assay be no more than 0.1%. It would be preferable to have the Lubrol concentration be 0.05% or less. Given the sensitivity of binding to both Lubrol and cholate, the detergent concentrations of samples eluting in a detergent gradient should be normalized if comparisons among samples are to be valid.

Magnesium Ion Concentrations and Rates of Binding

It has been known for some time that increasing concentrations of Mg^{2+} accelerate the rate of GTPγS binding.[3,4] At 2 mM Mg^{2+} concentration

G_{i2} binds GTPγS severalfold faster than G_{i1} or G_{i3} (Fig. 2A). Increasing the concentration of Mg^{2+} to 50 mM allows all three forms of G_i to bind GTPγS at much faster rates (Fig. 2B). Because most peripheral tissues contain mixtures of G_{i2} and G_{i3}, it is advisable to use at least 25 mM

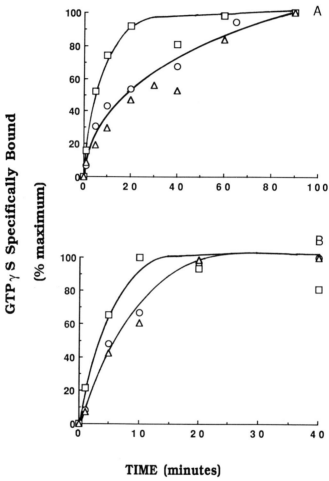

FIG. 2. [^{35}S]GTPγS binding to purified G_i proteins at 2 mM (A) and 50 mM (B) Mg^{2+}. Concentrations of the other assay components and assay conditions are described in the text. G_{i1} (○) was purified from bovine brain, whereas G_{i2} (□) and G_{i3} (△) were from human erythrocytes. Data are plotted as a percentage of maximum bound after subtraction of nonspecific binding measured in the presence of 10 μM GTPγS. (Adapted from Ref. 3 with permission of the American Society for Biochemistry and Molecular Biology.)

MgCl$_2$ and to incubate the binding assay for at least 30 min to obtain equilibrium binding.

In comparison to G$_i$, both forms of G$_o$ bind GTPγS much faster and at very low Mg^{2+} concentrations (Fig. 3). Consequently, assays for brain G proteins can generally be conducted with lower concentrations of Mg^{2+}

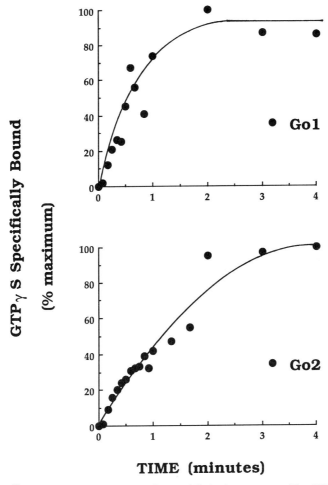

FIG. 3. [^{35}S]GTPγS binding to the two forms of G$_o$ in the presence of 1 mM EDTA and no added Mg^{2+}. The G$_{o1}$ and G$_{o2}$ proteins were purified from bovine brain as described in Ref. 4. Concentrations of the assay components and assay conditions are described in the text. Data are plotted as a percentage of maximum bound after subtraction of nonspecific binding measured in the presence of 10 μM GTPγS. (Adapted from Ref. 4 with permission of the American Society for Biochemistry and Molecular Biology.)

and for shorter time periods. However, in crude mixtures of brain G proteins, because there is a substantial amount of G_{i1} it is advisable to use 25 mM MgCl$_2$ and 30-min incubations.

Concluding Remarks

GTPγS binding is a quick and easy assay for G proteins if used under appropriate conditions. In general, it is most useful for the purification of G_i. In brain tissues it can be used for purification of G_o as well.

Acknowledgments

Research in our laboratories is supported by NIDA Grant DA-07622 to D.J.C. and CA-44998 and DK-38761 to R.I.

[5] Activation of Cholera Toxin by ADP-Ribosylation Factors: 20-kDa Guanine Nucleotide-Binding Proteins

By Joel Moss, Randy S. Haun, Su-Chen Tsai, Catherine F. Welsh, Fang-Jen Scott Lee, S. Russ Price, and Martha Vaughan

Introduction

Cholera toxin, a secretory product of *Vibrio cholerae,* is responsible in large part for the devastating fluid and electrolyte loss characteristic of cholera.[1] The toxin is an oligomeric protein of 84 kDa, consisting of one A subunit (~29 kDa) and five B subunits (11.6 kDa) (for review, see Ref. 2). The B oligomer is responsible for toxin binding to cell surface ganglioside G_{M1} [galactosyl-*N*-acetylgalactosaminyl-(*N*-acetylneuraminyl)galactosylglucosylceramide].[2] The A subunit is a latent ADP-ribosyltransferase; activation requires proteolysis near the carboxyl terminus in a domain between two cysteines. Reduction of the disulfide then releases the larger (22 kDa) catalytically active A1 protein (CTA1) and a smaller, carboxyl-terminal A2 protein (CTA2).[3] ADP-ribosylation is believed to be responsible for the effects of the toxin on cells.

The major ADP-ribose acceptor substrates for the A subunit are the

[1] M. T. Kelly, *Pediatr. Infect. Dis.* **5,** 5101 (1986).
[2] J. Moss and M. Vaughan, *Adv. Enzymol. Relat. Areas Mol. Biol.* **61,** 303 (1988).
[3] J. J. Mekalanos, R. J. Collier, and W. R. Romig, *J. Biol. Chem.* **254,** 5855 (1979).

regulatory guanine nucleotide-binding (G) proteins that couple membrane-associated cell surface receptors with their intracellular effectors.[2,4,5] The G proteins are heterotrimers consisting of α, β, and γ subunits (for review, see Refs. 2, 4, and 5). The α subunit possesses the guanine nucleotide-binding site, an intrinsic GTP hydrolytic activity, and an ADP-ribose acceptor site. The G protein is active when α with bound GTP is dissociated from the $\beta\gamma$ complex. Inactivation of α(GTP) results from hydrolysis of the GTP to GDP and reassociation of α with $\beta\gamma$. Receptor, when occupied by agonist, promotes GDP release from α and thereby facilitates GTP binding. Cholera toxin-catalyzed ADP-ribosylation inhibits the intrinsic GTPase activity of α and thereby prolongs its active state.[6,7] Cholera toxin selectively modifies the multiple forms of $G_{s\alpha}$, the α subunit of the G protein involved in the stimulation of adenylyl cyclase (adenylate cyclase) and the regulation of ion flux.[4] Other G proteins such as transducin (or G_t), G_o, and G_i can also serve as cholera toxin substrates.[8,9] It appears that occupancy of receptor by agonist promotes ADP-ribosylation of $G_{i\alpha}$.[9]

Cholera toxin catalyzes, in addition to the modification of G proteins, the ADP-ribosylation of (a) free arginine and other simple guanidino compounds, (b) unrelated proteins, presumably owing to the presence of an accessible arginine residue, and (c) its own A1 catalytic unit (auto-ADP-ribosylation).[10-14] The toxin also hydrolyzes NAD^+ to ADP-ribose, nicotinamide, and H^+.[15] The ADP-ribosyltransferase reaction is stereospecific and proceeds by an S_N2-like mechanism, with formation of α-ADP-ribosylarginine from β-NAD^+.[16]

---

[4] L. Birnbaumer, R. Mattera, A. Yatani, J. Codina, A. M. J. Van Dongen, and A. M. Brown, "ADP-Ribosylating Toxins and G Proteins: Insights into Signal Transduction" (J. Moss and M. Vaughan, eds.), p. 225. American Society for Microbiology, Washington, D.C., 1990.
[5] P. J. Casey and A. G. Gilman, *J. Biol. Chem.* **263**, 2577 (1988).
[6] D. Cassel and Z. Selinger, *Proc. Natl. Acad. Sci. U.S.A.* **74**, 3307 (1977).
[7] S. E. Navon and B. K.-K. Fung, *J. Biol. Chem.* **259**, 6686 (1984).
[8] C. Van Dop, M. Tsubokawa, H. R. Bourne, and J. Ramachandran, *J. Biol. Chem.* **259**, 696 (1984).
[9] T. Iiri, M. Tohkin, N. Morishima, Y. Ohoka, M. Ui, and T. Katada, *J. Biol. Chem.* **264**, 21394 (1989).
[10] J. Moss and M. Vaughan, *J. Biol. Chem.* **252**, 2455 (1977).
[11] J. B. Trepel, D. M. Chuang, and N. H. Neff, *Proc. Natl. Acad. Sci. U.S.A.* **74**, 5440 (1977).
[12] J. Moss and M. Vaughan, *Proc. Natl. Acad. Sci. U.S.A.* **75**, 3621 (1978).
[13] J. Moss, S. J. Stanley, P. A. Watkins, and M. Vaughan, *J. Biol. Chem.* **255**, 7835 (1980).
[14] J. K. Northup, P. C. Sternweis, M. D. Smigel, L. S. Schleifer, E. M. Ross, and A. G. Gilman, *Proc. Natl. Acad. Sci. U.S.A.* **77**, 6516 (1980).
[15] J. Moss, V. C. Manganiello, and M. Vaughan, *Proc. Natl. Acad. Sci. U.S.A.* **73**, 4424 (1976).
[16] N. J. Oppenheimer, *J. Biol. Chem.* **253**, 4907 (1978).

All of the toxin-catalyzed reactions are enhanced by a family of 20-kDa guanine nucleotide-binding proteins, known as ADP-ribosylation factors or ARFs.[17-21] In the presence of GTP or nonhydrolyzable GTP analogs, but not GDP, its analogs, or adenine nucleotides, ARFs stimulate ADP-ribosylation by the cholera toxin A1 catalytic unit.[20-22] The reaction is enhanced by certain phospholipids and detergents, which also may promote high-affinity binding of guanine nucleotides by ARFs.[19,23,24] Depending on the detergent and phospholipid, activation of cholera toxin by ARF occurs under conditions of both low and high affinity for GTP.[24]

At least six mammalian ARF genes have been described, which can be divided into three classes, based on deduced amino acid sequences, size, phylogenetic analysis, and gene structure.[25-33] Class I ARFs (ARFs 1–3) have 181 amino acids, differ from one another primarily near the amino and carboxyl termini, and have introns in identical locations within the coding region.[25-27,30,31,33] Class II ARFs (ARFs 4 and 5) have 180 amino

---

[17] I. M. Serventi, J. Moss, and M. Vaughan, *Curr. Top. Microbiol. Immunol.* **175**, 43 (1992).
[18] R. A. Kahn and A. G. Gilman, *J. Biol. Chem.* **259**, 6228 (1984).
[19] R. A. Kahn and A. G. Gilman, *J. Biol. Chem.* **261**, 7906 (1986).
[20] S.-C. Tsai, M. Noda, R. Adamik, J. Moss, and M. Vaughan, *Proc. Natl. Acad. Sci. U.S.A.* **84**, 5139 (1987).
[21] S.-C. Tsai, M. Noda, R. Adamik, P. Chang, H.-C. Chen, J. Moss, and M. Vaughan, *J. Biol. Chem.* **263**, 1768 (1988).
[22] M. Noda, S.-C. Tsai, R. Adamik, D. A. Bobak, J. Moss, and M. Vaughan, *Biochemistry* **28**, 7936 (1989).
[23] M. Noda, S.-C. Tsai, R. Adamik, J. Moss, and M. Vaughan, *Biochim. Biophys. Acta* **1034**, 195 (1990).
[24] D. A. Bobak, M. M. Bliziotes, M. Noda, S.-C. Tsai, R. Adamik, and J. Moss, *Biochemistry* **29**, 855 (1990).
[25] S. R. Price, M. S. Nightingale, S.-C. Tsai, K. C. Williamson, R. Adamik, H.-C. Chen, J. Moss, and M. Vaughan, *Proc. Natl. Acad. Sci. U.S.A.* **85**, 5488 (1988).
[26] J. L. Sewell and R. A. Kahn, *Proc. Natl. Acad. Sci. U.S.A.* **85**, 4620 (1988).
[27] D. A. Bobak, M. S. Nightingale, J. J. Murtagh, S. R. Price, J. Moss, and M. Vaughan, *Proc. Natl. Acad. Sci. U.S.A.* **86**, 6101 (1989).
[28] L. Monaco, J. J. Murtagh, K. B. Newman, S.-C. Tsai, J. Moss, and M. Vaughan, *Proc. Natl. Acad. Sci. U.S.A.* **87**, 2206 (1990).
[29] M. Tsuchiya, S. R. Price, S.-C. Tsai, J. Moss, and M. Vaughan, *J. Biol. Chem.* **266**, 2772 (1991).
[30] S.-C. Tsai, R. S. Haun, M. Tsuchiya, J. Moss, and M. Vaughan, *J. Biol. Chem.* **266**, 23053 (1991).
[31] C.-M. Lee, R. S. Haun, S.-C. Tsai, J. Moss, and M. Vaughan, *J. Biol. Chem.* **267**, 9028 (1992).
[32] R. S. Haun, I. M. Serventi, S.-C. Tsai, C.-M. Lee, E. Cavanaugh, L. Stevens, J. Moss, and M. Vaughan, *Clin. Res.* **40**, 148A (1992).
[33] I. M. Serventi, E. Cavanaugh, J. Moss, and M. Vaughan, *J. Biol. Chem.* **268**, 4863 (1993).

acids and are more similar to one another than to other ARFs, having differences near the amino terminus and in the carboxyl half of the protein.[28,29] ARF 6 (class III ARF) has 175 amino acids and differs from the other ARFs throughout the coding region but primarily in the carboxyl half of the protein.[29] Deduced amino acid sequences of all the ARFs have the consensus domains believed to be important for guanine nucleotide binding and GTP hydrolysis (phosphate binding).[34] The ARFs are myristoylated at the amino termini on a glycine that follows the initiator methionine, which is a specific acceptor site for the N-myristoyltransferase.[35] ARFs, of all three classes, synthesized as recombinant proteins in *Escherichia coli*, stimulate cholera toxin-catalyzed ADP-ribosylation.[36-38]

The ARFs are highly conserved proteins, and they have been found in all eukaryotic cells, including what is believed to be the most ancient eukaryote, *Giardia*.[26,35,39,40] The deduced amino acid sequence of *Giardia* ARF is 63-70% identical to those of mammalian ARFs, and the recombinant protein synthesized in *E. coli* stimulates cholera toxin-catalyzed ADP-ribosylation in a GTP-dependent reaction.[40] Two very similar ARF genes have been described in *Saccharomyces cerevisiae;* deletion of both is lethal.[26,41] Mammalian ARFs can complement the double ARF deletion mutant yeast, consistent with conservation of structure and function.[37,42] There are also closely related genes, termed *arl* (ARF-like), which produce proteins that have some similarities in structure but not in function.[43]

Here, we address first cholera toxin-catalyzed ADP-ribosylation. Because ARF promotes ADP-ribosylation and has not been covered in this series for several years, sections are included on the preparation of native ARF, the synthesis of recombinant ARF in *E. coli*, and the use of comple-

---

[34] S. R. Price, A. Barber, and J. Moss, "ADP-Ribosylating Toxins and G Proteins: Insights into Signal Transduction," p. 397. American Society for Microbiology, Washington, D.C., 1990.

[35] R. A. Kahn, C. Goddard, and M. Newkirk, *J. Biol. Chem.* **263**, 8282 (1988).

[36] O. Weiss, J. Holden, C. Rulka, and R. A. Kahn, *J. Biol. Chem.* **264**, 21066 (1989).

[37] R. A. Kahn, F. G. Kern, J. Clark, E. P. Gelmann, and C. Rulka, *J. Biol. Chem.* **266**, 2606 (1991).

[38] S. R. Price, C. F. Welsh, R. S. Haun, S. J. Stanley, J. Moss, and M. Vaughan, *J. Biol. Chem.* **267**, 17766 (1992).

[39] S.-C. Tsai, R. Adamik, M. Tsuchiya, P. P. Chang, J. Moss, and M. Vaughan, *J. Biol. Chem.* **266**, 8213 (1991).

[40] J. J. Murtagh, Jr., M. R. Mowatt, C.-M. Lee, F.-J. S. Lee, K. Mishima, T. E. Nash, J. Moss, and M. Vaughan, *J. Biol. Chem.* **267**, 9654 (1992).

[41] T. Stearns, R. A. Kahn, D. Botstein, and M. A. Hoyt, *Mol. Cell. Biol.* **10**, 6690 (1990).

[42] F.-J. S. Lee, J. Moss, and M. Vaughan, *J. Biol. Chem.* **267**, 24441 (1992).

[43] J. W. Tamkun, R. A. Kahn, M. Kissinger, B. J. Brizuela, C. Rulka, M. P. Scott, and J. A. Kennison, *Proc. Natl. Acad. Sci. U.S.A.* **88**, 3120 (1991).

mentation with yeast mutants to define ARFs functionally. The reader is referred to recent reviews on this and related subjects.[44]

Assays for ADP-Ribosylation Factor Stimulation of Cholera Toxin ADP-Ribosyltransferase Activity

Assays of ADP-ribosylation factor activity are based on its stimulation of cholera toxin ADP-ribosyltransferase activity. Two assays have been most widely used.

ADP-Ribosylation of $G_{s\alpha}$ or Other Proteins

The assay contains protein substrate ($G_{s\alpha}$, other proteins, or membrane fraction), 20 µl of 1 mM GTP, 20 µl of dimyristoylphosphatidylcholine (DMPC)/cholate (5 mM/0.5%), 10 µl of CTA (1 µg), previously activated by incubation at 30° for 10 min with 30 mM dithiothreitol (DTT) in either 60 mM glycine buffer or 50 mM phosphate buffer (pH 7.5), and 30 µl of [^{32}P]NAD$^+$ solution containing 167 mM phosphate buffer (pH 7.5), 16.7 mM MgCl$_2$, 1.67 mM ATP, 66.7 mM thymidine, 66.7 µM NAD$^+$, and 2 µCi of [^{32}P]NAD$^+$ with water to make a total volume of 100 µl. The reaction is incubated at 30° for 30 to 60 min and terminated by the addition of 1–2 ml of cold 7.5% trichloroacetic acid (TCA) and bovine serum albumin (BSA, 5 µg). The tubes are kept in an ice bath for at least 30 min before centrifugation (2800 g, 30 min). After decanting the supernatant and wiping off residual fluid, the precipitate is heated at 65°–70° for 10 min with 0.125 M Tris base, 1% sodium dodecyl sulfate (SDS)/10% mercaptoethanol/10% glycerol/0.002% bromphenol blue before separation of [^{32}P]ADP-ribosyl proteins ($G_{s\alpha}$) by electrophoresis in 12% SDS–polyacrylamide gels (SDS–PAGE).

ADP-Ribosylagmatine Formation Catalyzed by ADP-Ribosyltransferase of Cholera Toxin

One of the simplest methods to assess ARF activity quantitatively is to measure the stimulation of cholera toxin-catalyzed ADP-ribosylagmatine formation from agmatine and radiolabeled NAD$^+$. The reaction mixture (300 µl) contains ovalbumin (30 µg), 0.2 mM GTP, 5 mM MgCl$_2$, 10 mM agmatine, 0.5 mM ATP, 200 µM NAD$^+$ with approximately 10^5 counts/min (cpm) of [*adenine*-^{14}C]NAD$^+$, 20 mM DTT, 1–2 µg of CTA activated as described for the assay above, lipid (such as cardiolipin, ~300 µg) or detergent (SDS, 0.003%), 50 mM potassium phosphate, pH 7.5, and the

---

[44] J. Moss, S.-C. Tsai, and M. Vaughan, this series, Vol. 235, p. 648.

ARF fraction. Not all lipids and detergents are equally effective in enhancing stimulation of cholera toxin-catalyzed reactions by ARF. In addition, ARFs exhibit different lipid/detergent requirements.[38] The reaction is incubated for 60 min at 30° and terminated by transferring to an ice bath. Duplicate samples (100 μl) are transferred to columns (0.8 ml, 0.5 × 4 cm) of AG 1-X2 (Bio-Rad, Richmond, CA) equilibrated with water. Elution with 5 ml of water yields [^{14}C]ADP-ribosylagmatine, which is mixed with 10 ml of scintillation counting fluid for radioassay. It must be established that ARF activity (i.e., the increment in radiolabeled product above that produced by toxin alone) is proportional to the amount of ARF added. Activity is expressed as [^{14}C]ADP-ribosylagmatine formation in picomoles per minute.

Purification of Native ADP-Ribosylation Factors

ARF 1 and ARF 3 from Bovine Brain

sARF I and II (soluble ARFs) have now been identified, by partial amino acid sequences, as products of the ARF 1 and 3 genes, respectively.[21,45] A detailed purification procedure has been published in this series.[46] The description here, therefore, is abbreviated.

Preparation of Ammonium Sulfate Precipitate from Cytosol. All steps are carried out at 4°. Bovine brain cortex (fresh or frozen) is minced and homogenized in 4 volumes of buffer A [20 mM Tris buffer, pH 8.0/1 mM EDTA/1 mM DTT/1 mM NaN$_3$/0.5 mM phenylmethylsulfonyl fluoride (PMSF)/10% sucrose]. The homogenate is centrifuged (15,000 g, 1 hr), and solid ammonium sulfate to 25% saturation is added to the supernatant (I), which is kept at 4° for 2 hr after adjustment to pH 7.5 with cold 0.74 M NH$_4$OH. After centrifugation (15,000 g, 1 hr), solid ammonium sulfate to 70% saturation is added to the supernatant (II) with the pH maintained at pH 7.5 by addition of cold 0.74 M ammonium hydroxide. After 1 hr at 4° and centrifugation (15,000 g, 1 hr), the precipitate is dissolved in buffer B (20 mM potassium phosphate, pH 7.0/1 mM EDTA/1 mM DTT/1 mM NaN$_3$/1 mM benzamidine) containing 0.25 M sucrose and dialyzed overnight at 4° with three changes of buffer B, containing 0.5 M sucrose (to prevent an increase in volume of the dialyzed solution).

CM-Sepharose Fast-Flow Column Chromatography. After centrifugation (105,000 g, 2 hr), the supernatant (III) is applied to a column of CM-Sepharose equilibrated with buffer B containing 0.25 M sucrose. Fractions

---

[45] S.-C. Tsai, R. Adamik, R. S. Haun, J. Moss, and M. Vaughan, *Proc. Natl. Acad. Sci. U.S.A.* **89,** 9272 (1992).

[46] J. Moss, S.-C. Tsai, S. R. Price, D. A. Bobak, and M. Vaughan, this series, Vol. 195, p. 243.

containing unretained protein are pooled and titrated to precisely pH 5.35 with cold 0.1 M acetic acid while stirring on ice. After centrifugation (15,000 g, 1 hr), the supernatant (IV) is applied to a column of CM-Sepharose equilibrated with buffer C (20 mM potassium phosphate, pH 5.35/1 mM EDTA/1 mM DTT/1 mM NaN$_3$/1 mM benzamidine/0.25 M sucrose), which is washed with buffer C containing 25 mM NaCl and eluted with a linear gradient of 25 to 200 mM NaCl in buffer C (4.5 volumes of each). Fractions are immediately adjusted to pH 7.0 with 1 M K$_2$HPO$_4$. Fractions with the highest ARF activity are pooled and concentrated. Two peaks of activity are obtained, ARF 1 (sARF I) at approximately 33 mM NaCl and ARF 3 (sARF II) at about 77 mM NaCl. (If the pH is above 5.35, ARF 1 is eluted with the 25 mM NaCl/buffer C wash.)

Hydroxylapatite Column Chromatography. ARF 1 or 3 is desalted on Sephadex G-25 equilibrated with buffer D (20 mM Tris buffer, pH 8.0/1 mM EDTA/1 mM DTT/1 mM NaN$_3$/5 mM MgCl$_2$/0.25 M sucrose) and applied to a column of hydroxylapatite equilibrated with buffer D. (The presence of MgCl$_2$ prevents inactivation of ARF). After washing with two volumes of buffer D, proteins are eluted with a linear gradient of potassium phosphate, pH 8.0 (0 to 50 mM), in buffer D (3.5 column volumes of each). A sharp peak of ARF activity that emerges ahead of the protein peak is separately collected and concentrated in a Centriprep 10 apparatus (Amicon, Beverly, MA).

Ultrogel AcA 54 Column Chromatography. Concentrated ARF 1 or ARF 3 is added to a column (1.2 × 119 cm) of Ultrogel AcA 54 equilibrated and eluted with buffer D containing 0.1 M NaCl. Fractions (1 ml) are collected. The purity of ARF 1 or 3 after this step is approximately 90%.

Synthesis and Purification of Recombinant ADP-Ribosylation Factor Proteins

Expression of Bovine ARF 2 and Human ARF 6

Bovine ARF 2 and human ARF 6 cDNAs are inserted into the vector pRC23,[47] in which expression of protein is tightly controlled by the phage λ P$_L$ promoter and is thermoinducible. A 1607-base pair (bp) *Eco*RI restriction fragment derived from bovine retinal ARF 2 cDNA is inserted into the vector *Eco*RI site.[25] Because the ARF 6 lacks convenient restriction sites for insertion into pRC23, the vector is modified to facilitate directional insertion of amplified ARF 6 cDNA.[29,38] It is desirable to maintain the distance and nucleotide sequence between the Shine–Dalgarno se-

---

[47] R. Crowl, C. Seamans, P. Lomedico, and S. McAndrew, *Gene* **38**, 31 (1985).

quence in the vector and the initiator ATG codon of the cDNA, since levels of expression can be dramatically influenced by alterations in this region.[47] To introduce this region of the ARF 2 cDNA as well as NcoI and NotI restriction sites for directional insertion into the vector, the ARF 2 coding region, an EcoRI restriction fragment from bovine retinal ARF 2 cDNA, is amplified using a forward primer to introduce a NcoI restriction site at position −1 in the ARF 2 cDNA sequence (+1 position corresponds to A of start ATG) and 3' to the EcoRI site located at position −12 of the cDNA sequence; the reverse primer is designed to introduce a NotI restriction site 3' to the termination codon and 5' to an EcoRI restriction site. Amplification is performed using 35 cycles of 94°, 1 min/37°, 1 min/72°, 2 min, followed by final extension for 10 min at 72°. The amplified ARF 2 cDNA is cleaved with EcoRI and inserted into the EcoRI restriction site of pRC23 to produce pRCARF2Nco/Not.

The ARF 6 cDNA sequence is prepared for cloning by amplification as described above using primers to insert a NcoI restriction site 5' to the initiation codon and a NotI site 3' to the termination codon. After cutting with NcoI and NotI, the amplified ARF 6 cDNA is subcloned into the modified pRC23 vector prepared by cutting pRCARF2Nco/Not with NcoI and NotI and isolation of the vector DNA.

For protein synthesis, E. coli strain RR1 bearing plasmid pRK248cIts, which encodes a temperature-sensitive cI repressor protein, is transformed with the ARF 2 or ARF 6 construct.[48,49] Transformed bacteria are grown to an OD_{600} of 0.4 at 32°. Synthesis of recombinant ARF protein (rARF) is then induced by increasing the temperature to 42° for 2 hr. The bacterial pellet from 500 ml of culture is dispersed in 2.5 ml of 50 mM Tris-Cl, pH 8.1/63 mM EDTA containing lysozyme (1.5 mg/ml) and incubated for 20 min at 4°. Cells are lysed by the addition of 2.5 ml of 50 mM Tris-Cl, pH 8.1/63 mM EDTA/0.5% cholate and incubation for 20 min. The lysate supernatant (∼5 ml) is applied to a column (2 × 120 cm) of Ultrogel AcA 54 equilibrated and eluted with 20 mM Tris-Cl, pH 8.0/0.25 M sucrose/1 mM EDTA/1 mM NaN_3/100 mM NaCl/2 mM DTT/1 mM PMSF/1 mM benzamidine. The major peak of ARF activity is identified using the ADP-ribosylagmatine assay. Peak fractions are pooled, concentrated to approximately 1.0 mg/ml, and stored at −80°. When the preparation is judged by SDS-PAGE to be less than 90% pure, the principal 14-kDa contaminant (lysozyme) is removed, after overnight dialysis against 20 mM potassium phosphate, pH 7.0/0.25 M sucrose/1 mM EDTA/1 mM DTT/1 mM

---

[48] H.-U. Bernard and D. R. Helinski, this series, Vol. 68, p. 482.
[49] H.-U. Bernard, E. Remaut, M. V. Hershfield, H. K. Das, D. R. Helinski, C. Yanofsky, and N. Franklin, *Gene* **5**, 59 (1979).

benzamidine/1 mM NaN$_3$, by incubation with CM-Sepharose equilibrated with the same buffer. The 14-kDa protein is bound and rARF remains in solution.

Synthesis and Purification of Recombinant ADP-Ribosylation Factor as Fusion Protein with Maltose-Binding Protein

Expression of recombinant proteins fused to a specific carrier protein offers several advantages over expression as nonfusion proteins in bacteria. The presence of the carrier protein, in theory, provides an efficient mechanism for purification via affinity chromatography, and, as was true for the ARF proteins, high levels of production of soluble protein can be achieved. The ARFs are expressed fused to maltose-binding protein (MBP) by inserting the cDNA into the pMAL expression vector (New England Biolabs, Beverly, MA) in-frame with and 3' to the *malE* gene.[50] The ARF 6 cDNA (clone 65S)[29] is amplified with primers that incorporate a *Sma*I site at the 5' end and a *Hin*dIII site at the 3' end so that, after digestion with *Sma*I and *Hin*dIII and insertion of the product into the vector, the first codon represents a glycine, corresponding to the penultimate glycine in human ARF 6. The purified and digested polymerase chain reaction (PCR) product is ligated into pMAL that has been digested with *Stu*I and *Hin*dIII. The pMAL vector also encodes the recognition site for factor Xa protease (I-E-G-R), which cleaves carboxylterminal to the arginine. By cloning ARF 6 into the *Stu*I site, a blunt-ended restriction site, and using factor Xa protease, no vector-derived sequence is added to the native sequence. This is advantageous when performing structure–function studies.

Following insert ligation and transformation of competent *E. coli* DH5α cells, ampicillin-resistant colonies are picked and grown in 5 ml of LB broth. As transcription of the construct is under control of the strong *tac* promoter, protein synthesis is induced with 0.3 mM isopropyl-1-thio-β-D-galactopyranoside (IPTG). Colonies producing MBP–ARF 6 fusion protein are identified by the appearance of a 62-kDa protein on SDS–PAGE in the induced, but not uninduced, cell pellets.

For large-scale preparation of MBP-ARF 6, transformed cells are grown in 300 ml of LB with ampicillin (50 μg/ml) at 37° to an OD$_{600}$ of 0.5, and protein expression is induced with 0.3 mM IPTG for 2 hr. The cell pellet is suspended in phosphate-buffered saline (PBS) containing 1 mM EDTA, 1 mM DTT, 1 mM benzamidine, and 0.2 mM PMSF, lysed by sonication, and extracted with 0.25% Tween 20/0.3 M NaCl. Because

---

[50] C. F. Welsh, J. Moss, and M. Vaughan, *Clin. Res.* **40**, 215A (1992).

of inadequate purification using amylose resin affinity chromatography, rARF 6 is obtained using a two-step gel-filtration procedure. The lysate supernatant is applied to a column (2 × 120 cm) of Ultrogel AcA 54 (Sepracor, Marlborough, MA) equilibrated with 20 mM Tris-Cl, pH 8.0/1 mM EDTA/1 mM DTT/0.25 M sucrose/100 mM NaCl/5 mM MgCl$_2$/1 mM benzamidine/0.2 mM PMSF (buffer A). A peak of ARF activity (fusion protein) identified using the ADP-ribosylagmatine assay elutes in the 60-kDa size range. Peak fractions are combined and dialyzed overnight against factor Xa cleavage buffer (buffer A lacking NaCl, sucrose, and PMSF) and incubated with 5% factor Xa (weight/weight of fusion protein) for 24–48 hr at 4°. After completion of proteolysis (monitored by SDS–PAGE), the mixture is applied to a column of Ultrogel AcA 54 equilibrated with buffer A. Recombinant ARF 6 elutes in the 20-kDa size range and is greater than 90% pure (Coomassie blue staining of gels after SDS–PAGE).

Synthesis of ADP-Ribosylation Factor as Glutathione S-Transferase Fusion Protein

A modification of the ligation-independent cloning procedure of Aslanidis and de Jong[51] provides a novel method of cloning cDNAs in any reading frame in a modified pGEX-based vector.[52] The modified vector, pGEX-5G/LIC, permits expression of a cDNA as a fusion gene, following PCR amplification, without restriction endonuclease digestion of the amplified DNA. Recombinant human ARFs 3, 4, and 5 have been produced as fusion proteins with glutathione S-transferase (GST) in the pGEX5-G/LIC vector.[38,45,52] The recombinant proteins have been useful in characterization of the biochemical differences of the mammalian ARFs[38] and for the preparation of ARF-specific antiserum.[45] The preparation of recombinant ARF as a GST fusion protein is described here.

Polymerase Chain Reaction Amplification of ARF 4 cDNA for Ligation-Independent Cloning. The coding region of ARF 4 cDNA is amplified using the primers 5' GGCCTGGTTCCGCGGGGCCTCACTATC 3' and 5' CTGCGCCTCGCTCCAATTTCATTTAA 3'. Underlined nucleotides represent ARF 4 cDNA sequences; remaining nucleotides are complementary to single-stranded tails of T4 DNA polymerase-treated vector. The PCR amplifications are carried out in a total volume of 100 μl containing 25 ng of plasmid DNA, 100 ng of each primer, deoxynucleoside triphosphates (200 μM of each dNTP), *Taq* DNA polymerase buffer (Promega,

---

[51] C. Aslanidis and P. J. de Jong, *Nucleic Acids Res.* **18**, 6069 (1990).
[52] R. S. Haun and J. Moss, *Gene* **112**, 37 (1992).

Madison, WI) and 2.5 units of *Taq* DNA polymerase (Promega). Before addition of polymerase, samples are heated at 95° for 5 min to denature the plasmid DNA. After addition of polymerase, samples are overlaid with mineral oil and subjected to 30 cycles of denaturation (94°, 1 min), annealing (52°, 1 min), and extension (72°, 2 min) using a thermal cycler (Perkin-Elmer Cetus, Norwalk, CT), followed by incubation at 72° for 7 min to extend incomplete products. Products are precipitated by addition of 0.1 volume of 3 M sodium acetate (pH 5.5) and 2 volumes of ethanol and separated in 1% agarose gels containing ethidium bromide at 0.2 μg/ml. The PCR products are purified from gel slices using QIAEX (Qiagen, Chatsworth, CA) as described by the manufacturer.

Ligation-Independent Cloning Reactions. Amplified DNA (~100–200 ng) or *Sac*II-linearized plasmid vector is incubated in 33 mM Tris–acetate (pH 8.0), 66 mM potassium acetate, 10 mM magnesium acetate, 0.5 mM DTT, and 10 mM MgCl$_2$ with 3 units of T4 DNA polymerase (New England Biolabs) and 2.5 mM TTP (amplified DNA) or dATP (linearized vector). The two reaction mixtures (each 20 μl) are incubated at 37° for 20 min, heated at 75° for 10 min (to terminate reactions), and mixed. One-tenth volume of 3 M sodium acetate (pH 5.5) and 2 volumes ethanol are added. Precipitated DNA is suspended in 15 μl of 10 mM Tris-HCl, pH 8.0/1 mM EDTA/100 mM NaCl preheated to 65° and annealed at room temperature for 1–2 hr. A sample (5 μl) of the mixture is used to transform competent DH5α bacteria.

Screening Transformants for Recombinant Clones. Isolated ampicillin-resistant colonies are randomly selected and grown overnight in 1 ml of LB broth containing ampicillin, 50 μg/ml, in a 24-well dish. A sample (100 μl) of fresh overnight culture is added to 900 μl of LB medium, and growth is continued for 1 hr with vigorous agitation. Then IPTG is added (final concentration 0.3 mM), and cultures are grown for an additional 2–3 hr. A sample (100 μl) of the culture is analyzed by SDS–PAGE to permit identification of cultures producing the fusion protein.

Purification of Fusion Proteins and Thrombin Cleavage. Cleared lysates from 500-ml cultures induced with 0.3–0.5 mM IPTG are mixed with 2 ml of 50% glutathione–agarose beads on a rocking platform at 4° for 30 min. After centrifugation (~2000 g, 5 min, 4°), beads with bound fusion protein are suspended in 10 ml of Dulbecco's phosphate-buffered saline (DPBS, BioWhittaker, Walkersville, MD) and transferred to Poly Prep chromatography columns (Bio-Rad). Columns are washed with 30 ml of DPBS, and fusion proteins are eluted with 1.5 ml of 5 mM glutathione in 50 mM Tris-HCl, pH 8.0. Eluates are concentrated, and glutathione is removed using Centricon 10 microconcentrators (Amicon). Buffer is exchanged by suspending the retentates in 1.5 ml of thrombin cleavage buffer

(50 mM Tris-HCl, pH 8.0/2.5 mM CaCl$_2$) and concentrating three times. The retentates are then suspended in cleavage buffer (final volume 0.5 ml), 6 μg of human thrombin (Sigma, St. Louis, MO) is added, and mixtures are rocked on a platform at room temperature for 4 hr. A sample (10 μl) of each is subjected to electrophoresis in a 12.5% acrylamide, 0.1% SDS gel to verify digestion of the fusion protein. Recombinant ARF is separated from glutathione S-transferase and fusion protein using a 1-ml column of glutathione–agarose equilibrated with DPBS. The protein concentration is determined using a Bio-Rad assay kit with bovine serum albumin as a standard.

Synthesis of Myristoylated ADP-Ribosylation Factor

As native ARF is modified by amino-terminal myristoylation,[35] a procedure based on the yeast myristoyl-CoA : protein N-myristoyltransferase (yNMT) coexpression system of Duronio et al.[53] has been used to produce myristoylated and nonmyristoylated recombinant ARF.

Construction of Yeast Myristoyl-CoA : Protein N-Myristoyltransferase Expression Vector. Construction of the yeast NMT expression vector is shown in Fig. 1. Briefly, the coding region of the yeast *NMT1* gene is amplified from 2 μg of genomic DNA with primers that encode an *Nco*I restriction site at the initiator methionine, a *Bgl*II site after the termination codon, and ligation-independent cloning (LIC) sequences at the 5' ends. After cloning the amplified *NMT1* sequence into the LIC vector,[54] the coding region is excised with *Nco*I and *Bgl*II and ligated into a pACYC177 vector (ATCC 37031, American Type Culture Collection, Rockville, MD) modified to contain the bacteriophage T7 promoter and gene *10* ribosome-binding site from the expression plasmid pET-3d (Novagen, Madison, WI). The resulting plasmid, pACYC177/ET3d/yNMT, in which the *NMT* sequences are under the control of the T7 promoter, contains a kanamycin resistance gene for coselection and a p15A replicon for compatibility with ColE1-derived plasmids.

Construction of Human ARF5 Expression Vector. The coding region of human *ARF5* is amplified from 25 ng of cDNA clone 44[29] as described above using the primers 5' CTGGTTCCGGCGAA*CATATG*GGCCTCAC-CGTGTCC 3' and 5' CTCGCTCCGGCGAA*GGATCC*TGGCTGGTTAG-CGC 3'. Underlined sequences correspond to human ARF5 cDNA (acces-

---

[53] R. J. Duronio, E. Jackson-Machelski, R. O. Heuckeroth, P. O. Olins, C. S. Devine, W. Yonemoto, L. W. Slice, S. S. Taylor, and J. I. Gordon, *Proc. Natl. Acad. Sci. U.S.A.* **87**, 1506 (1990).
[54] R. S. Haun, I. M. Serventi, and J. Moss, *BioTechniques* **13**, 515 (1992).

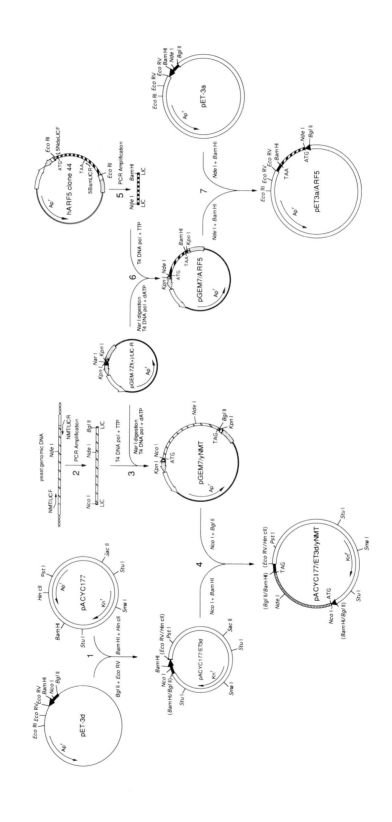

sion number M57567), double-underlined sequences to pGEM7Zf(+)/ LIC-R vector LIC sequences, and italicized sequences to *Nde*I and *Bam*HI restriction endonuclease sites. The gel-purified, amplified DNA cloned into the LIC vector results in plasmid pGEM7/ARF5 containing a *Nde*I restriction site at the initiation methionine codon. The *ARF5* coding sequences are placed under the control of a T7 promoter by ligating the *Nde*I/*Bam*HI fragment of pGEM7/ARF5 into *Nde*I- and *Bam*HI-digested pET-3a (Novagen), resulting in plasmid pET3a/ARF5.

Expression of Myristoylated ARF 5. To assess the ability of yNMT specifically to myristoylate recombinant ARF 5, 5-ml samples of prewarmed medium are inoculated with 100 µl of overnight cultures of BL21(DE3) (Novagen) transformed with either pET3a/ARF5, pACYC177/ET3d/yNMT, or pET3a/ARF5 and pACYC177/ET3d/yNMT with appropriate antibiotic(s), 100 µg/ml. Radiolabeled myristic acid ([9,10-^{3}H(N)]-myristic acid, Du Pont New England Nuclear, Wilmington, DE) is evaporated to dryness in a Savant Speed-Vac (Farmingdale, NY) and suspended in one-tenth the original volume of dimethyl sulfoxide (10 mCi/ml, final concentration). After cultures have reached an OD$_{600}$ of approximately 1, duplicate 1-ml samples of each are transferred to wells of a 24-well dish prewarmed to 37°; 10 µl of [^{3}H]myristic acid is added to each, and 10 µl of 100 m*M* IPTG is added to one of each pair. A sample of each culture (100 µl) is removed after 1 hr. Cells are pelleted by brief centrifugation, suspended in 30 µl of loading buffer, and subjected to electrophoresis in a 15% acrylamide/0.1% SDS (w/v) gel. The gel is then fixed in 10% acetic acid/10% methanol (v/v) for 15 min, washed twice

---

FIG. 1. Construction of yeast NMT and human ARF 5 expression vectors. In Step 1, the *Bgl*II + *Eco*RV fragment of pET-3d was excised and ligated into *Bam*HI- and *Hinc*II-digested pACYC177 resulting in plasmid pACYC177/ET3d. In Step 2, the yeast *NMT1* gene was amplified from genomic DNA using primers corresponding to the NMT coding region and containing ligation-independent cloning (LIC) sequences at the 5' ends with a *Nco*I restriction site at the initiation codon and a *Bgl*II restriction site after the stop codon. In Step 3, the amplified DNA was treated with T4 DNA polymerase in the presence of TTP and cloned into T4 DNA polymerase-treated LIC vector pGEM-7Zf(+)/LIC-R. In Step 4, the yeast NMT expression vector pACYC177/ET3d/yNMT was constructed by ligating the *Nco*I + *Bgl*II fragment from pGEM7/yNMT into *Nco*I- and *Bam*HI-digested pACYC177/ET3d. In Step 5, the coding region of human ARF 5 was amplified from a cDNA clone using primers that contain a *Nde*I restriction site at the initiation codon, a *Bam*HI restriction site 3' to the termination codon, and LIC sequences at the 5' ends. In Step 6, after T4 DNA polymerase treatment, the amplified ARF 5 sequence was cloned into T4 DNA polymerase-treated pGEM-7Zf(+)/LIC-R resulting in pGEM7/ARF5. In Step 7, the ARF 5 coding region was then excised by digestion with *Nde*I and *Bam*HI and ligated to *Nde*I- and *Bam*HI-digested pET-3a yielding the ARF 5 expression vector pET3a/ARF5.

for 3 min with water, treated with ISS Pro-mote (Integrated Separation Systems, Natick, MA) for 15 min, dried on Whatman (Clifton, NJ) 3MM paper, and exposed to Kodak (Rochester, NY) XAR-5 film overnight. A myristoylated protein is detected only in cells expressing both the yeast *NMT1* and human *ARF 5* sequences (Fig. 2, lanes 3 and 6). Comparison of the culture grown with IPTG (lane 6) with that grown without (lane 3) indicates that expression of myristoylated-ARF 5 is enhanced by addition of the inducer.

Purification of Myristoylated and Nonmyristoylated ARF 5. For production of nonmyristoylated ARF 5, competent BL21(DE3) bacteria are transformed with the plasmid pET3a/ARF 5 and selected for ampicillin resistance. Myristoylated ARF 5-producing bacteria are obtained by preparing competent bacteria from the pET3a/ARF5 transformed BL21(DE3) strain, transforming with the pACYC177/ET3d/yNMT plasmid, and selecting for both ampicillin and kanamycin resistance. For large-scale production of recombinant protein, 500 ml of prewarmed LB broth containing the appropriate antibiotic(s), 100 μg/ml, is inoculated with 5 ml from an

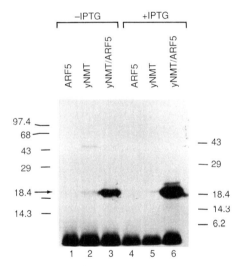

FIG. 2. Autoradiogram of *in vivo* [³H]myristic acid-labeled proteins. BL21(DE3) bacteria transformed with plasmid pET3a/ARF5 (expressing ARF 5, lanes 1 and 4), pACYC177/ET3d/yNMT (expressing yNMT, lanes 2 and 5), or pET3a/ARF5 and pACYC177/ET3d/yNMT (expressing both ARF 5 and yNMT, lanes 3 and 6) were grown in the presence of [³H]myristic acid with (+IPTG) or without (−IPTG) inducer. After 1 hr, the cells were pelleted, dispersed in loading buffer, and subjected to electrophoresis in a 12.5% acrylamide, 0.1% SDS gel. The gel was fixed with acetic acid/methanol, washed with water, treated with salicylate, dried, and exposed to film. Sizes of protein standards (kilodaltons) are indicated. The arrow indicates the position of 20-kDa ARF 5.

overnight culture of a transformed bacterial strain. Myristic acid [Research Organics, Inc., Cleveland, OH, 1% (w/v) in ethanol] is added (final concentration 100 μM), and cultures are shaken at 37°. After the culture reaches an OD_{600} of about 1, 0.01 volumes of 100 mM IPTG are added, cultures are grown for an additional 2 hr, bacteria are collected by centrifugation, and pellets are frozen at $-20°$. To pellets suspended in 5 ml of cold DPBS, 1.5 ml lysozyme (10 mg/ml) is added; suspensions are placed on ice for 5 min, and 1.5 ml of 0.5 M EDTA is added. Mixtures are transferred to a 45-ml cell disruption bomb (Parr Instrument Co, Moline, IL), which is pressurized to 1500 psi with N_2 and placed on ice for 15 min before cells are disrupted by decompression.

Batches of lysed cells containing myristoylated or nonmyristoylated recombinant ARF 5 are centrifuged at 98,000 g for 2 hr (Beckman Ti 50.3, 37,000 rpm). One-half of the supernatant is applied to a column (1.2 × 117 cm, V_t = 132 ml) of Ultrogel AcA 54 (IBF Biotechnics) equilibrated and eluted with 20 mM Tris-HCl, pH 8.0/1 mM EDTA/2 mM DTT/100 mM NaCl/0.25 M sucrose/5 mM $MgCl_2$/1 mM NaN_3. Fractions (1 ml) are collected, assayed for ARF activity (ADP-ribosylagmatine assay) and protein content (Bio-Rad assay with bovine serum albumin as standard), and analyzed for recombinant ARF 5 by SDS–PAGE (14% gel) followed by reaction of blots with rARF 5 polyclonal antibody. Fractions containing ARF are pooled and stored in small portions at $-20°$.

Assessment of Activity by Complementation of ADP-Ribosylation Factor Function in Saccharomyces cerevisiae

Disruption of the two ARF genes (*ARF 1* and *ARF 2*) in *Saccharomyces cerevisiae* is lethal.[41] Although yeast ARF is only about 65% identical to mammalian ARFs, the function of ARF seems to have been preserved during evolution, since expression of human ARF 1, ARF 4, ARF 5, ARF 6, and *Giardia* ARF in yeast could rescue the double *ARF 1/ARF 2* disrupted yeast.[26,37,41,42] Many cell metabolic regulators are conserved through evolution and can serve the same function in heterologous species. By functional complementation, one can determine whether ARFs from other species can perform a biological role the same as or similar to that of yeast ARFs in protein trafficking.

Strains and Culture Conditions. Yeast culture media are prepared, as described by Sherman *et al.*[55] YPD medium contains 1% Bacto-yeast extract, 2% Bacto-peptone, and 2% glucose; YPGal medium contains 1%

---

[55] F. Sherman, G. R. Fink, and J. B. Hicks, "Methods in Yeast Genetics." Cold Spring Harbor Laboratory, Cold Spring Harbor, New York, 1986.

Bacto-yeast extract, 2% Bacto-peptone, and 2% galactose; SD contains 0.7% Difco (Detroit, MI) yeast nitrogen base without amino acids and 2% glucose; and SGal contains 0.7% Difco yeast nitrogen base without amino acids and 2% galactose. Nutrients essential for auxotrophic strains are supplied at specified concentrations.[55]

Polymerase Chain Reaction. The PCR mixture contains 50 mM KCl, 10 mM Tris-Cl (pH 8.3), 1.5 mM MgCl$_2$, 0.01% gelatin, dNTPs (200 μM of each), 0.1% Tween 20, amplification primers (each 25 pmol), and *Taq* polymerase, 2.5 units (total volume, 100 μl). Amplification (Perkin-Elmer Cetus TCI thermal cycler) is performed for 35 cycles of 95°, 1 min/52°, 1 min/72°, 1 min, followed by extension at 72° for 10 min. Samples of reaction mixtures are analyzed by electrophoresis in 1.5% agarose gels.

Construction of ARF Expression Plasmid. For inducible expression of ARF cDNA in yeast, pYEUra3 expression vector (Clontech, Palo Alto, CA) carrying the *GAL1–GAL10* promoter is used. The vector is a low copy number CEN plasmid carrying the selectable *URA3* marker. Sense and antisense *ARF* coding regions are generated using specific cDNA clones and a pair of specific primers (Fig. 3). The PCR products are purified, cleaved with restriction enzymes, and inserted downstream of the *GAL1* promoter. The transcription terminator sequence from *yARF1* 3'-untranslated region was constructed by inserting a PCR fragment generated using two primers (5' AACTCAACTCGAGAATTCTCAGAATATG-

Forward primer	sense	Reverse primer	Restriction Enzyme	Expression plasmid
5'aagaaaggatccaaacaggttta3'	yeast ARF 1	3'acaaaatgtcgtggggagctcct5'	BamHI/XhoI	pYEUyR1s
5'ccgcgtcggatcccgcgccctc3'	human ARF 5	3'ggtccccgtccgagatctacggg5'	BamHI/XbaI	pYEUhR5s
5'ccggctctctagacgcgatggg3'	human ARF 6	3'caatgacaagagctcaaaccgcc5'	XbaI/XhoI	pYEUhR6s
5'ttcaaaatctagaaggctggcca3'	Giardia ARF	3'cgagtatacaaagagctcgcaca5'	XbaI/XhoI	pYEUgRs
	anti-sense			
5'taatagctcgagaaatgggtttgt3'	yeast ARF 1	3'taaacaaaatgtcctagggacgca5'	XhoI/BamHI	pYEUyR1a
5'cccctcgagggccccgccatg3'	human ARF 5	3'ggtccccgtccgagatctacggg5'	XhoI/XbaI	pYEUhR5a
5'aatgcctcgagccccggctcctc3'	human ARF 6	3'gaggtgggtaggagatcttcctct5'	XhoI/XbaI	pYEUhR6a
5'ggccctcgagaagaatgggccaa3'	Giardia ARF	3'cgagtatacaaagatctcgcacag5'	XhoI/XbaI	pYEUgRa

FIG. 3. Construction of ARF expression plasmids. Sense and antisense ARF coding regions were generated with pairs of specific primers as indicated. The PCR fragments were purified, digested with restriction enzymes, and subcloned immediately downstream of the *GAL1* promoter in the pYEUra3 vector.

GAT 3' and 5' TGATTCCTCGAGGGGTGCTGTAAAACAA 3') followed by digestion with XhoI. The DNA sequence and orientation of the ARF genes in the expression vector are confirmed by sequencing.[56] Yeast are transformed by the lithium acetate method.[57]

Sporulation and Complementation. Yeast are grown on YPD medium, transferred to sporulation solution (1% potassium acetate) with appropriate auxotrophic nutrients, and incubated at 30°, 250 rpm, for 1 day. Cells are harvested by centrifugation (1200 g, 5 min, 20°) and suspended in sporulation medium with appropriate auxotrophic nutrients. Random sporulation is carried out as described by Treco.[58] Briefly, the sporulated cells are treated with lyticase (Sigma) overnight with gentle shaking at 30°, then with Nonidet P-40 (NP-40, United States Biochemical, Cleveland, OH) on ice for 15 min, and sonified six times for 45 sec; spores are diluted with water to a concentration of 1000 spores/ml. Samples (100 μl) are placed on several YPGal plates and incubated for 3 days at 30°. Colonies are screened for the marker of interest by replica plating.

Recombinant human ARFs 5 and 6 and *Giardia* ARF proteins exhibit GTP-dependent ARF activity *in vitro*.[38,40] To determine whether these ARF proteins can functionally replace yeast ARFs, sense or antisense cDNAs of human ARF 5, human ARF 6, *Giardia* ARF, or yeast ARF 1 under the control of the *GAL1* promoter are constructed. Expression plasmids are transformed into the heterozygous diploid strain B2 (*MAT*a/ *MAT*α *ade2/* + , *arg4/* + , cyh_r/cys_h, *his3/* + , + /*leu2, trp1/trp1, ura3/ura3, arf1/ARF1, ARF2/arf2*) containing genomic disruptions of one of the *ARF 1* or *ARF 2* alleles. Transformants with the Ura$^+$ phenotype are isolated and sporulated. Tetrad spores are disrupted and plated on YPGal plates.

As yeast lacking both ARF genes are not viable,[41] spores lacking both ARF genes could grow only if a functional ARF protein is synthesized from the expression construct. Spores retaining one or both ARF genes can grow in either glucose or galactose medium. Spores lacking both ARF genes, however, can grow only when galactose induces synthesis of a functionally homologous ARF protein. In each experiment, approximately 300 colonies are screened by replicating colonies on plates of SD without uracil and SGal without uracil. Spores that grow on galactose plates, but not glucose plates, are isolated. Total DNA from each of the colonies is analyzed by Southern blotting to confirm that both yeast *ARF 1* and *ARF 2* genes are disrupted and to demonstrate that these colonies are rescued

---

[56] F. Sanger, S. Nicklen, and A. R. Coulson, *Proc. Natl. Acad. Sci. U.S.A.* **74,** 5463 (1977).
[57] H. Ito, Y. Fukuda, K. Murata, and A. Kimura, *J. Bacteriol.* **153,** 163 (1983).
[58] D. A. Treco, "Current Protocols in Molecular Biology," p. 13.2.9. Wiley, New York, 1990.

by a human or *Giardia* ARF gene. No spores derived from antisense cDNA plasmid transformants grow on galactose but not glucose plates.

Strains rescued by human ARF 5 and ARF 6 and *Giardia* grow much more slowly than wild-type cells on YPGal (Fig. 4). It appears that the properties of the homologous ARFs may not be identical to those of yeast

FIG. 4. Phenotype of complementation of yeast, human, and *Giardia* ARFs. Cells isolated from the viable spores lacking both ARF genes (*arf1, arf2*) and harboring yeast expression plasmids (pYEUyR1s for expressing yeast ARF 1, pYEUhR5s for expressing human ARF 5, pYEUhR6s for expressing human ARF 6, and pYEUgRs for expressing *Giardia* ARF) were streaked on both YPD and YPGal plates. After incubation for 2 days at 30°, plates were photographed. Cells (B2) with plasmid pYEUra3 were used as a control.

ARFs, although levels of expression of recombinant ARFs may have been different, despite growth under identical conditions on YP galactose medium with the identical *GAL1* promoter.

A growing number of ARF family members are being identified, and each presumably carries specific targeting information. Human ARF 5 and ARF 6 and *Giardia* ARF may have information that is not completely useful in the protein trafficking system of yeast Golgi. This functional complementation system may be valuable in evaluating structure–function relationships with ARFs from different species.

[6] Pertussis Toxin-Catalyzed ADP-Ribosylation of G Proteins

By DONNA J. CARTY

Introduction

Pertussis toxin (also known as islet-activating protein) is an A–B type bacterial toxin produced by certain *Bordetella pertussis* strains. Like other A–B type toxins, it has two components, an enzymatically active A component and a B component that binds to the surface of a cell and enables the A component to enter the cell.[1] The A, or active, component (also called S1) is a 28-kDa protein containing both NAD^+ glycohydrolase and ADP-ribosyltransferase activities. The B, or binding, component consists of five subunits.[2] Inside the cell, ATP and phospholipids bind to the B component and induce dissociation of the A and B components.[3-6] Then, when a disulfide bond within the A component between Cys-41 (which is included in the binding site for NAD^+) and Cys-200[7,8] is reduced by

---

[1] D. M. Gill, *in* "Bacterial Toxins and Cell Membranes" (J. Jeljaszewicz and T. Wadstrom, eds.), p. 291. Academic Press, New York, 1978.
[2] M. Tamura, K. Nogimori, S. Murai, M. Yajima, K. Ito, T. Katada, M. Ui, and S. Ishi, *Biochemistry* **21,** 5516 (1982).
[3] L. K. Lim, R. D. Sekura, and H. R. Kaslow, *J. Biol. Chem.* **260,** 2585 (1985).
[4] D. L. Burns and C. R. Manclark, *J. Biol. Chem.* **261,** 4324 (1986).
[5] J. Moss, S. J. Stanley, P. A. Watkins, D. L. Burns, C. R. Manclark, H. R. Kaslow, and E. L. Hewlett, *Biochemistry* **25,** 2720 (1986).
[6] H. R. Kaslow, L. K. Lim, J. Moss, and D. D. Lesikar, *Biochemistry* **26,** 123 (1987).
[7] D. L. Burns and C. R. Manclark, *J. Biol. Chem.* **264,** 564 (1989).
[8] H. R. Kaslow, J. D. Schlotterbeck, V. L. Mar, and W. N. Burnett, *J. Biol. Chem.* **264,** 6386 (1989).

intracellular reduced glutathione, the ADP-ribosyltransferase and NAD^+ glycohydrolase activities are expressed.[9,10]

Among the cellular targets for ADP-ribosylation by the active A component of pertussis toxin are several G proteins belonging to the G_i family.[11–14] (Although it belongs to this family, G_z is not a substrate.[15]) Because ADP-ribosylation functionally uncouples substrate G proteins from receptors,[16] sensitivity to pertussis toxin can serve as an indication of the involvement of a pertussis toxin substrate G protein in a signal transduction pathway. (Insensitivity to ADP-ribosylation does not, however, rule out the involvement of a nonsubstrate G protein.)

Substrate G proteins are ADP-ribosylated at a common conserved cysteine occurring at the fourth position from the carboxy termini of the α subunits.[17] However, only heterotrimeric ($\alpha\beta\gamma$) holo-G proteins are substrates for ADP-ribosylation.[14] Because G-protein activation involves the dissociation of the G-protein α subunit from the $\beta\gamma$ subunits, the extent of ADP-ribosylation (measurable using $[^{32}P]NAD^+$ in the reaction) of a substrate G protein can be used as an indication of the extent of ligand-induced G-protein activation.[18]

Experimental Procedures

Activation of Pertussis Toxin

Because the binding component is necessary for entry of the A component into the cell, only the holotoxin can be used for studies involving intact cells. Typically, for whole cell experiments, the toxin is added to the culture medium at a final concentration of 0.01–1.00 μg/ml. Effects of the toxin can be assayed 6 to 8 hr later.

We routinely purchase pertussis toxin from List Laboratories (Campbell, CA, Cat. No. 180). It arrives lyophilized with sodium chloride and

---

[9] J. Moss, S. J. Stanley, D. L. Burns, J. A. Hsai, D. A. Yost, G. A. Myers, and E. L. Hewlett, *J. Biol. Chem.* **258**, 11879 (1983).

[10] H. R. Kaslow and D. L. Burns, *FASEB J.* **6**, 2684 (1992).

[11] T. Katada and M. Ui, *J. Biol. Chem.* **257**, 7210 (1982).

[12] C. Van Dop, G. Yamanaka, F. Steinberg, R. D. Sekura, C. R. Manclark, L. Styer, and H. R. Bourne, *J. Biol. Chem.* **259**, 23 (1984).

[13] P. W. Sternweis and J. D. Robishaw, *J. Biol. Chem.* **259**, 13806 (1984).

[14] E. J. Neer, J. M. Lok, and L. G. Wolf, *J. Biol. Chem.* **259**, 14222 (1984).

[15] H. K. Fong, K. K. Yoshimoto, P. Eversole-Cire, and M. I. Simon, *Proc. Natl. Acad. Sci. U.S.A.* **85**, 3066 (1988).

[16] A. G. Gilman, *Annu. Rev. Biochem.* **56**, 615 (1987).

[17] R. E. West, J. Moss, M. Vaughan, T. Liu, and T. Y. Liu, *J. Biol. Chem.* **260**, 14428 (1985).

[18] G. S. Kopf, M. J. Woolkalis, and G. L. Gerton, *J. Biol. Chem.* **261**, 7327 (1986).

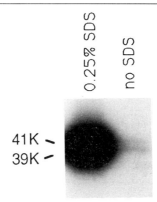

FIG. 1. Effect of including 0.25% SDS on pertussis toxin activation. Pertussis toxin was activated by a 30-min incubation at 32° in the presence of 100 mM DTT with or without 0.25% SDS. Rat brain membranes (25 μg protein) were then ADP-ribosylated using the protocol described in the text. Precipitated protein was electrophoresed on a polyacrylamide gel which was stained to confirm comparable recovery of protein and exposed to Kodak XAR-5 autoradiography film. The 30K–50K range of the gel is shown. SDS greatly improved pertussis toxin activation.

phosphate buffer. We reconstitute 50 μg of the toxin in 500 μl sterile distilled water, as suggested by List, which results in a concentration of 100 μg/ml in 10 mM sodium phosphate buffer, pH 7.0, with 50 mM sodium chloride. The toxin is stored in 5-μl aliquots (sufficient for a 10-sample assay) at −70° for later activation as needed. It tolerates freezing well in this buffer in its nonreduced state.

To obtain an enzymatically active A component in a cell-free system, the A and B components must be dissociated, and a reducing agent such as dithiothreitol (DTT) or 2-mercaptoethanol must be used to break the disulfide bonds of the A component. Both ATP[3,4] and small amounts of detergents[5,10,19] promote dissociation of the A and B components. Together, they are synergistic.[5] The cholic acid analog detergent CHAPS, Lubrol, and sodium dodecyl sulfate (SDS) have all been used for this purpose.[5,10,19] To activate a 5-μl aliquot, we add 5 μl of 200 mM DTT and 1 μl of 2.5% SDS and incubate for 20 min at 32°. The concentration of DTT must be at least 20 mM, but higher concentrations up to 250 mM are more effective. Figure 1 shows that even in the presence of 100 mM DTT, the addition of 0.25% SDS greatly increases ADP-ribosylation of partially purified bovine brain G proteins. Because our assay involves an eventual 100-fold dilution in the final assay medium, the concentration of

[19] K. Enomoto and D. M. Gill, *J. Biol. Chem.* **255**, 1252 (1980).

toxin in the final assay medium will be reduced to 1 μg/ml and that of SDS to 0.0025%. A concentration of SDS as low as 0.02% in the final assay medium can inhibit ADP-ribosylation, and, since CHAPS has also been shown to inhibit ADP-ribosylation of G proteins at higher concentrations,[5] we suggest that the detergent concentration in the activation step be kept to a minimum. However, it is probable that the inclusion of 1 mM ATP in the activation could improve ADP-ribosylation further.

After incubation, we add 50 μl of 1 mg/ml bovine serum albumin (BSA) to the activated toxin for a final assay concentration of approximately 0.1 mg/ml. This prevents loss of ADP-ribosylation substrate protein through absorption to the surface of the glass assay tube and ensures good recovery of substrate protein when the reaction is halted by sodium chloride/acetone precipitation.

Because our toxin aliquots are sufficient for only 10 assays, any unused activated toxin is discarded with little loss. Reduced toxin has been shown to lose activity rapidly.[8]

Preparation of Cell Homogenates and Fractions

Although cell homogenates can be used for ADP-ribosylation assays, it is important, especially in the case of secretory cells, to minimize proteolysis by including protease inhibitors (e.g., phenylmethylsulfonyl fluoride, leupeptin, aprotinin, p-aminobenzamidine, or soybean or lima bean trypsin inhibitors).

There is great variation in the proportion of G-protein substrate in total cellular protein. Figure 2 shows a comparison of ADP-ribosylation using equal amounts of protein (25 μg) from cell membranes of rat brain and liver. It is evident that rat brain contains proportionally more substrate G protein than rat liver.

Because holo-G proteins are, in general, membrane-associated proteins, fractionation to produce enriched membrane fractions will increase the proportion of G proteins in the total protein. Extraction of membrane fractions with a nonionic detergent such as Lubrol or cholate (1%) can often yield further enrichment, since G proteins, though membrane associated, are not integral membrane-crossing proteins. However, the inclusion of more protein in the assay does not always produce better results. Figure 3 shows a comparison of ADP-ribosylations done with 5, 25, and 125 μg of rat brain membranes. Note that 25 μg gives only a slightly stronger signal than 5 μg, and 125 μg gives much less! For this reason, a preliminary experiment varying the amount of protein used should be performed to determine the amount giving optimum results.

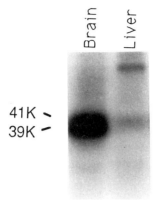

FIG. 2. Comparison of the amounts of G-protein pertussis toxin substrate in 25 μg of rat brain membranes and 25 μg rat liver membranes. The membranes were ADP-ribosylated as described in the text. Proteins were precipitated, electrophoresed, stained, and autoradiographed, also as described. The 30K–50K range of the gel is shown. Rat brain membranes contained proportionally more G-protein substrate than rat liver membranes.

Preliminary Considerations

Test Tubes. If the reaction will be halted by precipitation with −20° sodium chloride/acetone, ADP-ribosylation is best carried out in glass assay tubes. These should be labeled in a manner impervious to acetone.

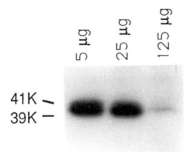

FIG. 3. Comparison of ADP-ribosylation performed using different amounts of rat brain membrane protein. For the experiment, 5, 25, or 125 μg of rat brain membranes were ADP-ribosylated using the protocol described in the text. The assays were identical except for the amount of protein. The proteins were precipitated, electrophoresed, stained, and autoradiographed as described. Coomassie staining (not shown) of BSA (included in pertussis toxin activation) indicated comparable recovery of protein from each assay. ADP-ribosylation was not linear with respect to rat brain membrane protein.

$[^{32}P]NAD^+$. The NAD^+ should be stored in the dark, and the assay solutions containing NAD^+ should be protected from light. We do this by covering containers with aluminum foil before and during the incubation. The $[^{32}P]NAD^+$ should not be purchased or synthesized more than 1 month before the experiment will be carried out. Otherwise, since ^{32}P has a half-life of only 2 weeks, it may be difficult to include the optimum amount of radioactivity in the reaction.

Controls

Three controls may be included among the individual assays to ensure correct interpretation of the results. (1) To discern ADP-ribosylation catalyzed by glycohydrolases other than the pertussis toxin active component, a sample may be incubated substituting water for pertussis toxin. Any proteins labeled equally in this sample and the same sample including pertussis toxin can be excluded from consideration. Thymidine is included in the reaction mixture to inhibit one such enzyme, poly(ADP-ribose) synthase (NAD^+ ADP-ribosyltransferase, EC 2.4.2.30).[20] (2) Excess nonradioactive NAD^+ (100 μM or greater) can be included in a companion sample to prove that ^{32}P-labeling of a particular protein is due to enzymatic transfer of ADP-ribose from $[^{32}P]NAD^+$. Such labeling, and only such labeling, would be eliminated or diminished. (3) To shift the equilibrium of G-protein α and $\beta\gamma$ subunits toward the trimeric state, a sample may be incubated in the presence of excess $\beta\gamma$ subunits, guanosine 5′-diphosphate (GDP), or guanosine 5′-triphosphate (GTP), which will be hydrolyzed to GDP. This should cause an increase in labeling of substrate G proteins, as shown in Fig. 4, but should not affect labeling of other proteins. Alternatively, nonhydrolyzable GTP analogs such as guanosine 5′-O-(3-thio)triphosphate (GTPγS) or guanylyl imidodiphosphate (GppNHp) may be included to induce dissociation of α and $\beta\gamma$ subunits and thus cause a decrease in G-protein labeling. The inclusion of an agonist for a receptor coupled to substrate G proteins should increase dissociation caused by GTPγS or GppNHp.

Final Assay Conditions

The final 50-μl assay mixture contains, in addition to the substrate material (usually in 5 μl) and the 1 μg/ml pertussis toxin, 1 mM DTT, 0.025% SDS, and 0.1 mg/ml BSA contributed by the activated pertussis toxin solution (5 μl of the activation mixture with BSA), the following reagents at the indicated concentrations:

---

[20] D. M. Gill and R. Meren, *Proc. Natl. Acad. Sci. U.S.A.* **75,** 3050 (1978).

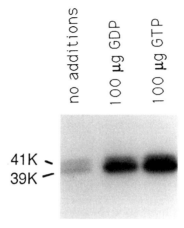

FIG. 4. Comparison of ADP-ribosylation of rat brain membrane G proteins in the absence or presence of 100 μM GDP or 100 μM GTP. For the experiment, 25 μg of rat brain membrane was ADP-ribosylated using the protocol described in the text. The assays were identical except that 5 μl of deionized distilled water, 1 mM GDP, or 1 mM GTP was included in the 50-μl assay mixture. The protein was precipitated, electrophoresed, stained, and autoradiographed as described in the text. ADP-ribosylation was increased by the inclusion of GDP or GTP.

Tris, pH 7.5	25 mM
EDTA	1 mM
Thymidine	10 mM
DTT	10 mM
Lubrol	0.1%
[^{32}P]NAD$^+$	5 μM [~20,000 counts/min (cpm/pmol) or 500,000 cpm/sample]

The samples are incubated at 32° for 40 min, and the reaction is terminated by protein precipitation. We precipitate protein by adding enough NaCl to bring the final concentration to 200 mM along with 10 times the reaction volume of acetone at −20°. The samples may be stored at −20° overnight after this precipitation or centrifuged at 0° and 2500 g for 30 min. Recovery of proteins is more complete and more consistent between samples when they are stored overnight before centrifugation. After centrifugation, supernatants, containing free [^{32}P]NAD$^+$ are removed by aspiration. One milliliter of 20% (w/v) trichloroacetic acid (TCA): 100 ml TCA + H$_2$O final is then added to each sample. This dissolves the NaCl which otherwise causes problems in running samples on polyacrylamide gels. After centrifugation and aspiration of the supernatant, as before, 1 ml of ethyl ether is added to each sample to dissolve labeled phospholipids, and a third centrifugation at the same conditions and aspiration are performed. The precipitates are then dissolved in sample buffer and electrophoresed on polyacrylamide gels.

Alternatively, the reaction may be terminated by simply boiling the samples in sample buffer before electrophoresis, but precipitation of protein gives a "cleaner" autoradiograph. To deal with any loss of protein occurring during precipitation, the polyacrylamide gel may be stained before drying, and Coomassie staining of BSA (from the pertussis toxin activation) may be used to normalize autoradiography results.

Polyacrylamide Gel Electrophoresis

The samples are electrophoresed on 9–12% (w/w) polyacrylamide gels according to the method of Laemmli[21] except that an altered ratio of bisacrylamide to acrylamide (0.4:30) is used to improve resolution.[22] For even better resolution of G-protein α subunits of similar molecular weights, a polyacrylamide gel containing 4 M deionized urea in the separating gel may be used. Gels are stained with Coomassie blue and dried in order to confirm consistent recovery of protein and placed in cassettes with Kodak (Rochester, NY) XAR-5 X-ray film. The duration of exposure necessary for optimal resolution of different G-protein α subunits varies from 15 min to 4 hr.

[21] U. K. Laemmli, *Nature (London)* **227**, 680 (1970).
[22] M. Toutant, D. Aunis, J. Bockaert, V. Homburger, and B. Rouot, *FEBS Lett.* **215**, 339 (1987).

[7] Synthesis and Use of Radioactive Photoactivatable NAD$^+$ Derivatives as Probes for G-Protein Structure

By RICHARD R. VAILLANCOURT, N. DHANASEKARAN, and ARNOLD E. RUOHO

Introduction

Heterotrimeric G proteins mediate the transfer of an extracellular signal from a cell surface receptor to an intracellular enzyme. The transfer of information is dependent on the binding and hydrolysis of guanine nucleotides. A number of methods have been used to probe G-protein structure and function, and each method has provided some information about the functional domains involved in GTP/GDP binding, GTPase activity, effector activation, receptor association, and $\beta\gamma$ subunit interaction. Because the heterotrimeric G proteins have not been crystallized and much of the structural information is indirectly derived from mutational studies,[1-3] epitope mapping,[4-6] and photocross-linking,[7,8] photoactivatable GTP analogs have been used to identify and map the guanine nucleotide binding site of G protein α subunits.[9,10] Utilizing a sulfhydryl derivatizing photoaffinity reagent we have shown the interaction of the C terminus of α_t with the GDP/GTP binding domain. Here we describe the synthesis and use of [^{125}I]iodoazidophenylpropionyl-NAD$^+$ ([^{125}I]AIPP-NAD$^+$) and 2-azido-[^{32}P]NAD$^+$ to probe the structure of G proteins, especially the α subunit. Transducin (G_t) is used as a prototypic model system to derivatize G proteins with the azido-NAD derivatives.

The basic experimental approach for the structural analysis of heterotrimeric G proteins by this method involves the ADP-ribosylation of G proteins with radiolabeled azido-NAD$^+$ and pertussis toxin (PT) [or cholera toxin (CT)]. The PT protein ADP-ribosylates specific cysteines whereas CT ribosylates arginine residues of Gα subunits.[11] The 2-azido-[^{32}P]ADP-ribose, which is "tethered" to the α subunit by such toxin-mediated ADP-ribosylation, is transferred to an "acceptor" polypeptide by cleavage of the thioglycosidic bond (in the case of PT) or Arg–glycosidic linkage (in the case of CT) with an appropriate reagent following photolysis. The "label transferred" polypeptides can then be purified, chemically cleaved, and the radiolabeled cleavage products purified. Microsequencing of the amino-terminal residues would then be matched with the cDNA sequence to identify the fragment to which label is transferred. This domain could only be radiolabeled if it is within reach of the azido-ADP-ribosyl moiety attached to the G-protein α subunit.

This chapter has been divided into three sections. The first section describes the synthesis and use of radioiodinated phenylazide NAD$^+$ de-

---

[1] S. Osawa, N. Dhanasekaran, C. W. Woon, and G. L. Johnson, *Cell* (*Cambridge, Mass.*) **63**, 697 (1990).
[2] S. Osawa, L. E. Heasley, N. Dhanasekaran, S. K. Gupta, C. W. Woon, C. Berlot, and G. L. Johnson, *Mol. Cell. Biol.* **10**, 2931 (1990).
[3] C. H. Berlot and H. R. Bourne, *Cell* (*Cambridge, Mass.*) **68**, 911 (1992).
[4] H. E. Hamm, D. Deretic, K. P. Hofman, A. Schleicher, and B. Kohl, *J. Biol. Chem.* **262**, 10831 (1987).
[5] D. Deretic and H. E. Hamm, *J. Biol. Chem.* **262**, 10839 (1987).
[6] V. N. Hingorani and Y.-K. Ho, *J. Biol. Chem.* **265**, 19923 (1990).
[7] N. Dhanasekaran, M. Wessling-Resnick, D. J. Kelleher, G. L. Johnson, and A. E. Ruoho, *J. Biol. Chem.* **263**, 17942 (1988).
[8] R. R. Vaillancourt, N. Dhanasekaran, G. L. Johnson, and A. E. Ruoho, *Proc. Natl. Acad. Sci. U.S.A.* **87**, 3645 (1990).
[9] D. J. Takemoto and L. J. Takemoto, *Biochem. J.* **225**, 227 (1985).
[10] V. N. Hingorani, L.-F. H. Chang, and Y.-K. Ho, *Biochemistry* **28**, 7424 (1989).
[11] G. L. Johnson and N. Dhanasekaran, *Endocr. Rev.* **10**, 317 (1989).

rivatives ([^{125}I]AIPP-NAD$^+$); the next section describes the synthesis and use of 2-azido-[^{32}P]NAD$^+$. The final section describes photocross-linking and label transfer using pertussis toxin 2-azido-[^{32}P]ADP-ribosylated transducin.

Synthesis of 3-[^{125}I]Iodo-4-azidophenylpropionyl-NAD$^+$

Photoactivatable derivatives of adenine nucleotides have been reported at the C-2,[12-14] C-8,[15,16] N-6,[17] and ribose hydroxyl groups.[18,19] A photoactivatable NAD$^+$ derivative suitable for the "tethered" molecule approach must have the following properties: (a) The compound must be a substrate for pertussis and cholera toxin ADP-ribosylation of Gα subunits. (b) The compound must ADP-ribosylate Gα subunits in reasonable yields. (c) The radiolabel must be easily detectable. (d) The position of the radiolabel must be near the photoactivatable atom so that on photolysis and reversal, facile transfer to the photoinserted polypeptide will occur. (e) The molecule must have bonds that will be susceptible to cleavage by chemical or enzymatic methods. (f) The excess NAD$^+$ photolabel, which has not been "tethered" by ADP-ribosylation, must be efficiently removed after ADP-ribosylation and before photolysis to avoid nonspecific labeling during photolysis.

Chen and Guillory[20] prepared a NAD$^+$ photoaffinity label by derivatizing a ribose hydroxyl group with a tritiated arylazido-β-alanine moiety. The compound, which was a substrate for yeast alcohol dehydrogenase (ADH), was used to photoaffinity label the NAD$^+$ binding site. Because specific NAD$^+$ binding to ADH was not compromised by derivatization on the ribose hydroxyl, one approach for the preparation of a photoactivatable NAD$^+$ derivative is to esterify the carboxylic acid 3-iodo-4-azidophenylpropionic acid (AIPP)[21] to the ribose hydroxyl (Fig. 1).

The nomenclature for NAD$^+$ derivatized on the AMP or nicotinamide

[12] P. T. Riquelme and J. J. Czarnecki, *J. Biol. Chem.* **258**, 8240 (1983).
[13] F. Boulay, P. Dalbon, and P. V. Vignais, *Biochemistry* **24**, 7372 (1985).
[14] J. R. Jefferson, J. B. Hunt, and G. A. Jamieson, *J. Med. Chem.* **11**, 2013 (1987).
[15] B. E. Haley and J. F. Hoffman, *Proc. Natl. Acad. Sci. U.S.A.* **71**, 3367 (1974).
[16] F. Marcus and B. E. Haley, *J. Biol. Chem.* **254**, 259 (1979).
[17] M. Lindberg, P.-O. Larsson, and K. Mosbach, *Eur. J. Biochem.* **40**, 187 (1973).
[18] B. P. Gottikh, A. A. Krayevsky, N. B. Tarussova, P. P. Purygin, and T. L. Tsilevich, *Tetrahedron* **26**, 4419 (1970).
[19] P. C. Carvalho-Alves, C. R. G. Oliveira, and S. Verjovski-Almeida, *J. Biol. Chem.* **260**, 4282 (1985).
[20] S. Chen and R. J. Guillory, *J. Biol. Chem.* **252**, 8990 (1977).
[21] J. M. Lowndes, M. Hokin-Neaverson, and A. E. Ruoho, *Anal. Biochem.* **168**, 39 (1988).

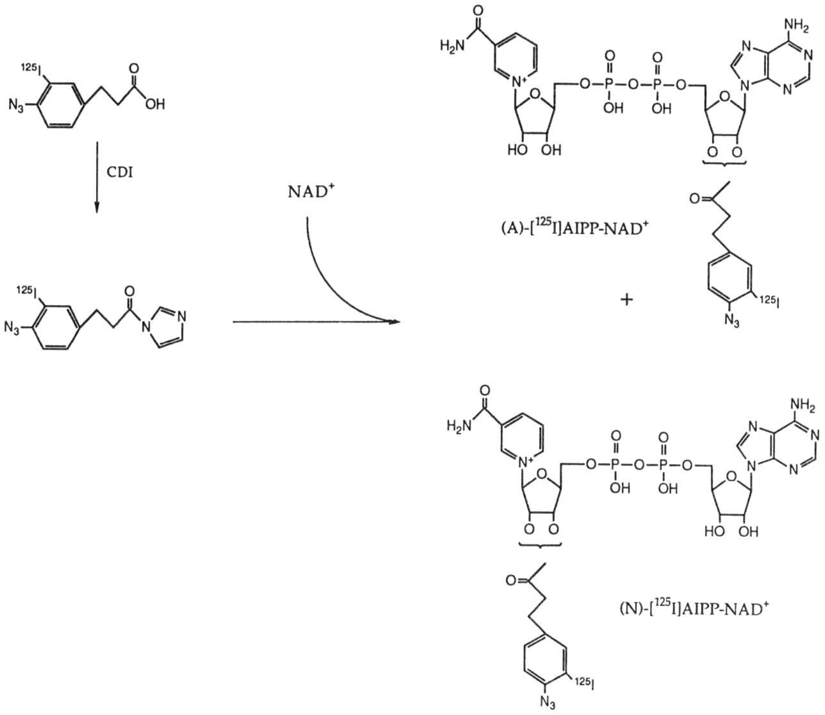

FIG. 1. Synthesis of [^{125}I]AIPP-NAD$^+$ isomers. [^{125}I]AIPP was activated with carbodiimidazole (CDI) forming the N-acylimidazole, which was then reacted with NAD$^+$. Synthesis of (A)-[^{125}I]AIPP-NAD$^+$ and (N)-[^{125}I]AIPP-NAD$^+$ is shown. Using C$_{18}$ reversed-phase TLC, the isomers were resolved using 1-butanol–water (5 : 1, v/v) as solvent. The relative mobilities for (A)-[^{125}I]AIPP-NAD$^+$ and (N)-[^{125}I]AIPP-NAD$^+$ were 0.37 and 0.44, respectively.

mononucleotide (NMN) hydroxyls is either (A) or (N), respectively.[20] (A)-[^{125}I]AIPP-NAD$^+$, which would be derivatized on an AMP hydroxyl, would meet requirements (c)–(e) listed above; that is, the radioiodine which is easily detectable would be ortho to the azide, and the molecule would have three readily reversible bonds: (1) the thioglycosidic bond between the cysteine residue and the 1' carbon of the ADP-ribose is cleavable with mercuric acetate,[22] (2) the ester bond, between [^{125}I]AIPP and the hydroxyl of NAD$^+$, could be cleaved with alkali or hydroxylamine, and (3) the phosphodiester bond could be enzymatically cleaved with

[22] M. K. Jacobson, P. T. Loflin, N. Aboul-Ela, M. Mingmuang, J. Moss, and E. L. Jobson, *J. Biol. Chem.* **265**, 10825 (1990).

snake venom phosphodiesterase.[23] Either of these methods would thus transfer the (A)-[^{125}I]AIPP-ADP-ribose moiety to an acceptor polypeptide.

Experimental Procedures

Materials. Sources of materials are as follows: Carbodiimidazole (CDI) and mercuric acetate are purchased from Aldrich Chemical Co. (Milwaukee, WI); carrier-free Na^{125}I from New England Nuclear (Boston, MA); dimethylformamide (DMF) from Pierce Chemical Co. (Rockford, IL); NAD$^+$, NMN, AMP, *Crotalus atrox* snake venom, and calf alkaline phosphatase from Sigma Chemical Co. (St.Louis, MO); Whatman 3MM paper from Whatman BioSystems Ltd. (Hillsboro, OR); precoated silica gel thin-layer chromatography (TLC) plates, type 60 F$_{254}$, prepared by EM Science (Germany) and obtained through VWR Scientific (Chicago, IL); Bio-Gel P-6 from Bio-Rad Chemical Division (Richmond, CA); pertussis toxin from List Biological Laboratories, Inc. (Campbell, CA); sodium nitrite from J. T. Baker Chemical Co. (Phillipsburg, NJ); X-Omat AR X-ray film from Eastman-Kodak Co. (Rochester, NY). All other chemicals and solvents used are reagent grade.

Synthesis of 3-[^{125}I]Iodo-4-azidophenylpropionylnicotinamide Adenine Dinucleotide. The synthesis of carrier-free [^{125}I]AIPP has been described by Lowndes *et al.*[21] For the preparation of [^{125}I]AIPP-NAD$^+$ (Fig. 1), 1.9 mCi of [^{125}I]AIPP is used, with the specific activity adjusted to 0.5 Ci/mmol with nonradioactive AIPP. Both [^{125}I]AIPP and AIPP in ethyl acetate are dried in a test tube under a stream of N$_2$. To the test tube is added 20 μl of 52 mg CDI/ml DMF, and the reaction is allowed to proceed at ambient temperature. After 15 min, 0.1 ml of 5.58 μg/ml aqueous NAD$^+$ is added to the activated [^{125}I]AIPP. The reaction is stirred overnight at ambient temperature. The DMF is removed under reduced pressure, and the remaining residue is washed three times with 2 ml of acetone to remove unreacted [^{125}I]AIPP. Following the acetone extractions, 0.2 ml of 1-butanol–water (5:1, v/v) is added to the residue, the solution is applied to a 10 × 20 cm C$_{18}$ reversed-phase TLC plate, and the plate is developed with 1-butanol–water (5:1, v/v) as solvent. The product is identified by autoradiography and extracted from the silica with the same solvent.

Two isomers of [^{125}I]AIPP-NAD$^+$ are obtained (Fig. 1). One isomer is derivatized on the adenosine ribose (denoted A), and the other isomer is derivatized on the NMN ribose (denoted N). The products are characterized by enzymatic cleavage using nucleotide pyrophosphatase to generate [^{125}I]AIPP-AMP and [^{125}I]AIPP-NMN followed by alkaline phosphatase

---

[23] C. Van Dop, G. Yamanaka, F. Steinberg, R. D. Sekura, C. R. Manclark, L. Stryer, and H. R. Bourne, *J. Biol. Chem.* **259**, 23 (1984).

treatment to generate [^{125}I]AIPP-adenosine and [^{125}I]AIPP-nicotinamide ribose (data not shown). The ratio of the two isomers is typically 65% (N)-[^{125}I]AIPP-NAD$^+$ and 35% (A)-[^{125}I]AIPP-NAD$^+$. This result is consistent with the C_{18} reversed-phase TLC analysis of a sample of (N)-arylazido-β-alanyl-NAD$^+$,[20] which is kindly provided by Dr. Guillory (University of Hawaii), which also showed two products with R_f values of 0.38 and 0.41.

Purification of Holotransducin. Transducin is purified from rod outer segment membranes isolated from frozen, dark-adapted bovine retinas with illumination as previously described.[24,25] Briefly, thawed retinas are placed in ice-cold 20 mM Tris, pH 7.4, 1 mM CaCl$_2$, 45% (w/w) sucrose and passed several times through a 60-ml syringe. Rod outer segments (ROS) which are disrupted in this manner are collected by flotation, washed in buffer without sucrose, and pooled before being layered over a step gradient of 25 and 35% (w/w) sucrose in 20 mM Tris, pH 7.4, 1 mM CaCl$_2$. The ROS are collected from the 25/35% interface after centrifugation at 100,000 g for 30 min at 4° and are subjected to a series of extensive washes at 100,000 g for 15 min at 4°. The first series of four isotonic washes are carried out in a buffer of 10 mM Tris, pH 7.4, 100 mM NaCl, 5 mM MgCl$_2$, 1 mM dithiothreitol (DTT), 0.1 mM EDTA, and then the ROS are washed four times in a hypotonic buffer consisting of 10 mM Tris, pH 7.4, 1 mM DTT, 0.1 mM EDTA.

To obtain transducin, washed ROS are extracted with 10 mM Tris, pH 7.4, 1 mM DTT, 0.1 mM EDTA, in the presence of 40 μM GTP followed by centrifugation at 100,000 g for 15 min at 4°. The solution is chromatographed on a hexylagarose column using a Waters (Milford, MA) 650E Advanced Protein Purification System. The column is washed at a flow rate of 20 ml/hr with 10 mM 3-(N-morpholino)propanesulfonic acid (MOPS), pH 7.5, 2 mM MgCl$_2$, 0.1 mM phenylmethylsulfonyl fluoride (PMSF), 1 mM DTT, and the protein is eluted with a gradient of 0–300 mM NaCl. The transducin is concentrated by vacuum dialysis and stored at $-20°$ at a concentration of 3 mg/ml in 5 mM Tris, pH 7.4, 2.5 mM MgCl$_2$, 50 mM NaCl, 2 mM DTT, 0.1 mM EDTA, and 50% glycerol.

The rhodopsin-catalyzed transducin binding to [^{35}S]GTPγS is monitored using the nitrocellulose filtration assay previously described.[25] A 300-μl reaction mixture at 4° typically contains 5.4 nM rhodopsin, [^{35}S]GTPγS [1.5–2 × 10^5 counts/min (cpm)/pmol], 10 mM Tris, pH 7.4, 100 mM NaCl, 5 mM MgCl$_2$, 1 mM DTT, 0.1 mM EDTA, and transducin. All measurements are made at ambient temperature. Rhodopsin is bleached under room light, mixed with [^{35}S]GTPγS, and incubated for 1 min. The

---

[24] B. K.-K. Fung and L. Stryer, *Proc. Natl. Acad. Sci. U.S.A.* **78,** 152 (1981).
[25] M. Wessling-Resnick and G. L. Johnson, *J. Biol. Chem.* **262,** 3697 (1987).

nucleotide exchange reaction is initiated by the addition of transducin, and 50-μl aliquots are withdrawn and filtered through nitrocellulose filters (Schleicher and Schuell, Keene, NH, BA85) and immediately washed with two 3.5-ml aliquots of ice-cold buffer consisting of 10 mM Tris, pH 7.4, 100 mM NaCl, 5 mM MgCl$_2$, 0.1 mM EDTA. The filters are dissolved in scintillation fluid, and the amount of [^{35}S]GTPγS associated with the filters is measured by liquid scintillation counting. Typically, the amount of [^{35}S]GTPγS binding sites is the same as the protein concentration as assessed by the Coomassie dye-binding method of Bradford.[26]

[^{125}I]AIPP-ADP-Ribosylation of Purified Holotransducin. Prior to ADP-ribosylation, 200 μl of 3 mg/ml purified transducin is passed through a 3-ml Bio-Gel P-6 column (0.7 × 10 cm, 3-drop fractions) to remove DTT and to reduce the glycerol concentration from 50 to 10% at 4°. Removal of DTT from the transducin preparation is essential in order to prevent reduction of the azide moiety of [^{125}I]AIPP-NAD$^+$ during the ADP-ribosylation procedure.[27] The Bio-Gel P-6 column is equilibrated with 5 mM Tris, pH 7.2, 2.5 mM MgCl$_2$, 50 mM NaCl, 50 μM EDTA, and 10% glycerol (P-6 buffer), and transducin is eluted with P-6 buffer. Fractions are assayed for protein using the Coomassie dye-binding method of Bradford[26] with bovine serum albumin (BSA) as the standard. In a volume of 50 μl is added 5 μg transducin, 0.5 mM ATP, 50 mM 2-mercaptoethanol, 0.5 μg pertussis toxin, and 20 μM [^{125}I]AIPP-NAD$^+$. After 2 hr at 30°, the incubation is stopped with 2× sample buffer and analyzed by sodium dodecyl sulfate–polyacrylamide gel electrophoresis (SDS–PAGE).[28]

For determination of the K_m for (N)-[^{125}I]AIPP-NAD$^+$ and NAD$^+$, concentrations are varied from 0.685 to 6.85 nmol and 0.25 to 5 nmol, respectively. For (N)-[^{125}I]AIPP-NAD$^+$ and NAD$^+$, the reactions are allowed to proceed at 30° for 2 hr and 20 min, respectively. The incubations are terminated with 2× sample buffer and analyzed by SDS–PAGE. The ^{125}I and ^{32}P content in the protein bands is determined using a Packard gamma counter or scintillation counter, respectively.

Proteolysis of [^{125}I]AIPP-ADP-Ribosylated Holotransducin. [^{125}I]AIPP-ADP-ribosylated transducin is treated with trypsin at an enzyme to substrate ratio of 1:10 for 90 min at 4°.[29] Soybean trypsin inhibitor is used to terminate the reaction at an inhibitor to trypsin ratio of 10:1 for 15 min at 4°. Analysis is performed by SDS–PAGE.

[26] M. Bradford, *Anal. Biochem.* **72**, 248 (1976).
[27] I. L. Cartwright, D. W. Hutchinson, and V. W. Armstrong, *Nucleic Acids Res.* **3**, 2331 (1976).
[28] S. P. Fling and D. S. Gregerson, *Anal. Biochem.* **155**, 83 (1986).
[29] B. K.-K. Fung and C. R. Nash, *J. Biol. Chem.* **258**, 10503 (1983).

Chemical Cleavage of Thioglycosidic Bond. Cleavage of thioglycosidic bonds has been described by Krantz and Lee.[30] Mercuric acetate is used to hydrolyze the thioglycosidic bond between Cys-347 of α_t and the 1' carbon of [^{125}I]AIPP-ADP-ribose as described by Jacobson *et al.*[22] Briefly, a final concentration of 10 mM mercuric acetate, 0.1% SDS, and 0.33% aqueous acetic acid is incubated for 15 min at 30° with [^{125}I]AIPP-ADP-ribosylated transducin. Radiolabeled transducin is analyzed by SDS–PAGE and autoradiography.

Electrophoresis. Electrophoresis of proteins is performed according to the method of Fling and Gregerson.[28] Following electrophoresis, gels are fixed in 10% acetic acid, 50% methanol for 30 min and stained with 0.25% (w/v) Coomassie blue in 10% acetic acid for 30 min. The gels are destained in 10% acetic acid and 5% glycerin, dried on a slab-dryer, and then exposed to Kodak X-Omat film with a Quanta III intensifier screen (Du Pont, Wilmington, DE) at −80°.

Results

Two iodinated, phenylazide derivatives of NAD$^+$ were synthesized to specific activities as high as 0.5 Ci/mmol. The synthesis involved the derivatization of the AMP- and NMN-ribose hydroxyl groups on the NAD$^+$ molecule with the *N*-acylimidazole derivative of 3-iodo-4-azidophenylpropionic acid (Fig. 1). As described below, the reaction favored derivatization of the ribose of NMN as opposed to that of AMP, with a ratio of 65:35. Characterization of [^{125}I]AIPP-NAD$^+$ isomers was performed enzymatically using snake venom nucleotide pyrophosphatase and alkaline phosphatase (data not shown).

Transducin could be ADP-ribosylated in a pertussis toxin-dependent manner with either [^{125}I]AIPP-NAD$^+$ isomer. For example, the time course of (N)-[^{125}I]AIPP-ADP-ribose incorporation is shown in Fig. 2. Quantitation of (N)-[^{125}I]AIPP-ADP-ribose incorporation showed 1.1% incorporation after 6 hr, whereas, under identical conditions, 33% of [^{32}P]ADP-ribose was incorporated into the α subunit of transducin after 2 hr using the natural substrate, [^{32}P]NAD$^+$ (data not shown).

To determine the relative substrate properties (i.e., K_m and V_{max}) of (N)-[^{125}I]AIPP-NAD$^+$ and NAD$^+$, transducin was ADP-ribosylated with [^{32}P]NAD$^+$ (Fig. 3A, a and b) and (N)-[^{125}I]AIPP-NAD$^+$ (Fig. 3B, a and b). The K_m values for pertussis toxin-dependent ADP-ribosylation of transducin were 6.2 and 51.5 μM with NAD$^+$ and (N)-[^{125}I]AIPP-NAD$^+$, respectively. The rate with (N)-[^{125}I]AIPP-NAD$^+$ produced a V_{max} of 1.66 pmol/min/mg.

[30] M. J. Krantz and Y. C. Lee, *Anal. Biochem.* **71,** 318 (1976).

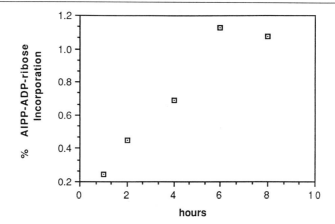

FIG. 2. Quantitation of ADP-ribosylation time course with (N)-[^{125}I]AIPP-NAD$^+$ using pertussis toxin. [^{125}I]AIPP-ADP-ribosylated transducin was excised from a gel which had been stained, destained, and dried onto Whatman 3MM paper. The ^{125}I content in the protein bands was determined using a Packard gamma counter.

The site for pertussis toxin-dependent ADP-ribosylation of transducin has been shown to be Cys-347.[31] A convenient method for demonstrating this fact is to perform trypsin cleavage of [^{125}I]AIPP-ADP-ribosylated holotransducin.[29,32] Trypsin treatment of ADP-ribosylated transducin using (N)-[^{125}I]AIPP-NAD$^+$ produced a 3-kDa peptide (Fig. 4), indicating that the ADP-ribosylation site for (N)-[^{125}I]AIPP-NAD$^+$ and [^{32}P]NAD$^+$ was on the carboxyl terminus of α_t, most likely Cys-347. The label in the 3-kDa polypeptide (or the intact α_t subunit) could be completely removed by mercuric acetate treatment as described by Jacobson *et al.*[22] (data not shown).

Discussion

It has been reported that the use of carbodiimidazole as a condensing reagent provides selectivity for reaction at the 3'-hydroxyl rather than the 2'-hydroxyl group of adenosine nucleosides and nucleotides.[33,34] In addition, the use of dimethylformamide–water (1 : 5, v/v) as a reaction solvent has been reported to favor reactivity at the ribose hydroxyls

---

[31] R. E. West, J. Moss, and M. Vaughn, *J. Biol. Chem.* **260**, 14428 (1985).
[32] J. B. Hurley, M. I. Simon, D. B. Teplow, J. D. Robishaw, and A. G. Gilman, *Science* **226**, 860 (1984).
[33] B. E. Griffin, M. Jarman, C. B. Reese, J. E. Sulston, and D. R. Trenthan, *Biochemistry* **5**, 3638 (1966).
[34] P. C. Zamecnik, *Biochem. J.* **85,** 257 (1962).

A. NAD+

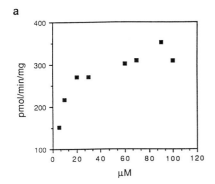
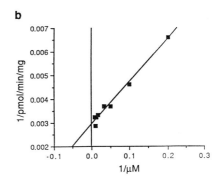

B. (N)-[125I]AIPP-NAD+

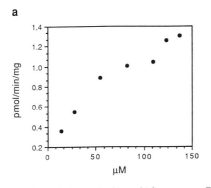
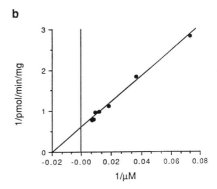

FIG. 3. Initial velocity and Lineweaver–Burk analysis of the data for the ADP-ribosylation of transducin by pertussis toxin using (A) [^{32}P]NAD$^+$ and (B) (N)-[^{125}I]AIPP-NAD$^+$. (A) Transducin (4 μg) was ADP-ribosylated for 20 min at 30° with pertussis toxin using NAD$^+$ from 0.25 to 5 nmol and 0.8 μCi [^{32}P]NAD$^+$. The protein was resolved by 10% SDS–PAGE, and the ^{32}P content in the protein bands was determined using a Packard scintillation counter. Plot (a) shows the concentration dependence of the ADP-ribosylation, and (b) shows a Lineweaver–Burk analysis of the data in (a). (B) Transducin (10 μg) was ADP-ribosylated for 2 hr at 30° with pertussis toxin using (N)-[^{125}I]AIPP-NAD$^+$ from 0.685 to 6.85 nmol at a specific activity of 383 mCi/mmol. The protein was resolved by 10% SDS–PAGE, and the ^{125}I content in the protein bands was determined using a Packard gamma counter. Plot (a) shows the concentration dependence of the ADP-ribosylation, and (b) shows a Lineweaver–Burk analysis of the data in (a).

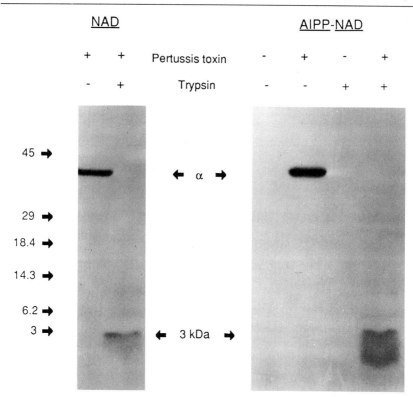

FIG. 4. Trypsin cleavage of transducin ADP-ribosylated with [^{32}P]NAD$^+$ and (N)-[^{125}I]AIPP-NAD$^+$. Autoradiograms were prepared of 8–25% SDS–PAGE gels of ADP-ribosylated transducin using [^{32}P]NAD$^+$ (5 mCi/mmol) or (N)-[^{125}I]AIPP-NAD$^+$ (500 mCi/mmol). After 90 min, ADP-ribosylated transducin was treated with or without trypsin for 90 min at 4°, and the reaction was terminated with soybean trypsin inhibitor as described in the text.

and not at the adenine ring amino group.[18] The carbodiimidazole-coupled [^{125}I]AIPP derivatization of NAD$^+$ was assumed to be on the 3'-hydroxyl, although an assessment of these compounds using ^{13}C nuclear magnetic resonance (NMR) was not performed owing to the low yields during chemical synthesis.

Both [^{125}I]AIPP-NAD$^+$ isomers were substrates for pertussis toxin-dependent ADP-ribosylation of transducin, and the site was the same as that for NAD$^+$ based on trypsin treatment of native transducin, which meets requirement (a) for tethering a photoactivatable molecule. The K_m for pertussis toxin-dependent ADP-ribosylation of transducin using (N)-[^{125}I]AIPP-NAD$^+$ and [^{32}P]NAD$^+$ was 51.5 and 6.2 μM, respectively. A K_m of 25 μM has been reported for ADP-ribosylation of membrane-bound

G proteins using pertussis toxin and [^{32}P]NAD$^+$.[35] Hydroxyl derivatization of NAD$^+$, with [^{125}I]AIPP, decreased the V_{max} 200-fold. Our results are similar to the data reported by Yamaguchi et al., who used an NAD$^+$ molecule derivatived at the hydroxyl position.[36,37] They showed that aryl-azido-β-alanyl-NAD$^+$ had a K_m twice that of NAD$^+$ and that the V_{max} of β-hydroxybutyrate dehydrogenase activity was one-tenth that of NAD$^+$.[36]

Because the K_m for (N)-[^{125}I]AIPP-NAD$^+$ was 51.5 μM, high concentrations of the compound would be necessary in order to ADP-ribosylate transducin at concentrations above the K_m. To overcome this problem, nonradioactive (N)-AIPP-NAD$^+$ could be added to the ADP-ribosylation reaction to obtain high concentrations of photolabel, but this would dilute the specific activity. The relatively low V_{max} also showed that a small percentage of the transducin molecules were [^{125}I]AIPP-ADP-ribosylated in relation to NAD$^+$; therefore, (N)-[^{125}I]AIPP-NAD$^+$ failed to meet criterion (b) listed above. Finally, experiments with transducin and pertussis toxin showed that (N)-[^{125}I]AIPP-NAD$^+$ was not efficiently removed by gel filtration chromatography after [^{125}I]AIPP-ADP-ribosylation (data not shown). Excess radiolabel, which would be present after [^{125}I]AIPP-ADP-ribosylation of transducin, would contribute nonspecific labeling. Therefore, (N)-[^{125}I]AIPP-NAD$^+$, as a substrate for ADP-ribosylation of transducin, failed to meet criterion (f) listed above.

Synthesis and Application of 2-Azido-[^{32}P]NAD$^+$

As reported in the previous section, adenine nucleotide derivatives at the C-2 position of the adenine ring have been synthesized.[12-14] Specifically, the C-2 position of the adenine ring of AMP has been derivatized with an azide moiety and the 5'-phosphate radiolabeled to produce 2-azido-[^{32}P]AMP.[13] This section describes the synthesis of 2-azido-[^{32}P]NAD$^+$ by a combined chemical and enzymatic methodology. An important feature of the use of enzymes for synthesis is the ability to produce 2-azido-[^{32}P]NAD$^+$ at a specific activity exceeding 200 Ci/mmol.

Experimental Procedures and Results

Materials. Sources of materials are as follows: 2-chloroadenosine, anhydrous hydrazine, polyethyleneimine (PEI)-cellulose TLC plates, and NMN are purchased from Sigma Chemical Co.; carrier-free ^{32}PO$_4$ from

---

[35] J. Moss, S. J. Stanley, D. L. Burns, J. A. Hsia, D. A. Yost, G. A. Myers, and E. L. Hewlett, *J. Biol. Chem.* **258**, 11879 (1983).

[36] M. Yamaguchi, S. Chen, and Y. Hatefi, *Biochemistry* **24**, 4912 (1985).

[37] M. Yamaguchi, S. Chen, and Y. Hatefi, *Biochemistry* **25**, 4864 (1986).

New England Nuclear; dimethyl sulfoxide (DMSO) from Pierce Chemical Co.; sodium nitrite from J. T. Baker Chemical Co.; ATP, phosphorus pentoxide, spectrophotometric grade acetone, 2,2-dimethoxypropane, di-*p*-nitrophenyl hydrogen phosphate, triethylamine, calcium hydride, trichloroacetonitrile, tetrabutylammonium hydrogen sulfate, and boric acid gel from Aldrich Chemical Co.; creatine phosphate, creatine kinase from rabbit muscle (EC 2.7.3.2), myokinase from rabbit muscle (EC 2.7.4.3, adenylate kinase; suspension in 3.2 M ammonium sulfate), and NAD^+ pyrophosphorylase from hog liver (EC 2.7.7.1, lyophilized with sucrose) from Boehringer Mannheim (Indianapolis, IN); X-Omat AR X-ray film from Eastman-Kodak Co.; DEAE-cellulose from Whatman BioSystems Ltd.; precoated C_{18} reversed-phase TLC plates, 0.25 mm thick, F_{254}, and precoated silica gel TLC plates, type 60 F_{254}, prepared by EM Science and obtained through VWR Scientific. All other chemicals and solvents used are of reagent grade.

Synthetic Procedures

2-Azido-[^{32}P]NAD$^+$ is prepared using two synthetic procedures. One method involves an entirely chemical approach (Fig. 5), and another combines a chemical and enzymatic approach (see Fig. 7). Both synthetic procedures involve the synthesis of 2-azido-[α-^{32}P]AMP. The chemical synthesis of 2-azido-[^{32}P]NAD$^+$ involves the coupling of 2-azido-[^{32}P]AMP with NMN. The enzymatic procedure involves enzymatic phosphorylation of 2-azido-[α-^{32}P]AMP at the β and γ positions, then enzymatic coupling of the product, 2-azido-[α-^{32}P]ATP, with NMN to produce 2-azido-[^{32}P]NAD$^+$.

Chemical Synthesis of 2-Azido-[^{32}P]NAD$^+$

Chemical Synthesis of 2-Azidoadenosine 5'-[α-^{32}P] Monophosphate

Synthesis of 2-hydrazinoadenosine. The following synthetic procedure (see Fig. 6) has previously been described and is used in this work with modifications.[38] Briefly, 3.31 mmol of 2-chloroadenosine (1.0 g) and 1.59 mol of anhydrous hydrazine (50 ml) are added to a 250-ml round-bottomed flask, and the flask is purged with argon. The reaction is stirred with a Teflon stirring bar for 16 hr at ambient temperature. Unreacted hydrazine is removed by rotary evaporation at 30°. The remaining residue is washed three times with 25 ml anhydrous isopropyl alcohol, followed by rotary evaporation to remove traces of hydrazine. The product is a heat-labile,

---

[38] H. J. Schaeffer and J. Thomas, *J. Am. Chem. Soc.* **80**, 3738 (1958).

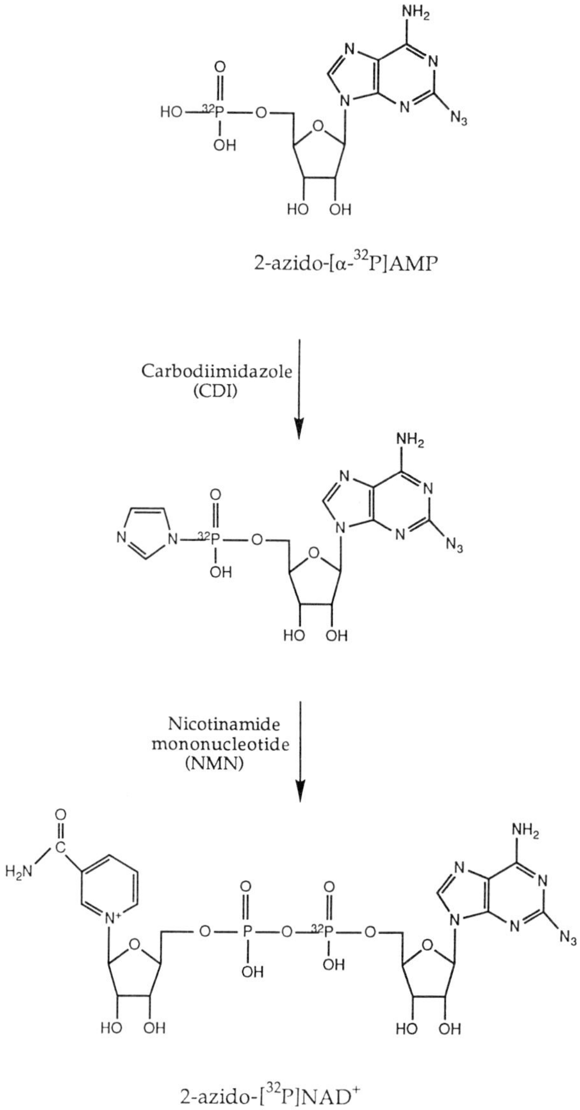

2-azido-[α-^{32}P]AMP

Carbodiimidazole (CDI)

2-azido-[^{32}P]NAD$^+$

FIG. 5. Summary of synthetic scheme for the chemical synthesis of 2-azido-[^{32}P]NAD$^+$. The activation of the phosphate of 2-azido-[α-^{32}P]AMP with carbodiimidazole forms the N-acylimidazole which reacts with the phosphate of NMN to produce 2-azido-[^{32}P]NAD$^+$. Details are described in the text.

gummy residue which is immediately taken to the next step in the synthesis without further characterization.

Synthesis of 2-azidoadenosine. The synthetic procedure has previously been described and is used in this work with modifications.[38] Briefly, 2-hydrazinoadenosine, as a gummy residue in a 250-ml round-bottomed flask, is dissolved in 5% ice-cold acetic acid (30 ml) until a clear solution is obtained and then chilled to 4°. To this solution is added 4.84 mmol of ice-cold, aqueous sodium nitrite (70 ml, 4.88 mg/ml). A precipitate begins to form after 30 min. The reaction is allowed to proceed for 1.5 hr. The precipitate is collected by vacuum on a 60-ml medium scintered glass funnel. The solid is dried under vacuum over phosphorus pentoxide. A yield of 888 mg of 2-azidoadenosine (87%) is obtained, starting from 2-chloroadenosine. The product is homogeneous as analyzed by TLC using C_{18} reversed-phase TLC plates and methanol–water (4 : 1, v/v; R_f 0.85) or normal-phase silica gel TLC plates with anhydrous ethyl ether–acetone–acetic acid (25 : 10 : 5, v/v; R_f 0.71). The IR spectrum of the (potassium bromide) salt shows a maximum at 2140 cm^{-1} (arylazide).

Synthesis of 2',3'-O-isopropylidene 2-azidoadenosine. The synthesis of 2',3'-O-isopropylidene 2-azidoadenosine is based on a modification of the procedure used to synthesize 2',3'-O-isopropylidene adenosine.[39] All subsequent steps are performed under dim light. Briefly, 2.87 mmol of 2-azidoadenosine (888 mg) and 391 mmol of spectrophotometric grade acetone (28.7 ml) are added to a 50-ml round-bottomed flask with 22.96 mmol 2,2-dimethoxypropane (2.82 ml). The reaction is initiated with 3.44 mmol di-*p*-nitrophenyl hydrogen phosphate (1.17 g) and stirred with a Teflon stirring bar at ambient temperature. After 5 hr, the reaction mixture is added to 0.1 M $NaHCO_3$ (71.7 ml) in a 250-ml round-bottomed flask. Acetone is removed by rotary evaporation. The concentration of the 2',3'-O-isopropylidene 2-azidoadenosine is estimated to be 12 mg/ml. This solution is divided so that 10 ml of 2',3'-O-isopropylidene 2-azidoadenosine (120 mg) is extracted twice with 60 ml of ethyl acetate and the organic phase collected using a 125-ml separatory funnel. After the extractions, the ethyl acetate is concentrated by rotary evaporation. The concentrated ethyl acetate solution is applied to 10 × 20 cm C_{18} reversed-phase TLC plates (~12 mg/plate), and the plate is developed in methanol–water (4 : 1, v/v). The product migrates with an R_f of 0.69 and is scraped from the glass plates with a razor blade. Silica containing 2',3'-O-isopropylidene 2-azidoadenosine is pooled into a 60-ml fine scintered glass funnel and extracted three times with 30 ml ethyl acetate. The ethyl acetate is concentrated by rotary evaporation and the remaining residue dissolved in metha-

---

[39] A. Hampton, *J. Am. Chem. Soc.* **83**, 3640 (1961).

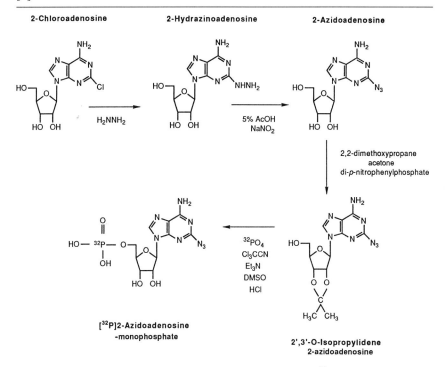

FIG. 6. Synthetic scheme for the chemical synthesis of 2-azido-[^{32}P]AMP. The intermediates are labeled and details are described in the text.

nol (1 volume) and then water (2 volumes). This solution is concentrated by lyophilization. The product is obtained as a dry powder in high yield and stored at $-20°$.

Removal of the isopropylidene protecting group is performed as follows: 2′,3′-O-isopropylidene 2-azidoadenosine is dissolved in 2 M aqueous acetic acid and heated for 45 min at 90°. Formation of 2-azidoadenosine (R_f 0.85) from 2′,3′-O-isopropylidene 2-azidoadenosine (R_f 0.64) is monitored by C_{18} reversed-phase TLC using methanol–water (4 : 1, v/v).

Synthesis of nonradioactive 2-azidoadenosine 5′-monophosphate as standard. The phosphorylation of 2′-3′-O-isopropylidene 2-azidoadenosine to produce 2-azido-AMP is performed according to the method of Yoshikawa and Kato[40] and the purification is performed according to Boulay *et al.*[13]

Synthesis of 2-azidoadenosine 5′-[^{32}P] monophosphate. The synthesis of 2-azido-[^{32}P]AMP (Fig. 6) is based on the procedure used to synthesize

[40] M. Yoshikawa and T. Kato, *Bull. Chem. Soc. Jpn.* **40**, 2849 (1967).

deoxy-[^{32}P]AMP and [^{32}P]AMP.[41] The specific activity of carrier-free (8500–9120 Ci/mmol) orthophosphoric acid in 0.02 N HCl (1 ml) is reduced (20 to 250 Ci/mmol) with aqueous 10 mM phosphoric acid in a 2.2-ml vial. To a 25-ml pear-shaped round-bottomed flask is added 17.2 μmol 2′,3′-O-isopropylidene 2-azidoadenosine (6 mg), 0.2 ml triethylamine (distilled and stored over CaH$_2$), and 25 mCi ^{32}PO$_4$, adjusted to the desired specific activity. This solution is lyophilized to remove hydrochloric acid and water. The phosphorylation reaction is initiated with 0.2 ml DMSO, 21.5 μmol triethylamine (3 μl), and 29.9 μmol trichloroacetonitrile (3 μl, distilled over phosphorus pentoxide and stored at $-80°$). The reaction is heated at 37° on a heating block containing paraffin oil for 30 min. The isopropylidene protecting group is removed by adding 0.8 ml of 0.1 N HCl and heating at 100° for 15 min on a heating block. The HCl is neutralized with 20 μl of triethylamine, and the solution is concentrated ($\sim$200 μl DMSO remains after lyophilization). 2-Azido-[^{32}P]AMP is purified using Dowex 50W-X4 H$^+$ ion-exchange chromatography, as described by Boulay et al.[13] A typical yield of 2-azido-[^{32}P]AMP from the phosphorylation of 2′,3′-O-isopropylidene 2-azidoadenosine is 25–30%.

Synthesis of 2-Azido-[^{32}P]NAD$^+$

Chemical synthesis of nonradioactive 2-azidonicotinamide adenine nucleotide as standard. The synthesis of 2-azido-NAD$^+$ has been reported by Kim and Haley.[42] The procedure is described here with modifications. Briefly, the trioctylammonium salt of NMN is prepared in the following manner. Nicotinamide mononucleotide (50.9 mg, 148 μmol) is refluxed in 1.04 ml methanol with 64.8 μl trioctylamine (148 μmol) in a cold-finger flask for 15 min at 65°. The solution is then transferred to a 25-ml round-bottomed flask and the methanol removed by rotary evaporation. To the residue is added 1 ml DMF, and the solvent is removed under vacuum. The reaction is performed as follows: the trioctylammonium salt of NMN is dissolved in a mixture of 1.04 ml dioxane (12.2 mmol), 0.37 ml hexamethylphosphoamide (2.12 mmol), 0.37 ml of 124.7 mg benzyltributylammonium chloride/ml DMF (150 μmol), 61.6 μl diphenyl phosphochloridate (297 μmol), and 88.3 μl tributylamine (370 μmol). The reaction is stirred for 4 hr at ambient temperature, and then approximately 6 ml diethyl ether is added to precipitate the unreacted diphenyl phosphochloridate; the residue is dissolved in 0.74 ml dioxane.

[41] R. H. Symons, *Nucleic Acids Res.* **4**, 4347 (1977).
[42] H. Kim and B. E. Haley, *J. Biol. Chem.* **265**, 3636 (1990).

The tributylammonium salt of 2-azido-AMP is prepared, based on a modification of the method of Hoard and Ott,[43] in the following manner. 2-Azido-AMP (free acid, 46.4 mg, 120 μmol) is dissolved in 1 ml water with 10 μl pyridine to make the pyridinium salt, and the solution is lyophilized. To the residue is added 1 ml of water with 28.6 μl (120 μmol) tributylamine to make the tributylammonium salt of 2-azido-AMP. The solvent is removed under vacuum, and 0.75 ml pyridine is added to the tributylammonium salt of 2-azido-AMP. The tributylammonium salt of 2-azido-AMP is then added to the activated NMN, and the reaction is allowed to proceed overnight at ambient temperature with stirring. After the solvent is removed under vacuum, the residue is dissolved in 50 ml of 0.1 M ammonium formate, pH 8.5, and applied to a 3-ml bed volume boronyl column. The column is washed with 25 ml of 0.1 M ammonium formate, pH 8.5, and 2-azido-NAD^+ is eluted with 15 ml of water. Fractions are analyzed for product by TLC using isobutyric acid—water—NH_4OH (66:1:33, v/v) and lyophilized. Authentic 2-azido-NAD^+ and 2-azido-[^{32}P]NAD^+ (see next section) comigrated on TLC in two different solvent systems: isobutyric acid–water–NH_4OH (66:1:33, v/v; R_f 0.49) and 1-propanol–NH_4OH–water (20:12:3, v/v; R_f 0.27). The yield of the chemical synthesis is 20% with respect to 2-azido-AMP.

Chemical synthesis of [adenylate-^{32}P]2-azidonicotinamide adenine dinucleotide. The synthesis of [adenylate-^{32}P]2-azido-NAD^+ (2-azido-[^{32}P]NAD^+), using two different chemical coupling procedures with 2-azido-[^{32}P]AMP and NMN, has been reported.[42,44] The chemical coupling of nucleotide phosphates using carbonyldiimidazole has been described by Hoard and Ott.[43] 2-Azido-[^{32}P]AMP, at a specific activity of 1 Ci/mmol, is coupled to NMN using this method (Fig. 5) and purified by preparative silica gel TLC using either of the solvent systems listed above.[42] 2-Azido-[^{32}P]NAD^+ is stored in methanol at $-20°C$.

Enzymatic Synthesis of 2-Azidoadenosine [α-^{32}P]Triphosphate and [adenylate-^{32}P]2-Azidonicotinamide Adenine Dinucleotide

Synthesis of 2-Azidoadenosine [α-^{32}P]Triphosphate. The β and γ phosphates are coupled to 2-azido-[α-^{32}P]AMP, based on a modification of the method described by Symons,[41] to produce 2-azido-[α-^{32}P]ADP and 2-azido-[α-^{32}P]ATP in the following manner (see Fig. 7). To the 2-azido-[α-^{32}P]AMP reaction mixture, described above, is added 612 μl of 50 mM HEPES, pH 8.0, 4 μl of 1 M $MgCl_2$, 16 μl of 1 mM ATP, 64 μl of creatine

---
[43] D. E. Hoard and D. G. Ott, *J. Am. Chem. Soc.* **87**, 1785 (1965).
[44] R. R. Vaillancourt, N. Dhanasekaran, G. L. Johnson, and A. E. Ruoho, *Proc. Natl. Acad. Sci. U.S.A.* **87**, 3645 (1990).

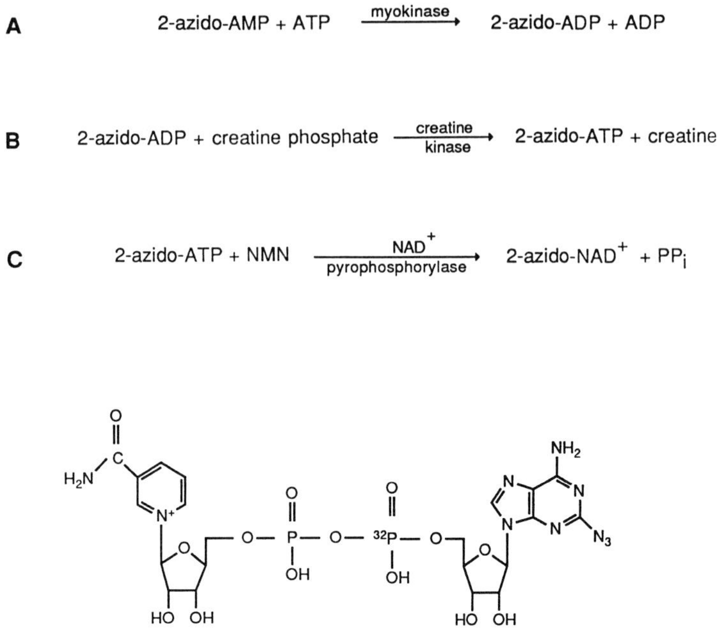

A 2-azido-AMP + ATP $\xrightarrow{\text{myokinase}}$ 2-azido-ADP + ADP

B 2-azido-ADP + creatine phosphate $\xrightarrow{\text{creatine kinase}}$ 2-azido-ATP + creatine

C 2-azido-ATP + NMN $\xrightarrow[\text{pyrophosphorylase}]{\text{NAD}^+}$ 2-azido-NAD$^+$ + PP$_i$

2-Azido-[adenylate-^{32}P]NAD$^+$

FIG. 7. Enzymatic reactions for the synthesis of 2-azido-[α-^{32}P]ATP and 2-azido-[*adenylate*-^{32}P]NAD$^+$ and chemical structure of 2-azido-[*adenylate*-^{32}P]NAD$^+$. The myokinase (A) and creatine kinase (B) reactions were performed simultaneously. Following 2-azido-[α-^{32}P]ATP synthesis, NAD$^+$ pyrophosphorylase and NMN (C) were added to the reaction vessel to complete the reaction. Details are described in the text.

phosphate (14.7 mg/ml), 64 μl of creatine kinase (21.4 mg/ml), and 40 μl of myokinase (2 mg/ml). The level of ammonium sulfate in the preparation of myokinase is decreased before use by centrifugation at 8000 g for 10 min. The reaction is allowed to proceed for 30 min at 37°. 2-Azido-[α-^{32}P]ATP (R_f 0.26) and 2-azido-[α-^{32}P]AMP (R_f 0.48) are identified by PEI-cellulose TLC using 0.5 M NH$_4$HCO$_3$. The enzymatic reaction mixture is lyophilized and applied to a C$_{18}$ reversed-phase high-performance liquid chromatography (HPLC) column (Supelco, 0.5 ml injection loop, 15 cm × 4.6 mm, Supelcosil LC-18-T, 3 μm packing), and fractions are collected at 1-min intervals. The elution conditions are summarized in Table I. ATP elutes at 20 min and 2-azido-[α-^{32}P]ATP elutes as a broad peak between 24 and 29 min, presumably owing to equilibration of the

TABLE I
CHROMATOGRAPHY AND ION-PAIR CONDITIONS
FOR 2-AZIDO-[α-^{32}P]ATP PURIFICATION[a]

Time (min)	Solvent A (%)
0	100
3.3	100
6.7	70
13.3	40
17.3	0
22.7	0
24.0	100

[a] Mobile phases for C_{18} HPLC: solvent A was 0.1 M KH_2PO_4, 4 mM tetrabutylammonium hydrogen sulfate, pH 6.0, and solvent B was solvent A–methanol (70:30, v/v), pH 7.2. The flow rate was 1.0 ml/min, and detection was at 254 nm.

azide moiety into tetrazolo isomers.[45] 2-Azido-[α-^{32}P]ATP purified from the HPLC column is desalted using DEAE-cellulose (HCO_3^- form), as described by Symons,[41] and concentrated using a Savant Speed-Vac (Farmingdale, NY). 2-Azido-[α-^{32}P]ATP is stored in 0.1 mM EDTA at $-20°$. Conversion of 2-azido-[α-^{32}P]AMP to 2-azido-[α-^{32}P]ATP is quantitative as assessed by TLC using a PEI-cellulose system as described in the preparation of 2-azido-[^{32}P]NAD$^+$. Coupling of the 2-azido-[α-^{32}P]ATP to NMN is performed following purification of the 2-azido-[^{32}P]ATP. Coupling could also be performed without prior purification of the 2-azido-[α-^{32}P]ATP.

Synthesis of [adenylate-^{32}P]-Azidonicotinamide Adenine Nucleotide. In all subsequent steps, 2-azido-[^{32}P]NAD$^+$ is identified by comigration on TLC or reversed-phase HPLC with authentic nonradioactive 2-azido-NAD$^+$. 2-Azido-[^{32}P]NAD$^+$ is synthesized, based on a modification of the method described by Alvarez-Gonzalez,[46] in the following manner, without prior purification of 2-azido-[α-^{32}P]ATP. To the 2-azido-[α-^{32}P]ATP reaction mixture is added 40 μl of 48 mM NMN and 120 μl of NAD$^+$ pyrophosphorylase (24 mg/ml). The reaction is complete after 4 hr, although the product is stable to overnight incubation at 37°. 2-Azido-[^{32}P]NAD$^+$ is purified by first diluting the reaction with 1 ml of 1 M HEPES, pH 8.5, and then applying the reaction mixture to a 1-ml boronyl column

[45] J. J. Czarnecki, *Biochim. Biophys. Acta* **800,** 41 (1984).
[46] R. Alvarez-Gonzalez, *J. Chromatogr.* **444,** 89 (1988).

TABLE II
CHROMATOGRAPHY CONDITIONS FOR
2-AZIDO-[α-^{32}P]NAD$^+$ PURIFICATION[a]

Time (min)	Solvent A (%)
0	100
12	100
20	75
23	10
25	0
34	0
36	100

[a] Mobile phases for C_{18} HPLC: solvent A was 0.1 M KH$_2$PO$_4$, pH 6.0, and solvent B was solvent A–methanol (80:20, v/v). The flow rate was 1.0 ml/min, and detection was at 254 nm.

equilibrated with 0.1 M ammonium formate, pH 8.5, collecting 30-drop fractions (~0.5 ml). The boronyl column is used to remove the enzymes and DMSO. The column is subsequently washed with 5 ml of 0.1 M ammonium formate, pH 8.5, and the product, 2-azido-[^{32}P]NAD$^+$, is eluted with water. Fractions which contain 2-azido-[^{32}P]NAD$^+$ (R_f 0.64) are identified using PEI-cellulose TLC developed with a step gradient of 0.2 M LiCl for 2 min, 1.0 M LiCl for 6 min, and 1.6 M LiCl to the top of a 10-cm plate.[47]

Lyophilized fractions are pooled and applied to a C_{18} reversed-phase HPLC column (Supelcosil LC-18-T, 15 cm × 4.6 mm, 3 μm packing). The elution conditions are summarized in Table II. NAD$^+$ elutes as a sharp peak at 22 min, whereas 2-azido-[^{32}P]NAD$^+$ elutes as a broad peak between 28 and 31 min, presumably owing to equilibration of the azide moiety into tetrazolo isomers.[45] The content of any single tube from the peak of the fractionation profile of the HPLC-purified 2-azido-[^{32}P]NAD$^+$ shows reequilibration to tetrazolo isomers when chromatographed on the same C_{18} HPLC system. Fractions containing 2-azido-[^{32}P]NAD$^+$ are identified using PEI-cellulose and comigration with authentic 2-azido-NAD$^+$, as described above. Fractions 28–31 (4 ml) containing 2-azido-[^{32}P]NAD$^+$ are diluted with 5 ml of 1.0 M HEPES, pH 8.5, to adjust the pH of the sample for boronyl affinity column chromatography. The diluted sample is desalted on a 1-ml boronyl column, as described above. Fractions

[47] K. Randerath, "Thin-Layer Chromatography," 2nd Ed., p. 219. Academic Press, New York, 1966.

containing 2-azido-[^{32}P]NAD$^+$ are identified on PEI-cellulose TLC with the LiCl step gradient described above and concentrated by means of a Savant Speed-Vac. High specific activity (250 Ci/mmol) 2-azido-[^{32}P]NAD$^+$ is adjusted to 0.1 mM EDTA and is stable to radiodecomposition when stored at $-20°$.

Analysis of 2-Azido-[^{32}P]NAD$^+$ by Enzyme Cleavage. 2-Azido-[^{32}P]NAD$^+$ is characterized using nucleotide pyrophosphatase and alkaline phosphatase. The assay is performed at 37° in 50 μl containing 100 mM Tris, pH 8.0, 20 mM MgCl$_2$, 0.11 units of nucleotide pyrophosphatase, 0.11 units alkaline phosphatase, and 116 pmol 2-azido-[^{32}P]NAD$^+$ [~195,000 disintegrations/min (dpm)]. After 1 hr, 1 μl from the assay is analyzed by silica TLC using (A) isobutyric acid–NH$_4$OH–water (66:1:33, v/v) and (B) 1-propanol–NH$_4$OH–water (20:12:3, v/v). Hydrolysis of 2-azido-[^{32}P]NAD$^+$ (R_f 0.44, A; 0.34, B) with nucleotide pyrophosphatase is complete after 1 hr, forming 2-azido-[^{32}P]AMP, which comigrates with nonradioactive 2-azido-AMP (R_f 0.50, A; 0.37, B). Subsequent treatment of 2-azido-[^{32}P]AMP with alkaline phosphatase produces ^{32}PO$_4$ (R_f 0.18, A; 0.05, B) and 2-azidoadenosine.

Time Course 2-Azido-[^{32}P]ADP-Ribosylation of Transducin. The method used here is similar to that described above for [^{125}I]AIPP-NAD$^+$ derivatives. Prior to ADP-ribosylation, 200 μl of 3 mg/ml purified transducin, stored at $-20°$, is passed through a 3-ml Bio-Gel P-6 column at 4° (0.7 × 10 cm, 3-drop fractions) to remove DTT and to reduce the glycerol concentration from 50 to 10%. Removal of DTT from the transducin preparation is essential in order to prevent reduction of the azide moiety of 2-azido-[^{32}P]NAD$^+$ during the ADP-ribosylation procedure. The Bio-Gel P-6 column is equilibrated, and transducin is eluted with 5 mM Tris, pH 7.2, 2.5 mM MgCl$_2$, 50 mM NaCl, 50 μM EDTA, and 10% glycerol (P-6 buffer). Transducin is ADP-ribosylated in 50 μl with either 25 μM NAD$^+$ or 25 μM 2-azido-NAD$^+$ under the following conditions: 0.5 mM ATP, 50 mM 2-mercaptoethanol, 1 μg pertussis toxin, 2.5 μg transducin, 0.63 μCi [^{32}P]NAD$^+$, or 2.87 μCi 2-azido-[^{32}P]NAD$^+$. After incubating at 30°, the reactions are stopped with 25 μl of 4× sample buffer and 25 μl of 8 mM NAD$^+$. The samples are analyzed by 9% SDS–PAGE.[28] The α subunits are excised from the stained and destained gel with a razor blade and the amount of ^{32}P incorporation determined in a Packard liquid scintillation counter using 4 ml Poly-Fluor (Packard) scintillation fluid.

Determination of Kinetic Constants for NAD$^+$ and 2-Azido-NAD$^+$

Transducin (5 μg, 1.25 μM) is ADP-ribosylated, as described above with varying concentrations of nonradioactive coenzyme from 0.1 to 50

μM for 20 min at 30°. The reaction is quantitated as described for the ADP-ribosylation time course using SDS–PAGE separation of the α subunit from β and γ.

Analytical Methods

The following molecular weight standards (from Sigma) are used for molecular weight calibration: carbonate dehydratase (29,000), ovalbumin (45,000), bovine serum albumin (66,000), phosphorylase b (97,400), β-galactosidase (116,000), and myosin (205,000).

Discussion

The successful synthesis of high specific activity 2-azido-[α-^{32}P]ATP and 2-azido-[^{32}P]NAD$^+$ has been described. The synthesis involved the derivatization of 2-chloroadenosine with anhydrous hydrazine and then nitrous acid, to form 2-azidoadenosine. The 2'- and 3'-hydroxyls of 2-azidoadenosine were reacted with acetone under acid conditions, forming 2',3'-O-isopropylidene 2-azidoadenosine, thus leaving the 5'-hydroxyl unprotected for phosphorylation with ^{32}PO$_4$. The final synthetic step involved a condensation reaction using trichloroacetonitrile to couple 2',3'-O-isopropylidene 2-azidoadenosine with ^{32}PO$_4$ (Fig. 6). The yield of 2-azido-[α-^{32}P]AMP formation was approximately 30% with respect to inorganic phosphate. 2-Azido-[α-^{32}P]AMP was purified for the chemical coupling with NMN, whereas the purification of 2-azido-[α-^{32}P]AMP was not required for the enzymatic synthesis of 2-azido-[α-^{32}P]ATP (see below).

The chemical coupling of 2-azido-[α-^{32}P]AMP with NMN to produce 2-azido-[^{32}P]NAD$^+$ was performed using carbodiimidazole. For this reaction to succeed, the reactants must be present in micromolar quantities, which limited the specific activity that could be obtained to approximately 1 Ci/mmol.

To synthesize high specific activity 2-azido-[^{32}P]NAD$^+$, an enzymatic method was needed since the chemical scale was reduced 250 times. Myokinase was used to phosphorylate 2-azido-[α-^{32}P]AMP to 2-azido-[α-^{32}P]ADP (Fig. 7) using ATP to initiate the reaction (16 nmol). Creatine kinase was then used to convert the ADP derivative to 2-azido-[α-^{32}P]ATP. The myokinase and creatine kinase reactions were performed simultaneously. The myokinase reaction was inefficient unless the reaction was coupled with creatine kinase, since the equilibrium constant for the myokinase reaction favors the formation of AMP and ATP, not ADP.[48] No detectable 2-azido-[α-^{32}P]ADP was synthesized when myokinase and

---

[48] L. Noda, *J. Biol. Chem.* **232**, 237 (1958).

2-azido-[α-^{32}P]AMP were incubated together (data not shown). Conversion of 2-azido-[α-^{32}P]AMP to 2-azido-[α-^{32}P]ATP was quantitative, and the reaction was complete after 30 min. Nonradioactive ATP was separated from 2-azido-[α-^{32}P]ATP by C$_{18}$ reversed-phase HPLC (Table I).

The 2-azido-[α-^{32}P]ATP was used with NMN and NAD$^+$ pyrophosphorylase to synthesize 2-azido-[^{32}P]NAD$^+$ (Fig. 7). The conversion of 2-azido-[α-^{32}P]ATP to 2-azido-[^{32}P]NAD$^+$ was quantitative. (It was not necessary to purify the ATP from the 2-azido-[α-^{32}P]ATP during the myokinase and creatine kinase reactions, since the C$_{18}$ HPLC column was capable of purifying NAD$^+$ from 2-azido-[^{32}P]NAD$^+$.) The entire reaction from the chemical synthesis of 2-azido-[^{32}P]AMP to the enzymatic synthesis of 2-azido-[^{32}P]NAD$^+$ can be performed in the same reaction vessel with the single caveat that the myokinase and creatine kinase reaction was allowed to proceed to completion (as assessed by PEI-cellulose TLC) before the addition of the NAD$^+$ pyrophosphorylase and NMN. When the NAD$^+$ pyrophosphorylase reaction was performed simultaneously with the myokinase and creatine kinase reactions, 2-azido-[α-^{32}P]AMP appeared to decompose (data not shown).

A time course of pertussis toxin-dependent ADP-ribosylation of transducin with 25 μM NAD$^+$ or 2-azido-NAD$^+$ indicated that the amount of ADP-ribose or 2-azido-ADP-ribose incorporation into α$_t$ was the same for both compounds and reached saturation after 1 hr (data not shown). Twenty minutes was chosen as a measure of initial velocity of the ADP-ribosylation reaction. Using the 20-min time point and a concentration of 1.25 μM transducin, similar kinetics of pertussis toxin-dependent ADP-ribosylation of α$_t$ were observed with either NAD$^+$ (Fig. 8A, a) or 2-azido-NAD$^+$ (Fig. 8B, a). Lineweaver–Burk analysis of the data determined the K_m for NAD$^+$ to be 1.29 μM (Fig. 8A, b) and 1.43 μM for 2-azido-NAD$^+$ (Fig. 8B, b). The maximal velocity, V_{max}, was essentially the same for both compounds, 120 pmol/min/mg for NAD$^+$ and 126 pmol/min/mg for 2-azido-NAD$^+$. The similarity of the kinetic constants for NAD$^+$ and 2-azido-NAD$^+$ shows that derivatization of the adenine base at the C-2 position did not affect the substrate recognition site for pertussis toxin.

The preferred structural conformation for adenine nucleotides with regard to the position of the adenine base in relation to the ribose sugar is the anti conformation.[49-51] It has been suggested that derivatization of the adenine base at the C-2 position favors the anti conformation, whereas

---

[49] S. S. Tavale and M. Sobell, *J. Mol. Biol.* **48**, 109 (1970).
[50] M. Ikehara, S. Uesugi, and K. Yoshida, *Biochemistry* **11**, 830 (1972).
[51] R. H. Sarma, C. H. Lee, F. E. Evans, N. Yathrinda, and M. Sundralingam, *J. Am. Chem. Soc.* **96**, 7337 (1974).

A. NAD^+

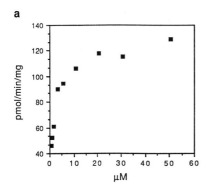
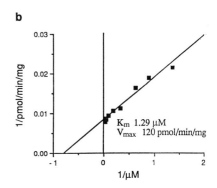

B. 2-azido-$[^{32}P]NAD^+$

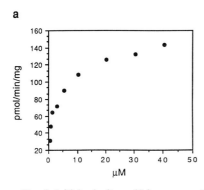
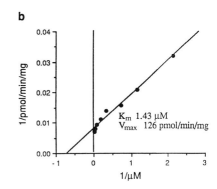

FIG. 8. Initial velocity and Lineweaver–Burk analysis of the data for the ADP-ribosylation of transducin by pertussis toxin using (A) $[^{32}P]NAD^+$ and (B) 2-azido-$[^{32}P]NAD^+$. Transducin (5 μg, 1.25 μM) was ADP-ribosylated, as described in the first section of the text, with varying concentrations of nonradioactive coenzyme from 0.1 to 50 μM for 20 min at 30° and either $[^{32}P]NAD^+$ or 2-azido-$[^{32}P]NAD^+$. The reaction was quantitated as described for the ADP-ribosylation time course using SDS–PAGE analysis. (A) A representative analysis of three different experiments with $[^{32}P]NAD^+$ is shown. The correlation coefficient was 0.967. Plot (a) shows the concentration dependence of ADP-ribosylation, and (b) shows a Lineweaver–Burk analysis of the data in (a). (B) A representative analysis of three different experiments with 2-azido-$[^{32}P]NAD^+$ is shown. The correlation coefficient was 0.983. Plot (a) shows the concentration dependence of ADP-ribosylation, and (b) shows a Lineweaver–Burk analysis of the data in (a).

derivatization at the C-8 position favors the syn conformation.[38] For example, fructose 1,6-bisphosphatase has a 100-fold higher affinity for 2-azido-AMP[12] versus 8-azido-AMP,[52] suggesting a difference in conformation between the C-2 and C-8 isomers. Based on the data reported here, the C-2 azido derivative of NAD^+ (presumably in the anti conformation) was an excellent substrate for pertussis toxin-dependent ADP-ribosylation of α_t. Clearly, ribose derivatization of NAD^+ decreased the pertussis toxin-dependent ADP-ribosylation of α_t (see above), which may indicate an important role for the ribose moiety in the pertussis toxin active site.

In summary, a synthetic procedure for the preparation of a high specific activity and photoactivatable NAD^+ molecule has been described. 2-Azido-$[^{32}P]NAD^+$ meets the following criteria for the "tethered" molecule, as was established previously: (1) 2-Azido-$[^{32}P]NAD^+$ was shown to be a substrate for pertussis toxin as assessed by ADP-ribosylation of α_t. (2) The amount of ADP-ribose incorporated into α_t was the same for NAD^+ and 2-azido-NAD^+. (3) The radiolabel was easily detected by autoradiography and liquid scintillation counting. (4) The position of the isotope is near the photoactivatable moiety such that facile transfer of radiolabel could occur to a photoinserted polypeptide. (5) The compound has readily reversible bonds which would be susceptible to cleavage by mercuric acetate or snake venom phosphodiesterase. Five of the six criteria (a)–(e) thus are met with 2-azido-$[^{32}P]NAD^+$. In contrast, $[^{125}I]AIPP-NAD^+$ met only criteria (a), (c), (d), and (e) listed above and failed to meet criteria (d) and (f).

Intramolecular and Intermolecular Transfer of 2-Azido-$[^{32}P]ADP$-Ribose from Carboxy Terminus of G-Protein α Subunits

The synthesis of high specific activity 2-azido-$[^{32}P]NAD^+$ has been described in the previous section. NAD^+ and 2-azido-$[^{32}P]NAD^+$ were shown to have the same K_m and V_{max} values for pertussis toxin-dependent ADP-ribosylation of transducin. This section illustrates the "tethering" of 2-azido-$[^{32}P]ADP$-ribose at the carboxyl terminus of the α subunit of transducin using pertussis toxin and 2-azido-$[^{32}P]NAD^+$. Photocross-linking and reversal with mercuric acetate (to transfer the radiolabel to the "acceptor" intramolecular and/or intermolecular domains) and SDS–PAGE are shown.

[52] F. Marcus and B. E. Haley, *J. Biol. Chem.* **254**, 259 (1979).

Experimental Procedures

Photolysis of 2-Azido-[^{32}P]ADP-Ribosylated Transducin. ADP-ribosylation of transducin is carried out as described above. Prior to photocross-linking the 2-azido-[^{32}P]ADP-ribosylated transducin, free 2-azido-[^{32}P]NAD$^+$ is removed by gel filtration chromatography using a Bio-Gel P-6 column. The ADP-ribosylated G protein elutes in the void volume, between fractions 8 and 10, whereas free radiolabel is included in the column and fractionates between fractions 21 and 30. Fractions containing ADP-ribosylated G proteins are assayed for protein using the Bradford method.[26] Pooled fractions are photolyzed at 4° through 2-mm thick Pyrex tubes for 5 sec at a distance of 10 cm from a 1-kW mercury lamp, Model AH-6 (Advanced Radiation, Santa Clara, CA).[53]

Results

As described above, an important feature of 2-azido-[^{32}P]NAD$^+$ is that the photoactive moiety and the radioactive atom are positioned on the same side of the phosphodiester bond. Thus, ADP-ribosylation of a G protein, followed by a photolytic reaction of the azide moiety with a neighboring polypeptide, can be used to transfer the molecule to the site of azide insertion by cleavage of the thioglycosidic linkage using mercuric acetate or cleavage of the phosphodiester bond using snake venom phosphodiesterase.

To confirm that the 2-azido-[^{32}P]ADP-ribosylated group was "tethered" to Cys-347 at the carboxyl terminus of transducin, the tryptic pattern of the 2-azido-[^{32}P]ADP-ribosylated α_t was examined. Figure 9 shows the tryptic cleavage pattern of the ADP-ribosylated α subunit of transducin. After 5 min at 4°, the 3-, 12-, and 38-kDa peptides were radiolabeled (Fig. 9, lane b). After 90 min, only the 3-kDa peptide was radiolabeled, which is consistent with ADP-ribosylation at Cys-347 (Fig. 9, lane c).

The stoichiometry of covalent modification of α_t with either [^{32}P]NAD$^+$ or 2-azido-[^{32}P]NAD$^+$ was 1 : 1, and 30% of the transducin molecules were ADP-ribosylated (data not shown). When the 2-azido-[^{32}P]ADP-ribosylated G_t was photolyzed and the photocross-links analyzed by SDS–PAGE, three major cross-linked polypeptides, which contained [^{32}P]ADP-ribosylated α_t, were detected with apparent molecular weights of 47,000, 83,000, and 105,000 (Fig. 10, lane b). A minor photocross-linked band at 92,000 was also produced. The molecular weights (as estimated from protein standards) for the photocross-linked polypeptides were consistent with the following species: α–γ (47K), α–α (83K), α–α–γ (92K),

---

[53] A. Rashidbaigi and A. E. Ruoho, *Proc. Natl. Acad. Sci. U.S.A.* **78**, 1609 (1981).

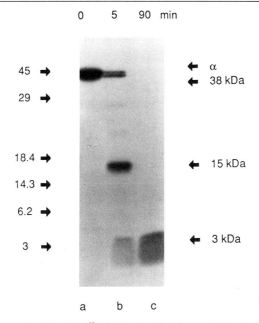

FIG. 9. Trypsin treatment of 2-azido-[^{32}P]ADP-ribosylated transducin. An autoradiogram was prepared of a 15% SDS–PAGE separation of 2-azido-[^{32}P]ADP-ribosylated transducin which was proteolyzed with trypsin for the indicated times as described in the text. After 5 min of trypsin treatment three major trypsin cleavage peptides were formed consisting of 38, 15, and 3 kDa, and the digest was complete after 90 min, producing a 3-kDa peptide derived from the carboxyl terminus.

and α–α–α (105K). The 47,000–105,000 cross-linked proteins required azide activation for their appearance (compare lanes a and b of Fig. 10) and were highly reproducible in several experiments involving multiple G_t preparations.

Two control experiments were performed to determine the pertussis toxin and azide dependence of photocross-linking (data not shown). First, transducin was ADP-ribosylated using 2-azido-[^{32}P]NAD$^+$ in the presence or absence of pertussis toxin. Free radiolabel was removed by gel filtration. The transducin was treated with and without light to activate the azide moiety. The results showed that radiolabel incorporation into α_t was dependent on pertussis toxin treatment (i.e., ADP-ribosylation) and that photoincorporation of radiolabel into transducin subunits (i.e., α_t dimers and trimers) was dependent on prior treatment with pertussis toxin. Second, transducin was ADP-ribosylated with pertussis toxin and [^{32}P]NAD$^+$ or 2-azido-[^{32}P]NAD$^+$, then free radiolabel was removed, and the samples

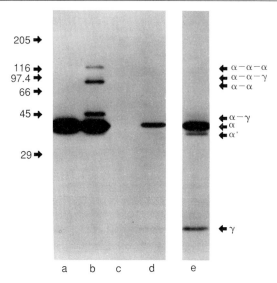

FIG. 10. Autoradiogram of ADP-ribosylated transducin and photocross-linked oligomers. G_t was ADP-ribosylated with 2-azido-[^{32}P]NAD$^+$ (lane a) and photolyzed for 5 sec (lane b). Cross-linked products were generated (right-hand arrows) with azide activation by light. Mercuric acetate removed the ^{32}P radiolabel from α_t in the absence of prior light exposure (land c). After photolysis, the cross-linked species were reversed, thereby transferring radiolabel from α_t to either the α or γ subunit (lanes d and e). The autoradiograms were obtained from the same gel. Lanes a through d were exposed for 17 hr, whereas lane e is a 75-hr exposure of lane d.

were photolyzed. Photocross-linking was observed into α and γ subunits with 2-azido-[^{32}P]NAD$^+$, whereas no photocross-linking was observed into α or γ subunits when [^{32}P]NAD$^+$ was used, indicating an azide-dependent photoincorporation of radiolabel. Therefore, tethering the photoactivatable compound onto the G protein was dependent on pertussis toxin, and photocross-linking was dependent on an azide moiety.

Cleavage of the thioglycosidic bond with mercuric acetate removed greater than 99% of the 2-azido-[^{32}P]ADP-ribose from nonphotolyzed α_t (Fig. 10, lane c). In contrast, mercuric acetate treatment following photolysis resulted in photoinsertion and label transfer into the α and γ subunits (Fig. 10, lanes d and e). The majority of the radiolabel was covalently attached to the α subunit, from either intermolecularly photocross-linked α subunits (i.e., α_t dimers and trimers) or intramolecularly photocross-linked α_t. Assessment of the stoichiometry of label transfer from the α–γ cross-link to the γ subunit was difficult because of loss of γ subunit during gel destaining (data not shown). One additional ^{32}P-radiolabeled band that was detected on reversal migrated immediately below the α subunit (Fig. 10, lane e). This ^{32}P-radiolabeled band was not superimposable on the Coomassie-stained β subunit and is most likely an intramolecular cross-linked α subunit (referred to as α' by Ho and colleagues[54]) which was not completely reversed with mercuric acetate and migrated on SDS–PAGE at a lower apparent molecular weight.

Evidence to support the identity of the cross-linked polypeptides was provided by immunoblotting using antisera specific for either α_t or β and γ subunits (data not shown). These data demonstrate that azide activation results in (a) the insertion of the 2-azido-[^{32}P]ADP-ribose moiety into α_t by an intramolecular reaction and (b) three types of intermolecular insertions, that is, α–α (dimers) and α–α–α (trimers) and an α–γ cross-link. The α–α–γ cross-link was not readily detected in this experiment by either autoradiography or immunoblotting, consistent with its low abundance relative to the major cross-linked products.

These findings are consistent with the following conclusions. (1) A significant fraction of G_t exists as dimers and/or trimers in solution. (2) The γ subunit of the G_t $\alpha\beta\gamma$ heterotrimer is positioned within 2 nm of the α_t Cys-347 residue, assuming a fully extended structure for the azido-ADP-ribosyl molecule (determined by the distance of a molecular space filling model from the 1' carbon on the ribose sugar to the C-2 carbon on the adenine ring of 2-azido-ADP-ribose). (3) 2-Azido-ADP-ribosyl-α_t Cys-347 is not oriented intermolecularly toward the β subunit. In addition, the G_t oligomeric molecules in solution interact in such a way that the α subunit carboxyl terminus of one heterotrimer is in close proximity with both the α and γ subunits of a second G_t heterotrimer.

Acknowledgments

This work was supported by National Institutes of Health Grant GM33138 to A.E.R. and partially funded by a Pharmaceutical Manufacturers Association Foundation Predoctoral Fellowship to R.R.V. We wish to thank Michael Sievert for assistance in preparing the manuscript.

---

[54] V. N. Hingorani, D. T. Tobias, J. T. Henderson, and Y.-K. Ho, *J. Biol. Chem.* **263**, 6916 (1988).

[8] Photoaffinity Guanosine 5'-Triphosphate Analogs as a Tool for the Study of GTP-Binding Proteins

By MARK M. RASENICK, MADHAVI TALLURI, and WILLIAM J. DUNN III

Introduction

Guanine nucleotide-binding proteins (G proteins) mediate myriad cellular functions from cell growth and division to protein synthesis and secretion. Heterotrimeric G proteins couple extracellular receptors to a variety of intracellular events, including the generation of second messengers and the transport of molecules across the cell membrane. A highlight of the "receptor coupling" process involves the ability of a specific agonist to bind to the receptor and to elicit the binding of GTP to the α subunit of a heterotrimeric G protein. The binding of GTP creates an activated G protein, and, once activated, the heterotrimeric G protein transmits and amplifies the message of a hormone or neurotransmitter through the invocation of a variety of intracellular processes. The mechanism whereby hormones or neurotransmitters activate G proteins and their intracellular effectors can be studied in reconstituted systems, where purified components appear to reconstruct certain aspects of the receptor–G-protein coupling process. Nonetheless, the complexity of biological membranes renders it likely that several additional components (such as membrane or cytoskeletal elements) participate in this event.

To study the relationship among G proteins, receptors, and effector molecules in complex systems, such as membranes or permeable cells, it is necessary to design probes which will allow, selectively, the examination of G-protein activation. One such probe is the hydrolysis-resistant, photoaffinity GTP analog P^3-(4-azidoanilido)-P^1-guanosine 5'-triphosphate (AAGTP, Fig. 1), originally synthesized by Pfeuffer.[1] In this chapter we describe the usefulness of this compound for both the examination of G-protein action in complex systems and for the purification of G proteins. Further, we describe an improved synthesis of this compound and its precursors (Fig. 2, I–IV).

[1] T. Pfeuffer, *J. Biol. Chem.* **252**, 7224 (1977).

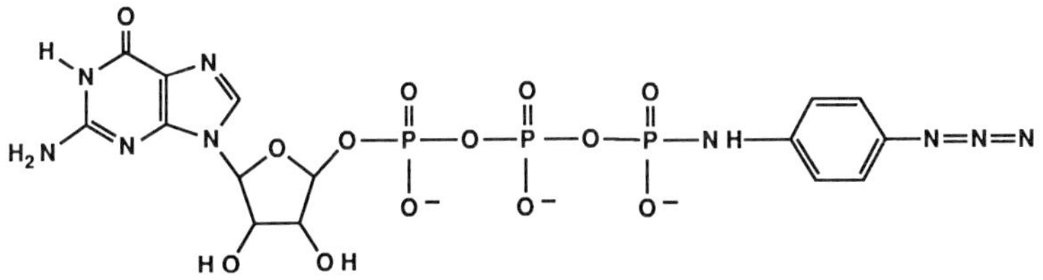

FIG. 1. Structure of 4-azidoanilido-GTP.

Preparation of P^3-(4-Azidoanilido)-P^1-guanosine 5'-Triphosphate

AAGTP (Fig. 1) is prepared by coupling 4-azidoaniline to GTP. 4-Azidoaniline is available from commercial vendors. However, high-performance liquid chromatography (HPLC) shows the compound obtained from Fluka (Ronkonkoma, NY) to be a four-component mixture, and thus of questionable purity. Furthermore, AAGTP made from the commercial compound contains a gray precipitate. Even after purification, AAGTP prepared from this source appears cloudy. The previous synthesis of 4-azidoaniline dates to 1906 and is unwieldy.[2] A revision of that synthesis has provided a high-quality precursor for AAGTP.

Synthesis of 4-Azidoaniline

4-Azidoaniline (**IV**) is prepared via the reaction scheme shown in Fig. 2, which is a variation of the procedure of Silberrad and Smart.[2] 4-Aminoacetanilide (**I**) is obtained from Fluka and used without further purification. A solution of 20 ml of dionized water and 11.5 ml of concentrated hydrochloric acid is prepared in a 250-ml three-necked flask equipped with an overhead stirrer and a 50-ml pressure-equalizing addition funnel. The solution is cooled to 0° with an external ice bath. It is also necessary to cool the mixture for the duration of the reaction. To the flask is added, in small portions, 7.5 g (50 mmol) of **I**. After 3–5 min when the temperature is at 0°, a solution of 3.7 g of NaNO$_2$ in 12.5 ml of deionized water is added dropwise over a period of 1 hr. The solution of the diazo cation (**II**) that results is allowed to stand with stirring for approximately 1 hr. About 25 ml of ice is added, and to this is added slowly, with vigorous stirring, a solution of 3.25 g NaN$_3$ in 10 ml of deionized water. As soon as the azide solution contacts the solution of **II**, a gray precipitate forms and gas evolves. It is necessary to stir the reaction mixture to suspend the solid as it is formed. Addition is complete in about 1 hr. The mixture

[2] O. Silberrad and F. J. Smart, *J. Chem. Soc.* **89**, 170 (1906).

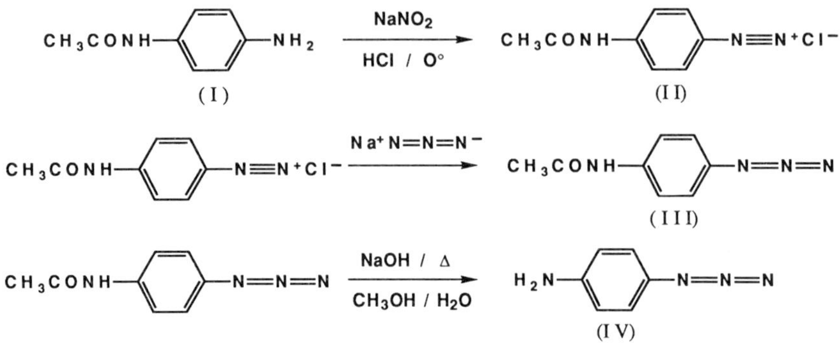

FIG. 2. Synthesis of 4-azidoaniline.

is then allowed to warm with stirring to room temperature and then to sit for 1 hr until gas evolution ceases. The resulting azide is then filtered and air-dried overnight. The dried product is taken up in ethanol and decolorized with activated charcoal. It is filtered, and, while hot, water is added dropwise until the solution becomes cloudy. On cooling the product crystallizes. It is filtered and air-dried overnight. The yield is 5.5 g (30 mmol) or 60% based on **II**. The melting point is 119°–120° with decomposition.

Three grams of **III** (20 mmol) is then dissolved in 10 ml of methanol in a 100-ml three-necked flask equipped with a reflux condenser and a magnetic stirrer. To this is added a solution of 3 g of NaOH in 50 ml of deionized water, and the mixture is refluxed with stirring for 4 hr. The mixture is cooled to room temperature and the condenser replaced with a short column distillation apparatus. By this time, the mixture will turn dark. It is distilled slowly, with the methanol coming off initially. When the temperature reaches 95° the receiver is changed, and approximately 30 ml of distillate is collected in a round-bottomed flask which is submersed in an ice bath. An additional 10 ml of water is added to the distillation flask, and a further 10 ml of distillate is collected and the distillation discontinued. (*Note:* Be sure that there is liquid in the distillation flask and do not attempt to distill further, as azides are thermally unstable.) On standing in the ice bath, the product crystallizes as a light, straw-colored solid. The material is filtered and air-dried, with a yield of 300 mg (2 mmol) or 10% based on **III**. The melting point is 62° in agreement with that reported.[2]

Preparation of GTP Analog from 4-Azidoaniline

[^{32}P]GTP (2 mCi; 410 Ci/mmol) is evaporated under reduced pressure and dissolved in 400 μl of 100 mM triethanolamine hydrochloride, pH 7.2, containing 400 μM *N*-ethyl-*N'*-(3-dimethylaminopropyl)carbodiimide. To this, 200 μl of 4-azidoaniline in peroxide-free dioxane is added, and the mixture is incubated at room temperature for 16 hr in the dark, with gentle rocking. Unreacted 4-azidoaniline is extracted three times with 2 ml peroxide-free ether. The resulting mixture is about 85% pure, and this is further purified by chromatography over Dowex anion-exchange Bio-Rad (Richmond, CA) AG 50W-X4 resin, preequilibrated with HCl, pH 1.15, and eluted with the same solution. Alternately, purification can be accomplished over polyethyleneimine-cellulose, eluting with 800 mM triethylammonium bicarbonate, pH 7.5,[3] or HPLC over a C_{18} column.[4]

General Considerations for Photolabeling with GTP Analog

Photolabeling of membranes, permeable cells, or protein solutions is accomplished in a room illuminated with gold fluorescent lights (Phillips F40 GO). Ideally, buffers used are devoid of sulfhydryl reagents and employ protease inhibitors such as benzamidine, which do not interfere with labeling. Many exceptions are possible, and they are noted below. Unless otherwise noted, exposure to AAGTP is at 23° for 3 min (membranes) or 30° for 10 min (soluble proteins). Exposure to UV light is for 2 to 5 min at a distance of 4 cm. Samples (in volumes not exceeding 50 μl) are in 1.8-ml screw-topped microcentrifuge tubes (Sarstedt, Newton, NC) and are placed in aluminum blocks, on ice, for the irradiation step. Using a hand-held 9-W lamp (Spectroline, Westbury, NY), 12 samples can be illuminated simultaneously.

The AAGTP-labeled proteins are analyzed by sodium dodecyl sulfate–polyacrylamide gel electrophoresis (SDS–PAGE). For analysis of column fractions, 10% polyacrylamide–0.27% bisacrylamide gels are used routinely. Where multiple G proteins in the 40-kDa range are labeled, better separation is obtained with 10% polyacrylamide–0.13% bisacrylamide gels containing 4 M urea. Gels are 8 × 10 cm and are run for 2 hr at 100 V.

Controls for AAGTP labeling include prior exposure of membranes to hydrolysis-resistant GTP analogs, followed by washing. For soluble G proteins, GTP analogs are allowed to bind to G proteins, and those proteins are passed through desalting columns (Bio-Gel P-6DG, Bio-Rad) prior to incubation with AAGTP. Although [^{32}P]AAGTP is labeled in the α position and the nucleotide is resistant to hydrolysis, it is still advisable to control for phosphorylation. This is accomplished by preserving some samples from exposure to UV irradiation. Inclusion of large excesses of nucleotide in the incubation buffer is not a good control, as this may act merely as a UV sink and provide a physical barrier to photoincorporation.

[3] R. Thomas and T. Pfeuffer, this series, Vol. 195, p. 280.
[4] S. Offermans, G. Schultz, and W. Rosenthal, this series, Vol. 195, p. 286.

Uses of P^3-(4-Azidoanilido)-P^1-guanosine 5'-Triphosphate for Identification of G Proteins

Labeling of Soluble G Proteins

The AAGTP analog is a useful probe for the purification of G proteins. Because the content of various chromatographic fractions is often established using SDS-PAGE, AAGTP allows the verification of both the total protein composition of a given fraction as well as the G proteins resident in that fraction (Fig. 3). Further, proteins labeled covalently with [^{32}P]AAGTP can be transferred to nitrocellulose and hybridized with antibody (Western blotting) (Fig. 4). This allows both colorimetric and radiographic detection and verification of a given G protein. When an appropriate antibody is available, AAGTP-labeled G-protein α subunits can be immunoprecipitated.

Generally, 50-μl aliquots of various column fractions [in the elution buffer containing 20 mM Tris-HCl, 1 mM EDTA, 1 mM dithiothreitol (DTT) and 0.2 mM Phenylmethylsulfonyl fluoride (PMSF) plus salt or detergent appropriate to the given column and fraction] are incubated with 40 nM [^{32}P]AAGTP for 10 min at 30° in the presence of 10 mM MgCl$_2$.

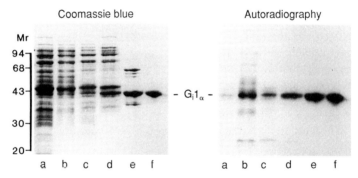

FIG. 3. AAGTP as a marker in the purification of G proteins. Recombinant $G_i\alpha_1$ (from M. Linder, Washington University, St. Louis, MO) was expressed in *Escherichia coli*, and proteins were extracted and purified as described [M. E. Linder, D. A. Ewald, R. J. Miller, and A. G. Gilman, *J. Biol. Chem.* **265,** 8243 (1990)]. Fractions from each column step were incubated with [^{32}P]AAGTP as described in the text, then subjected to SDS-PAGE on 10% polyacrylamide gels (0.27% bisacrylamide, without urea). Coomassie blue staining of protein is shown at left, and the resulting autoradiograph (16 hr exposure) is at right. Lanes a are the *E. coli* extract; lanes b represent the fractions eluted from DEAE-Sephacel which were loaded onto phenyl-Sepharose; lanes c are the phenyl-Sepharose fractions which were loaded onto hydroxyapatite; lanes d are the hydroxyapatite fractions which were loaded onto heptylamine-Sepharose; lanes e are the fractions eluted from the first heptylamine-Sepharose chromatography; lanes f show pure $G_i\alpha_1$ after a second heptylamine-Sepharose step.

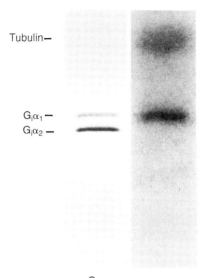

FIG. 4. Transfer of AAGTP from tubulin to synaptic membrane G proteins. Synaptic membranes (80 μg) were incubated with 15 μg tubulin–AAGTP at 23° for 3 min in a reaction volume of 50 μl. After UV irradiation and quenching (as described in the text), the membranes were subjected to SDS–PAGE, and the proteins were transferred to nitrocellulose. Nitrocellulose was probed with anti-$G_i\alpha$ (No. 8730 from D. Manning, Philadelphia, PA), and the immune complex was detected with alkaline phosphatase. The presence of urea and the low concentration of bisacrylamide combine to give good resolution of the G-protein α subunits. Note that, under these conditions, tubulin transfers [^{32}P]AAGTP only to $G_i\alpha_1$.

UV irradiation is for 2–5 min, and reactions are stopped with the addition of 25 μl sample buffer containing 150 mM Tris-HCl, pH 6.8, 150 mM DTT, 9% SDS, 9 mM EDTA, and 30% glycerol.

Clearly, the conditions of labeling of column fractions violate some of the tenets stated above for optimal photolabeling. Both DTT and PMSF, which scavenge free radicals, are present in the buffer along with AAGTP. As can be seen from Fig. 3, however, clear photoincorporation of AAGTP into $G_i\alpha_1$ (and some other proteins) is obtained. Under these conditions it is likely that the extent of photoincorporation is reduced. Because AAGTP is being used to follow the G protein through the purification process, a reduction of the extent of total photolabeling in exchange for preservation of the activity of that protein (provided by both DTT and PMSF) represents a viable compromise.

Labeling of Tubulin

Incorporation of AAGTP into tubulin has provided a mechanism to follow the activation of G proteins by tubulin, a process which appears to involve the direct transfer of GTP (or a GTP analog) from tubulin to $G_i\alpha_1$ or $G_s\alpha$. Tubulin is extracted by polymerization–depolymerization as described by Shelanski et al.,[5] and bound nucleotide is removed from tubulin by extraction with activated charcoal (Norit A) as described.[6] [^{32}P]AAGTP (140 mM) is then added, and tubulin is incubated at 37° for 20 min. Polymerized microtubules are collected by centrifugation at 120,000 g for 20 min in a Beckman TL-100 ultracentrifuge set at 37°. Microtubule pellets are rinsed, resuspended in 2 mM HEPES (pH 7.4), 1 mM MgCl$_2$, 5% glycerol, and stored in aliquots under liquid N$_2$ until use (within 3 weeks). Prior to use, microtubule-associated proteins are removed by phosphocellulose chromatography,[7] and tubulin is desalted twice through Bio-Gel P6-DG resin (Bio Rad) to remove unbound AAGTP. Although tubulin initially incorporates 0.85 mol AAGTP/mol tubulin dimer, after storage and desalting, 0.5 mol AAGTP/mol tubulin remains tightly bound (for at least 3 days on ice).[7] Although tubulin binds 2 mol GTP/mol, it appears that α-tubulin binds this nucleotide nonexchangeably and β-tubulin is the site of exchangeable GTP binding. It is also the site of AAGTP binding.

An alternate procedure to incorporate AAGTP into tubulin (at a higher specific activity) involves incubation of charcoal-treated, phosphocellulose-purified tubulin with [^{32}P]AAGTP. Immediately after charcoal treatment, phosphocellulose-purified tubulin (2 mg/ml) in 100 mM PIPES, pH 6.8, 1 mM MgCl$_2$, 2 mM EDTA is incubated with [^{32}P]AAGTP (100 μM, 26 Ci/mmol) for 20 min on ice followed by two Bio-Gel P6-DG centrifugation–desalting steps.

Tubulin–AAGTP is stored in aliquots under liquid N$_2$ at a concentration of at least 1 mg/ml, and it appears to be stable for at least 3 weeks. Once generated, the reagent can be used for studies involving the transfer of GTP analog between tubulin and various G proteins (see Fig. 4). Tubulin can transfer AAGTP to Gα under conditions where that Gα does not bind to the AAGTP present in the medium.[8] Tubulin also appears to be an effective activator of G$_s$ and G$_{i1}$ and can bypass a tightly coupled receptor

---

[5] M. Shelanski, F. Gaskin, and C. Cantor, *Proc. Natl. Acad. Sci. U.S.A.* **70**, 765 (1973).
[6] M. Weingarten, A. Lockwood, S. Hwo, and M. Kirschner, *Proc. Natl. Acad. Sci. U.S.A.* **17**, 1858 (1975).
[7] M. M. Rasenick and N. Wang, *J. Neurochem.* **51**, 300 (1988).
[8] S. Roychowdhury, N. Wang, and M. M. Rasenick, *Biochemistry* **32**, 4955 (1993).

to activate those proteins.[8-10] This process may represent one interface between the cytoskeleton and G-protein-mediated signal transduction.

Labeling of Membrane G Proteins

It is likely that G proteins in membranes are subjected to a multitude of regulatory factors. To examine the behavior of specific proteins within this complex milieu, AAGTP can provide a useful tool (Fig. 4). Membranes from NG108 neuroblastoma × glioma hybrid cells have been useful for studying the coupling between δ-opiate receptors and $G_o\alpha$.[11] Rat liver membranes have been used to study the coupling of somatostatin receptors to G_q and/or $G_{11}\alpha$.[12]

Labeling of membranes from rat cerebral cortex is accomplished by incubating the membranes with 0.1 μM [^{32}P]AAGTP for 3 min at 23°, followed by centrifugation–washing (15,000 g, 4°, 10 min) twice with 20 volumes of buffer. Such treatments allow only tightly bound AAGTP to remain associated with membranes prior to photolabeling.[13] In synaptic membranes, receptors are uncoupled from G proteins subsequent to the disruption of cells. Thus, even in the absence of an agonist, at least four G proteins in rat cerebral cortex synaptic membranes bind AAGTP.[14] For G proteins linked to receptors, binding of AAGTP in the absence of agonist is significantly lower.[4,11] In either case, background binding of AAGTP can be reduced considerably by the inclusion of 1 mM mercaptoethanol in the buffer. Another technique which reduces background AAGTP binding while increasing agonist-specific AAGTP incorporation involves inclusion of GDP in the binding buffer.[11] Although 10–30 μM GDP was found to reduce the overall incorporation of AAGTP by 39- to 41-kDa proteins by 50- to 100-fold, Offermanns *et al.*[11] were able to show a 2-fold increase in labeling induced by agonist in membranes from NG108-15 cells. In other systems, the amount of GDP required to increase agonist-specific labeling could be much lower. In permeable C_6 glioma cells, it has been observed that 10–50 nM GDP will give a similar enhancement of isoproter-

---

[9] K. Yan and M. M. Rasenick, *in* "Biology of Cellular Transducing Signals" (J. Vanderhoek, ed.), p. 163. Plenum, New York, 1990.

[10] N. Wang and M. M. Rasenick, *Biochemistry* **30**, 10957 (1991).

[11] S. Offermans, G. Schultz, and W. Rosenthal, *J. Biol. Chem.* **266**, 3365 (1991).

[12] G. Bernstein, J. Blank, A. Smrcka, T. Higashijima, P. Sternweis, J. Exton, and E. Ross, *J. Biol. Chem.* **267**, 8081 (1992).

[13] S. Hatta, M. M. Marcus, and M. M. Rasenick, *Proc. Natl. Acad. Sci. U.S.A.* **83**, 5439 (1986).

[14] J. H. Gordon and M. M. Rasenick, *FEBS Lett.* **235**, 201 (1988).

enol-induced labeling in $G_s\alpha$ (M. Lazarevic and M. M. Rasenick, unpublished observations, 1992).

Studying GTP-Modulated Enzymes

It is noteworthy that unlabeled AAGTP is an activator of G proteins. Intracellular effectors such as adenylyl cyclase (adenylate cyclase, EC 4.6.1.1) are activated or inhibited by AAGTP with an EC_{50} comparable to that seen for guanylyl imidodiphosphate (GppNHp).[7,13] Because GppNHp and AAGTP have similar affinities for the α subunits of G_s, G_i, and G_o,[14] this suggests that AAGTP is a useful hydrolysis-resistant analog for studies which compare the activation of a G protein with the recruitment of a subsequent signal transduction process.[13,15]

Labeling of G Proteins in Permeable Cells

Coupling between receptors, G proteins, and intracellular effector molecules is often lost when cells are disrupted. This loss of coupling is a complex phenomenon which takes different forms in different cell types. Generally, receptors from these disrupted cells bind agonist normally and also bind agonist with reduced affinity in response to GTP or GTP analogs. Nonetheless, agonist-induced activation or inactivation of intracellular effectors (e.g., adenylyl cyclase) is compromised. Hydrolysis-resistant analogs of GTP are capable of activating G proteins in the absence of agonist under conditions where receptors are uncoupled from those G proteins. In a complex system, such as the permeable C_6 cell, AAGTP allows the investigator to distinguish between those G proteins which are tightly coupled to a receptor and those which are uncoupled.[16] In this system, the β-adrenergic receptor is tightly coupled to G_s, and AAGTP, added to the cells in the absence of isoproterenol or any other β-adrenergic agonist, does not label $G_s\alpha$. In the same cells and under the same conditions, $G_i\alpha$ incorporates AAGTP quite well. No receptor has been reported to couple to this G protein in C_6 cells. In the presence of both isoproterenol and AAGTP, $G_s\alpha$ is labeled by the later compound. Thus, AAGTP can be used as a probe to ascertain the cognate receptors of various G proteins in a system where the coupling between those components resembles that seen in the native state.

AAGTP labeling of G proteins in permeable cells (Fig. 5) is carried out on monolayers of cells which have been made permeable with saponin.[16,17] The monolayers of cells (in 24-well tissue culture plates) are then

[15] H. Ozawa and M. M. Rasenick, *J. Neurochem.* **56**, 330 (1991).
[16] M. M. Rasenick, M. Lazarevic, M. Watanabe, and H. E. Hamm, *Methods* **5**, 252 (1993).
[17] M. M. Rasenick and R. Kaplan, *FEBS Lett.* **207**, 296 (1986).

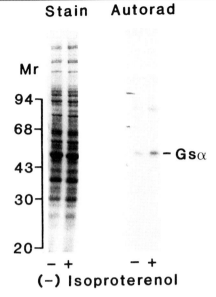

FIG. 5. AAGTP labeling of $G_s\alpha$ in permeable C_6 glioma cells. The C_6 2B cells were grown to about 80% confluence in 24-well tissue culture plates and made permeable as described in the text. (−) Isoproterenol (1 μM) was incubated with the indicated sample for 3 min, and UV irradiation was performed as described. Soluble proteins were precipitated with TCA and subjected to SDS–PAGE and autoradiography. TCA-precipitated samples were boiled for 3 min in SDS sample buffer to improve solubilization. This caused dimerization of $G_s\alpha_4$, and the higher molecular weight band was recognized by an anti-$G_s\alpha$ antibody on Western blots.

covered with 100 μl of Hanks' balanced salt solution including ATP (500 μM), [^{32}P]AAGTP (0.1 μM), GDP (10–50 nM), and the agonist to be tested. After 3 to 5 min of incubation at 37°, cell monolayers (in the 24-well plates) are exposed to UV light (4 min, 3 cm), and the reaction is stopped with 150 μl of 2 mM HEPES, pH 7.4, containing 5 mM MgCl$_2$ and 4 mM DTT. The wells are scraped individually, decanted into microcentrifuge tubes, and washed with another 100 μl of the HEPES buffer. Each tube is then homogenized with a Teflon pestle and centrifuged at 800 g for 5 min. The pellet is discarded, and the resulting supernatant is centrifuged at 100,000 g for 45 min. The membrane pellets are resuspended in SDS sample buffer and subjected to SDS–PAGE as described above. Protein from the 100,000 g supernatants is collected by trichloroacetic acid (TCA) precipitation and analyzed by SDS–PAGE. We had reported that $G_s\alpha$ is released from synaptic membranes,[18] and a similar phenomenon

---

[18] M. M. Rasenick, G. L. Wheeler, M. W. Bitensky, C. Kosack, and P. J. Stein, *J. Neurochem.* **43**, 1447 (1984).

was seen for S49 cells.[19] If similar experiments are done with C_6 membranes, little G-protein α subunit is found in the wash buffer (K. Yan and M. M. Rasenick, unpublished observations, 1992). Thus, in order to evaluate the spectrum of activated G proteins, both membranes and cytosol must be examined, especially if $G_s\alpha$ is the protein being investigated.

Acknowledgments

The authors thank Mr. Ferdinand Belga as well as Drs. Kun Yan and Milenko Lazarevic for providing data for this chapter. Gratitude is also expressed to Drs. Maurine Linder and David Manning for gifts of material. The work described herein is supported by grants from the U.S. National Science Foundation (IBN 91-21540) and the National Institute of Mental Health (MH 39595). M.M.R. is the recipient of a Research Scientist Development Award from the U.S. National Institute of Mental Health.

[19] L. Rasnas, P. Svoboda, J. Jasper, and P. Insel, *Proc. Natl. Acad. Sci. U.S.A.* **86**, 7900 (1989).

[9] Preparation of Activated α Subunits of G_s and G_is: From Erythrocyte to Activated Subunit

By LUTZ BIRNBAUMER, DAGOBERTO GRENET, FERNANDO RIBEIRO-NETO, and JUAN CODINA

Introduction

A detailed description of a procedure for the purification of G_s and G_i (mixture of G_{i2} plus G_{i3}) from human erythrocytes is presented first, followed by an example of separate purification of G_{i2} from G_{i3}. This is then followed by a general protocol for activation and isolation of the activated α subunits of the three G proteins. We have applied the activation and isolation procedure for α subunits to purified G_{i1} and G_o proteins as well, and we suspect that it can also be used for α subunits of the G_q family of G proteins.

Our procedures for G-protein purification[1,2] avoid the use of stabilizing

[1] J. Codina, J. D. Hildebrandt, R. D. Sekura, M. Birnbaumer, J. Bryan, C. R. Manclark, R. Iyengar, and L. Birnbaumer, *J. Biol. Chem.* **259**, 5871 (1984).
[2] J. Codina, W. Rosenthal, J. D. Hildebrandt, R. D. Sekura, and L. Birnbaumer, *J. Recept. Res.* **4**, 411 (1984).

ligands such as NaF/Mg and nonhydrolyzable GTP analogs [e.g., guanylyl imidodiphosphate (GMP-P(NH)P) or GTPγS] which are known to affect functionally the behavior and activity of G proteins. While eliminating many possible alterations of the subunit composition of these proteins as might result from the effect of these ligands to induce subunit dissociation, our methods are more cumbersome and more tedious than those described by others using NaF/Mg.[3] Whether one should choose to use this methodology for purification of G_s and G_i depends on how the proteins are to be used. Our experience coupled to a comparison with published data for G proteins obtained by other published methods is that our approach yields proteins which appear to have higher specific activities (activity of α subunits saturating at 250–500 pM, with EC_{50} values ranging from 20 to 100 pM). This may be related to the stability of the purified proteins.

As purified, both G_s and G_i are detergent–protein complexes, with the proteins being heterotrimers of αβγ composition and approximate M_r 96,000. The human erythrocyte G proteins differ in the α subunits: $α_s$ has an M_r of 42,000 and is a substrate for the ADP-ribosyltransferase activity of cholera toxin; $α_{i2}$ and $α_{i3}$ have M_r values of 40,000 and are substrates for the ADP-ribosyltransferase activity of pertussis toxin (PTX). There are at least two classes of β subunits [$β_{35}$ and $β_{36}$ by sodium dodecyl sulfate–polyacrylamide gel electrophoresis (SDS–PAGE)] which may be a mixture of up to four gene products ($β_1$, $β_2$, $β_3$, and $β_4$). The relative abundance of the $β_{35}$ and $β_{36}$ bands varies from tissue to tissue. In human erythrocyte membrane-derived G proteins, the $β_{35}$ and $β_{36}$ bands are equally represented. The γ subunits of G_s and G_i comigrate on urea and polyacrylamide gel electrophoresis with an apparent M_r of about 5000, and using two-dimensional peptide mapping they appear to be the same for G_s and a pool of G_{i2} plus G_{i3}.[4]

Materials

Chemicals and Reagents

Reagents for polyacrylamide gel electrophoresis are from Bio-Rad (Richmond, CA) and are used without further purification. DEAE-Sephacel as well as Sephadex G-50 (coarse) and Sepharose 4B-CL are from Pharmacia (Piscataway, NJ); Ultrogel AcA 34 is from LKB Instruments;

---

[3] J. K. Northup, P. C. Sternweis, M. D. Smigel, L. S. Schleifer, E. M. Ross, and A. G. Gilman, *Proc. Natl. Acad. Sci. U.S.A.* **77**, 6516 (1980).

[4] J. D. Hildebrandt, J. Codina, W. Rosenthal, L. Birnbaumer, A. Yamazaki, M. W. Bitensky, and E. J. Neer, *J. Biol. Chem.* **260**, 14867 (1984).

hydroxylapatite (HAP) is from Bio-Rad, and ultrafiltration filters are from Amicon Corporation (Danvers, MA). 2-Mercaptoethanol, 4-(2-hydroxyethyl)-1-piperazineethanesulfonic acid (HEPES; Sigma, St. Louis, MO) and ethylene glycol (Fisher, Pittsburgh, PA), are used without further treatments. Heptylamine-Sepharose 4B-CL is synthesized as described by Northup et al.[3] Bovine serum albumin (BSA) is Cohn fraction V from Sigma.

Cholate is from Sigma and is recrystallized six times from ethanol before use. First, 80 g of cholic acid is dissolved in 1 liter of 95% ethanol by bringing to a boil; then 4 liters of cold water is added, and the mixture is allowed to stand at room temperature for 3 hr. The clear supernatant is decanted off and discarded. The remaining suspension is brought to 1 liter with 95% ethanol, and the above extraction steps are repeated five times. The final precipitate is dried in a convection oven at 60°–70° overnight.

Lubrol-PX is from Sigma and is purified by ion-exchange chromatography. A 10% (w/v) solution of Lubrol-PX is made, and a 1.0 × 18 cm column of Bio-Rex RG 501-X8 resin (mixed bed resin of Dowex 50 and Dowex 1 with indicator dye) is prepared. The column is subjected to a regeneration cycle by washing sequentially with 2 liters of water, 500 ml of 2 N NaOH, 2 liters water, 500 ml of 2 N HCl, and 4 liters of water (final pH of effluent > 5.0). The Lubrol-PX solution is passed over the Bio-Rex column at room temperature, and the percolate is neutralized to pH 8.0 with NaOH and used.

All other reagents and chemicals are of the highest grade commercially available and are used as obtained.

Biologicals

Blood. Outdated blood is obtained from local blood banks. The blood can be stored at 0°–4° for up to 2 weeks before processing. It has become more difficult to work with human materials because of potential human immunodeficiency virus (HIV)-related complications. We have reproduced the complete purifications with comparable results using pig blood instead of human blood.

cyc⁻ S49 Mouse Lymphoma Cell Membranes. cyc⁻ S49 cells are grown as suspension cultures in Dulbecco's modified Eagle's medium (DMEM) and supplemented with 10% heat-inactivated (50°, 20 min) horse serum according to the procedures of Coffino et al.[5] Cells can be obtained from the University of California at San Francisco Cell Culture Center

---

[5] P. Coffino, H. R. Bourne, and G. M. Tomkins, *J. Cell. Physiol.* **85,** 603 (1975).

after having obtained consent from Dr. Henry R. Bourne (Department of Pharmacology, School of Medicine, UCSF). cyc^- membranes are prepared as described by Ross et al.,[6] using Mg^{2+}-free buffers throughout.

For 1 liter of cell suspension, cells are grown at 37° to a density of $2-3 \times 10^6$ cells/ml and are harvested by centrifugation for 20 min at 3000 g at 4°. All subsequent steps are at 4°. The medium is discarded, and the cell pellet is weighed (yield should be ~0.75-1.0 g wet weight). Cells are washed twice with 0.5 liter Puck's balanced salt solution without divalent cations (GIBG/BRL, Gaithersburg, MD) and then resuspended in 20 volumes (with respect to wet weight) of lysis medium [25 mM sodium HEPES, 120 mM NaCl, 2 mM $MgCl_2$, and 1 mM dithiothreitol (DTT), pH 8.0]. Cells are lysed by nitrogen cavitation/decompression using a Parr bomb after equilibration with N_2 at 400 psi for 20 min. The lysate is centrifuged at 1500 g for 5 min, and the pellet is discarded. The supernatant is centrifuged at 40,000 g for 20 min, and the supernatant is discarded. The pellet is resuspended in 20 volumes (with respect to starting wet weight of cells) of wash buffer (25 mM sodium HEPES and 1 mM DTT, pH 8.0) and recentrifuged at 40,000 g for 20 min. This pellet is resuspended in 0.75-1.0 ml of wash buffer (15-25 mg protein/ml), fractionated into 100-μl aliquots, and stored frozen at $-70°$ until used.

Analytical Procedures and Routine Assays

Protein Determinations

Owing to the presence of high concentrations of 2-mercaptoethanol throughout the purification procedure and because fluorometry is more sensitive than the classic Lowry procedure, proteins are determined fluorometrically by addition of fluorescamine (Fluram, Roche Diagnostics, Branchburg, NJ) to the samples according to instructions of the manufacturer. Buffers such as Tris and other compounds containing primary amines are therefore not used.

Electrophoresis

Sample Preparation. Samples to be electrophoresed and containing up to 1% (w/v) cholate, 1% (w/v) Lubrol-PX, 30% (v/v) ethylene glycol, and/or 20 mM 2-mercaptoethanol are made 50 mM in NaCl and then precipitated quantitatively by addition of 9 volumes of ice-cold acetone. The precipitated proteins are then resuspended in Laemmli's sample

---

[6] E. M. Ross, M. E. Maguire, T. W. Sturgill, R. L. Biltonen, and A. G. Gilman, *J. Biol. Chem.* **252**, 5761 (1977).
[7] U. K. Laemmli, *Nature (London)* **227**, 680 (1970).

buffer[7] containing 1% (w/v) sodium dodecyl sulfate (SDS) and Pyronin Y as the tracking dye and incubated at room temperature for at least 1 hr prior to electrophoresis. Boiling of samples should be avoided for it leads to aggregation of the α subunit of G_s and the G_i proteins. To prevent formation of insoluble potassium salts of SDS, K^+ is avoided throughout.

Sodium Dodecyl Sulfate–Polyacrylamide Gel Electrophoresis. Analysis of proteins by SDS–PAGE is carried out in 10% (w/v) acrylamide slabs, 13.8 by 17.7 cm and 0.75 or 1.5 mm thick, according to Laemmli.[7] Electrophoresis is at room temperature at a constant voltage of 100–150 V (~30 mA of initial current).

Sodium Dodecyl Sulfate–Discontinuous Urea and Polyacrylamide Gradient Gel Electrophoresis. Electrophoresis is in 0.75 mm thick, 13.8 by 17.7 cm slabs using Laemmli's running buffer.[7] The gels contain a stacking gel of approximately 3 cm and a separating gel of approximately 10 cm. The stacking gel is 6.25% (w/v) acrylamide/0.172% (w/v) bisacrylamide made in 62.5 mM Tris-HCl (pH 6.9), 0.1% (w/v) SDS, and 0.1% (v/v) N,N,N',N'-tetramethylethylenediamine (TEMED), polymerized at 0.02% (w/v) ammonium persulfate. The separating gel is discontinuous in composition and formed of two sections. The top half (5 ml) is 12.5% (w/v) acrylamide/0.344% (w/v) bisacrylamide made in 0.375 M Tris-HCl (pH 8.9), 0.1% (w/v) SDS, and 0.01% (v/v) TEMED (buffer A) with 0.02% (w/v) ammonium persulfate. The bottom half of the separating gel (5 ml) is a gradient (top to bottom) from 12.5% (w/v) acrylamide/0.344% (w/v) bisacrylamide/4 M urea in buffer A with 0.02% (w/v) ammonium persulfate to 25% (w/v) acrylamide/0.688% (w/v) bisacrylamide/8 M urea in buffer A with 0.01% (w/v) ammonium persulfate. Gels are run at 100 V for 4 to 6 h.

Sodium Dodecyl Sulfate–Urea Gradient Polyacrylamide Gel Electrophoresis. Electrophoresis is in 9% (w/v) polyacrylamide gel slabs polymerized in a 4 to 8 M urea gradient dissolved in Laemmli's separation buffer (urea gradient/SDS–PAGE) as described by Scherer et al.[8] For construction of urea gradients, the Laemmli separation buffer ingredients are dissolved in 9 M urea (ultrapure from ICN, Costa Mesa, CA, or BRL, Gaithersburg, MD), freshly prepared and deionized by passing up to 500 ml three times over an initially dry 50-ml column of 20–50 mesh Bio-Rad AG 501-X8(D) resin (Bio-Rad, 142-6425). Thus, two 100-ml separating gel buffers are made: one is 8 M urea, 9% acrylamide, and the other is 4 M urea, 9% acrylamide prepared as follows. Tris base, 4.537 g (Sigma, T-1503), 100 mg SDS (Bio-Rad, 161-0301), 9 g acrylamide (BioRad, 161-0101), and 260 mg bisacrylamide (BioRad, 161-0201) are dissolved

---

[8] N. M. Scherer, M.-J. Toro, M. L. Entman, and L. Birnbaumer, *Arch. Biochem. Biophys.* **259**, 431 (1987).

together with either 89 ml (8 M solution) or 44.4 ml (4 M solution) of 9 M urea and adjusted to a volume of 90–95 ml with water. The volumes are then brought to 100 ml after adjusting to pH 8.8 with concentrated HCl, and the solutions are filtered through 0.45-μm nitrocellulose filters (e.g., Millipore, Bedford, MA, SLHA O250S). The gel slabs are cast between 17 by 17 cm glass plates, using 1 cm wide spacers, so as to be 1.5 mm thick and 150 mm wide, using 16.5 ml each of the 8 and 4 M urea/acrylamide solutions that have each been mixed with 15 μl of 8.4% (v/v) TEMED (BioRad, 161-0801) and 30 μl 12.5% (w/v) ammonium persulfate (BioRad, 161-0700).

Separating and stacking gels are cast on afternoons and used between 15 and 36 hr afterwards.

Laemmli's Sample Buffer. Laemmli's 2× sample buffer contains 50 ml water, 1.51 g Tris base, 2 g SDS, and 25 g glycerol. Take to 90 ml and adjust to pH 6.8 with a meter using concentrated HCl (5–6 drops). Add 2 mg Pyronin Y and filter through Millipore 0.45 μm filters. Before using, mix 1 part 14.1 M 2-mercaptoethanol with 9 parts of the above solution. For use, mix 1 volume of sample with 1 volume of 2× Laemmli's sample buffer, or dilute 2× Laemmli's sample buffer to 1× and add to pelleted samples.

Staining and Autoradiography. After completion of electrophoresis, gels are stained for at least 6 hr in acetic acid–methanol–water (1:5:5, v/v) containing 0.1% (w/v) Coomassie Brilliant Blue and then destained overnight in 10% (v/v) acetic acid. Coomassie blue-stained gels can subsequently be stained with silver by the method of Poehlig and Neuhoff[9] or Wray *et al.*[10] For autoradiography, gels are dried on Bio-Rad filter paper or between two sheets of cellophane and juxtaposed at −70° to Kodak (Rochester, NY) XR-5 film for 1–3 days.

Molecular Weight Standards. Phosphorylase b (M_r 97,400), BSA (M_r 67,000), ovalbumin (M_r 43,000), α-actin (M_r 41,800), carbonic anhydrase (M_r 30,000), soybean trypsin inhibitor (M_r 21,000), and α-lactalbumin (M_r 14,000) are obtained from Pharmacia. Additional low molecular weight standards can be prepared by cyanogen bromide cleavage of horse heart myoglobin, yielding five polypeptides with molecular weights of 16,900, 14,400, 8200, 6200, and 2500. For the cleavage reaction, 100 mg myoglobin is reacted with 50 mg cyanogen bromide in 13 ml of 0.1 N HCl at 24° for 24 hr. The solution is diluted with 130 ml of water, lyophilized, and resuspended at a final concentration of 10 mg/ml. Standards are prepared for electrophoresis by dissolving in Laemmli's sample buffer and heating to 100° for 3–5 min.

[9] H. M. Poehlig and V. Neuhoff, *Electrophoresis* **2**, 141 (1981).
[10] G. W. Wray, T. Boulikas, V. P. Wray, and R. Hancock, *Anal. Biochem.* **118**, 197 (1981).

Adenylyl Cyclase Assays, G_s Assay, and Preactivation of G_s

Adenylyl Cyclase. The enzyme (adenylyl, EC4.6.1.1) is assayed in a final volume of 50 μl containing, unless specified otherwise, 0.1 mM ATP, 5–8 × 10^6 counts/min (cpm) of [α-^{32}P]ATP (specific radioactivity >200,000 cpm/pmol), 10 mM MgCl$_2$, 1.0 mM EDTA, 1 mM cAMP, 10,000–12,000 cpm [^{3}H]cAMP (specific radioactivity >15Ci/mmol), a nucleoside triphosphate regenerating system consisting of 20 mM creatine phosphate, 0.2 mg/ml creatine phosphokinase, and 0.02 mg/ml myokinase, 25 mM Tris-HCl, pH 8.0, and, when present, 1–10 μM GMP-P(NH)P, 10 mM NaF or 100 μM GTP without or with 10^{-4} M isoproterenol, 10 μl of cyc^- membranes (10–20 μg protein, diluted in 10 mM sodium HEPES, pH 8.0, and 1.0 mM EDTA), and 10 μl of various media containing G_s activity. Incubations are at 32.5°, and [^{32}P]cAMP formed is quantitated by double chromatography over Dowex-55 and aluminum oxide.[11,12] In some instances, formation of [^{32}P]cAMP can be monitored by postaddition of 10 μl containing the [α-^{32}P]ATP and [^{3}H]cAMP to 40 μl of a mixture that contains cyc^- membranes, fractions with G_s activity, and all of the above-mentioned reagents and that had been subjected to a preliminary incubation at 32.5° lasting between 5 and 40 min. These incubations (e.g., 5–15 min or 40–50 min assays of activity) are then stopped after 10 min, and [^{32}P]cAMP formed is determined as described above.

"Reconstitution" of cyc^- Membrane Adenylyl Cyclase. Reconstitution of cyc^- adenylyl cyclase activity is obtained by mixing equal volumes of cyc^- membranes (1–3 mg/ml diluted in 10 mM sodium HEPES, pH 8.0, and 1.0 mM EDTA) and G_s activity suitably diluted in 1.0% BSA, 1.0 mM EDTA, 20 mM 2-mercaptoethanol, and 10 mM Tris-HCl, pH 8.0, with less than 0.001% cholate, less than 0.001% Lubrol-PX, with or without 1.5 M KCl plus 20 mM 2-mercaptoethanol, letting the mixtures stand for 10–30 min on ice and then proceeding with the assay of adenylyl cyclase activity on 20-μl aliquots of these mixtures by either of the two methodologies described above.

Treatment of Fractions Containing G_s Activity. G_s proteins present in buffer containing 0.4–1.0% (w/v) sodium cholate or 0.1–1.0% (v/v) Lubrol-PX, 10 mM sodium HEPES, pH 8.0, and up to 20 mM 2-mercaptoethanol, 30% (v/v) ethylene glycol, 150 mM NaCl, and 1 mM EDTA are suitably diluted in media containing 1.0% BSA, 1.0 mM EDTA, and treated at 32.5° in the presence of additives such as up to 100 mM MgCl$_2$, 15 mM NaF, or 100 μM GMP-P(NH)P or GTPγS for appropriate times. At the end of the treatments, the mixtures are diluted 10 to 200-fold with ice-

---

[11] Y. Salomon, C. Londos, and M. Rodbell, *Anal. Biochem.* **58**, 541 (1974).
[12] J. Bockaert, M. Hunzicker-Dunn, and L. Birnbaumer, *J. Biol. Chem.* **251**, 2653 (1976).

cold medium containing 1.0% (w/v) BSA, 1.0–5.0 mM EDTA, 10 mM sodium HEPES, pH 8.0, with or without 1.5 M KCl and 20 mM 2-mercaptoethanol. cyc^- reconstituting activity is then assayed as described in the previous paragraph.

All incubations are carried out in 12 × 75 mm polypropylene test tubes (Walter Sarstedt, St. Louis, MO, Cat. No. 55.526), to avoid losses of G_s protein due to adsorption to walls.

Pertussis Toxin-Catalyzed ADP-Ribosylations

Prior to use PTX is activated by incubation at 250 μg/ml for 30 min at 32° in the presence of 50 mM DTT and 1 mM adenyl-5'-yl imidodiphosphate. The mixture is then diluted 10-fold in 10 mM Tris-HCl, pH 8.0, and 0.025% (w/v) BSA and kept on ice until use. The ADP-ribosylation reactions contain 2 volumes of 300 μM GDPβS, 30 mM thymidine, 3 mM EDTA, 3 mM ATP, 6 mM DTT, 0.5% Lubrol PX, and 45 mM Tris-HCl, pH 8.0; 1 volume of diluted toxin; 1 volume of 1 μM [^{32}P]NAD$^+$ (2–5 × 10^6 cpm); and 2 volumes of additions containing up to 50 nM native G proteins, in 10 mM sodium HEPES, pH 8.0, 1 mM EDTA, 20 mM 2-mercaptoethanol, 30% (v/v) ethylene glycol, 0.1% (w/v) Lubrol PX, and 0.1% (w/v) BSA. Final volumes vary between 10 and 60 μl. Incubations are for 30 min at 32° or 30–36 hr at 4°, and the reactions are terminated by addition of 2 to 3 times the incubation volume of 2× Laemmli's sample buffer containing 4 mM unlabeled NAD. The ADP-ribosylated proteins are then subjected directly to SDS–PAGE or urea gradient/SDS–PAGE. The slabs are stained by Coomassie blue, to monitor efficiency of sample (BSA) transfer onto the gels, and autoradiographed to monitor [^{32}P]ADP-ribosylation.

Purification Procedures

Purification of G_s and G_i

A summary of the procedures used to purify G proteins is given in steps 1–9. Unless stated otherwise, all procedures and manipulations are carried out at 0°–4°, and sodium salts of both cholic acid and EDTA are used.

Step 1: Preparation of Membranes. Red cells from 14 units of blood are washed 4 times by centrifugation (2000 g, 15 min) with 4-liter batches of 5 mM potassium phosphate buffer, pH 8.0 (P_i buffer), containing 150 mM NaCl. The upper layer of white cells is discarded by aspiration after each centrifugation. Washed cells (1500 ml) are lysed with 40 liters of P_i buffer, and the membranes are collected in eight 50-ml stainless steel tubes in a Sorvall SS-34 (Wilmington, DE) rotor fitted with a KSR-R Szent Georgy-Blume continuous flow adaptor by centrifuging at 35,000 g and pumping the lysate at 110–130 ml/min with the aid of a Cole-Palmer (Chicago, IL) Masterflex pump. After pumping an additional 2.0 liters of P_i buffer through the system, the collected membranes are removed from the stainless steel tubes, taking care to discard dark pellets of nonlysed cells that accumulate at the bottom of the tubes.

The collected membranes are resuspended in 2.4 liters of P_i buffer and washed three times in 250-ml polycarbonate bottles by centrifugation at 16,500 g for 30 min in large six-place fixed-angle rotors. The supernatants are discarded, as are the tight pellets that accumulate at the bottom of the centrifuge bottles. Typically, the light pink washed membranes from 14 units of human blood are recovered in a final volume of 500 ml at 15–20 mg protein/ml. They are made 10 mM in sodium HEPES, pH 8.0, and stored frozen at $-70°$ until use.

Steps 2 and 3: Preextraction and Extraction of Washed Human Erythrocyte Membranes. Membranes (650 ml, ~12 g protein) are thawed under continuous stirring at room temperature and preextracted three times by incubating them in a final volume of 700 ml for 30–40 min at 4° with 0.1% cholate, 10 mM MgCl$_2$, 20 mM 2-mercaptoethanol, 300 mM NaCl, and 10 mM sodium HEPES, pH 8.0, followed by separation of the extracted proteins from the membranes by centrifugation at 70,000 g_{av} for 30 min in a Beckman type 45 Ti rotor (Palo Alto, CA). The preextraction procedure is then completed by washing the membranes once with 700 ml of 1.0 mM EDTA in 10 mM sodium HEPES, pH 8.0. G proteins are then extracted from the pellet of the fourth centrifugation by resuspending in 700 ml of Millipore-filtered (0.45-μm porosity, type HA filters) 1% cholate, 10 mM MgCl$_2$, 20 mM 2-mercaptoethanol, and 10 mM sodium HEPES, pH 8.0, and stirring at 4° for 45 min. The extracted G proteins are separated from nonextracted proteins by centrifugation, and the supernatants are collected and made 11 mM in sodium EDTA and 30% in ethylene glycol. This mixture is referred to as "cholate extract." Typically, a 4-fold repetition of the above procedure yields from 70 units of blood approximately 3000 ml of cholate extract containing about 3 g of protein.

Steps 4 through 9: Chromatography. All buffers and solutions used throughout are filtered through 0.45-μm HA Millipore filters. All procedures are carried out in buffer A, which is composed of 1.0 mM sodium EDTA, 20 mM 2-mercaptoethanol, 30% ethylene glycol, and 10 mM sodium HEPES, pH 8.0. Amicon PM-10 membrane filters used to concentrate by ultrafiltration are pretreated with 1% BSA and washed extensively with distilled water prior to use.

Step 4: First DEAE Ion-Exchange Chromatography. Cholate extract from 70 units of blood is loaded at 1 ml/min onto a DEAE-Sephacel column (5 × 60 cm) preequilibrated with 0.9% cholate in buffer A. The column is then washed with 1.0 liter of 0.9% cholate in buffer A, and proteins are eluted with a linear gradient (2 × 2 liters) between 0.9% cholate in buffer A and 0.9% cholate in buffer A plus 500 mM NaCl. G_s activity elutes as a single peak in a volume of 650 to 725 ml at approximately 125 mM NaCl, and G_i, assessed by ADP-ribosylation with pertussis toxin, elutes as an overlapping peak slightly after G_s, at approximately 150 mM NaCl. The double peak of G_s plus G_i containing 25–30% of the applied protein, and better than 90% of the G proteins, is collected. The concentration of NaCl in the eluate is reduced by dialysis against 2 × 8 liters (12 hr each) of 0.9% cholate in buffer A. The resulting sample is then concentrated over a 2-day period to 35–40 ml by ultrafiltration over an Amicon PM-10 filter, diluted to 80 ml with 0.9% cholate in buffer A, and again concentrated to 35–40 ml. The approximate flow rate is 20 ml/hr.

Step 5: Ultrogel AcA 34 Exclusion Chromatography. The concentrated material in buffer A containing 0.9% cholate (~750 mg protein) is applied at a flow rate of 1.0 ml/min to a 5 × 55 cm column of Ultrogel AcA 34 equilibrated with 0.9% cholate in buffer A plus 100 mM NaCl. As illustrated in Fig. 1, the G_s activity elutes from the column in two separate peaks called A and B, which are individually processed further. G_i comigrates with G_s. Typically, peak A is in about 180 ml containing approximately 30% of the applied protein and 60% of the recovered G_s activity, and peak B is in about 170 ml containing 6% of the applied protein and 25% of the recovered activity. Combination of peaks A and B leads to aggregation of materials and permits no further purification of G proteins.

Steps 6A and B: Heptylamine-Sepharose 4B-CL Chromatography

Step 6A: Peak A from the Ultrogel AcA 34 step is diluted 2.25-fold in buffer A containing 100 mM NaCl so that final concentrations are 0.4% cholate and 100 mM NaCl in buffer A, and the mixture is applied at a flow rate of 0.5 ml/min to a 5 × 40 cm column of heptylamine-Sepharose 4B-CL equilibrated with buffer A plus 0.4% cholate and 100 mM NaCl. The column is then washed first with 1500 ml of buffer A plus 0.4% cholate and 100 mM NaCl and then with 1000 ml of buffer A plus 0.4% cholate and 500 mM NaCl. Proteins are eluted with a linear gradient (2 × 5000 ml) of cholate from 0.4 to 4.5% in buffer A at a flow rate of 100 ml/hr. G_s activity (25–30% of the applied activity) elutes as a single peak at approximately 2.25% cholate in a total volume of around 1200 ml. This is the source of most of the G_s. The protein recovery of the G_s part of

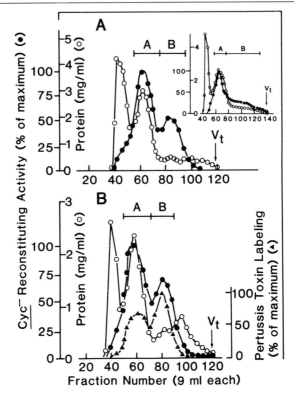

FIG. 1. Elution profiles of G_s and G_i from the Ultrogel AcA 34 column. Results of three experiments are shown. (A) Distribution of G_s activity if the eluate of Step 4 is dialyzed prior to concentration for gel filtration. (*Inset to A*) Distribution of G_s activity if the eluate of Step 4 is concentrated for gel filtration without prior dialysis. (B) Results from a preparation where the distribution of both G_s (cyc^- reconstituting activity) and G_i (ADP-ribosylation with [^{32}P]NAD$^+$ and pertussis toxin) was determined. Note that both G_s and G_i subfractionate into two peaks, of which the first is called peak A and the second peak B.

the purification at this stage is about 40–45 mg. The bulk of G_i is eluted prior to G_s as well as partially overlapping with it. G_i is localized by ADP-ribosylation with [^{32}P]NAD$^+$ and pertussis toxin and pooled for further processing.

Step 6B: Peak B from the Ultrogel AcA 34 step is treated identically to peak A except that the size of the heptylamine-Sepharose column is smaller (2.5 × 10 cm). The column is first washed with 250 ml of buffer A plus 0.4% cholate and 100 mM NaCl and then with 250 ml of buffer A plus 0.4% cholate and 500 mM NaCl. Proteins are eluted with a double reciprocal gradient (2 × 300 ml) of cholate from 0.4 to 2.0% and NaCl from 300 to 0 mM in buffer A at a flow rate of 1 ml/min. G_s activity (50–60% of the applied activity) elutes as a single peak at approximately 1.2% cholate and 150 mM NaCl in a total volume of about 100 ml. Polypeptides of M_r 40,000 start to eluate at around 0.7% cholate. However, a peak of material susceptible to ADP-ribosylation by pertussis toxin (i.e., the α subunit of G_i), elutes at 1.0% cholate. It is part of the bulk of G_i (for further purification, see below), the trailing edge of which elutes with G_s. At this point, a typical peak B preparation of G_s contains about 5–6 mg protein.

Purification of G_s from Preparations A and B

Step 7: Second DEAE Ion-Exchange Chromatography. The pooled fractions from preparations A and B obtained after heptylamine-Sepharose chromatography are diluted 6-fold and 4-fold, respectively, with buffer A containing 0.5% Lubrol-PX and are applied to a 2 × 10 cm column of DEAE-Sephacel equilibrated with 0.5% Lubrol-PX in buffer A at a flow rate of 0.5 ml/min. The column is then washed with 100 ml of 0.5% Lubrol-PX in buffer A and eluted with a linear 0–4% cholate gradient (2 × 300 ml) in buffer A with 0.5% Lubrol-PX (flow rate 1 ml/min). In each instance, G_s activity elutes at 1.5% cholate, and recovery of activity is about 75% of the applied peak A material and essentially 100% of the applied peak B material. At this point, a peak A preparation contains approximately 2 mg protein and a peak B preparation contains around 2.5 mg. In addition, by SDS–PAGE analysis the fractions eluting after G_s activity from the DEAE-Sephacel column that had received the peak B preparation are found to contain material resembling purified G_i and M_r 35,000 peptides (not shown) and constitute the B-II preparation. The pooled fractions containing G_s activity and the peak B preparation constitute the B-I preparation.

Step 8: Third DEAE Ion-Exchange Chromatography. The A and B-I preparations obtained after the second DEAE chromatography are diluted 1:4 with 0.5% Lubrol-PX in buffer A and applied to a 2 × 10 cm column of DEAE-Sephacel equilibrated with 0.5% Lubrol-PX in buffer A. The column is washed with 100 ml of equilibration buffer and then eluted at a flow rate of 0.5 ml/min with a linear 0–300 mM NaCl gradient (2 × 300 ml) in buffer A containing 0.5% Lubrol-PX. Typically, a B-I preparation yields three discrete peaks of protein: the first peak elutes at about 40 mM NaCl, the second minor protein peak elutes at around 75 mM NaCl, and the third peak elutes at about 120 mM NaCl. This last peak contains approximately 70% of the applied G_s activity and around 0.7 mg protein. In contrast, an A preparation yields only two protein peaks eluting in the positions of the first and third peaks of the B-I preparation. The second peak of the A preparation contains approximately 50% of the applied G_s activity, around 0.6 mg protein, and constitutes purified G_s. The first peaks of each preparation constitute the M_r 35,000 containing protein that has been characterized as the $\beta\gamma$ complex.

Step 9: Hydroxylapatite Chromatography. The third peak of the B-I preparation obtained in Step 8 is fractionated further by chromatography over hydroxylapatite (HAP). To this end, the pool of G_s activity from Step 8, which by SDS–PAGE analysis of polypeptide composition contains approximately equal amounts of G_s and G_i, is applied to a 1.0×2.0 cm column of HAP preequilibrated with 100 mM NaCl and 0.1% Lubrol-PX in buffer A made with 0.1 mM EDTA (buffer B). Prior to elution of the protein, the column is washed with buffer B. Proteins are then eluted with buffer B containing 30 or 300 mM potassium phosphate, pH 8.0. First, on elution (flow rate 0.25 ml/min) with 30 mM phosphate buffer, a HAP I fraction is obtained containing in 7 ml around 40% of the recovered G_s activity and 70% of the recovered protein. This constitutes G_i with some contaminating G_s. On subsequent elution (flow rate 0.25 ml/min) with 300 mM phosphate buffer, a HAP II fraction is obtained containing in 7 ml approximately 60% of the recovered G_s activity and 30% of the recovered protein. This constitutes purified G_s. HAP I and HAP II fractions are desalted by chromatography over Sephadex G-50 equilibrated with 0.1% Lubrol-PX and 50 mM NaCl in buffer A.

Purification of G_i. There are four sources of purified G_i: (*a*) G_i in the HAP I fraction from Step 9 of the purification of G_s; (*b*) G_i in the B-II fraction from Step 7 of the purification of preparation B of G_s; and (*c*) G_i present in fractions that precede G_s in the heptylamine-Sepharose chromatography steps for preparations A and B.

a. Separation of G_i by hydroxylapatite chromatography. See Step 9 of G_s purification.

b. Purification of G_i from fraction B-II of Step 7 of B preparation of G_s. The pooled fractions constituting the B II preparation obtained in Step 7 are diluted 4-fold with buffer A containing 0.5% Lubrol-PX and are treated exactly as in Step 8 of the G_s purification. Two peaks of protein are eluted from the DEAE-Sephacel column: the first constitutes the M_r 35,000 containing the $\beta\gamma$ complex of the G proteins, and the second constitutes pure G_i.

c. Purification of G_i separated from G_s on heptylamine-sepharose chromatography. The elution profiles of G_i from the heptylamine-Sepharose A and B columns are determined by ADP-ribosylation of 10-μl aliquots of the eluates. The fractions containing G_i, which as mentioned above elutes prior to G_s from both the A and B heptylamine-Sepharose columns, are pooled, diluted 6-fold (preparation A) or 4-fold (preparation B) with 0.5% Lubrol-PX in buffer A, and then subjected to two successive DEAE-

Sephacel chromatography steps developed in the same manner as described for Steps 7 and 8 of the purification of preparations A and B of G_s. After each of the purification steps, the position of elution of G_i is determined either by ADP-ribosylation with pertussis toxin or by SDS–PAGE analysis of the polypeptide composition of proteins in the column eluates, identifying the polypeptides by silver or Coomassie blue staining. The final pooled and concentrated fractions constitute the H-A and H-B preparations of G_i.

Figure 2 provides an overview of these purification schemes.

Comments. Several factors contribute to the successful isolation of the unaltered and unactivated form of G_s by the above procedure. The first is that membranes could be extensively preextracted by repeated washings first with buffer containing low (0.1%) cholate, 10 mM MgCl$_2$, 300 mM NaCl, 1.0 mM EDTA, and 10 mN sodium HEPES without loss of G_s activity, even though extraneous proteins were removed. This leads,

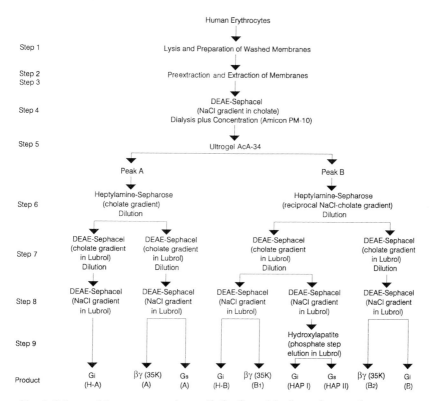

FIG. 2. Scheme of the strategy used to purify G_s, G_i, and $\beta\gamma$ dimers from erythrocyte membranes.

in addition, to changes in the membranes such that G_s can then be extracted in a rather selective manner using 1% cholate, provided that 10 mM MgCl$_2$ is included. Without the "preextraction" treatment, 1% cholate leads to extraction of up to 60% of G_s activity without MgCl$_2$. The second factor is that addition to solubilized G_s of 1 mM EDTA, 20 mM 2-mercaptoethanol, 30% ethylene glycol, and 10 mM sodium HEPES, pH 8.0 (buffer A), results in a stabilization of G_s activity. Although the strategy used is based on the methodology published by Northup et al.[3] for the purification of G_s from liver membranes in the presence of a stabilizing mixture that included NaF, ATP, and Mg^{2+}, some of the properties and chromatographic behaviors of G_s derived from human erythrocyte membranes in the absence of these stabilizing agents differ from those of the protein obtained using NaF/Mg/ATP mixtures.[8,9,12] Thus, (1) Mg^{2+} has to be avoided throughout because its presence in the absence of ATP and NaF leads invariably to loss of activity, and (2) the elution pattern of G_s activity differs in several respects.

The chromatographic procedures are consequently adapted and/or modified to give the method described above. As developed, the first DEAE chromatography poses no special problems. However, the Ultrogel AcA 34 exclusion chromatography step shows variability depending on the way the pooled fractions obtained from the first DEAE column are handled (Fig. 1). Figure 1A shows the elution profile of G_s activity and proteins when the pooled fractions are sequentially dialyzed, concentrated, diluted, reconcentrated, and then applied onto the Ultrogel AcA 34 column as described. The inset to Fig. 1A shows results obtained when the pooled fractions are simply concentrated by ultrafiltration without first decreasing the salt concentration. Clearly, G_s activity distributes into two classes: peak A, which is not affected by the history of the fraction applied onto the Ultrogel AcA 34 column, and what we now call peak B, which is not recognizable as such when the DEAE fractions are kept at approximately 120 mM NaCl throughout the concentration step. These differences in behavior may be related to the hydrophobic nature of G_s and/or its contaminants such that, at "high" salt, a proportion of G_s tends to stick to the Ultrogel AcA 34 matrix or to be associated with a protein that does so. Sternweis et al.[13] found G_s to elute as a single peak from Ultrogel AcA 34 which, by its position, seems to elute between our peaks A and B. Neer and collaborators, using cholate extraction of NaF/Mg-solubilized G_s from brain, found that it, too, fractionated into two peaks on gel filtration.[14]

[13] P. C. Sternweis, J. K. Northup, M. D. Smigel, and A. G. Gilman, *J. Biol. Chem.* **256**, 11517 (1981).
[14] E. J. Neer, J. M. Lok, and L. G. Wolf, *J. Biol. Chem.* **259**, 14222 (1984).

Attempts to continue purification of a pool of peaks A plus B are unsuccessful in yielding a meaningful resolution of G_s from the bulk of contaminants by either heptylamine chromatography, developed according to Northup et al.,[3] or various forms of DEAE ion-exchange chromatography. G_s activity is therefore purified separately from peak A and peak B using a sequence of heptylamine, DEAE, and, for material derived from peak B only, hydroxylapatite chromatographies.

Heptylamine-Sepharose tends to have a poor capacity for G_s activity applied as part of the peak A pool obtained from the Ultrogel Aca 34 column, and this capacity may vary up to 2-fold from batch to batch of heptylamine-Sepharose synthesized by us. Consequently, the capacity of heptylamine-Sepharose to retain peak A G_s needs to be titrated. Peak B G_s activity is retained without difficulty by heptylamine-Sepharose, and no calibration of retention capacity is needed. Although peak B G_s is eluted from heptylamine-Sepharose columns by a minor modification of the procedure of Sternweis et al.[13] for liver G_s (i.e., by using a double reciprocal gradient of 300–0 mM NaCl and 0.4–3% cholate), peak A G_s is eluted from the much larger heptylamine-Sepharose column by a simple linear 0.4–4% cholate gradient. The eluted fractions from the heptylamine-Sepharose columns are then pooled and processed as outlined in Fig. 2 and described in detail above.

Peak A G_s is purified to better than 95% purity after only two successive DEAE ion-exchange chromatographies. Peak B G_s requires an additional hydroxylapatite treatment, followed by Sephadex G-50 chromatography to remove the phosphate buffer with which G_s had been eluted from the hydroxylapatite. The only additional feature worth commenting on here is that, as described above, the concentration of the salt with which G_s activity is eluted from the heptylamine and the second DEAE column is reduced by simple dilution, and the fractions are then directly applied onto the next respective chromatography columns. This was found necessary to avoid irreversible formation of aggregates that seem to form on concentration of the pooled fractions over Amicon filters.

Yield of G_s and G_i. A typical procedure that starts with 60 units (pints) of outdated blood yields about 600–750 µg from G_s from Step 8 and 150–200 µg of G_s from Step 9. The overall purification is about 5000-fold with respect to starting washed human erythrocyte membranes, with a yield of about 10%. The isolated proteins appear to have preserved all the properties expected from studies with crude cholate extracts.

A typical purification procedure also yields about 60–70 µg of G_i at Step 8 from peak A, 300–400 µg of G_i at Step 8 from peak B-I (fraction H-B), 100–150 µg of G_i at Step 8 from preparation B-II (fraction B), and 200–250 µg at Step 9 (Fig. 2). It is possible to increase the quantities of G_i, but not G_s, by 50–80% if a broader region of the fractions from the

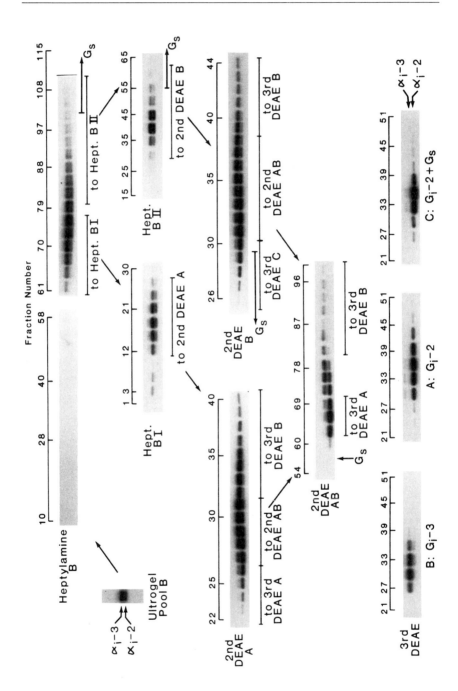

first DEAE-Sephacel chromatography are pooled. If this is done, however, the preparations of G_s and G_i which are obtained are significantly less pure. In particular, glycoproteins of M_r between 60,000 and 100,000 and a polypeptide of about M_r 28,000 tend to contaminate all preparations.

Finally, about 1600 μg of the $\beta\gamma$ complex is also obtained, free of G_s or G_i. They are distributed in three fractions (35Ks of Fig. 2). Approximately 800 μg is obtained from the peak A preparation, about 400 μg is obtained from Step 8 of the peak B preparation of G_s before hydroxylapatite chromatography, and around 400 μg is obtained as a side product of the B-II preparation of G_i.

FIG. 3. Progress of the purification of human erythrocyte G_{i2} and G_{i3}. Areas of autoradiograms of urea gradient/SDS-PAGE gels that correspond to α subunits of PTX-sensitive G proteins (M_r range 35,000–45,000) are shown. Pool B (230 ml) from an Ultrogel AcA 34 column was chromatographed over heptylamine-Sepharose [variant B: 2.5 cm diameter, 100 ml bed volume; 600-ml reciprocal 0.4–2% cholate/200–0 mM NaCl gradient in buffer B (20 mM 2-mercaptoethanol, 1 mM EDTA, 10 mM sodium HEPES, pH 8.0, 30% ethylene glycol); 6-ml fractions]. As shown, the "upper" (leading) PTX substrate was partially separated from the "lower" (trailing) PTX substrate, the last part of which in turn overlapped with G_s. Fractions were pooled as shown and rechromatographed over heptylamine-Sepharose (heptylamine B-I and B-II chromatographies: columns of 0.9 cm diameter, 10 ml bed volume; 600-ml reciprocal gradient as above; 6-ml fractions). The fractions with PTX substrates from these two steps were then subjected to separate second DEAE chromatography steps over DEAE-Toyopearl (0.9 cm diameter, 10 ml bed volume; 600-ml 0–4% cholate gradient in buffer B plus 0.5% Lubrol-PX; 6-ml fractions). This resulted in a further partial separation of "lower" (now leading) from "upper" (now trailing) PTX substrate, with the difference that whereas the leading "lower" PTX substrate eluting from the second DEAE-A step was free of G_s, that eluting from the second DEAE-B step contained an about equimolar amount of G_s. The eluates from each of the second DEAE columns were combined to give three pools, of which the two center pools (~50/50 "upper" and "lower" PTX substrates) were combined, diluted, and rechromatographed over DEAE-Toyopearl (second DEAE-AB: 0.9 cm diameter, 50 ml bed volume; 600-ml 0–4% cholate gradient as above; 6-ml fractions). The fractions with the leading PTX substrate ("lower") from the second DEAE-AB step were combined with those from the second DEAE-A step, and the fractions with the trailing "upper" PTX substrate were combined with those from both the second DEAE-A and -B steps. These and the pool of leading "lower" PTX substrate plus contaminating G_s from the second DEAE-B pools were then subjected to a third DEAE chromatography (columns designated A, B, and C as shown; 0.9 cm diameter, 10 ml bed volume; 600-ml 0–300 mM NaCl gradient in buffer B plus 0.5% Lubrol-PX; 6-ml fractions). "Upper" PTX substrate (G_{i3}) eluted from the third DEAE column slightly earlier that the "lower" PTX substrate (G_{i2}), which in turn essentially cochromatographed with G_s. Fractions from the third DEAE A, B, and C columns were pooled to give pools A (G_{i3}, lane A), B (G_{i2}, lane B), and C (G_s plus G_{i2}, lane C). Flow rates were 1.0–1.1 ml/hr. All other conditions and materials, as well as methods for performing PTX labeling and urea gradient/SDS-PAGE, were as described in earlier sections of the chapter.

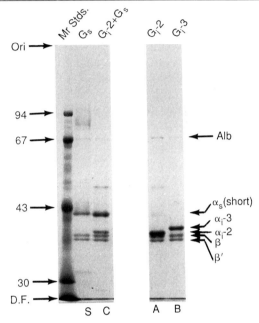

FIG. 4. Urea gradient/SDS–PAGE analysis of fractions from the purification shown in Fig. 3. Photographs of Coomassie blue-stained gels are shown. Lanes S, A, B, and C correspond to aliquots of proteins obtained, respectively, from purifying G_s activity from the Ultrogel AcA 34 pool A and G_{i2}, G_{i3}, and a mixture of G_s plus G_{i2} from the Ultrogel AcA pool B. Note presence of two β subunits in each of the proteins. The γ subunits (M_r 5000–8000) were not resolved on these gels and migrated with the dye front (D.F.). Lanes S, C, A, and B received 250, 350, 250, and 500 ng protein, respectively.

Purification of G_{i2} Separate from G_{i3}

A typical procedure is illustrated in Fig. 3. Human erythrocyte G proteins are extracted from erythrocyte membranes (55 g protein) and subjected to the first three chromatography steps [first DEAE (DEAE-Sephacel), Ultrogel AcA 34, and heptylamine-Sepharose] as described for pool B of the Ultrogel AcA 34 eluate, which contains approximately 65% of the PTX substrates (Fig. 1B). The fractions are assayed for G_s (cyc^- reconstituting activity) and for the presence of PTX substrates as seen by [^{32}P]ADP-ribosylation followed by urea gradient/SDS–PAGE and autoradiography. The latter shows two major PTX-labeled bands, denoted as "upper" and "lower," which correspond to the ADP-ribosylated forms of α_{i3} and α_{i2}, respectively. Resolution of G_{i2} from G_{i3}, as well as separation from other proteins, is accomplished by sequential chromatographies, first over a second heptylamine-Sepharose column and then over two or three

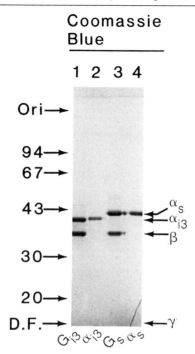

FIG. 5. Coomassie blue staining of human erythrocyte G proteins G_s and G_{i3} and their resolved and purified GTPγS-activated α subunits. Lanes 1 and 3 each received approximately 500 ng of protein that was subjected to activation by GTPγS, followed by dialysis as described in the text. Lanes 2 and 4 received aliquots of the DEAE eluates with the activated α subunits prior to being subjected to the final dialysis step. For details, see text.

DEAE-Toyopearl columns as shown in Fig. 3. G_s is purified from the Ultrogel AcA pool A, taking care to pool the eluates so as to minimize contamination with PTX substrates.

A total of four pools of G proteins are obtained with the following protein yields: 81 μg of G_s (Fig. 4, lane S), 360 μg of G_{i2} (Fig. 4, lane A), 414 μg of G_{i3} (Fig. 4, lane B), and a mixture of 450 μg of G_s plus 220 μg of G_{i2} (Fig. 4, lane C). The G_s, G_{i2}, and G_{i3} proteins prepared in this way are at least 90% pure in terms of unrelated contaminants, and they contain less than 5% cross-contaminating G proteins as assessed by PTX labeling and cyc^- reconstitution assays (not shown).

Preparation of Activated G-Protein α Subunits

The GTPγS-activated forms of α_s (α_s^*), α_{i2} (α_{i2}^*), and α_{i3} (α_{i3}^*) are prepared from human erythrocyte G_s, G_{i2}, and G_{i3} as follows. An aliquot of the pooled heterotrimeric G protein [0.5 to 2 μM in 1.0 ml of 1 mM EDTA, 20 mM 2-mercaptoethanol, 10 mM sodium HEPES, pH 8.0, 30% (v/v) ethylene glycol (buffer A) plus 0.5% Lubrol-PX and around 125 mM NaCl] is made 20 μM in [^{35}S]GTPγS (3700 cpm/pmol) and 100 mM in MgCl$_2$ (in that order) and incubated for 1 hr at 32°. The mixture is then diluted 13.3-fold with 1 mM EDTA, 1 mM DTT, and 10 mM Tris-HCl, pH 8.0 (buffer B), plus 0.5% Lubrol-PX and applied to a DEAE-Toyopearl column of 100 μl bed volume prepared in a disposable 1-ml syringe that had been prewashed at room temperature with 5 ml of 1% BSA, 10 ml of 1 M NaCl, and 10 ml of buffer B plus 0.5% Lubrol-PX and 7.5 mM MgCl$_2$. The column with the sample is then washed with 1 ml of buffer B plus 0.5% Lubrol-PX and 7.5 mM MgCl$_2$, 10 ml of buffer B with 7.5 mM MgCl$_2$, 1.5 ml of buffer B plus 7.5 mM MgCl$_2$ and 60 mM NaCl, and ten 100-μl aliquots of buffer B plus 7.5 mM MgCl$_2$ and 200 mM NaCl. All the above operations (dilution of the GTPγS-activated G protein and chromatographic steps) are performed at room temperature.

The final eluates are immediately cooled to 0°–4°, and all subsequent steps are performed in the cold. Fractions 2, 3, and 4, containing the Lubrol-free GTPγS-activated α subunits plus around 5% of the initially added [^{35}S]GTPγS are dialyzed two times against 650 ml of buffer B plus 7.5 mM MgCl$_2$ and 20 mM KCl in an MRA Multichamber Dialysis Apparatus (MRA Corporation, Clearwater, FL). The bottoms of the dialysis chambers are made with a Spectrapor membrane having a molecular weight cutoff of 6000–8000 that had been boiled for 5 min in 1 mM EDTA. To minimize loss of α subunits, the sample chambers are first filled with 1% BSA and then rinsed with dialysis buffer prior to sample addition. After removing aliquots for protein determination, electrophoretic analysis, and quantification of radioactivity, the dialyzed fractions, containing [^{35}S]GTPγS and α subunits at molar ratios that range from 1.5 to 2.0, are aliquoted into 0.5-ml Eppendorf tubes and stored at $-70°$ until used. The α-subunit concentrations are calculated on the basis of protein content assuming a molecular weight of 40,000. Figure 5 shows Coomassie blue-stained α and β subunits of G$_{i3}$ and G$_s$ that were subjected to this procedure and the resolved GTPγS-activated α_{i3} and α_s subunits recovered from the DEAE-Toyopearl column before they were subjected to dialysis in the presence of BSA.

Acknowledgments

Supported in part by National Institutes of Health Grant DK-19318 to L.B. and Diabetes and Endocrinology Research Center Grant DK-27685.

[10] Purification and Separation of Closely Related Members of Pertussis Toxin–Substrate G Proteins

By Toshiaki Katada, Kenji Kontani, Atsushi Inanobe, Ichiro Kobayashi, Yoshiharu Ohoka, Hiroshi Nishina, and Katsunobu Takahashi

Introduction

G proteins, which have a common heterotrimetric structure consisting of $\alpha\beta\gamma$ subunits, carry signals from membrane-bound receptors to effectors such as enzymes or ion channels. The α subunits have binding sites specific for GTP (or GDP) with a high affinity and sites for NAD^+-dependent ADP-ribosylation which is catalyzed by bacterial toxins such as pertussis toxin (islet-activating protein, IAP) or cholera toxin. Functions of G proteins as signal transducers are profoundly affected by toxin-induced modifications of the α subunits. Thus, these bacterial toxins, together with radiolabeled guanine nucleotides, have been widely used as tools to identify and analyze G proteins. The procedures described here have been used in our laboratories for the purification of IAP–substrate G proteins from bovine and rat brains. The basic protocol used is similar to those described previously.[1-7]

Materials

All reagents used during the purification are analytical grade. Cholate and Lubrol PX (purchased from Sigma, St. Louis, MO, or Nacalai Tesque, Kyoto, Japan) are purified by the method of Codina et al.[8] before they can be used. Typically a 10% (w/v) solution of the purified cholic acid is

[1] T. Katada, M. Oinuma, and M. Ui, J. Biol. Chem. **261,** 8182 (1986).
[2] T. Katada, M. Oinuma, K. Kusakabe, and M. Ui, FEBS Lett. **213,** 353 (1987).
[3] I. Kobayashi, H. Shibasaki, K. Takahashi, S. Kikkawa, M. Ui, and T. Katada, FEBS Lett. **257,** 177 (1989).
[4] I. Kobayashi, H. Shibasaki, K. Takahashi, K. Tohyama, Y. Kurachi, H. Ito, M. Ui, and T. Katada, Eur. J. Biochem. **191,** 499 (1990).
[5] A. Inanobe, H. Shibasaki, K. Takahashi, I. Kobayashi, U. Tomita, M. Ui, and T. Katada, FEBS Lett. **263,** 369 (1990).
[6] H. Shibasaki, T. Kozasa, K. Takahashi, A. Inanobe, Y. Kaziro, M. Ui, and T. Katada, FEBS Lett. **285,** 268 (1991).
[7] K. Kontani, K. Takahashi, A. Inanobe, M. Ui, and T. Katada, Arch. Biochem. Biophys. **294,** 527 (1992).
[8] J. Codina, D. J. Carty, L. Birnbaumer, and R. Iyengar, this series, Vol. 195, p. 177.

made by adding 10 N NaOH to a suspension of cholic acid in water at room temperature until a pH of 7–8 is reached, and the solution is stored at 4°. Final 10% (v/v) Lubrol PX which has been purified by passing through a mixed-bed resin (AG 501-X8, Bio-Rad, Richmond, CA) may also be stored at 4° for several months. CHAPS {3-[(3-Cholamidopropyl)dimethylammonio]propanesulfonic acid} is obtained from Dojindo Laboratories (Tokyo, Japan) and used without purification. Aprotinin is a generous gift from Hoechst Japan Ltd. (Tokyo). The IAP reagent (Kakenseiyaku, Tokyo, Japan), is stored at 4° in 100 mM sodium phosphate (pH 7.0) and 2 M urea at a final concentration of 0.5–1.0 mg of protein/ml.

Assays for G Proteins

[^{35}S]GTPγS Binding Assay

Column fractions to be assayed for [^{35}S]GTPγS-binding activity are diluted with TED [20 mM Tris-HCl (pH 8.0), 1 mM sodium EDTA, and 1 mM dithiothreitol (DTT)] in order to reduce the concentration of Lubrol-PX present in the fractions. The diluted samples (5 μl; containing less than 0.1% of Lubrol-PX) are incubated at 30° for 30–60 min in a reaction mixture (15 μl) such that the final concentrations of reagents are 20 mM Tris-HCl (pH 7.5), 1 mM EDTA, 1 mM DTT, 25 mM MgCl$_2$, 0.02% Lubrol-PX, 250 mM (NH$_4$)$_2$SO$_4$, and 2.5 μM [^{35}S]GTPγS [3–8 × 10^3 counts/min (cpm)/pmol]. The reaction is terminated by the addition of an ice-cold buffer (300 μl) consisting of 20 mM Tris-HCl (pH 7.5), 25 mM MgCl$_2$, and 100 mM NaCl. The [^{35}S]GTPγS-bound G proteins are collected on a Millipore (Bedford, MA) HA plate (96 wells of nitrocellulose filter; 0.6-cm diameter). The filters, after being washed four times with 300 μl of the ice-cold buffer, are dried and counted for ^{35}S in a liquid scintillation counter.

Pertussis Toxin-Catalyzed [^{32}P]ADP-Ribosylation of G Proteins

The diluted samples (5 μl; containing less than 0.1% of Lubrol-PX) are added to a reaction mixture (17.5 μl) such that the final concentrations of reagents are 50 mM Tris-HCl (pH 7.5), 1 mM EDTA, 10 mM thymidine, 1 mM L-α-dimyristoylphosphatidylcholine, 1 μM GDP, and 2.5 μM [α-^{32}P]NAD (3–8 × 10^3 cpm/pmol). The reaction is started by the addition (2.5 μl) of 100 μg/ml of IAP which has been preactivated by an incubation with 100 mM DTT and 0.1 mM ATP in 50 mM Tris-HCl (pH 7.5) at 30° for 15 min. After incubation at 30° for 30–60 min, the reaction is terminated by the addition of 150 μl of 0.2 M Tris-HCl (pH 7.5) containing 0.4%

sodium dodecyl sulfate (SDS) and 10 μg/ml of bovine serum albumin, followed by further addition of 150 μl of 30% trichloroacetic acid. The precipitated proteins are collected on the Millipore HA plate. The filters, after being washed four times with 300 μl of 4% trichloroacetic acid, are dried and counted for ^{32}P in a liquid scintillation counter.

Preparation of Membrane Extract

Cerebral cortices from five bovine brains are washed with TSA [10 mM Tris-HCl (pH 7.5), 10% (w/v) sucrose and 10 kallikrein inhibitory units/ml of aprotinin] and homogenized with an equal volume of TSA using a Polytron. The homogenate, after being filtered through two layers of cheesecloth, is diluted to a total volume of 6 liters with TSA and centrifuged at 10,000 rpm for 30 min at 4° in a Beckman JA-10 rotor. The pellets are again resuspended with a total volume of 4.5 liters of TSA by a Potter–Elvehjem homogenizer and then centrifuged at 14,000 rpm for 30 min in a Beckman JA-14 rotor. The pellets are further resuspended with a total volume 4.5 liters of TEDA (TED plus 10 kallikrein inhibitory units/ml of aprotinin) fortified with 100 mM NaCl and centrifuged at 14,000 rpm for 30 min. The washed membranes (~1 liter of pellet volume) are resuspended with an equal volume of TEDA and poured into liquid N_2 with vigorous shaking. The frozen membranes having a popcornlike shape are stored at $-80°$ until use. This amount is enough for 5–6 runs of the following series of G-protein purification.

For a single run of the first DEAE-Sephacel step, 380 g of the frozen brain membrane fraction is thawed in a glass beaker and mixed with 760 ml of TEDA containing 65 mM NaCl and 1.5% sodium cholate to give final concentrations of NaCl and sodium cholate of about 60 mM and 1%, respectively. The mixture is maintained at 0° for 1 hr with stirring, then centrifuged at 40,000 rpm for 80 min in a Beckman 45Ti rotor. The clear supernatant (900 ml of the extract) is subjected to the purification procedure for G proteins.

Purification Procedure

Purification of Mixture of Closely Related Pertussis Toxin–Substrate G Proteins

The extract is applied to a column (4.4 × 25 cm) of DEAE-Sephacel (Pharmacia-LKB, Piscataway, NJ), which has been equilibrated with 1.5 liters of TEDA, 1% sodium cholate, and 60 mM NaCl at a flow rate of about 150 ml/hr. The column is washed with 500 ml of TEDA, 1% sodium cholate, and 60 mM NaCl, and IAP–substrate G proteins are then eluted from the column with 400 ml of TEDA, 1% sodium cholate, and 160 mM NaCl. Botulinum C_3 substrate (*rho*) GTP-binding proteins could be eluted from the column with 400 ml of TEDA, 1% sodium cholate, and 275 mM NaCl.[9,10] The column is finally washed with 400 ml of TEDA, 1% sodium cholate, and 1000 mM NaCl.

The protein peak fractions (200–250 ml; ~400 mg of protein) eluted with 160 mM NaCl from the DEAE-Sephacel column are concentrated to about 20 ml by an ultrafiltration system with Minimodule NM-3 (HC membrane; Asahikasei, Tokyo, Japan), and the concentrate is applied to a column (4.4 × 80 cm) of Sephacryl S-300 HR (Pharmacia-LKB) which has been equilibrated with 2.5 liters of TEDA, 0.8% sodium cholate, 100 mM NaCl, and 1 μM GDP and eluted at a flow rate of about 100 ml/hr. The IAP–substrate G proteins are eluted in a single symmetrical peak from the column, and the G-protein-rich fractions (80–100 ml; ~100 mg of protein) are pooled.

The above sequential column chromatography steps are usually repeated twice, and the thus obtained fractions (160–200 ml; 200 mg of protein) are pooled and diluted with 2.2 volumes of TED fortified with 318 mM NaCl and 1 μM GDP. The diluted sample is applied to a column (3.2 × 12 cm) of phenyl-Sepharose CL-4B (Pharmacia-LKB) which has been equilibrated with 400 ml of TED, 0.25% sodium cholate, 250 mM NaCl, and 1 μM GDP and eluted with a linear gradient (360 ml) starting with 0.25% sodium cholate and 250 mM NaCl and ending with 1% sodium cholate and 0 mM NaCl in TED at a flow rate of 80–100 ml/hr (see Fig. 1). The column is further washed with 100 ml of TED, 1% sodium cholate, 500 mM NaCl, and 7 M urea. The mixture of $\alpha\beta\gamma$ trimeric IAP–substrate G proteins (25–30 mg of protein with a specific activity of 6–8 nmol/mg) is thus obtained from the cholate extract with a recovery of 30–40%. The G-protein-rich fraction can be stored at $-80°$ with a great stability (more than a couple of months) and is further subjected to purification and separation of closely related members of IAP–substrate G proteins and their constituent α and $\beta\gamma$ subunits.

Separation of Closely Related Members of Pertussis Toxin–Substrate G Proteins

Various $\alpha\beta\gamma$ trimeric G proteins serving as the substrate of IAP can be purified from the phenyl-Sepharose fraction by means of sequential

[9] T. Maehama, Y. Ohoka, T. Ohtsuka, K. Takahashi, K. Nagata, Y. Nozawa, K. Ueno, M. Ui, and T. Katada, *FEBS Lett.* **263**, 376 (1990).

[10] T. Maehama, K. Takahashi, Y. Ohoka, T. Ohtsuka, M. Ui, and T. Katada, *J. Biol. Chem.* **266**, 10062 (1991).

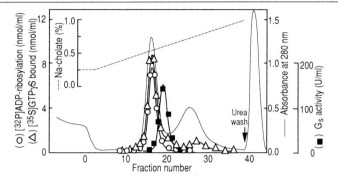

Fig. 1. Phenyl-Sepharose CL-4B column chromatography of IAP–substrate G proteins. The Sephacryl S-300 HR eluate was chromatographed on a phenyl-Sepharose column as described in the text. G_s activity was assayed by the method described by Sternweis et al. [P. C. Sternweis, J. K. Northup, M. D. Smigel, and A. G. Gilman, J. Biol. Chem. **256**, 11517 (1981)].

chromatography steps with DEAE-Toyopearl 650S, hydroxyapatite HCA-100S, and Mono Q columns.[2,3] The G-protein-rich fraction is concentrated to 5 ml (5–6 mg of protein/ml) by the Minimodule NM-3 apparatus and then diluted with 3 volumes of TED containing 1% Lubrol-PX. The diluted sample is applied to a column (1 × 8 ml) of DEAE Toyopearl 650S (Tosoh, Tokyo, Japan) which has been equilibrated with 50 mM NaCl in TED containing 0.75% Lubrol-PX and is eluted with a linear gradient of NaCl (50–200 mM; 60 ml) in the same buffer. Three major peaks of G proteins are recovered from the DEAE column (Fig. 2); the first (I), second (II),

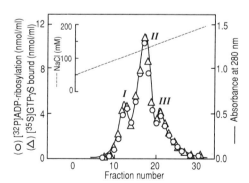

Fig. 2. DEAE-Toyopearl 650S column chromatography of IAP–substrate G proteins. The phenyl-Sepharose CL-4B eluate was chromatographed on a DEAE-Toyopearl column as described in the text.

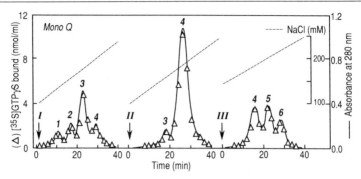

FIG. 3. Mono Q column chromatography of IAP–substrate G proteins. The three peaks (I–III) of G proteins eluated from the DEAE-Toyopearl column were chromatographed on a Mono Q column as described in the text.

and third (III) peaks mainly contain the $\alpha\beta\gamma$ trimeric forms of G_{i1}, G_o, and G_{i2}, respectively.[2,3,11]

Each of the three peaks (I–III) is applied and eluted from a column (1 × 6 cm) of HCA 100S hydroxyapatite (Mitsui Toatsu Chemicals, Inc., Tokyo, Japan) to concentrate and to purify further the G proteins.[12] Each fraction thus obtained is applied to a column of Mono Q HR5/5 (Pharmacia-LKB) which has been equilibrated with TED containing 100 mM NaCl and 0.7% CHAPS and eluted at a flow rate of 0.75 ml/min with the following series of NaCl gradients using a Pharmacia FPLC (fast protein liquid chromatography) system: 100 mM (or 150 mM) for 2 min; 100–250 mM (or 150–250 mM) over 40 min; 250–500 mM over 5 min. Figure 3 shows typical elution patterns from the Mono Q columns, to which the peaks I, II, and III from the DEAE Toyopearl column has been applied. Six peaks of G proteins (0.5–3 mg of each protein) are purified to near homogeneity; peaks 1, 2, 3, 4, 5, and 6 are eluted at about 145, 165, 175, 195, 205, and 215 mM NaCl, respectively, under the present conditions. Based on findings of the immunoreactivities of the peaks to specific antibodies raised against various α subunits, peaks 1–6 are identified as free α_{o2}, $\alpha_{o2}\beta\gamma$, $\alpha_{i1}\beta\gamma$, $\alpha_{o1}\beta\gamma$, $\alpha_{i2}\beta\gamma$, and $\alpha_{o1}\beta\gamma$, respectively.[3,5,6] The G proteins are concentrated using an ultrafiltration system employing a Centriflo CF-25 or Centricon 10 (Amicon, Danvers, MA) to 0.5–1 ml and filtered through a column (0.8 × 10 cm) of Sephadex G-50 fine (Pharmacia-LKB) in 50 mM HEPES–NaOH (pH 7.4) and 0.7% CHAPS. The purified $\alpha\beta\gamma$ trimeric G

[11] H. Itoh, T. Katada, M. Ui, H. Kawasaki, K. Suzuki, and Y. Kaziro, *FEBS Lett.* **230**, 85 (1988).

[12] K. Nagata, T. Katada, M. Tohkin, H. Itoh, Y. Kaziro, M. Ui, and Y. Nozawa, *FEBS Lett.* **237**, 113 (1988).

proteins can be stored at $-80°$ for more than 6 months. Repeated freezing and thawing of the proteins, however, are not recommended.

Separation of α and $\beta\gamma$ Subunits of Pertussis Toxin–Substrate G Proteins

The first phenyl-Sepharose fraction is concentrated to about 5 ml (5–6 mg of protein/ml) by the Minimodule NM-3 apparatus and incubated with 50 μM GTPγS (if required, plus [^{35}S]GTPγS; ~2,000 cpm/nmol) at 30° for 30 min in 10 ml of TED containing 25 mM MgCl$_2$ and 250 mM (NH$_4$)$_2$SO$_4$. The reaction mixture is diluted with 10 volumes of TMD [20 mM Tris-HCl (pH 8.0), 5 mM MgCl$_2$, and 1 mM DTT] containing 300 mM NaCl. The diluted sample is applied to a column (2.2 × 25 cm) of phenyl-Sepharose CL-4B which has been equilibrated with 250 mM NaCl and 0.25% sodium cholate in TMD and is eluted with a linear gradient (200 ml) starting with 0.25% sodium cholate and 250 mM NaCl and ending with 1.2% sodium cholate and 0 mM NaCl in TMD. Two major peaks of protein are recovered from the column (Fig. 4). The first peak contains GTPγS-bound α subunits (with the ^{35}S radioactivity of GTPγS), whereas the second has $\beta\gamma$ subunits resolved from the α subunits. The purified α and $\beta\gamma$ subunits are subjected to purification of heterogeneous α and $\beta\gamma$ subunits as described below.

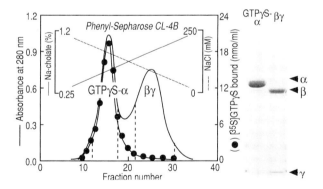

FIG. 4. Resolution of G proteins into constituent α and $\beta\gamma$ subunits by phenyl-Sepharose CL-4B column chromatography. The first phenyl-Sepharose eluate was incubated with GTPγS, MgCl$_2$, and (NH$_4$)$_2$SO$_4$, and the resultant GTPγS-bound α subunits and $\beta\gamma$ subunits were recovered from a phenyl-Sepharose CL-4B column as described in the text. The right-hand side shows Coomassie blue staining of the eluted proteins (3 μg of protein), which were separated by SDS–polyacrylamide gel electrophoresis (12% gel).

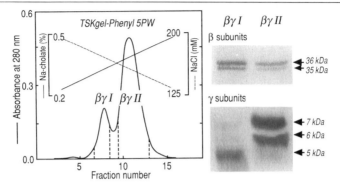

FIG. 5. Elution profile of βγ subunits from the TSKgel-Phenyl 5PW column. The βγ subunits resolved from α subunits were applied to and eluted from a TSKgel-Phenyl 5PW column as described in the text. The right-hand side shows silver staining of the eluted βγ subunits (0.5 μg of protein in the pooled peaks of βγI and βγII) which were separated by SDS–polyacrylamide gel electrophoresis (12 and 16.5% gels for β and γ subunits, respectively).

Separation of Heterogeneous α Subunits

The GTPγS-bound α subunits (10–15 mg of protein) are concentrated to 5 ml by the Minimodule NM-3 apparatus and diluted with 3 volumes of TMD containing 1% Lubrol-PX. The diluted sample is applied to a column (1 × 8 ml) of DEAE Toyopearl 650S which has been equilibrated with 50 mM NaCl in TMD containing 0.75% Lubrol-PX, and the column is eluted with a linear gradient of NaCl (50–200 mM; 60 ml) in the same buffer. Three major peaks of α subunits (with ^{35}S radioactivity) are recovered from the DEAE column[4] as had been observed in Fig. 2.

Each of the three peaks (I–III) is applied and eluted from a column (1 × 6 cm) of hydroxyapatite HCA 100S to concentrate and to purify further the GTPγS-bound α subunits.[12] Each fraction (1–5 mg of protein) thus obtained is applied to a column of Mono Q HR5/5 which has been equilibrated with TMD containing 100 mM NaCl and 0.7% CHAPS, and the column is eluted at the flow rate of 0.75 ml/min with the following series of NaCl gradients using a Pharmacia FPLC system: 100 mM for 2 min; 100–250 mM over 40 min; 250–500 mM over 5 min. Five peaks of GTPγS-bound α subunits are purified to near homogeneity.[4] Based on the findings of immunoblot analyses using specific antibodies raised against various α subunits, the five peaks are identified as α_{i1}, α_{o2}, α_{i3}, α_{o1}, and α_{i2}, respectively, in the order of elution from the Mono Q column.[4,6] Each GTPγS-bound α subunit is concentrated and filtered through a column (0.8 × 10 cm) of Sephadex G-50 fine in 50 mM HEPES–NaOH (pH 7.4), 50 μM MgCl$_2$, and 0.7% CHAPS prior to use or storage at $-80°$.

Separation of Heterogenous βγ Subunits

An aliquot (4–5 mg of protein) of the βγ subunits eluted from the second phenyl-Sepharose CL-4B column is diluted with TMD and further applied to a column (0.5 × 5 cm) of TSKgel-Phenyl 5PW (Tosoh) which has been equilibrated with 187.5 mM NaCl and 0.25% sodium cholate in TMD, and the column is eluted at a flow rate of 0.75 ml/min with the following gradients of cholate and NaCl using the Pharmacia FPLC system: 0.25% cholate/187.5 mM NaCl for 5 min, 0.25–0.6% cholate/187.5–100 mM NaCl over 30 min, and 0.6–1.0% cholate/100–0 mM NaCl over 5 min. Two major peaks of βγ subunits, the first (βγI; fractions 6–8) and second (βγII; fractions 10–13) peaks, are recovered from the column (Fig. 5). There are 36- and 35-kDa β subunits in both the βγI and βγII fractions. However, the apparent molecular sizes of γ subunits are different between the two fractions; there are mainly 5-kDa and 7/6-kDa γ subunits in fractions βγI and βγII, respectively. Each of the two βγ subunits (βγI and βγII) could be further separated into five peaks via column chromatography with HRLC MA7Q (Bio-Rad).[7] The resolved peaks are concentrated and filtered through Sephadex G-50 fine in 50 mM HEPES–NaOH (pH 7.4) and 0.7% CHAPS prior to use or storage at −80°.

[11] Purification of Transducin

By JOËLLE BIGAY *and* MARC CHABRE

Introduction

Transducin is the heterotrimeric GTP-binding protein (G protein) which, in vertebrate retinal rods, couples the membrane photoreceptor rhodopsin to the intracellular effector enzyme cGMP phosphodiesterase. Transducin is specifically localized in the disk membranes of the retinal rod outer segments (ROS), where it represents about 10% of the total protein content. Although it remains preferentially membrane bound *in situ* and when the ROS preparations are kept in isotonic buffer (e.g., with 150 mM KCl or NaCl and 1 to 5 mM MgCl$_2$), inactive GDP-bound transducin, Tα_{GDP}–Tβγ, can be fully solubilized from dark-adapted ROS membranes on lowering the buffer ionic strength, in the absence of detergent. On illumination of the ROS membrane, if no guanine nucleotide is added to the medium, transducin undergoes tight binding to photoexcited rhodopsin (R*). Addition of GTP or a nonhydrolyzable analog can then induce the dissociation of active GTP-bound transducin from R* and quantitative release of the $T\alpha_{GTP}$ subunit from the membrane, on which $T\beta\gamma$ still remains bound if the buffer is of isotonic ionic strength. The dependence of transducin solubility on ROS membrane illumination and on the presence of GTP, first described by Kühn,[1] is the basis of very simple and efficient techniques of isolation and purification of transducin subunits. We first describe the preparation of cattle ROS membranes, the most common material for large-scale extraction of transducin.

Purification of Bovine Rod Outer Segment Membranes

One hundred fresh dark-adapted bovine retinas, dissected under dim red light, are shaken in 400 ml sucrose buffer [45% (w/v) sucrose in buffer X: 12 mM KCl, 5 mM MgCl$_2$, 20 mM HEPES, pH 7.4, 0.1 phenylmethylsulfonyl fluoride (PMSF), 5 mM 2-mercaptoethanol (2-ME)]. The mixture is sedimented at 20,000 g for 20 min (JA 14, Beckman) in two 200-ml tubes after covering each tube with an additional 5 ml of buffer X without sucrose. The floating ROS membranes are then sedimented after dilution 1/2 in buffer X and purified by centrifugation (53,000 g for 45 min at 4° in a SW 28 rotor, Beckman) on a stepwise sucrose gradient (density 1.105, 1.115, and 1.135). Purified ROS membranes are collected at the 1.135/1.115 interface. The rhodopsin concentration is assayed spectroscopically on an aliquot solubilized in 1% lauryldimethylamine oxide (LDAO). The absorbance at 500 nm ($\varepsilon = 40,300/M$ cm) is measured in the dark in the presence of 0.1 mM hydroxylamine and controlled after illumination. The purified ROS membranes are sedimented after dilution 1/2 in buffer X, and the pellets are stored in the dark at $-80°$. Typically 40 to 50 mg rhodopsin is obtained from 100 retinas.

Extraction of Transducin Subunits

The ROS pellets are first resuspended at 0.5 mg rhodopsin/ml in Iso buffer (120 mM KCl, 0.1 mM MgCl$_2$, 20 mM Tris, pH 7.5, 0.1 mM PMSF, 5 mM 2-ME) and centrifuged for 5 min at 400,000 g (TL100.3 rotor, Beckman) or 45 min at 31,000 g (JA 20 rotor, Beckman) to eliminate soluble proteins. The pellets are fully illuminated, resuspended at 0.5 mg rhodopsin/ml in Hypo buffer (0.1 mM MgCl$_2$, 5 mM Tris, pH 7.5, 0.1 mM PMSF, 5 mM 2-ME) which solubilizes cGMP-phosphodiesterase, and sedimented.

For extraction of inactive holotransducin ($T\alpha_{GDP}$—$T\beta\gamma$), the illuminated and washed ROS membranes are suspended at 4 mg rhodopsin/ml

---

[1] H. Kühn, *Curr. Top. Membr. Transp.* **15**, 171 (1981).

in Hypo buffer and centrifuged after the addition of 100 μM GTP or 200 μM GDP. The transducin-containing supernatant is carefully removed and recentrifuged to eliminate traces of membrane contamination.

Preferential extraction of the Tα subunit in its permanently active (T$\alpha_{GTP\gamma S}$) or inactive (Tα_{GDP}) form is achieved by suspending the illuminated and washed ROS membranes in Iso buffer and sedimenting immediately after the addition of 100 μM GTPγS or 200 μM GTP. In the later case Tα_{GTP} is solubilized and decays rapidly to Tα_{GDP} in the supernatant. The T$\beta\gamma$ subunit can then be solubilized by resuspending the Tα-depleted ROS membranes in Hypo buffer. Aliquots of the successive ROS extracts are then analyzed on polyacryalmide gels, without heating the samples. About 50 μg Tα or 100 μg Tα–T$\beta\gamma$ is extracted from ROS membranes containing 1 mg rhodopsin.

Sodium dodecyl sulfate (SDS)–polyacrylamide gel electrophoresis is performed according to Laemmli,[2] using a SE 280 Tall Mighty Small Slab Unit (gel size 80 × 110 × 0.75 mm, Hoeffer Scientific, San Francisco, CA), run at constant current (25–30 mA, corresponding to voltages of 100 V at the beginning and 250 V at the end of the run). The separating gel contains 12% acrylamide, 0.32% bisacrylamide, and 0.1% SDS. Standard procedures are used for Coomassie blue staining and drying the gels.

Figure 1 sums up the different steps of extraction and shows the protein content in the respective supernatants. As seen in lane 4 (Fig. 1), the inactive holotransducin extract (Tα_{GDP}–T$\beta\gamma$) is generally pure enough to be used without further purification. If necessary, purification is performed as described below for the Tα_{GDP} extract.

Purification of Holotransducin or Separated Subunits from Extracts

Free Nucleotide Removal

For large-scale purification of transducin subunits on ion-exchange columns, the excess free nucleotide must be eliminated from the protein extract, as it can compete with the protein for binding on the charged residues of the anion-exchange resin. The protein extract, filtered through a Millex-GV 0.22-μm unit (Millipore, Bedford, MA), is applied on Sephadex G-25 columns XK 16/20 (Pharmacia, Piscataway, NJ) for extract volumes up to 3 ml, or XK 16/70 for extract volumes up to 25 ml. The columns are equilibrated with buffer A (20 mM Tris, pH 7.5, 0.1 mM MgCl$_2$, 0.1 mM PMSF, 5 mM 2-ME). The protein is eluted at 2 ml/min

---

[2] U. K. Laemmli, *Nature (London)* **227**, 680 (1970).

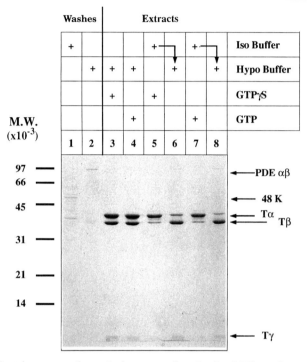

FIG. 1. Protein content of transducin extracts from bovine ROS membranes, as assayed by Coomassie blue staining on a 12% polyacrylamide gel. Lanes 1 and 2 show results of preliminary washing steps, successively in Iso buffer in the dark to eliminate soluble proteins and in Hypo buffer after illumination, to extract the cGMP phosphodiesterase. Lanes 3 to 8 represent subsequent transducin extracts in media supplemented with GTPγS or GTP, as indicated: lane 3, separated $T\alpha_{GTP\gamma S}$ and Tβγ subunits; lane 4, holoenzyme $T\alpha_{GDP}$–Tβγ; lane 5, TαGTPγS with a small fraction of Tβγ and, lane 6, Tβγ with the remaining TαGTPγS in the subsequent Hypo buffer extract; lane 7, TαGDP with a small fraction of Tβγ and, lane 8, Tβγ with the remaining TαGTPγS in the subsequent Hypo buffer extract.

and appears in the first peak, with a second peak corresponding to the free nucleotide.

An alternative method is to concentrate/dilute (in buffer A) and reconcentrate the transducin extract using a Centricon 30 microconcentrator (Amicon, Danvers, MA) until free nucleotides are diluted below 10 μM. Note that with inactive holoenzyme $T\alpha_{GDP}$–Tβγ extract or $T\alpha_{GDP}$ extract, 0.1% Lubrol PX (Sigma, St. Louis, MO) (or equivalent detergent such as Thesit from Boehringer Mannheim, Mannheim, Germany) must be added to avoid the binding of $T\alpha_{GDP}$ to the membrane of the microconcentrator.

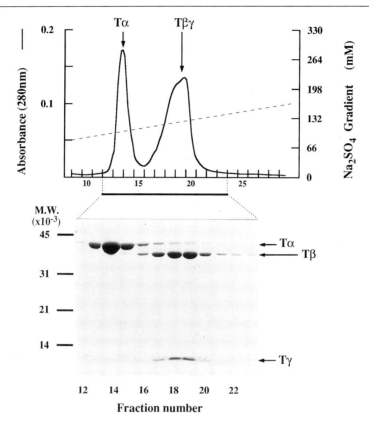

FIG. 2. Anion-exchange chromatography on a Polyanion SI HR 5/5 column (Pharmacia) of a transducin extract in Hypo buffer supplemented with GTP. The excess free nucleotide had been removed on a Centricon 30 microconcentrator (Amicon). The protein contents of the peak fractions are visualized on a 12% polyacrylamide gel.

Ion-Exchange Chromatography

Our standard purification procedure[3] is based on the use of the weak anion-exchange column Polyanion SI HR5/5 (polyamine charges bound on silica bed, from Pharmacia), with the Pharmacia FPLC (fast protein liquid chromatography) system. This anion-exchange support, however, is no longer available from Pharmacia. Other anion-exchange columns from Pharmacia (Mono Q or Mono P), which are strong anion exchangers (tertiary or quaternary amines bound on hydrophilic support), do not separate Tα properly from T$\beta\gamma$, and they even lead to partial (Mono Q)

[3] P. Deterre, J. Bigay, C. Pfister, and M. Chabre, *FEBS Lett.* **191**, 181 (1984).

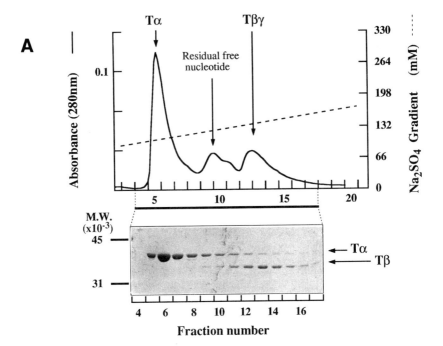

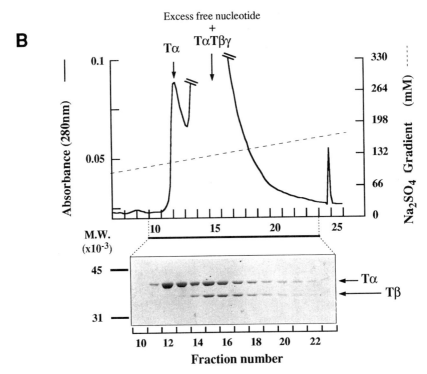

or complete (Mono P) retention of Tα on the column. We have tested the weak anion-exchange gel Synchropak AX 300, (Synchrom, Linden, IN), which is also composed of polyamine charges bound on silica bed. It can efficiently replace the Polyanion SI for separations with NaCl or MgCl$_2$ gradients, but not with Na$_2$SO$_4$ gradients.

Other protocols have been described, using Blue-Sepharose CL-6B columns (Pharmacia)[4,5] or ω-aminooctyl agarose columns.[6,7] We have not tested these columns for purification of transducin from bovine retinal extracts, but we have purified recombinant Tα_{GDP}, expressed in a baculovirus system,[8] by elution on a Blue-Sepharose CL-6B column with a KCl gradient. Recombinant Tα_{GDP} elutes at about 320 mM KCl, as described by Yamazaki et al.[5] for purification of Tα_{GDP} extracted from frog retina. One should note that the reported purifications of inactive Tα_{GDP}–T$\beta\gamma$ holoenzyme using Blue-Sepharose CL-6B yield Tα_{GDP} separated from T$\beta\gamma$ as in the case of active T$\alpha_{GTP\gamma S}$. Such a separation of subunits of the Tα_{GDP}–T$\beta\gamma$ holoenzyme is not observed on Polyanion SI.[3]

Purification of Inactive Tα_{GDP} or Activated T$\alpha_{GTP\gamma S}$

The Tα subunit is best purified from an extract in Iso buffer with GTP, which is already enriched in Tα_{GDP} (Fig. 1, lane 5), or from an extract in Iso buffer with GTPγS, which is enriched in T$\alpha_{GTP\gamma S}$ (Fig. 1, lane 7), but good purification is also obtainable from extracts in Hypo buffer with GTPγS, which contain equal amount of Tα and T$\beta\gamma$ (Fig. 1, lane 3). The extract is injected on a Polyanion SI column equilibrated with buffer A. The proteins are then eluted with a linear Na$_2$SO$_4$ gradient (0 to 0.66 M, 4.4 mM/min, 0.5 ml/min). Typical Na$_2$SO$_4$ elution profiles for various types of transducin extracts are shown in Figs. 2 and 3. With this salt gradient, the first major peak always corresponds to the Tα subunit. When present, isolated T$\beta\gamma$ subunit is eluted as a second peak, but, as seen on

[4] C. Kleuss, M. Pallat, S. Brendel, W. Rosenthal, and G. Schultz, *J. Chromatogr.* **407**, 281 (1987).
[5] A. Yamazaki, M. Tatsumi, and M. W. Bitensky, this series, Vol. 159, p. 702.
[6] B. K. K. Fung, *J. Biol. Chem.* **258**, 10495 (1983).
[7] A. B. Fawzi, D. S. Fay, E. A. Murphy, H. Tamir, J. J. Erdos, and J. K. Northup, *J. Biol. Chem.* **266**, 12194 (1991).
[8] E. Faurobert, personal communication (1992).

FIG. 3. Anion-exchange chromatography on a Polyanion SI HR 5/5 column (Pharmacia) of (A) transducin extract in Iso buffer supplemented with GTPγS, after removal of excess free nucleotide on a Centricon 30 microconcentrator (Amicon), and (B) transducin extract in Iso buffer with GTP, loaded onto the column without removal of excess free nucleotide.

Fig. 1, this Tβγ peak is always contaminated by Tα. To obtain pure Tβγ subunit, another salt gradient must be used (see [34], this volume). If free nucleotides have not been removed, an additional peak between Tα and Tβγ or superimposed on Tβγ is observed (Fig. 3). With extracts containing holotransducin $T\alpha_{GDP}$–Tβγ, the first peak contains only a small fraction of separated $T\alpha_{GDP}$, and the major fraction of $T\alpha_{GDP}$ elutes together with Tβγ in the second peak (not shown, see Ref. 3). With this protocol, it is possible to purify 30 to 50 μg Tα from ROS membranes corresponding to 1 mg rhodopsin.

Slightly modified Na_2SO_4 gradient programs (4.4 mM/min from 0 to 0.13 M, and 8.8 mM/min above 0.13 M, flow rate 0.5 ml/min) also allow purification of $T\alpha_{GTP\gamma S}$–PDEγ complexes as well as separation of PDEαβγ2, PDEαβγ, and PDEαβ from extracts which contain both activated cGMP phosphodiesterase and actived transducin,[9,10] as eluted together from the illuminated ROS pellet by Hypo buffer containing GTPγS.

[9] P. Deterre, J. Bigay, M. Robert, C. Pfister, H. Kühn, and M. Chabre, *Proteins* **1**, 188 (1986).
[10] P. Deterre, J. Bigay, F. Forquet, M. Robert, and M. Chabre, *Proc. Natl. Acad. Sci. U.S.A.* **85**, 2424 (1988).

[12] Expression of G-Protein α Subunits in *Escherichia coli*

By ETHAN LEE, MAURINE E. LINDER, and ALFRED G. GILMAN

Introduction

The ability to express cloned eukaryotic genes in *Escherichia coli* is useful to assign function to a particular protein in the absence of a contaminating background, especially when it is difficult to resolve potential candidates from one another. In addition, the large quantity of protein that can often be obtained using prokaryotic expression systems permits studies that would otherwise be impractical.

G proteins are a family of guanine nucleotide-binding regulatory proteins that link a large number of cell surface receptors to regulation of several intracellular effectors (reviewed in Refs. 1 and 2). G proteins are heterotrimers, consisting of an α polypeptide and a tightly associated complex of β and γ subunits. The α subunit, which differs in each G-protein oligomer, binds guanine nucleotides with high affinity and possesses an

[1] J. R. Hepler and A. G. Gilman, *Trends Biochem. Sci.* **17**, 383 (1992).
[2] M. I. Simon, M. P. Strathmann, and N. Gautam, *Science* **252**, 802 (1991).

intrinsic GTP hydrolase activity. Binding of GTP to α leads to dissociation of α-GTP from $\beta\gamma$, and both α-GTP and $\beta\gamma$ are then free to interact with downstream effectors and regulate their activity.

To date, 17 genes that encode G-protein α subunits have been identified. Several of these can be expressed in *E. coli*, and the characteristics of the recombinant proteins are similar, but not identical, to those of counterparts purified from mammalian sources.[3-5] The differences are attributable to myristoylation,[6,7] palmitoylation,[8,9] and perhaps other covalent modifications of Gα proteins that are not normally performed by *E. coli*. However, projects requiring very large amounts of purified protein (e.g., structural studies) have been hampered by the relatively low levels of expression of G-protein α subunits in previously described prokaryotic systems. In this chapter, we describe a general method for expressing several G-protein α subunits in *E. coli* at levels 10–100 times higher than achieved previously. In addition, we describe a method for purification of the recombinant proteins by affinity chromatography on a resin containing chelated Ni^{2+} after addition of an amino-terminal hexahistidine tag to the recombinant protein. Such purification is rapid and results in the isolation of highly purified protein in a single step. Furthermore, the introduction of a tobacco etch virus (TEV) polyprotein cleavage site between the hexahistidine tag and the G-protein α subunit permits the efficient removal of the tag by recombinant TEV protease.

Methods

Construction of Expression Vectors

Construction of the plasmids used for expressing G-protein α subunits is performed using methods described by Sambrook and colleagues.[10] The prokaryotic expression vector pQE-6 (Qiagen, Chatsworth, CA) belongs

---

[3] M. P. Graziano, M. Freissmuth, and A. G. Gilman, *J. Biol. Chem.* **264**, 409 (1989).
[4] M. E. Linder, D. A. Ewald, R. J. Miller, and A. G. Gilman, *J. Biol. Chem.* **264**, 8243 (1990).
[5] P. J. Casey, H. K. W. Fong, M. I. Simon, and A. G. Gilman, *J. Biol. Chem.* **265**, 2383 (1990).
[6] J. E. Buss, S. M. Mumby, P. J. Casey, A. G. Gilman, and B. M. Sefton, *Proc. Natl. Acad. Sci. U.S.A.* **84**, 7493 (1987).
[7] M. E. Linder, I.-H. Pang, R. J. Duronio, J. I. Gordon, P. C. Sternweis, and A. G. Gilman, *J. Biol. Chem.* **266**, 4654 (1991).
[8] M. E. Linder, P. Middleton, J. R. Hepler, R. Taussig, A. G. Gilman, and S. M. Mumby, *Proc. Natl. Acad. Sci. U.S.A.* **90**, 3675 (1993).
[9] M. Parenti, V. Alessandra, C. M. H. Newman, G. Milligan, and A. I. Magee, *Biochem. J.* **291**, 349 (1993).
[10] J. Sambrook, E. F. Fritsch, and T. Maniatis, "Molecular Cloning: A Laboratory Manual." Cold Spring Harbor Laboratory, Cold Spring Harbor, New York, 1989.

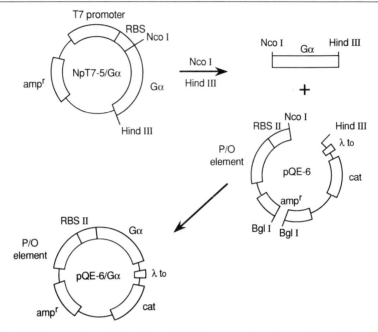

Fig. 1. Construction of plasmids used for the expression of recombinant Gα subunits. For explanation, see text.

to the pDs family of vectors.[11,12] The important features of this vector are outlined in Fig. 1. The pQE-6 vector confers ampicillin resistance (amp^r) and contains a very strong coliphage T5 promoter upstream of two *lac* operators [the promoter/operator (P/O) element N25OPSN25OP29]. Transcription of genes subcloned into pQE-6 is, therefore, repressed by the *lac* repressor. We have taken advantage of the fact that regulation of the rate of transcription can be achieved by adjusting the concentration of the inducer, isopropyl-β-D-thiogalactopyranoside (IPTG). This is presumably due to increasing clearance from the promoter with increasing concentrations of IPTG.[13] Ribosomal binding is mediated by the synthetic ribosome binding site, RBSII, and efficient transcriptional termination is mediated by the terminator t_o from phage λ. The pQE-6 vector also contains the gene for chloramphenicol acetyltransferase (CAT) with its own ribosomal

---

[11] H. Bujard, R. Gentz, M. Lanzer, D. Stueber, M. Mueller, I. Ibrahimi, M.-T. Haeuptle, and B. Dobberstein, this series, Vol. 155, p. 416.
[12] D. Stuber, H. Matile, and G. Garotta, *Immunol. Methods* **4**, 121 (1990).
[13] J. Crowe, "The QIAexpressionist." Qiagen, Chatsworth, California, 1992.

binding site downstream of t_o. The *cat* gene, however, does not contain its own promoter.

Construction of the pQE6/Gα plasmid is illustrated in Fig. 1. The pQE-6 vector contains an *NcoI* site encoding the initiation ATG sequence and a *Hin*dIII site within a downstream multicloning region. This permits direct subcloning of G-protein α subunits previously harbored in the NpT7-5 expression vectors into pQE-6.[3,4] To circumvent an internal *Nco*I restriction site within the pQE-6 vector, the *Bgl*I–*Nco*I fragment containing the authentic ATG sequence and the *Bgl*I–*Hin*dIII fragment containing t_o are purified and ligated to the isolated *Nco*I–*Hin*dIII fragment encoding the desired G-protein α subunit cDNA.

The pQE-6 plasmid must be maintained in a host strain that expresses *lac* repressor; thus, the appropriate constructs are identified by restriction digestion after transformation into such a cell strain (e.g., JM109). The pQE-6/Gα plasmid is then transformed into the appropriate expression host. In cases where the host strain expresses either low levels or no *lac* repressor, cotransformation with the pREP4 plasmid,[12] which contains the *laqI* gene, is performed. The pREP4 plasmid contains a kanamycin resistance marker and is compatible with pQE-6. Double transformants containing both plasmids are selected with Luria–Bertani (LB) plates containing 50 μg/ml of both kanamycin and ampicillin.

The pQE-6 plasmid contains the pBR322 replication region (pBM1 replicon) and is therefore compatible with plasmids carrying pSC101 and p15A replicons.[10] Previous studies have shown that correctly modified myristoylated G-protein α subunits can be synthesized when yeast protein *N*-myristoyltransferase is coexpressed with the appropriate G-protein α subunits in *E. coli*[7]; however, yields were low. It is now known that coexpression of *Saccharomyces cerevisiae* protein *N*-myristoyltransferase in a p15A replicon-based plasmid with the appropriate Gα subunit in a pQE-6 vector results in a substantial increase in the yield of soluble myristoylated protein.[14]

Time Course of Expression

When induced with a low concentration of IPTG (30 μM) at an early time point in the growth curve (OD_{600} of 0.4), cells harboring a pQE6/Gα plasmid continue to accumulate protein in the cytosolic fraction long after induction. As shown in Fig. 2 with BL21/DE3 cells harboring pQE-6/$G_i\alpha_1$, there is no immunoreactive $G_i\alpha_1$ protein prior to induction (0 hr). On addition of IPTG, $G_i\alpha_1$ expression rises slowly, with peak accumulation

---

[14] S. M. Mumby and M. E. Linder, this volume [20].

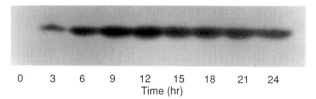

FIG. 2. Time course of induction of BL21/DE3 cells harboring pREP4 and pQE-6/$G_i\alpha_1$. A 100-ml culture of BL21/DE3 cells harboring pREP4 and pQE-6/$G_i\alpha_1$ plasmids was grown from a single colony in enriched medium (see section on culture of *E. coli*) containing 50 μg/ml of both ampicillin and kanamycin. The culture was maintained at 30° with shaking at 150 rpm. At an OD_{600} of 0.4, the cells were induced with 30 μM IPTG plus 1 μg/ml of chloramphenicol. At the time points indicated, 5-ml samples were withdrawn, and the cells were harvested by centrifugation at 4° in a Beckman J-6B centrifuge for 10 min at 3500 rpm. The pellets were washed with TEDP [50 mM Tris-HCl (pH 8.0), 1 mM EDTA, 2 mM dithiothreitol, and 0.1 mM phenylmethylsulfonyl fluoride]. The cells were lysed by freezing in liquid nitrogen and thawing in 200 μl of TEDP containing 0.1 mg/ml of lysozyme. The lysates were centrifuged at 4° in a Beckman Ti 70.1 rotor for 30 min at 40,000 rpm, and the supernatants were recovered. Protein was quantified as described by Bradford [M. M. Bradford, *Anal. Biochem.* **72,** 248 (1976)], and 10 μg of each sample was subjected to SDS–PAGE in an 11% polyacrylamide gel and transferred to nitrocellulose. Immunoblotting was performed as previously described [S. M. Mumby and J. E. Buss, *Methods* **1,** 216 (1990)] using an affinity-purified $G_i\alpha_1$/$G_i\alpha_2$-specific antibody, BO87.[8]

occurring after 9–12 hr. Peak expression times for different α subunits are shown in Table I.

The growth curve of BL21/DE3 cells[15] harboring the pQE-6/$G_i\alpha_1$ plasmid before and after induction is shown in Fig. 3. The cells continue to grow long after induction with IPTG. Cells are harvested soon after the accumulation of Gα protein is maximal since proteolysis is a problem for some α subunits when incubation is continued for an extended period of time.

Host Strains Used for Expression

Various cell strains have been tested for the ability to accumulate high levels of G-protein α subunits in the soluble fraction. An *E. coli* strain that expresses a particular G-protein α subunit at high levels may not necessarily behave similarly with other α subunits. However, the *E. coli* strain BL21/DE3,[15] which is protease deficient, is able to accumulate high levels of several different G-protein α subunits ($G_i\alpha_1$, $G_i\alpha_2$, $G_i\alpha_3$, and $G_s\alpha_s$). Although this strain normally grows vigorously and expresses the *lacI* gene product, the introduction of certain pQE-6/Gα plasmids into the strain inhibits growth. This observation probably reflects the insuffi-

[15] F. W. Studier and B. A. Moffatt, *J. Mol. Biol.* **189,** 113 (1986).

TABLE I
PEAK TIME OF ACCUMULATION OF
RECOMBINANT α SUBUNITS[a]

Protein	Time (hr postinduction)	OD_{600}[b]
$G_s\alpha_s$	12–15	2.0
$G_i\alpha_1$	9–12	2.4
$G_i\alpha_2$	16–18	3.1
$G_i\alpha_3$	16–18	3.0
$G_o\alpha_A$[c]	16–18	4.8

[a] Cells were cultured in enriched medium, harvested, and processed as described in the text (see section on culture of *E. coli*). Unless otherwise indicated, the host strain was BL21/DE3 harboring pREP4.
[b] For comparison, the OD_{600} of cells harboring NpT7-5/Gα plasmids ranged from 1.4 to 1.7 at the time of harvest.
[c] The host strain was M15 harboring pREP4.

cient levels of *lac* repressor expressed in these cells, since growth inhibition can be prevented by introduction of the pREP4 plasmid. Expression of $G_o\alpha_A$ is higher in strain M15[16] than in BL21/DE3. Because *lac* repressor is absent in M15, cotransformation with pREP4 is required to maintain the pQE-6/$G_o\alpha_A$ plasmid.

Culture of Escherichia coli

Culture of cells harboring the pQE-6 vector is described below. A single colony from a plate or a stab from a freezer stock is used to inoculate LB medium containing the appropriate antibiotics (50 μg/ml ampicillin ± 50 μg/ml kanamycin). The overnight culture is then diluted 1000-fold in 110 ml of enriched medium (2% tryptone, 1% yeast extract, 0.5% NaCl, 0.2% glycerol, and 50 mM KH_2PO_4, pH 7.2) containing the appropriate antibiotics and grown overnight at 30° with shaking at 150–200 rpm. This culture is used to inoculate 10 liters (10 × 1 liter in 2-liter Erlenmeyer flasks) of enriched medium containing 50 μg/ml ampicillin. Kanamycin is omitted at this stage. The cultures are maintained at 30° in a rotary air shaker (150–200 rpm). When the OD_{600} reaches 0.4, IPTG (30 μM) and chloramphenicol (1 μg/ml) are added. Chloramphenicol increases the expression of some, but not all, α subunits. Because chloramphenicol did not inhibit the expression of any α subunit tested, it was added routinely. Induced cell cultures are maintained at 30° with shaking at 150–200 rpm

[16] M. R. Villarejo and I. Zabin, *J. Bacteriol.* **120**, 466 (1974).

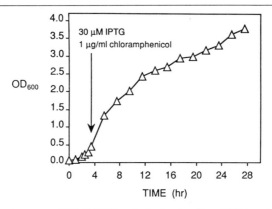

FIG. 3. Growth curve of BL21/DE3 cells harboring the pREP4 and the pQE-6/$G_i\alpha_1$ plasmids. Cells were cultured as described in the text. At the indicated times, 2 ml of cells were withdrawn and the OD_{600} determined.

until harvesting. Bacteria are collected by centrifugation at 4° in a Beckman JA-10 rotor for 15 min at 8000 rpm. The cell pellets are washed (without resuspension) with ice-cold TEDP [50 mM Tris-HCl (pH 8.0), 1 mM EDTA, 2 mM dithiothreitol, and 0.1 mM phenylmethylsulfonyl fluoride]. The resulting cell paste is frozen by direct immersion in liquid nitrogen and stored at $-80°$. Lysis of certain cell strains, particularly BL21/DE3, is facilitated by freezing and thawing; cells are routinely frozen prior to lysis and further processing.

A comparison of the levels of expression of Gα proteins obtained with this method and that utilized previously is shown in Fig. 4. Cells harboring pQE-6/Gα accumulate more recombinant protein than cells harboring NpT7-5/Gα (previous expression system) in all cases. Furthermore, cell density is now higher at the time of harvest (Table I). The increased yield reflects a combination of increased accumulation and greater cell density. The greater accumulation also permits more efficient purification of the recombinant proteins.

Purification of Recombinant Proteins

The high level of protein expression obtained with this method mandates alterations of prior protocols for protein purification because of limitations of column capacity and the tendency of some proteins to precipitate at high concentrations. The protocol described below is a modification of the purification scheme developed by Linder et al.[4] for purification of recombinant $G_i\alpha_1$, $G_i\alpha_2$, $G_i\alpha_3$, and $G_o\alpha_A$.[4] The method is applicable to these proteins and to $G_s\alpha$.

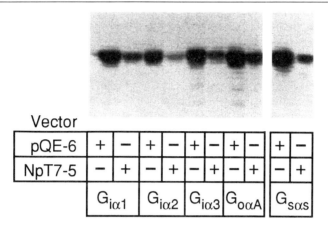

FIG. 4. Comparison of expression levels of recombinant $G_i\alpha_1$, $G_i\alpha_2$, $G_i\alpha_3$, $G_o\alpha_A$, and $G_s\alpha_s$ using previous and current expression systems. Cultures expressing recombinant proteins using the current system were prepared as described in the text. The previous expression system[3-5] utilized a T7 RNA polymerase-driven vector, NpT7-5 [S. Tabor and C. C. Richardson, *Proc. Natl. Acad. Sci. U.S.A.* **84**, 4767 (1987)]. BL21/DE3 cells harboring NpT7-5/Gα constructs were grown and harvested as described previously [H. Itoh and A. G. Gilman, *J. Biol. Chem.* **266**, 16226 (1991)]. Crude lysates and the supernatant fractions were prepared as described in the text. Crude supernatant fractions (10 μg) of $G_i\alpha_1$, $G_i\alpha_2$, $G_i\alpha_3$, $G_o\alpha_A$, and $G_s\alpha_s$ were processed as described in the legend to Fig. 2. An affinity-purified Gα common antibody, P-960 (S. M. Mumby and A. G. Gilman, this series, Vol. 195, p. 215), was used to probe $G_i\alpha_1$, $G_i\alpha_2$, $G_i\alpha_3$, and $G_o\alpha_A$. $G_s\alpha_s$ was visualized with the $G_s\alpha_s$ specific affinity-purified antibody, 584 (S. M. Mumby and A. G. Gilman, this series, Vol. 195, p. 215).

Lysis of Escherichia coli. Frozen cell paste (from 10 liters of culture) is suspended in 1 liter of TEDP plus 0.1 mg/ml of lysozyme (Sigma, St. Louis, MO) using an 18-gauge cannula and syringe and is incubated on ice for 30 min. To facilitate lysis, the cells are then sonicated (5 times 30 sec each time, on ice) using a probe-tip sonicator (Heat Systems Ultrasonics, Farmingdale, NY). Alternatively, frozen cell paste is resuspended in TEDP plus lysozyme using a Polytron homogenizer (Brinkman, Westburg, NY), followed by incubation on ice for 30 min. MgSO$_4$ (5 mM) and DNase I (20 mg) are then added to the lysate, which is incubated on ice for 30 min; the viscosity of the lysate decreases dramatically during this time. The lysate is centrifuged at 4° in a Beckman JA-14 rotor for 1 hr at 14,000 rpm, and the supernatant is recovered. All subsequent steps are performed at 4°.

DEAE-Sephacel Extraction. The supernatant is incubated for 30 min with occasional stirring with 300 ml of DEAE-Sephacel anion-exchange resin (Pharmacia, Piscataway, NJ) that had been equilibrated with TEDP. This slurry is then collected in a Büchner funnel lined with Whatman

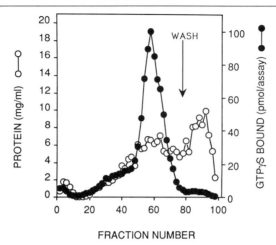

FIG. 5. Phenyl-Sepharose chromatography of recombinant $G_{i\alpha_1}$. The DEAE eluate was chromatographed on a phenyl-Sepharose column as described in the text. Samples from the indicated fractions were assayed for protein (2.5 μl) and GTPγS binding (5 μl). The GTPγS binding assay was performed at a final nucleotide concentration of 2 μM for 2 hr at 30°.

(Clifton, NJ) No. 4 paper. The resin is washed with 1 liter of TEDP and eluted with three 250-ml volumes of TEDP containing 300 mM NaCl.

Phenyl-Sepharose Chromatography. GDP (25 μM) and $(NH_4)_2SO_4$ (final concentration 1.2 M) are added to the DEAE eluate, and the mixture is incubated on ice for 10 min prior to centrifugation at 4° in a Beckman JA-14 rotor for 10 min at 8500 rpm. The supernatant is applied to a 200-ml column of phenyl-Sepharose (Pharmacia, Piscataway, NJ) that had been equilibrated with TEDP containing 1.2 M $(NH_4)_2SO_4$ and 25 μM GDP (Fig. 5). The protein is eluted with 1.2 liters of TEDP containing linearly decreasing concentrations of $(NH_4)_2SO_4$ (1.2–0 M) and increasing concentrations of glycerol (0–35%, v/v); 15-ml fractions are collected. To complete the elution, the column is washed with 250 ml of TEDP. Samples (5 μl) of the fractions are assayed for GTPγS binding activity[17] and are analyzed by Western blotting.[18] Desalting of the phenyl-Sepharose pool by successive dilution and concentration frequently results in the formation of protein precipitates. To minimize precipitation, peak fractions from the phenyl-Sepharose column are pooled and concentrated to 50–100 ml using an Amicon (Danvers, MA) ultrafiltration device with a PM30 membrane. The concentrated pool is dialyzed overnight against 4 liters of Q-Sepharose

---

[17] J. K. Northup, M. D. Smigel, and A. G. Gilman, *J. Biol. Chem.* **257**, 11416 (1982).
[18] S. M. Mumby and J. E. Buss, *Methods (San Diego)* **1**, 216 (1990).

buffer A [50 mM Tris-HCl (pH 8.0), 2 mM dithiothreitol] containing 2 μM GDP using Spectra/Por dialysis tubing (Spectrum Medical Industries, Houston, TX) with a molecular weight cutoff of 3500.

Q-Sepharose Chromatography. The dialyzed phenyl-Sepharose pool is chromatographed on a Pharmacia HiLoad 26/10 Q-Sepharose FPLC (fast protein liquid chromatography) column (Fig. 6). This column behaves similarly to the Mono Q column used previously but has the advantages of much higher capacity and lower cost. After loading the protein onto HiLoad Q-Sepharose, the column is washed with 150 ml of Q-Sepharose buffer A. The protein is eluted at 2 ml/min with 760 ml of Q-Sepharose buffer A containing linearly increasing concentrations of NaCl (0–300 mM for $G_s\alpha_s$; 0–250 mM for other Gα proteins); 80 fractions (9.5 ml each) are collected. Peak fractions from the HiLoad Q-Sepharose are pooled and assayed for GTPγS binding activity. Recombinant α subunits from this pool often represent more than 90% of total protein [as judged by Coomassie blue staining of sodium dodecyl sulfate (SDS)-containing polyacrylamide gels] and may be sufficiently pure for certain studies (see Fig. 8).

Hydroxyapatite Chromatography. Peak fractions from the Q-Sepharose column are pooled, and potassium phosphate (pH 8.0) is added to a final concentration of 10 mM. The pool is loaded (Fig. 7) onto a Bio-Gel HTP hydroxyapatite column (Bio-Rad, San Francisco, CA; 5 mg protein/

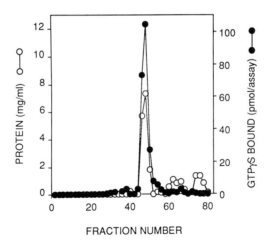

FIG. 6. Q-Sepharose chromatography of recombinant $G_i\alpha_1$. The dialyzed phenyl-Sepharose pool was chromatographed on a HiLoad 26/10 Q-Sepharose FPLC column as described in the text. Assay conditions were the same as those described in the legend to Fig. 5. Aliquots (2.5 μl) were assayed for protein and GTPγS binding.

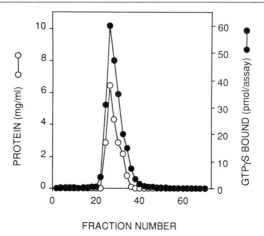

FIG. 7. Hydroxyapatite chromatography of recombinant $G_i\alpha_1$. The Q-Sepharose pool was chromatographed on a hydroxyapatite column as described in the text. Assay conditions and volumes were the same as those described in the legend to Fig. 6.

ml resin) that has been equilibrated with $T_{10}P_{10}D_2$ [10 mM Tris HCl (pH 8.0), 10 mM potassium phosphate (pH 8.0), and 2 mM dithiothreitol]. The column is washed with 2.5 volumes of $T_{10}P_{10}D_2$ and eluted with 20 volumes of 10 mM Tris-HCl (pH 8.0) and 2 mM dithiothreitol containing linearly increasing concentrations of potassium phosphate (10–500 mM for $G_s\alpha_s$; 10–200 mM for other Gα proteins). Recombinant protein in peak pools often represents more than 95% of the total protein (determined by Coomassie blue or silver staining). Electrophoretic analysis of the pools obtained during purification of $G_i\alpha_1$ is shown in Fig. 8. Activated G-protein α subunits are digested by trypsin to a lower molecular weight species that is resistant to further proteolysis, whereas nonactivated α subunits are degraded extensively.[19] In the presence of the nonhydrolyzable guanine nucleotide analog GTPγS, a substantial amount of the trypsin-resistant species is observed (Fig. 8, lane F). Thus, a majority of the recombinant $G_i\alpha_1$ can bind GTPγS. Quantitative estimation of binding of [^{35}S]GTPγS to purified preparations of Gα subunits routinely gives stoichiometries in excess of 0.8 mol/mol.

An additional chromatographic step is needed occasionally, typically with recombinant α subunits that are expressed at lower levels. In these cases, the protein is further purified on a high-resolution phenyl-Superose hydrophobic FPLC column.

[19] B. K.-K. Fung, *J. Biol. Chem.* **258,** 10495 (1983).

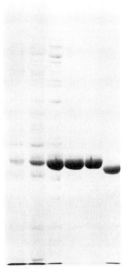

FIG. 8. Sodium dodecyl sulfate–polyacrylamide gel electrophoresis (SDS–PAGE) of fractions obtained from the purification of $G_i\alpha_1$. Samples of fractions from various stages in the purification of $G_i\alpha_1$ were resolved by SDS–PAGE in an 11% polyacrylamide gel, and the proteins were visualized by staining with Coomassie blue. Lane A contained 20 μg of the supernatant fraction; lane B, 20 μg of the DEAE-Sephacel eluate; lane C, 20 μg of the phenyl-Sepharose pool; lane D, 10 μg of the Q-Sepharose pool; lane E, 10 μg of the hydroxyapatite pool. Lane F shows trypsin protection analysis of the hydroxyapatite pool. An aliquot (10-μg) of the hydroxyapatite pool was incubated with 200 μM GTPγS and 20 mM MgSO$_4$ for 2.5 hr at 30°. After incubation, 1 μg of trypsin was added to the sample, and the reaction was continued for 15 min on ice. The reaction was terminated by addition of sample buffer, followed by boiling.

Phenyl-Superose Chromatography. Peak fractions from the hydroxyapatite column are pooled, and GDP is added to a final concentration of 50 μM. Ammonium sulfate (3.6 M) is added to a final concentration of 1.2 M, and the sample is applied to a high-resolution phenyl-Superose HR 10/10 hydrophobic column (Pharmacia) that had been equilibrated with phenyl-Superose buffer A [50 mM Tris HCl (pH 8.0), 1 mM EDTA, 2 mM dithiothreitol, and 50 μM GDP] containing 1.2 M (NH$_4$)$_2$SO$_4$. The protein is eluted at 1 ml/min with a 120-ml gradient containing linearly descending concentrations of (NH$_4$)$_2$SO$_4$ (1.2–0 M) in phenyl-Superose buffer A; 80 fractions (1.5 ml each) are collected.

Final pools are exchanged into HED [50 mM sodium HEPES (pH 8.0), 1 mM EDTA, and 2 mM dithiothreitol] with a Centricon 30 filtration unit

TABLE II
YIELDS OBTAINED FOR VARIOUS
RECOMBINANT α SUBUNITS[a]

Protein	Yield (mg)
$G_s\alpha_s$	35
$G_i\alpha_1$	400
$G_i\alpha_2$	40
$G_i\alpha_3$	ND[b]
$G_o\alpha_A$[c]	65

[a] Ten liters of *E. coli* harboring pREP4 and pQE-6/Gα were cultured and the recombinant Gα subunit purified as described in the text. Unless otherwise indicated, the host strain was BL21/DE3 harboring pREP4. Yields are protein obtained after hydroxyapatite chromatography.
[b] Not determined.
[c] The host strain was M15 harboring pREP4.

(Amicon) by successive concentration and dilution. Proteins are aliquoted, snap-frozen in liquid nitrogen, and stored at $-80°$ at a concentration greater than 2 mg/ml. GDP may also be added to a final concentration of 50 μM to stabilize the protein during prolonged storage. Table II shows the expected yields for various constructs.

Construction of Vectors to Express Histidine-Tagged Proteins

The vectors H_6pQE-60 and H_6TEVpQE-60 derived from the Qiagen vector pQE-60 are shown in Fig. 9. The Qiagen vector pQE-60 is essentially identical to the pQE-6 vector utilized above except that it lacks one of two internal *Nco*I sites (other than the *Nco*I site that encodes the initiator ATG). For the construction of H_6pQE-60 and H_6TEVpQE-60, the pQE-60 vector is digested with *Nco*I and *Hin*dIII (Fig. 9) and ligated to an adaptor sequence that encodes the hexahistidine [Met-(His)$_6$-Ala] or the hexahistidine TEV [Met-(His)$_6$-Ala-Glu-Asn-Leu-Tyr-Phe-Gln-Gly-Ala] sequences. The adaptor sequences are given in the legend to Fig. 9. Cleavage of the H_6TEVGα fusion protein by TEV protease results in the removal of the hexahistidine sequence and most of the TEV cleavage sequence. A glycine that forms part of the TEV recognition

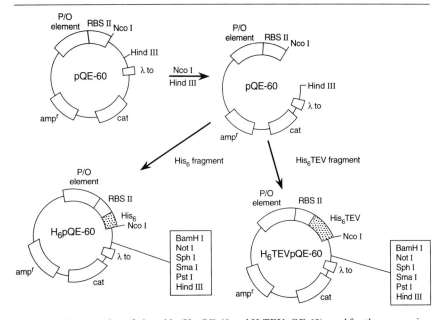

FIG. 9. Construction of plasmids (H₆pQE-60 and H₆TEVpQE-60) used for the expression of recombinant H₆Gα and H₆TEVGα subunits. For explanation, see text. The adaptor sequences utilized were as follows:

H₆pQE-60: CATGCATCACCATCACCATCACGCCATGGTATAGGATCCGCGGCCGCAT
GTAGTGGTAGTGGTAGTGCGGTACCATATCCTAGGCGCCGGCGTA
GCCCGGGCTGCAGA
CGGGCCCGACGTCTTCGA

and

H₆TEVpQE-60: CATGCATCACCATCACCATCACGCTGAGAATCTTTATTTTCAGGGC
GTAGTGGTAGTGGTAGTGCGACTCTTAGAAATAAAAGTCCCG
GCCATGGTATAGGATCCGCGGCCGCATGCCCGGGCTGCAGA
CGGTACCATATCCTAGGCGCCGGCGTACGGGCCCGACGTCTTCGA

The constructs were confirmed by dideoxynucleotide sequencing.

sequence remains on the recombinant protein. In addition, an alanine that was introduced to facilitate cloning is present on the cleaved α subunit (between the glycine and the initiator methionine).

The adaptor sequences are chosen such that the initiation ATG codon is preserved, but the *Nco*I site that encodes the ATG is destroyed. A new *Nco*I site is introduced within the adaptor sequence downstream of the His₆ or His₆ TEV coding sequence; this permits subcloning using the newly generated *Nco*I site such that the G-protein α subunits are in-frame with the tag. To further facilitate manipulations using these two vectors, various other unique restriction sites are engineered downstream from the NcoI site (see Fig. 9).

Expression and Purification of Histidine-Tagged Proteins

Optimal expression (i.e., cell strain and induction conditions) of histidine-tagged G-protein α subunits occurs under conditions that are identical to those for the nontagged counterparts. Ten-liter cultures of cells harboring H_6pQE-60/Gα are grown, harvested, and lysed as described for the nontagged constructs. Because EDTA and dithiothreitol interfere with chromatography on the Ni^{2+}-containing resin, $T_{50}\beta_{20}P_{0.1}$ [50 mM Tris-HCl, (pH 8.0), 20 mM 2-mercaptoethanol, and 0.1 mM phenylmethylsulfonyl fluoride] is substituted for TEDP. The lysate is centrifuged at 4° in a Beckman Ti 45 ultracentrifuge rotor for 30 min at 30,000 rpm, and the supernatant is collected. (A high-speed spin of the lysate is necessary for the histidine-tagged proteins to bind efficiently to the chelated Ni^{2+} resin.)

The crude supernatant is then applied directly to a 50-ml Ni^{2+} NTA column (Qiagen) that had been equilibrated with $T_{50}\beta_{20}P_{0.1}$ containing 100 mM NaCl. The column is washed with 2.5 column volumes of $T_{50}\beta_{20}P_{0.1}$ containing 500 mM NaCl and 10 mM imidazole, pH 8.0. Proteins are eluted with $T_{50}\beta_{20}P_{0.1}$ containing 100 mM NaCl and 10% glycerol with a 600-ml linear gradient of imidazole (10–150 mM); 80 fractions (7.5 ml each) are collected (Fig. 10). In general, the histidine-tagged proteins have a lower solubility than the nontagged proteins, and glycerol is added to the buffer to prevent precipitation. (Solubility is a particular problem following elution from the Ni^{2+} NTA column.) The pooled sample (typically representing >90% of the eluted protein) is desalted using an Amicon positive pressure ultrafiltration device by dilution with $T_{50}E_1D_2$ [50 mM Tris-HCl (pH 8.0), 1 mM EDTA, and 2 mM dithiothreitol] plus 10% glycerol and was further chromatographed on a Q-Sepharose column, as described above (Fig. 11). The recombinant protein in the peak pool often represents more than 99% of the eluted protein. Despite the greater ease of purification of the histidine-tagged constructs, yields are comparable to those for the nonhistidine-tagged counterparts. Explanations include the decreased solubility and slightly lower levels of expression of the histidine-tagged proteins.

$H_6TEVG\alpha$ is purified using Ni^{2+} affinity chromatography, followed by exchange into TEDP containing 10% glycerol as described above for $H_6G\alpha$. An expression vector encoding TEV protease fused to glutathione S-transferase (GST) was generously provided by Dr. Stephen Johnston (University of Texas Southwestern Medical Center, Dallas). The fusion

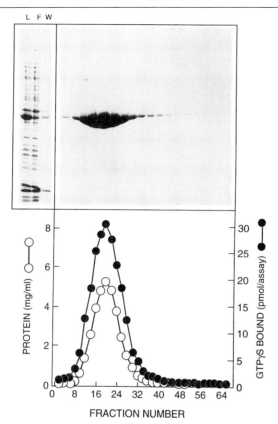

FIG. 10. Chromatography of $H_6G_i\alpha_1$ over Ni^{2+} NTA. Crude supernatant was chromatographed on a Ni^{2+} NTA column as described in the text. Samples of fractions (5 μl) were resolved by SDS–PAGE on an 11% polyacrylamide gel, and the proteins were visualized by staining with Coomassie blue. Also shown are the load, flow-through, and wash in lanes L, F, and W, respectively. Aliquots (1.5-μl) were assayed for protein and GTPγS binding as described in the legend to Fig. 5.

protein is expressed and purified using glutathione-Sepharose 4B resin (Pharmacia). A more detailed description of the construction and uses of this protease is presented elsewhere.[20] Both GST- and histidine-tagged TEV protease are available commercially from Life Technologies (Bethesda Research Laboratories, Gaithersburg, MD). The TEV protease is added to the pooled protein in a ratio of 1 : 20 (w/w), and the mixture is incubated at 30° for 30 min. GDP (50 μM) is added to stabilize the G

[20] T. D. Parks, K. K. Leuther, E. D. Howard, S. A. Johnston, and W. G. Dougherty, *Anal. Biochem.* **216,** 413 (1994).

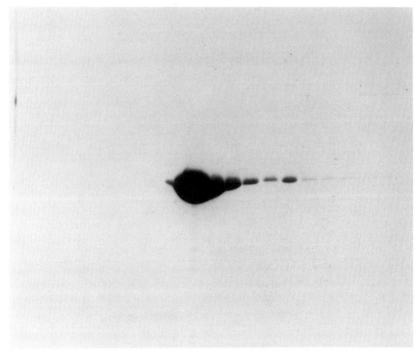

FIG. 11. Q-Sepharose chromatography of recombinant $H_6G_i\alpha_1$. The Ni^{2+} NTA pool was chromatographed on a HiLoad 26/10 Q-Sepharose FPLC column as described in the text. Samples of fractions (5 μl) were resolved by SDS–PAGE in an 11% polyacrylamide gel, and the proteins were visualized by staining with Coomassie blue.

protein during the incubation. The mixture is then chromatographed on a Q-Sepharose column. Elution from Q-Sepharose is the same as described above, except that $T_{50}\beta_{20}$ is substituted for $T_{50}D_2$. Peak fractions are pooled, concentrated, and applied sequentially to Ni^{2+} NTA and glutathione columns (equilibrated with $T_{50}\beta_{20}$) to remove uncleaved $H_6G\alpha$ protein and residual GST–TEV protease. Final yields of the cleaved proteins are typically 25–50% lower than those of the uncleaved proteins. A comparison of the various purified forms of $G_i\alpha_1$ is shown in Fig. 12.

Discussion

Previous work has shown that soluble G-protein α subunits can be expressed in *E. coli*; however, yields were relatively low, and much of

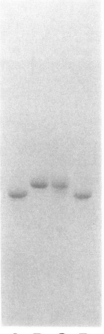

A B C D

FIG. 12. Analysis by SDS–PAGE of $G_i\alpha_1$, $H_6G_i\alpha_1$, $H_6TEVG_i\alpha_1$, and TEV protease-cleaved $H_6TEVG_i\alpha_1$. Samples (2 μg) of purified proteins were resolved in a 9% polyacrylamide gel containing 4 M urea, and the proteins were visualized by staining with Coomassie blue. Lane A, $G_i\alpha_1$; lane B, $H_6G_i\alpha_1$; lane C, $H_6TEVG_i\alpha_1$; and lane D, cleaved $H_6TEVG_i\alpha_1$.

the expressed protein was insoluble. We hypothesized that the proper folding of the newly synthesized recombinant polypeptide was the rate-limiting step in the production of functional protein, rather than the rate of synthesis of the nascent polypeptide. The fact that large amounts of soluble protein could be produced by slow induction over a long period of time supports this idea. Key features of the expression system described above include the use of the regulated expression plasmid pQE-6, low concentrations of inducer (IPTG), slow growth of the culture at 30°, and long induction time. The use of this expression system permits large quantities of highly purified recombinant G-protein α subunits to be obtained quickly without the use of large fermenters. Recombinant α subunits purified using this method are indistinguishable from those described previously. The high accumulation of soluble recombinant protein facilitates purification, and the resultant protein has been used successfully for crystallization.

Acknowledgments

We thank Dr. Henry Barnes for helpful discussions, Christy Jones for excellent technical assistance, and Begonia Ho for assistance in construction of the histidine-tagged vectors. Work from the authors' laboratory was supported by National Institutes of Health Grant GM34497 and GM07062, American Cancer Society Grant BE30-O, The Perot Family Foundation, The Lucille P. Markey Charitable Trust, and The Raymond Willie Chair of Molecular Neuropharmacology.

[13] Synthesis and Applications of Affinity Matrix Containing Immobilized βγ Subunits of G Proteins

By IOK-HOU PANG, ALAN V. SMRCKA, and PAUL C. STERNWEIS

Introduction

The heterotrimeric G proteins are critical for many cellular signal transduction systems. They serve to mediate signaling from a variety of receptors into regulation of downstream effector molecules, such as adenylyl cyclase (adenylate cyclase), cGMP phosphodiesterase, phospholipases, and ion channels.

All G proteins are composed of α, β, and γ subunits. Stimulation of the proteins is effected by exchange of GTP for GDP. This activation can be mimicked, *in vitro,* by the binding of AlF_4^- in concert with GDP. Activation promotes dissociation of α and βγ subunits. Because generic purified complexes of βγ subunits interact with a wide variety of unique α subunits, it is possible to use βγ as an affinity reagent for the study of α subunits. The development of a functional immobilized βγ resin has provided a novel method for isolating and purifying α subunits of G proteins[1,2] and a unique means for studying the interaction between α and βγ subunits.[3]

Synthesis of βγ-Affinity Matrix

The immobilization of βγ subunits on ω-aminobutyl-agarose takes advantage of the relative insensitivity of the function of βγ subunits to modification by maleimides. Thus, the βγ subunits are cross-linked to agarose through their cysteines. The synthesis involves the preparation

---

[1] I.-H. Pang and P. C. Sternweis, *Proc. Natl. Acad. Sci. U.S.A.* **86,** 7814 (1989).
[2] I.-H. Pang and P. C. Sternweis, *J. Biol. Chem.* **265,** 18707 (1990).
[3] M. E. Linder, I.-H. Pang, R. J. Duronio, J. I. Gordon, P. C. Sternweis, and A. G. Gilman, *J. Biol. Chem.* **266,** 4654 (1991).

of the purified $\beta\gamma$ subunits and the actual attachment of the subunits to the matrix with a bifunctional cross-linking reagent.

Preparation of $\beta\gamma$ Subunits

G proteins can be readily isolated from bovine brain.[4] The holoproteins are activated by AlF_4^- to effect dissociation of subunits. The subunits are separated by chromatography through heptylamine-Sepharose or phenyl-Sepharose. Fractions enriched with $\beta\gamma$ can be concentrated by pressure filtration with PM10 filters (Amicon Corporation, Danvers, MA) and stored for future use at $-80°$. Specific and detailed procedures for purification of G proteins and their subunits are given by Sternweis and Pang[5] and Codina *et al.*[6]

Preparation of $\beta\gamma$ for synthesis involves chromatography through hydroxylapatite. This procedure serves as a final purification step for the $\beta\gamma$, removes the sulfhydryl reducing agent dithiothreitol (DTT) in which $\beta\gamma$ is usually purified, and switches the $\beta\gamma$ protein into the desired solution for the synthetic reaction. The detailed procedure follows; all steps are carried out at $0°$ to $4°$.

1. Dilute concentrated preparations of $\beta\gamma$ (~50 mg of concentrated protein) at least 20-fold with solution A [sodium HEPES, pH 7.5, 20 mM; NaCl, 25 mM; sodium cholate, 0.5% (w/v)].

2. Load the diluted solution onto a column of hydroxylapatite (100 ml, Bio-Gel HTP, Bio-Rad, Richmond, CA) which has been equilibrated with solution B (sodium HEPES, pH 7.5, 20 mM; NaCl, 50 mM; sodium cholate, 1%).

3. Wash the column with 100 ml of solution B. Then elute the hydroxylapatite with a 500-ml linear gradient of 0 to 80 mM potassium phosphate, pH 7.5, in solution B. Collect the eluate in 8- to 10-ml fractions. Finally, further elute the column with solution C (sodium HEPES, pH 7.5, 20 mM; NaCl, 50 mM; sodium cholate, 1%; potassium phosphate, pH 7.5, 400 mM).

4. Define the eluted peak of $\beta\gamma$ by assay for protein and by visualization of polypeptide content with silver staining after separation by sodium dodecyl sulfate–polyacrylamide gel electrophoresis (SDS–PAGE). The latter is most important if removal of contaminating proteins is a goal. The $\beta\gamma$ subunits elute in the range of 20 to 60 mM phosphate.

---

[4] P. C. Sternweis and J. D. Robishaw, *J. Biol. Chem.* **259**, 13806 (1984).
[5] P. C. Sternweis and I.-H. Pang, in "Receptor–Effector Coupling: a Practical Approach" (E. C. Hulme, ed.), p. 1. Oxford Univ. Press, Oxford, 1992.
[6] J. Codina, D. J. Carty, L. Birnbaumer, and R. Iyengar, this series, Vol. 195, p. 177.

5. Pool fractions that contain the desired βγ subunits. Concentrate the subunits to approximately 2 mg protein/ml by pressure filtration with an Amicon PM10 membrane (~20-fold concentration). This procedure also partially concentrates the detergent, sodium cholate. The increased concentration of detergent appears to be helpful for more efficient coupling of the protein to the matrix.

Synthesis of βγ-Agarose

The following protocol is designed to synthesize 10 ml of the βγ matrix. The procedure can be adjusted proportionally to accommodate smaller or larger syntheses. During the synthesis, it is important to minimize the exposure of the solutions to air or oxygen. Therefore, all solutions are deaerated prior to use, and nitrogen or argon is used to displace air during incubation.

1. Place 10 ml of ω-aminobutyl-agarose (Sigma, Cat. No. A-6142) in a column that can be easily sealed at both ends (~1.5 × 20 cm). Equilibrate the agarose by washing at room temperature (20° to 22°) with 50 ml of 50 mM sodium phosphate, pH 7. Close the outlet when washing is complete.

2. Dissolve 2 mg (5 μmol) of *m*-maleimidobenzoylsulfosuccinimide ester (sulfo-MBS, Pierce, Rockford, IL, Cat. No. 22312) in 1 ml of 50 mM sodium phosphate, pH 7. Add the sulfo-MBS solution together with 9 ml of 50 mM sodium phosphate, pH 7, to the agarose gel. Displace the air above the agarose suspension with nitrogen or argon. Cap the column and incubate at room temperature for 30 min with gentle rocking to allow thorough mixing. This couples the succinimidyl group of the cross-linker with the amino groups of the substituted agarose.

3. Open the column, drain the reaction solution, and wash the agarose with 50 ml of 50 mM sodium phosphate, pH 7, to remove any remaining free sulfo-MBS.

4. Add 10 ml of the dithiothreitol-free βγ preparation (~2 mg/ml) to the sulfo-MBS-modified agarose. Purge the column with nitrogen or argon, cap, and incubate at room temperature for 60 min with constant gentle rocking. The βγ subunits are immobilized by reaction of their free sulfhydryl groups with the maleimide group of the attached cross-linker.

5. Drain and wash the agarose with 50 ml of solution D (sodium phosphate, pH 7, 50 mM; EDTA, 1 mM; sodium cholate, 1%; NaCl, 300 mM). The presence of detergent and higher ionic strength minimizes nonspecific binding of βγ to the matrix. Collect 5-ml fractions.

6. Determine the amount of protein in the collected fractions. The difference between the total protein applied and the total protein recovered

in the washes is interpreted to be the amount of βγ covalently linked to the matrix.

7. The remaining unreacted active groups on the matrix are quenched by adding 10 ml of solution E (sodium phosphate, pH 7, 50 mM; EDTA, 1 mM; sodium cholate, 1%; 2-mercaptoethanol, 50 mM) and incubation at room temperature for 60 min with gentle rocking.

8. Wash the column with 50 ml of solution F [sodium HEPES, pH 8, 20 mM; EDTA, 1 mM; Lubrol PX, 0.5% (v/v); NaCl, 300 mM; dithiothreitol, 3 mM]. The matrix is stored at 0° to 4° or on ice.

Comments

The optimal concentration of sulfo-MBS for cross-linking is determined empirically. The attachment of βγ to the matrix increases with higher concentrations of cross-linker, but the proportion of immobilized βγ that interacts with α subunits declines at the higher concentrations (Fig. 1). A procedure for assessment of α-interacting sites is presented below. The amount of βγ that is cross-linked to the agarose is linearly proportional to the concentration of βγ present during the conjugation reaction (Fig. 2).

When the synthesis does not work, it is usually due to a failure to immobilize the βγ. This problem is almost always a result of using aged and ineffective cross-linker. Another cross-linker, sulfo-MPB (Pierce, Cat.

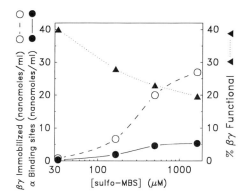

FIG. 1. Effect of the concentrations of sulfo-MBS on conjugation of βγ to butylamine-agarose. Several batches of βγ-agarose were synthesized using various concentrations of sulfo-MBS and 40 nmol of βγ for each milliliter of agarose. The total amount of βγ covalently bound to the matrix was analyzed by assay of protein as described in the text. The amount of functional βγ was determined by measurement of the capacity of the resin to bind α subunits. The ratio of α-subunit binding sites to total amount of βγ conjugated defined the fraction of immobilized βγ that was functional.

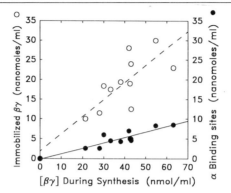

FIG. 2. Effect of the concentration of $\beta\gamma$ during synthesis on the total and functional $\beta\gamma$ immobilized. The plot summarizes 11 syntheses which utilized 500 μM sulfo-MBS and the indicated concentrations of $\beta\gamma$.

No. 22318) has also been used for immobilization by $\beta\gamma$ subunits. Effective concentrations are the same as found for sulfo-MBS. The sulfo-MPB appears to give a more consistent, higher yield of functional sites immobilized. This may be due to the extended chain length between the functional groups of the cross-linker.

The detergent Lubrol PX has become generally unavailable. An effective substitute is polyoxyethylene 10 lauryl ether (Sigma, Cat. No. P9769). Other detergents that allow subunit association and retention of activity may also be used.

Characterization of $\beta\gamma$-Agarose

Determination of α Binding Sites

To determine the binding capacity of the $\beta\gamma$ matrix for an α subunit, an excessive amount of functional α subunits is incubated with the agarose. The amount of α that binds to and is subsequently eluted from the matrix defines the functional α binding sites on the $\beta\gamma$ resin. The exact identity of the α subunits used in this procedure is not critical. Either purified α_o, which is abundant in brain, or a mixture of α subunits will serve well for this characterization. This is accomplished as follows.

1. Place 0.3 ml of $\beta\gamma$-agarose in each of two columns (0.8 cm, i.d.) and wash each with 3 ml of solution G (sodium HEPES, pH 8, 20 mM; EDTA, 1 mM; dithiothreitol, 3 mM; Lubrol PX, 0.1%; NaCl, 100 mM; GDP, 5 μM).

2. To the first sample, add a sufficient quantity of purified α subunits to achieve a stoichiometry of 0.5 to 1 relative to the total βγ attached to the agarose (i.e., if there are 8 nmol of βγ linked to 0.3 ml of agarose, add 4 to 8 nmol of α). This amount of α will be in excess of the functional binding sites on the matrix, because the yield of functional subunits under the conditions used for synthesis is usually about 25% of those which were immobilized (see Figs. 1 and 2). To the second sample of βγ-agarose, add one-tenth the amount of α added to the first sample. In the latter case, the functional βγ should be saturating, and this sample will either verify that all of the added α subunits can interact with βγ or quantify the functional subpopulation of α subunits in a partially active preparation. Sufficient amounts of solution G are added to both samples to bring the final volume of solution to 0.3 ml in addition to the 0.3 ml of agarose. Incubate these at room temperature for 60 min with gentle mixing.

3. Drain and wash the agarose 9 times with 0.3 ml of solution G. Wait 2 to 3 min between washes to collect the initial 0.3 ml and the 9 subsequent washes as fractions 1 through 10.

4. To release the bound α, elute the columns 5 times with 0.3 ml of solution H [sodium HEPES, pH 8, 20 mM; EDTA, 1 mM; dithiothreitol, 3 mM; Lubrol PX, 0.1% (v/v); NaCl, 100 mM; GDP, 5 μM; AlCl$_3$, 30 μM; MgCl$_2$, 50 mM; NaF, 10 mM]. Wait 2 to 3 min between washes to allow time for activation with AlF$_4^-$ and dissociation. Collect as fractions 11 through 15.

5. Determine the content of α subunits by assaying fractions for binding of GTPγS.[7] The total amount of α subunits eluted from the first sample by solution H (containing AlF$_4^-$) quantifies the functional α binding sites on the βγ-agarose. Essentially all of the functional α subunits added to the second sample should bind to the agarose and be eluted only by solution H. Additionally, the α subunits in the various wash and eluate fractions can be visualized by staining after SDS–PAGE.

Note. When making solutions with aluminum and fluoride, it is important to add the sodium fluoride and magnesium chloride prior to aluminum chloride; this will prevent association of Al^{3+} with chelators such as EDTA.

Comparison of Binding Affinities of Free and Immobilized βγ for α Subunits

After the functional α-binding capacity of a batch of βγ-agarose is determined, one can test the relative binding affinity of the immobilized

---
[7] J. K. Northup, M. D. Smigel, and A. G. Gilman, *J. Biol. Chem.* **257**, 11416 (1982).

$\beta\gamma$ for α subunits.[1] This is especially recommended for initial syntheses, or where the matrix is to be used for measuring quantitative aspects of subunit association.

1. Calculate the amount of functional α binding sites in 0.3 ml of $\beta\gamma$-agarose.
2. In six columns, mix 0.3 ml of $\beta\gamma$-agarose with various amounts of purified $\beta\gamma$ such that the concentrations of the free $\beta\gamma$ equal 0, 25, 50, 100, 200, and 400% of the concentration of the functional immobilized $\beta\gamma$.
3. Add a submaximal amount of purified α subunits (10 to 15% of the amount of functional immobilized $\beta\gamma$) to each column. Adjust all samples to the same volume with solution G. Incubate at room temperature for 60 min with gentle mixing.
4. Drain and wash the samples of agarose with 9 aliquots of 0.3 ml of solution G. Then elute 5 times with 0.3 ml of solution H. Collect all fractions as described in the previous section.
5. Analyze the fractions for the content of α subunits by measuring GTPγS-binding activity. The amount of α subunit in the fractions eluted with solution H represents α bound to the matrix. Increasing amounts of free $\beta\gamma$ should reduce the bound α with an IC_{50} approximately equal to the amount of the immobilized functional $\beta\gamma$. An example of such an analysis is shown in Fig. 3.

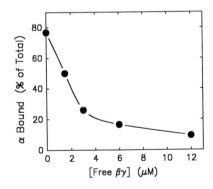

FIG. 3. Characterization of the efficacy of immobilized $\beta\gamma$. Purified α_o subunit (0.25 nmol) was mixed with 0.5 ml of $\beta\gamma$-agarose containing 3 μM α-binding sites and various amounts of free $\beta\gamma$ as indicated. The samples were washed, and α subunits that bound to the $\beta\gamma$-agarose were eluted. The percentage of α_o that bound to the $\beta\gamma$-agarose is shown as the percentage of total α subunit activity recovered from all of the fractions collected for each sample.

Application of $\beta\gamma$-Agarose Affinity Chromatography

Applications which utilize $\beta\gamma$-affinity chromatography take advantage of the relative ease in the manipulation of the affinity between α and $\beta\gamma$ subunits. Early studies have shown that binding of GTP to the α subunits of G proteins decreases their affinity for $\beta\gamma$ subunits and, thus, induces dissociation of the subunits. Moderately rapid hydrolysis of the bound GTP results in inactivation of the protein and promotes reassociation of α with $\beta\gamma$. This limits the utility of GTP as a reagent for elution of α subunits from the $\beta\gamma$-agarose. Nonhydrolyzable analogs of GTP, such as GTPγS, induce prolonged activation of the G-protein subunits and are, therefore, more efficacious agents for the elution of the bound α from the immobilized $\beta\gamma$ subunits. AlF_4^- is another activator of the G proteins. This complex ion, together with Mg^{2+} and the GDP that is already bound to the α subunit, serves functionally as a GTP analog and provides for an increase in the steady-state level of activated and dissociated G-protein subunits. Because the action of AlF_4^- does not require the exchange of nucleotide, it has proved especially useful for activating some of the G proteins, such as G_t and G_q, that demonstrate poor nucleotide exchange when isolated. The differential elution of α subunits from immobilized $\beta\gamma$ due to divergent rates of guanine nucleotide exchange has proved advantageous for the isolation of the $G_q\alpha$ subunits.[2]

We have successfully applied these principles in the following applications. The procedures invariably involve (1) binding of α subunits to the $\beta\gamma$-agarose either by basal exchange in a GDP-containing buffer or after promotion of dissociation of α subunits from accompanying $\beta\gamma$ with AlF_4^-, (2) washing of the matrix with GDP-containing buffers to remove nonassociated proteins, and (3) specific elution of bound α subunits from the matrix with one or a series of activating agents, such as GTP, GTPγS, or AlF_4^-.

Removal of α_o from Mixture of α_o and α_i Subunits

Bovine brain is a rich source of several G proteins, such as G_{i1}, G_{i2}, G_q, and especially G_o; the latter constitutes about 1% of the membrane protein. Conventional column chromatography can easily yield highly purified α_o, but preparations of α_{i1} or α_{i2} generally contain α_o as the most abundant contaminant. Because $\beta\gamma$ subunits have an apparent lower affinity for α_o than for α_{i1} or α_{i2},[8] $\beta\gamma$-agarose can be used as a final

---

[8] P. C. Sternweis, *J. Biol. Chem.* **261,** 631 (1986).

step of purification to remove the α_o subunit and other contaminating polypeptides.[1] The following procedures are performed at 4°.

1. Equilibrate an appropriate amount of $\beta\gamma$-agarose (containing functional α binding sites that equal approximately 50% of the expected total amount of α to be loaded) with solution G.

2. Incubate the mixture of α_i and α_o with $\beta\gamma$-agarose in solution G with constant and gentle mixing for 1 hr. Place the gel suspension in a column.

3. Wash the column with 2 to 5 bed volumes of solution G. Elute α_o with 10 to 20 bed volumes of solution I (sodium HEPES, pH 8, 20 mM; EDTA, 1 mM; dithiothreitol, 3 mM; Lubrol PX, 0.1%; NaCl, 100 mM; GTP, 5 μM). GTP is sufficient to dissociate most of the α_o but is not very effective for eluting α_i subunits.

4. Elute α_i with 10 to 20 bed volumes of solution J (sodium HEPES, pH 8, 20 mM; EDTA, 1 mM; dithiothreitol, 3 mM; Lubrol PX, 0.1%; NaCl, 100 mM; GDP, 5 μM; AlCl$_3$, 30 μM; MgCl$_2$, 10 mM; NaF, 10 mM). Release of remaining α_i can be effected with a final elution using solution 1 but with the concentrations of MgCl$_2$ and Lubrol increased to 50 mM and 0.5%, respectively.

5. The collected fractions are quantified by assay of GTPγS-binding activity and analysis of the subunit profiles by SDS–PAGE and staining.

A modification of this procedure is useful as the final purification step of other nonabundant α subunits such as α_q.[9,10] The $\beta\gamma$-affinity matrix is effectively used to remove any remaining contaminants after conventional chromatographic separations.

Isolation of α Subunits from Solubilized Membranes

Affinity chromatography with $\beta\gamma$-agarose is an expedient and specific technique to isolate and purify α subunits from detergent extracts of membranes from various tissues or cultured cell lines. The general steps are very similar to those of the previous application. To ensure isolation of a maximal amount of α subunits, the use of an activation–inactivation cycle during loading of the extracts onto the $\beta\gamma$ matrix is recommended. This cycle involves the activation of G proteins by AlF$_4^-$ to induce dissociation of the α subunits from the accompanying $\beta\gamma$. In the presence of the $\beta\gamma$-agarose, inactivation is then achieved by adding EDTA (which removes free Mg^{2+} and Al^{3+} by chelation) to the mixture. This activation step should improve the extent of exchange of α subunits from endogenous

---

[9] A. V. Smrcka, J. R. Hepler, K. O. Brown, and P. C. Sternweis, *Science* **251**, 804 (1991).
[10] A. V. Smrcka and P. C. Sternweis, *Methods Neurosci.* **18**, 72 (1993).

free $\beta\gamma$ to the immobilized $\beta\gamma$. In addition, a saturating amount of immobilized $\beta\gamma$ relative to the expected amount of α subunits is required. Generally, this can be largely achieved by use of a 5- to 10-fold excess of functional, immobilized $\beta\gamma$ relative to the endogenous $\beta\gamma$ contained in the extracts. Examples using the following procedures have been published.[1]

1. Extract plasma membranes at 5–10 mg protein/ml with solution K (sodium HEPES, 20 mM, pH 8; EDTA, 1 mM; dithiothreitol, 1 mM; NaCl, 100 mM; GDP, 5 μM; Lubrol PX, 1%). Combine the extract with an appropriate amount of $\beta\gamma$-agarose. To activate the G proteins, add sufficient amounts of $AlCl_3$, $MgCl_2$, NaF, and GDP to achieve final concentrations of 30 μM, 6 mM, 10 mM, and 5 μM, respectively. Solution H can be used to adjust to higher volumes if desired. Incubate at room temperature for 1 hr with gentle mixing.

2. Add EDTA to a final concentration of 20 mM for inactivation. Incubate the suspension at room temperature for 1 hr with gentle mixing.

3. Place the mixture in a column of appropriate size. Wash with a solution containing GDP such as solution G (9–10 bed volumes) and elute with solutions containing AlF_4^- such as solution H (5 bed volumes) as described above, except at room temperature.

This procedure can be modified for the isolation of α_q or other α subunits that have a slow rate of guanine nucleotide exchange.[2] An elution with a GTPγS-containing solution, such as solution G containing GTPγS rather than GDP, will differentially release the α subunits that have faster rates of exchange for guanine nucleotides. A second elution with solution H (AlF_4^-) will then elute α subunits with poor rates of nucleotide exchange. Neutral detergents (such as Lubrol) have yielded the best results for this procedure. More efficient extraction of G proteins can be obtained with 25 mM sodium cholate. Such extracts can be diluted with solution H for application to the matrix.

Study of Interactions among G Protein Subunits

Because immobilized $\beta\gamma$ subunits provide an easy means for separation of α subunits which are free or associated with $\beta\gamma$, the affinity matrix is obviously a helpful tool in studying subunit interactions. Many interesting aspects of the interactions can be explored. For example, this technique can be used to explore the functional behavior of expressed proteins (either wild type or mutated). Thus, Linder *et al.*[3] used association with $\beta\gamma$-agarose to provide evidence that myristoylation of α_o increased the affinity of α for $\beta\gamma$.

The matrix should also be useful for investigating the effects of ions or potential modulatory proteins on the interaction between α and βγ. It should be feasible to quantify the kinetics of subunit interaction by direct measurement of rates of association and dissociation using radiolabeled α subunits. Finally, the βγ matrix will be helpful in evaluating the specificity of α and βγ subunit interaction. This can be achieved by experiments measuring competition between radiolabeled α with various purified α subunits for binding to the βγ-agarose. The potencies in displacement will define the relative affinities for the immobilized βγ. By comparing affinity profiles obtained from different batches of βγ-agarose synthesized with purified βγ from various tissues or organisms or with specific species of β and γ obtained from expression systems, one can potentially determine the specificity of the interaction of a particular α with specific βγ subunits.

Acknowledgments

We thank Elizabeth Nowak and Steve Gutowski for superb technical assistance in the development of this technique. This work was made possible by funding from the National Institute of General Medical Science (Grant GM31954) and salary support from the American Heart Association.

[14] Purification of Activated and Heterotrimeric Forms of G_q Proteins

By JONATHAN L. BLANK and JOHN H. EXTON

Introduction

Guanine nucleotide-binding proteins (G proteins) of the G_q family transduce the signal from Ca^{2+}-mobilizing receptors to β-isozymes of phosphoinositide phospholipase C.[1-3] The products of phosphatidylinositol 4,5-bisphosphate (PIP_2) hydrolysis, namely, inositol 1,4,5-trisphosphate and 1,2-diacylglycerol, cause the release of Ca^{2+} from internal stores and activate protein kinase C, respectively. This results in the phosphorylation of regulatory proteins and the generation of physiological responses.

[1] S. J. Taylor, H. Z. Chae, S. G. Rhee, and J. H. Exton, *Nature* (*London*) **350**, 516 (1991).
[2] A. V. Smrcka, J. R. Hepler, K. O. Brown, and P. C. Sternweis, *Science* **251**, 804 (1991).
[3] M. Strathman and M. Simon, *Proc. Natl. Acad. Sci. U.S.A.* **87**, 9113 (1990).

Preparation of Activated G_q Proteins

Activation and Solubilization of $\alpha_{q/11}$. The α subunits of G_q and G_{11}, designated $\alpha_{q/11}$ were the first proteins of the G_q family to be isolated. The procedure first involves the activation of the G proteins present in bovine liver plasma membranes.[4] Membranes are prepared at 4° by a modification[4] of the method of Prpic et al.[5] Bovine liver (1.5 kg) is homogenized in 6 liters of isolation medium (5 mM HEPES, pH 7.5, 1 mM EGTA, 250 mM sucrose) using a Waring blendor. The homogenate is centrifuged at 350 g for 10 min, and the supernatant is collected, brought up to 6 liters with isolation medium, and recentrifuged at 1500 g for 10 min. The pellets are combined and resuspended with thorough mixing in 6.5 liters of isolation medium containing 11.9% (v/v, final) Percoll (Pharmacia LKB Biotechnology Inc., Piscataway, NJ). The mixture is centrifuged at 13,700 g for 90 min, and the membrane layer, which lies just below the surface of the self-forming Percoll gradient, is collected using a pipette. Membranes (~1 liter) are diluted to 5 liters with washing buffer [25 mM HEPES, pH 7.5, 1 mM EGTA, 0.5 mM dithiothreitol (DTT)] and centrifuged at 13,700 g for 20 min. The pellets are combined, resuspended in approximately 800 ml of washing buffer, and the membrane fraction collected by centrifugation at 13,700 g for 20 min. Membranes from 3 kg bovine liver are incubated at 30° for 30 min in buffer A (25 mM HEPES, pH 7.5, 1 mM EGTA, 10 mM MgCl$_2$, 0.5 mM DTT, 10 μg/ml leupeptin, 10 μg/ml aprotinin) containing 50 μM guanosine 5'-o-(thio)triphosphate (GTPγS), 3 μM [Arg^8]vasopressin, 300 μM 5'-adenylyl imidodiphosphate, and 50 mM NaCl. Buffer A containing 50 mM NaCl and 10% (w/v) sodium cholate (Calbiochem, La Jolla, CA) is added to give a final concentration of 0.8% (w/v) cholate, and the mixture is kept at 4° with agitation for 1 hr before centrifugation at 100,000 g for 60 min.

Chromatography on Heparin-Sepharose/Q-Sepharose. This and subsequent steps are performed at 4° using a fast protein liquid chromatograph (FPLC) system (Pharmacia). The supernatant (~640 ml) is loaded at 2.5 ml/min on a 160 ml (30 × 2.6 cm) heparin-Sepharose column connected in series with a 75 ml (14 × 2.6 cm) Q-Sepharose fast flow column. Both columns are equilibrated with buffer A containing 0.8% cholate (buffer B) and 50 mM NaCl, and are washed with this buffer until the A_{280} returns to baseline. The columns are then disconnected, and PIP$_2$–phospholipase C is eluted from the heparin-Sepharose column and may be further purified as described.[6] The Q-Sepharose column is developed with a linear gradient

---
[4] S. J. Taylor, J. A. Smith, and J. H. Exton, *J. Biol. Chem.* **265**, 17150 (1990).
[5] V. Prpic, K. C. Green, P. F. Blackmore, and J. H. Exton, *J. Biol. Chem.* **259**, 1382 (1984).
[6] K. Shaw and J. H. Exton, *Biochemistry* **31**, 6347 (1992).

in buffer B of 0.05–0.43 M NaCl over 480 ml and then 0.43–1 M NaCl over 120 ml. The flow rate is 4 ml/min and fraction size 12 ml. Fractions are pooled on the basis of activation of the β_1-isozyme of PIP$_2$–phospholipase C. The purification of this enzyme from bovine liver and brain are described elsewhere.[4,6,7]

Phospholipase C Activation Assay. The phospholipase assay is based on that described by Taylor and Exton.[8] The substrate, 100 μM [^{3}H]PIP$_2$ [400 counts/min (cpm)/nmol] from New England Nuclear (Boston, MA), is incorporated into lipid vesicles with PIP$_2$, phosphatidylethanolamine (PE), and phosphatidylserine (PS) in a molar ratio of 1:4:1. The PIP$_2$ can be prepared from mixed phosphoinositides (Sigma, St. Louis, MO) by chromatography on neomycin-linked glass beads (glyceryl-CPG-240 Å, 200–400 mesh, Fluka, Ronkonkoma, NY) as described by Schacht.[9] The PE (bovine liver) and PS (bovine brain) are supplied by Avanti Polar Lipids (Birmingham, AL). The assay is performed at 37° for 15 min or less in a final volume of 200 μl. The phospholipids are dried under a flow of nitrogen and prepared at twice the desired final concentration by sonication into assay buffer containing 75 mM HEPES, pH 7.0, 150 mM NaCl, 4 mM EGTA, and 1 mg/ml bovine serum albumin (BSA). Forty microliters of a 20 mM MgCl$_2$/5 mM CaCl$_2$ solution is added to each assay tube containing 100 μl of the 2× substrate mixture. The CaCl$_2$/EGTA buffer system maintains the final free Ca^{2+} concentration at 220 nM, as determined by the COMICS program.[10] Either 40 μl of water or 500 μM GTPγS can then be added, depending on whether preactivated or trimeric G$_{q/11}$ is to be assayed. In general, activation is determined by adding column fractions to the mixture prior to addition of phospholipase with a total volume of 20 μl and a final concentration of 0.08% (w/v) cholate [or 0.1% (w/v) β-octylglucoside]. The reaction is terminated by adding 200 μl of 10% (w/v) trichloroacetic acid, followed by 100 μl of 1% (w/v) BSA. After 5 min on ice, the mixture is centrifuged (900 g for 4 min), and the radioactivity in 400 μl of supernatant is determined.

Sephacryl S-300 Gel Filtration and Octyl-Sepharose Chromatography. Active fractions are pooled and concentrated to 20 ml using an Amicon (Danvers, MA) stirred cell with a YM30 membrane. This pool is applied to a 400 ml Sephacryl S-300 column (82 × 2.5 cm) equilibrated with buffer B containing 100 mM NaCl at a flow rate of 2 ml/min. Fractions (5 ml) pooled on the basis of phospholipase activation (~30 ml) are diluted with

---

[7] S. H. Ryu, K. S. Cho, K. Y. Lee, P. G. Suh, and S. G. Rhee, *J. Biol. Chem.* **262**, 12511 (1987).
[8] S. J. Taylor and J. H. Exton, *Biochem. J.* **248**, 791 (1987).
[9] J. Schacht, *J. Lipid Res.* **19**, 1063 (1978).
[10] O. D. Perrin and I. G. Sayce, *Talanta* **14**, 833 (1967).

2 volumes of buffer A containing 400 mM NaCl and loaded (2 ml/min) onto octyl-Sepharose (60 ml, 30 × 1.6 cm) equilibrated with buffer A containing 0.26% cholate and 300 mM NaCl. The column is washed with this buffer and developed with a 600-ml linear double gradient of cholate (0.26–1.2%) and NaCl (300–0 mM). It is then washed with 40 ml of buffer A containing 1.2% cholate. A broad peak of stimulatory activity is eluted and concentrated to approximately 3 ml using a YM30 membrane.

Mono Q Chromatography. The concentrate is diluted 1 : 4 with buffer C [25 mM HEPES, pH 7.25, 5 mM MgCl$_2$, 1 mM EGTA, 0.5 mM DTT, 1% (w/v) β-octylglucoside (Calbiochem), 10 μg/ml leupeptin, and 10 μg/ml aprotinin] and applied at 0.8 ml/min to a Mono Q HR 5/5 column equilibrated with buffer C containing 50 mM NaCl. After washing with buffer C containing 80 mM NaCl, the column is developed with 80–316 mM NaCl in 24 ml of buffer C and fractions of 0.6 ml collected. The leading (pool A) and trailing (pool B) edges of the peak of activity are pooled separately and each subjected to ADP-ribosylation[11] by pertussis toxin (PTX) (List Biological Laboratories, Campbell, CA), to remove α_i contamination. The pools are diluted 1 : 1 in incubation mix to give final concentrations of 25 mM HEPES, pH 7.25, 100 μM ATP, 10 μM GTPγS, 2 mM NAD$^+$, 10 μg/ml leupeptin, 10 μg/ml aprotinin, and 10 μg/ml DTT-activated PTX. Following incubation at 30° for 20 min, the pools are diluted 1 : 1 with buffer C and rechromatographed on Mono Q as described above. The peak of activity on Mono Q (B) can be pooled in its entirety. The peak of activity from Mono Q (A) is divided symmetrically into two pools (leading half, pool A1; trailing half, pool A2). This procedure allows partial resolution of α_{11} (pool A1) and α_q (pool B) as determined by Western blot analysis using antisera, for example, W082 and E976 which specifically recognize α_q and α_{11}, respectively, and W083 and X384 which recognize both α_q and α_{11} (Fig. 1).[12]

Sodium Dodecyl Sulfate–Polyacrylamide Gel Electrophoresis. Using the following procedure, α_q and α_{11} can be electrophoretically resolved in 13% polyacrylamide–sodium dodecyl sulfate (SDS) gels and display apparent molecular weights of 42,000 and 43,000, respectively.

Solution A: 30% (w/v) acrylamide, 0.8% (w/v) bisacrylamide (supplied by Protogel, National Diagnostics, Atlanta, GA)
Solution B: 1% (w/v) Tris–base, 0.33% (w/v) SDS, 4.8% (w/v) glycine
Solution C: 1.29% (w/v) Tris-HCl, 4.75% (w/v) imidazole, 0.33% (w/v) SDS; adjust to pH 6.8 with concentrated HCl

---

[11] T. Sunyer, B. Monastrinsky, J. Codina, and L. Birnbaumer, *Mol. Endocrinol.* **3**, 1115 (1988).
[12] S. J. Taylor and J. H. Exton, *FEBS Lett.* **286**, 214 (1991).

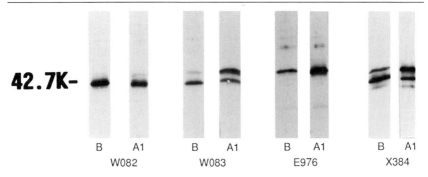

FIG. 1. Partial resolution of GTPγS-activated α_q and α_{11} by Mono Q chromatography. (Reproduced from Taylor and Exton[12] with permission of The American Society for Biochemistry and Molecular Biology.)

Solution D: 0.2% (w/v) Tris–base, 0.96% (w/v) imidazole, 3.33% (w/v) SDS, 5.33% (w/v) sucrose, 0.02% (w/v) bromphenol blue

All solutions are filtered after preparation. The 13% polyacrylamide resolving gel is prepared by mixing 13 ml solution A, 10 ml solution B, and 7 ml water. This solution is degassed for 10 min and polymerized by the addition of 40 μl N,N,N',N'-tetramethylethylenediamine (TEMED) and 160 μl of 10% (w/v) ammonium persulfate immediately prior to casting the gels. This is sufficient for 6 minigels of resolving gel dimensions 83 × 50 × 0.75 mm ($W \times H \times D$). Gels are allowed to polymerize for at least 45 min under 0.1% (w/v) SDS.

The stacking gel is prepared by mixing 1.6 ml of solution A, 4 ml of solution C, and 6.4 ml water. The stacking gel solution is degassed and polymerized with 16 μl TEMED and 100 μl of 10% (w/v) ammonium persulfate. The SDS is removed from the resolving gel, and the stacking gel poured to the top of the inner glass plate, overall dimensions 83 × 73 × 0.75 mm ($W \times H \times D$), and the comb inserted.

Sample buffer (2×) is prepared by first mixing 20 ml glycerol, 24 ml solution B, and 6 ml 45 mM EDTA, and then adding 50 ml of solution D. Sample buffer is stored at $-20°$, and 50 μl of 1 mM DTT is added to a 1-ml aliquot prior to use. Samples are diluted 1:1 with this buffer and heated for 5 min at 100°.

Gels are electrophoresed at 200 V using solution B diluted 1:2 (v/v) with water as reservoir buffer. Gels are run until the prestained ovalbumin 47-kDa marker (Bio-Rad, Richmond, CA) has migrated halfway through the resolving gel (2.5 cm, ~1.5 hr).

Preparation of Heterotrimeric G Proteins

Membranes prepared from 1.5 kg of bovine liver are resuspended in 25 mM HEPES, pH 7.5, containing 1 mM EGTA, 0.5 mM DTT, 10 μg/ml leupeptin, 10 μg/ml aprotinin, 1% cholate (w/v, final concentration), and 50 mM NaCl.[13] After mixing for 1 hr at 4°, the detergent extract is collected by centrifugation at 100,000 g for 1 hr.

Heparin/DEAE-Sephacel Chromatography. Membrane extract (300–500 ml) is applied at 4 ml/min to a 140-ml heparin-Sepharose column (2.6 × 26 cm) connected in series with a 300-ml DEAE-Sephacel column (5 × 15 cm), equilibrated with buffer D (20 mM Tris-HCl, pH 8.0, 1 mM EDTA, 1 mM DTT) containing 0.8% (w/v) cholate and 50 mM NaCl. After washing overnight with 600 ml of buffer D, the columns are disconnected, and the DEAE-Sephacel is washed at 5 ml/min with a further 600 ml of this buffer. The column is eluted with a 1000-ml linear gradient from 50 to 300 mM NaCl in buffer D containing 0.8% (w/v) cholate. Fractions of 10 ml are collected. Guanine nucleotide-dependent activation of the β_1 isozyme of phospholipase C by DEAE-Sephacel fractions is determined by performing the phospholipase C assay described above in the presence and absence of 100 μM GTPγS.

Octyl-Sepharose Chromatography. Fractions that reconstitute GTPγS-dependent activation of purified brain phospholipase C-β_1 are pooled (200–300 ml), diluted with 4 volumes of buffer D containing 300 mM NaCl, and applied overnight to a 60-ml column (1.6 × 30 cm) of octyl-Sepharose equilibrated in buffer D containing 300 mM NaCl and 0.3% (w/v) cholate. The column is washed with 100 ml of the column equilibration buffer and eluted with a 600-ml dual gradient in buffer D of increasing cholate from 0.3 to 1.5% (w/v) and decreasing NaCl from 300 to 0 mM. The flow rate is 2 ml/min, and 6-ml fractions are collected.

Hydroxylapatite Chromatography. Peak fractions are pooled (140–180 ml), taking care to avoid the leading part of the peak that is contaminated by G_i species. The pool is applied to a 24-ml column (1 × 30 cm) of high-performance liquid chromatography (HPLC)-grade hydroxylapatite (Calbiochem) equilibrated in 20 mM Tris-HCl, pH 8.0, containing 0.1 mM EDTA, 1 mM DTT, 0.8% (w/v) cholate, 100 mM NaCl, and 5 mM K_2HPO_4. After washing with 50 ml of this buffer, the column is developed with a linear gradient of K_2HPO_4 from 5 to 100 mM in 360 ml of buffer. Fractions of 3.6 ml are collected at a flow rate of 2 ml/min.

---

[13] J. L. Blank, A. H. Ross, and J. H. Exton, *J. Biol. Chem.* **266**, 18206 (1991).

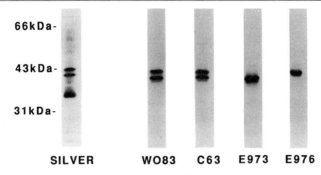

FIG. 2. Western blot analysis of heterotrimeric $G_{q/11}$.

Mono Q Chromatography. The leading fractions from the peak of GTPγS-dependent activity are pooled, thereby removing G_s from $G_{q/11}$ at this step. The pool (~30 ml) is diluted with 2 volumes of buffer E [20 mM Tris-HCl, pH 7.5, 1 mM EDTA, 1 mM DTT, 1% (w/v) β-octylglucoside] and applied to a Mono Q HR 5/5 column equilibrated in buffer E containing 80 mM NaCl. The flow rate for Mono Q chromatography is 1 ml/min. The column is washed with buffer E containing 80 mM NaCl until the A_{280} returns to baseline. The column is eluted with a linear gradient from 80 to 300 mM NaCl in 25 ml of buffer E, followed by a 4-ml gradient to 1 M NaCl. Fractions of 0.5 ml are collected.

Sephacryl S-300 Gel Filtration. Fractions containing the peak of activity (5–9 ml) are pooled and applied to a 335-ml Pharmacia HiLoad 26/60 Sephacryl S-300 HR column (2.6 × 63 cm) equilibrated in buffer D containing 0.8% (w/v) cholate and 100 mM NaCl. The column is eluted at 0.75 ml/min, and fractions of 4 ml are collected. The peak of activity elutes in 4 to 6 fractions which are pooled, divided into aliquots, and frozen in liquid nitrogen. The pool may be concentrated to less than 1 ml using a Centricon 30 cartridge (Amicon) prior to storage. The G proteins are stable for several months at −70°.

This procedure is usually performed over 6 days and gives a mixture of trimeric G_q and G_{11} in roughly equimolar proportions. The yield of proteins from this procedure has been improved from that described previously[13] by using 1% (w/v) cholate in the membrane extraction buffer. Final yields of α subunits vary between preparations, ranging from 14 to 67 μg α_q (33 ± 9 μg, mean ± SEM, $n = 6$) and 9 to 53 μg α_{11} (27 ± 7 μg, mean ± SEM, $n = 6$). The β-subunit content of the preparation is stoichiometrically equivalent to the total amount of α_q and α_{11}, indicating the isolation of the G proteins as heterotrimers.

In the absence of receptor, the $G_{q/11}$ preparation does not detectable GTPγS binding[13,14] and therefore cannot be quantified on this basis. Instead, the α-subunit content is routinely determined by densitometric scanning of silver-stained gels, using purified β subunit as a reference protein. Quantification of $G_{q/11}$ is also possible by measuring the total amount of stoichiometrically bound GDP to α_q and α_{11}, as described by Berstein et al.[14] Western blot analysis (Fig. 2) of the preparation using antisera raised to peptide sequences unique to α_q (E973) and α_{11} (E976) or common to α_q and α_{11} (W083 and C63) identify the 42- and 43-kDa proteins as α_q and α_{11}, respectively. This has been confirmed by sequence analysis of tryptic digests (J. L. Blank, D. Runge, C. Slaughter, and J. H. Exton, unpublished results).

Discussion

Preactivation of liver plasma membranes by GTPγS allows for partial resolution of α_q and α_{11} on Mono Q by the procedure described above, whereas attempts to resolve heterotrimeric G_q and G_{11} have been unsuccessful. However, preparation of G_q and G_{11} in the unresolved trimeric form has several advantages over purification of the GTPγS-liganded α subunits. The yield of GTPγS-liganded α subunits is always lower,[4] partly due to their incomplete activation prior to extraction from the plasma membrane. The preactivated α subunits of $G_{q/11}$ cannot be isolated from those of G_i without covalent modification by PTX treatment prior to Mono Q chromatography. This procedure requires complete ADP-ribosylation, and it is expensive and inconvenient. Chromatography of trimeric $G_{q/11}$ on octyl-Sepharose and hydroxylapatite removes contaminating G_i and G_s species, respectively, obviating the need for toxin-dependent ADP-ribosylation. The binding of GTPγS to the α subunits is essentially irreversible, and the preactivated preparation is, therefore, unsuitable for reconstitution studies with purified receptors. Preparation of the heterotrimers is therefore necessary in order to study guanine nucleotide binding and GTPase activities of $G_{q/11}$ that are regulated by receptors and phospholipases.[14,15]

---

[14] G. Berstein, J. L. Blank, A. V Smrcka, T. Higashijima, P. C. Sternweis, J. H. Exton, and E. M. Ross, *J. Biol. Chem.* **267**, 8081 (1992).

[15] G. Berstein, J. L. Blank, D.-Y. Jhon, J. H. Exton, S. G. Rhee, and E. M. Ross, *Cell (Cambridge, Mass.)* **70**, 411 (1992).

[15] Purification of Phospholipase C-Activating G Protein, G_{11}, from Turkey Erythrocytes

By GARY L. WALDO, JOSÉ L. BOYER, and T. KENDALL HARDEN

Introduction

The inositol lipid signaling cascade represents a major second messenger pathway that involves at least three molecular components.[1] Cell surface receptors for a broad range of stimuli couple to the activation of phospholipase C-β family isozymes through members of the G_q class of guanine nucleotide-binding proteins (G proteins).[2-6] Based on their essentially ubiquitous distribution, G_q and G_{11} apparently are the most prominent G-protein activators of phospholipase C (PLC), although G_{14} and G_{16} also are likely important for activation in a more restricted number of tissues.

Only G_{11} and G_q have become available in purified form, and purification of G_{14} and G_{16} has not yet been reported. Although expression and purification of recombinant $G\alpha_q$, $G\alpha_{11}$, $G\alpha_{14}$, and $G\alpha_{16}$ will represent major advances for the study of G-protein-regulated PLC, a need for purified native proteins of the G_q class will remain. Turkey erythrocytes are an easily obtained homogeneous cell type which apparently primarily express G_{11} but not the similar (88% homologous) G_q.[4,7,8] These cells thus provide an excellent starting source for purification of a member of the G_q family. The turkey erythrocyte offers the further advantage that other protein components of the phosphoinositide-specific signal transduction pathway, including a G-protein-regulated PLC and G-protein $\beta\gamma$ subunits, can be purified by relatively straightforward procedures.[9-11] The purifica-

[1] M. J. Berridge, Annu. Rev. Biochem. **56,** 159 (1987).
[2] S. J. Taylor, J. A. Smith, and J. H. Exton, J. Biol. Chem. **265,** 17150 (1990).
[3] A. V. Smrcka, J. R. Hepler, K. O. Brown, and P. C. Sternweis, Science **251,** 804 (1991).
[4] G. L. Waldo, J. L. Boyer, A. J. Morris, and T. K. Harden, J. Biol. Chem. **266,** 14217 (1991).
[5] M. I. Simon, M. P. Strathmann, and N. Gautam, Science **252,** 802 (1991).
[6] S. G. Rhee and K. D. Choi, J. Biol. Chem. **267,** 12393 (1992).
[7] D. H. Maurice, G. L. Waldo, A. J. Morris, R. A. Nicholas, and T. K. Harden, Biochem. J. **290,** 765 (1993).
[8] G. Milligan, this volume [21].
[9] A. J. Morris, G. L. Waldo, C. P. Downes, and T. K. Harden, J. Biol. Chem. **265,** 13501 (1990).
[10] J. L. Boyer, G. L. Waldo, T. Evans, J. K. Northup, C. P. Downes, and T. K. Harden, J. Biol. Chem. **264,** 13917 (1989).
[11] J. L. Boyer, G. L. Waldo, and T. K. Harden, J. Biol. Chem. **267,** 25451–25456 (1992).

tion of avian G_{11} described here does not utilize $\beta\gamma$-affinity chromatography, a highly effective but specialized method of G-protein α-subunit purification that is not available to all laboratories.[12] Rather, the procedure relies on conventional chromatographic techniques and monitors the capacity of column fractions to reconstitute AlF_4^- sensitivity to purified turkey erythrocyte PLC.

Reconstitution Assay

The reconstitution assay is performed in 12 × 75 mm conical polypropylene tubes. Prepare substrate solution by mixing at 4° 10 nmol phosphatidylserine (PS), 40 nmol phosphatidylethanolamine (PE), and 10 nmol phosphatidylinositol 4,5-bisphosphate [PtdIns(4,5)P_2] with 0.02–0.03 μCi of Ptd[^{3}H]Ins(4,5)P_2 per assay. Dry the phospholipid mixture under nitrogen and resuspend with a probe type sonicator in 50 μl/assay of cold buffer containing 20 mM 4-(2-hydroxyethyl)-1-piperazineethanesulfonic acid (HEPES), pH 7.4, 1 mM MgCl$_2$, 100 mM NaCl, 1 mM dithiothreitol (DTT), 0.1 mM benzamidine, and 0.1 mM phenylmethylsulfonyl fluoride (PMSF). Thus, for 100 assays combine in a single tube 1 μmol PS, 4 μmol PE, 1 μmol PtdIns(4,5)P_2, and 2–3 μCi Ptd[^{3}H]Ins(4,5)P_2. Dry the lipids as described above and resuspend in 5 ml of cold buffer.

The activity of each G-protein sample is measured in the absence of PLC (blank), in the presence of 10–20 ng turkey erythrocyte PLC (vehicle), and in the presence of 10–20 ng turkey erythrocyte PLC and 20 μM AlCl$_3$ and 10 mM NaF (AlF$_4^-$). Therefore, for each G-protein sample combine in a conical polypropylene microcentrifuge tube on ice 0.325 ml resuspended lipids and 6.5 μl of G-protein sample, mix, and transfer 50 μl to duplicate assay tubes for blank, vehicle, and AlF$_4^-$ which contain 10 μl of 10 mM HEPES or 10 μl of a 10× AlF$_4^-$ solution (200 μM AlCl$_3$ and 100 mM NaF) in 10 mM HEPES. Add 25 μl of 4× assay buffer [150 mM HEPES, pH 7.0, 300 mM NaCl, 16 mM MgCl$_2$, 8 mM EGTA, 7.2 mM CaCl$_2$, and 2 mg/ml fatty acid-free bovine serum albumin (BSA)]. Add 15 μl of 10 mM HEPES or 15 μl of 10 mM HEPES containing 10–20 ng turkey erythrocyte PLC or purified bovine brain PLC-β (see Discussion), mix, transfer the samples to a 30° bath, and incubate for 10 min. The final assay volume is 100 μl containing (final concentrations) 50 mM HEPES, pH 7.0, 5 mM MgCl$_2$, 125 mM NaCl, 2 mM EGTA, 0.5 mM DTT, 0.5 mg/ml BSA, 100 μM PS, 400 μM PE, 100 μM PtdIns(4,5)P_2, 15,000–20,000 counts/min (cpm) Ptd[^{3}H]Ins(4,5)P_2, 1.8 mM CaCl$_2$ to give approximately 3 μM free Ca^{2+}, and G-protein-containing sample (1 μl/assay). The reac-

[12] I. H. Pang and P. C. Sternweis, *Proc. Natl. Acad. Sci. U.S.A.* **86**, 7814 (1989).

tion is terminated by the addition of 0.375 ml of $CHCl_3:CH_3OH:HCl$ (40:80:1, v/v) followed by 0.125 ml $CHCl_3$ and 0.125 ml of 0.1 M HCl. The samples are mixed and phases separated by centrifugation for 5 min at 3250 g at room temperature (Beckman J-6, 4000 rpm). Four hundred microliters of the upper aqueous phase is removed and the water-soluble [^{3}H]Ins(1,4,5)P_3 quantitated in a liquid scintillation counter.

Comments. A description of the preparation of radioactive and nonradioactive polyphosphoinositide substrates is given elsewhere.[13] The assay is linear from 2 to 30 min under conditions where less than 15% of the substrate is hydrolyzed. PtdIns(4)P may be substituted for PtdIns(4,5)P_2. The substrate concentration may be varied between 25 and 200 μM, always in a 1:4:1 molar ratio of PS:PE:PtdIns(4,5)P_2 or PtdIns(4)P. Optimal divalent cation concentrations are 5–6 mM Mg^{2+} and 0.3–30 μM free Ca^{2+}.

The volume of G-protein sample may range from 0.2 to 2 μl/assay. Do not exceed a final concentration of 0.02% cholate in the assay. Bovine brain PLC-β (β_1) may be substituted for turkey erythrocyte PLC in the reconstitution assay for the purification of $G\alpha_{11}$ (see Discussion). Finally, low levels of activity are occasionally observed if volumes of G protein must exceed 2 μl/assay. In such instances, protocols utilizing Sephadex G-50 columns or dialysis may be preferred to form G-protein-containing phospholipid vesicles from lipids dispersed in a sodium cholate-containing buffer.[4,11]

Preparation of Turkey Erythrocyte Plasma Membranes

Turkey erythrocyte plasma membranes for the purification of G_{11} may be prepared from the same blood used for the preparation of cytosol for the purification of G-protein-regulated PLC.[13] A flowchart for the preparation of turkey erythrocyte plasma membranes is presented in Fig. 1. All steps are at 4°. Prepare washed erythrocytes from 40 liters of whole blood and disrupt the erythrocytes by nitrogen cavitation exactly as described in Ref. 13. Centrifuge the disrupted cells at 3250 g for 10 min with maximum brake (Beckman J-6, 4000 rpm). Carefully decant and save the resulting dark red supernatant, which contains the plasma membranes, and discard the predominately nuclear pellet. Centrifuge the low-speed supernatant at 28,000 g for 20 min with the brake set at half-maximum (Beckman JA-14, 13,500 rpm). Centrifugation of the membrane fraction from nitrogen-cavitated turkey erythrocytes results in a two-layered pellet that is difficult to visualize at this stage because of the dark

---

[13] G. L. Waldo, A. J. Morris, and T. K. Harden, this series, Vol. 238 [15].

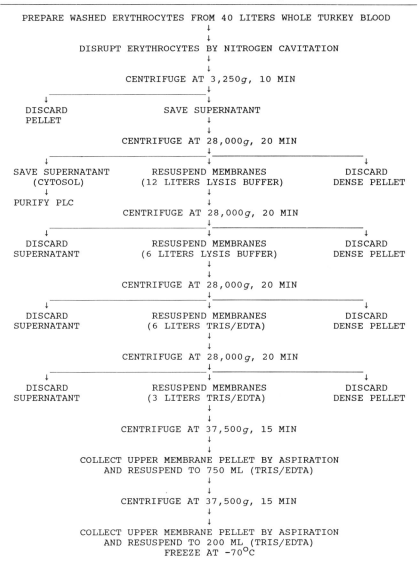

FIG. 1. Purification of turkey erythrocyte plasma membranes. For the preparation of washed erythrocytes from whole turkey blood, see Ref. 13.

red supernatant. In all centrifugation steps described below only the easily resuspended plasma membrane-rich upper portion of the pellet is collected; the denser lower portion of the pellet which adheres to the walls of the centrifuge bottles is discarded. Remove by carefully decanting

60–70% of the supernatant from this step and discard or use for the purification of PLC. Gently swirl the remaining supernatant to resuspend the loose membrane pellet. The yield of plasma membranes can be increased by gently rinsing the surface of the dense pellet with buffer and combining the rinse with the resuspended plasma membrane fraction.

Dilute the resuspended membranes to 12 liters with 5 mM Tris, pH 7.4, 1 mM [ethylenebis(oxyethylene nitrilo)]tetraacetic acid (EGTA), 5 mM $MgCl_2$, 0.1 mM PMSF, and 0.1 mM benzamidine (lysis buffer). Centrifuge the diluted membranes at 28,000 g, carefully aspirate and discard the supernatant, resuspend the loose membrane pellets with 6 liters of lysis buffer, and centrifuge as before. Discard the supernatant and resuspend the loose membrane pellets with 6 liters of 20 mM Tris, pH 7.4, 5 mM (ethylenedinitrilo)tetraacetic acid (EDTA), 0.1 mM benzamidine, and 0.1 mM PMSF (Tris/EDTA). Centrifuge the samples as before, collect the membrane pellets, resuspend the pellets with 3 liters Tris/EDTA, and centrifuge the samples as described. Resuspend the membrane pellets from this step to 750 ml with Tris/EDTA and centrifuge at 37,500 g (Beckman JA-17, 16,500 rpm) for 15 min. Discard the supernatant and carefully collect the white upper portion of the pellets by aspiration into a clean flask; dilute to 250 ml with Tris/EDTA. The dark lower portion of the pellet is discarded. Centrifuge the samples again at 37,500 g, harvest the white upper portion of the pellet, dilute to 200 ml with Tris/EDTA, and store frozen at $-70°$. The yield of erythrocyte plasma membrane protein from 40 liters of whole blood is 3–4 g. Using one J-6 and two J2-21 Beckman centrifuges and two 1-liter cell-disruption bombs, two investigators will require approximately 16 hr to prepare membranes from the washed erythrocytes obtained from 40 liters of whole blood.

Purification of G_{11} Protein

Membrane Solubilization. Rapidly thaw 3–4 g of turkey erythrocyte membranes and centrifuge at 37,500 g for 20 min. Discard the supernatant and resuspend the membrane pellet in cold 25 mM Tris, pH 7.4, 1 mM DTT, 50 μM $AlCl_3$, 10 mM NaF, 10 mM $MgCl_2$, 0.1 mM benzamidine, 0.1 mM PMSF, 2 μg/ml leupeptin, and 2 μg/ml aprotinin (buffer A) containing 1.2% sodium cholate. Dilute the membranes with this buffer to a final detergent to protein ratio of 6:1 (w/w). Thus, dilute 4 g membranes to 2000 ml and stir at 4° for 2 hr. Centrifuge the solution for 45 min at 140,000 g (Beckman type 35 rotor, 35,000 rpm) and collect the supernatant as soluble extract. Approximately 15–20% of the total membrane protein and greater than 80% of G_{11} is extracted by this procedure.

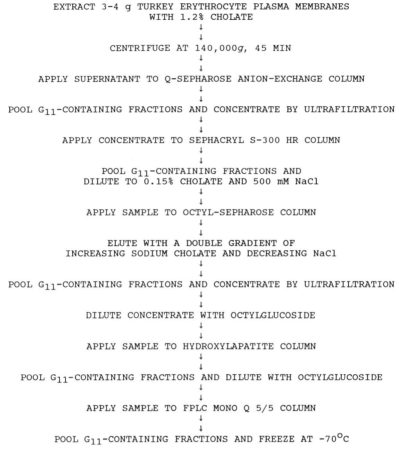

FIG. 2. Purification of G_{11} from turkey erythrocyte plasma membranes.

Q-Sepharose Chromatography. Apply the cholate extract at a flow rate of 4 ml/min to a 2.5 × 25 cm (125 ml) column of Q-Sepharose equilibrated with buffer A containing 0.8% sodium cholate. Wash the column with 150 ml of equilibration buffer, elute with a 500-ml linear gradient from 0 to 600 mM NaCl in buffer A containing 0.8% sodium cholate, and wash with 200 ml of 1 M NaCl in the same buffer. Collect the eluate in 7-ml fractions. Fractions conferring AlF_4^- sensitivity to PLC in reconstitution assays elute at approximately 250–300 mM NaCl (see Discussion). Pool the fractions of peak activity (50–100 ml) and concentrate to approximately 25 ml in an Amicon stirred-cell concentrator (Amicon Corp., Danvers, MA) using a PM30 membrane.

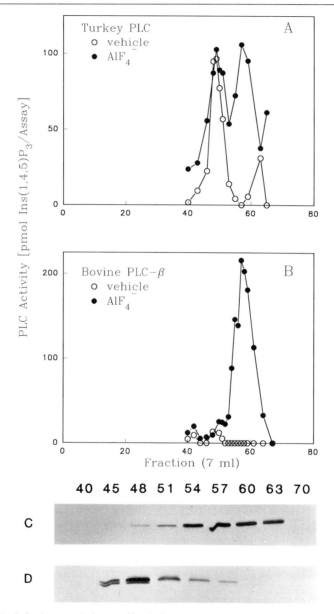

FIG. 3. Q-Sepharose elution profile. A detergent extract of turkey erythrocyte plasma membranes was prepared and applied to a Q-Sepharose column as described in the text. Two-microliter aliquots of fractions eluting from the column were reconstituted in substrate-containing phospholipid vesicles with purified turkey erythrocyte PLC (A) or purified bovine brain PLC-β1 (B). Assays were for 20 min at 30° in the absence (○) or presence (●) of 20

Sephacryl S-300 HR Chromatography. Apply the concentrated pool from the Q-Sepharose column at 2 ml/min to a 2.5 × 92 cm (450 ml) Sephacryl S-300 HR column equilibrated with buffer A containing 100 mM NaCl and 0.8% sodium cholate. Collect the eluate in 4.6-ml fractions. The elution volume (V_e) of partially purified G_{11} is 280–310 ml. Pool the fractions conferring AlF_4^- sensitivity to PLC (pool volume 40–50 ml).

Octyl-Sepharose Chromatography. Dilute the Sephacryl S-300 pool with buffer A containing 592 mM NaCl (4.3 ml of buffer/ml of pool) to give final concentrations of 0.15% cholate and 500 mM NaCl. Apply the diluted pool at 2 ml/min to a 1.6 × 18 cm (36 ml) column of octyl-Sepharose equilibrated with buffer A containing 0.15% sodium cholate and 500 mM NaCl. Wash the column with 50 ml of equilibration buffer and elute with a 200-ml linear gradient increasing from 0.15 to 1.2% sodium cholate and decreasing from 500 to 0 mM NaCl. Wash the column with 50 ml of 1.2% cholate followed by 50 ml of 2% cholate in buffer A. Collect the eluate in 3-ml fractions. Fractions containing G_{11} elute as a broad peak at approximately 200 mM NaCl and 0.8% cholate, although some G_{11} may continue to elute from the column during the 1.2 and 2% sodium cholate washes. Pool the fractions conferring AlF_4^- sensitivity to PLC (70–80 ml) and concentrate through an Amicon PM30 membrane to approximately 10 ml.

Hydroxylapatite Chromatography. The primary purpose of the hydroxylapatite chromatography step is the exchange of a nonionic detergent for sodium cholate. Dilute the concentrated pool from the octyl-Sepharose column 10-fold with 25 mM Tris, pH 7.4, 1 mM DTT, 50 μM $AlCl_3$, 3 mM $MgCl_2$, 3 mM NaF, 0.1 mM benzamidine, 0.1 mM PMSF, 2 μg/ml leupeptin, 2 μg/ml aprotinin, and 0.8% *n*-octylglucoside (buffer B) and apply at 0.5 ml/min to a 1 × 10 cm (7.8 ml) column of hydroxylapatite (Bio-Gel HTP; Bio-Rad Laboratories, Richmond, CA) equilibrated with the same buffer. Wash the column with 25 ml of buffer B and elute with 25 ml of 300 mM K_2HPO_4, pH 7.4, in buffer B. Collect the eluate in 1-ml fractions and pool the fractions conferring AlF_4^- sensitivity to PLC (5–6 ml).

Mono Q HR 5/5 Chromatography. Dilute the hydroxylapatite pool 10-fold with buffer B and apply at 1 ml/min to a Mono Q HR 5/5 (Pharmacia) column (1 ml) equilibrated with the same buffer. Wash the column

---

μM $AlCl_3$ and 10 mM NaF. The indicated fractions from the Q-Sepharose column were subjected to electrophoresis through 11% sodium dodecyl sulfate–polyacrylamide gels, transblotted onto nitrocellulose, and immunoreacted with antisera 118 (C) raised against the carboxyl-terminal dodecapeptide common to G_q and G_{11} or with β-8 antisera (D) raised against purified human placental $\beta\gamma$ subunits.

with 4 ml of equilibration buffer and elute with a 20-ml linear gradient from 0 to 400 mM NaCl. Collect the eluate in 0.5-ml fractions, pool the fractions conferring AlF$_4^-$ sensitivity to PLC (1–3 ml), and freeze in small aliquots at −70°. G$_{11}$ elutes at approximately 250 mM NaCl. Substitution of 0.8% 3-[(3-cholamidopropyl)dimethylammonio]-1-propane sulfonate (CHAPS) for n-octylglucoside in buffer B for both the hydroxylapatite and Mono Q columns produces similar results.

A flowchart for G$_{11}$ purification from turkey erythrocyte plasma membranes is presented in Fig. 2. The final yield of G$_{11}$ α-subunit protein from 3–4 g of turkey erythrocyte plasma membrane protein is 30–100 μg. This represents approximately a 5% recovery of Gα_{11} with greater than a 1000-fold purification. Actual purification (-fold) and recovery of activity are difficult to determine absolutely due to problems in establishing quantitatively reliable values for specific activity of G$_{11}$ in the initial sodium cholate extract from turkey erythrocyte membranes. The G$_{11}$ preparation is stable for at least 6 months when stored frozen at −70°. Repeated freeze–thaw cycles of the G$_{11}$ preparation should be avoided.

Discussion

The purified avian G protein has been unambiguously identified as Gα_{11} based on complete homology of the internal amino acid sequence with the sequence predicted from avian Gα_{11} cDNA.[7] Electrophoresis of purified avian G$_{11}$ on sodium dodecyl sulfate-polyacrylamide gels capable of resolving G$_q$ from G$_{11}$ offer additional evidence that the primary α subunit purified by the above procedure is G$_{11}$.[8] This conclusion is further supported by immunological data using antisera selective for G$_q$ and G$_{11}$.[4,7]

Purified avian G$_{11}$ confers AlF$_4^-$ and guanosine 5'-O-(γ-thiotriphosphate) (GTPγS) sensitivity to both purified turkey erythrocyte PLC and bovine brain PLC-β. GTPγS activation ranges from 10 to 50% of that observed in the presence of AlF$_4^-$. The relatively low response to GTPγS may be a consequence of a low rate of dissociation of GDP from members of the G$_q$ class of G proteins. Using the above procedure, Gα_{11} copurifies with variable amounts of $\beta\gamma$ subunit (ratio of α to $\beta\gamma$ ranges from 10:1 to 1:1) even with AlF$_4^-$ present during the purification. Inclusion of AlF$_4^-$ in all purification buffers stabilizes the capacity of G$_{11}$ fractions to reconstitute AlF$_4^-$ sensitivity to purified turkey erythrocyte PLC in phosphoinositide substrate-containing phospholipid vesicles. Some preparations have resulted in significant resolution of $\beta\gamma$ subunits from Gα_{11} on the Q-Sepharose column (Fig. 3). We previously have reported that $\beta\gamma$ subunits purified from bovine brain or turkey erythrocytes directly activate turkey erythrocyte PLC, but not bovine brain PLC-β1.[11] The results shown in Fig. 3 are consistent with this observation and suggest the utility of using PLC-β1 when screening fractions for α-subunit activity and the necessity of comparing vehicle with AlF_4^--stimulated activity when fractions are screened in reconstitution assays with turkey erythrocyte PLC. In other words, because both $G\alpha_{11}$ and $\beta\gamma$ subunits activate the turkey erythrocyte PLC, samples need to be assayed with and without AlF_4^- to resolve α versus $\beta\gamma$ subunit-promoted activity. It should be emphasized that we have not proved whether all AlF_4^- independent increases in vehicle result from $\beta\gamma$-promoted activation of the enzyme.

The purification protocol described in this chapter relies on identification of G_{11}-containing fractions by monitoring the capacity of column fractions to confer AlF_4^- sensitivity to purified PLC. Purified PLC may not be available to some investigators, or the high cost of phosphoinositide substrates may impose limitations. In such cases immunoblot analysis of column fractions using antisera specific for the carboxyl-terminal dodecapeptide common to G_q and G_{11} or antisera specific for G_{11} is a viable alternative to reconstitution assays for the identification of G_{11}.

[16] Purification of Recombinant $G_q\alpha$, $G_{11}\alpha$, and $G_{16}\alpha$ from Sf9 Cells

By JOHN R. HEPLER, TOHRU KOZASA, and ALFRED G. GILMAN

Introduction

Heterotrimeric guanine nucleotide-binding regulatory proteins (G proteins) couple many cell surface receptors to regulation of specific effector proteins at the plasma membrane.[1-5] G proteins consist of three subunits, designated α (39–46 kDa), β (37 kDa), and γ (8 kDa); to date, at least 21 unique α, 5 β, and 6 γ subunits have been identified.[2,3] The identity of the α subunit defines the individual G protein oligomer. The α subunits have been classified into four major subfamilies (G_s, G_i, G_q, and G_{12}), based on amino acid sequence relationships and functional similarities.[2] The $G_q\alpha$ subfamily consists of five members: $G_q\alpha$, $G_{11}\alpha$, $G_{14}\alpha$, $G_{15}\alpha$, and

[1] A. G. Gilman, *Annu. Rev. Biochem.* **56**, 615 (1987).
[2] M. I. Simon, M. P. Strathmann, and N. Gautam, *Science* **252**, 802 (1991).
[3] J. R. Hepler and A. G. Gilman, *Trends Biochem. Sci.* **17**, 383 (1992).
[4] Y. Kaziro, H. Itoh, T. Kozasa, M. Nakafuku, and T. Satoh, *Annu. Rev. Biochem.* **60**, 349 (1991).
[5] H. R. Bourne, D. A. Sanders, and F. McCormick, *Nature (London)* **348**, 125 (1990).

$G_{16}\alpha$. The $G_q\alpha$, $G_{11}\alpha$, and $G_{14}\alpha$ proteins are closely related (79% amino acid identity), whereas $G_{15}\alpha$ and $G_{16}\alpha$ are more distantly related to G_q (58 and 57% amino acid sequence identity, respectively). The $G_{15}\alpha$ and $G_{16}\alpha$ proteins are similar to one another (85% identity) and may represent the mouse and human homologs of the same gene.[6] All members of the G_q class lack the cysteine residue near the carboxyl terminus that is covalently modified (ADP-ribosylated) by pertussis toxin in members of the G_i class of α subunits. Thus, G_q family members are thought to couple cell surface receptors to signaling pathways that are insensitive to this toxin. A wide variety of hormones, neurotransmitters, and growth factors activate phospholipase C (PLC) which, in turn, generates two second messengers, inositol 1,4,5-trisphosphate [Ins(1,4,5)P$_3$] and diacylglycerol.[7,8] Members of the G_q subfamily have been shown to link cell surface receptors to stimulation of the three known β isoforms of PLC.[9–13]

Native "G_q" purified from mammalian sources actually consists of a mixture of G_q and G_{11},[14] and their close structural similarity (88% amino acid identity) makes biochemical resolution difficult.[15] G_{14} has not yet been purified from native sources. G_{16} and G_{15} have been identified only in tissues of hematopoietic origin[16] and also have not been purified as native proteins. The availability of full-length cDNAs for members of the G_q subclass of α subunits[16,17] has allowed expression and purification of resolved recombinant proteins for functional studies and biochemical characterization.[12,13] Here we describe methods for the baculovirus-driven expression in Sf9 insect cells and purification of $G_q\alpha$, $G_{11}\alpha$, and $G_{16}\alpha$. Much of this work has been reported elsewhere.[12,13]

Special Materials

βγ-Agarose. The βγ-agarose reagent is synthesized as described previously.[18] The G-protein βγ complex is an abundant brain protein, and

---

[6] T. M. Wilkie, P. A. Scherly, M. P. Strathmann, V. Z. Slepak, and M. I. Simon, *Proc. Natl. Acad. Sci. U.S.A.* **88,** 10049 (1991).

[7] M. J. Berridge, *Annu. Rev. Biochem.* **56,** 159 (1987).

[8] Y. Nishizuka, *Nature (London)* **334,** 661 (1988).

[9] A. V. Smrcka, J. R. Hepler, K. O. Brown, and P. C. Sternweis, *Science* **251,** 804 (1991).

[10] S. J. Taylor, H. Z. Chae, S. G. Rhee, and J. H. Exton, *Nature (London)* **350,** 516 (1991).

[11] G. L. Waldo, J. L. Boyer, A. J. Morris, and T. K. Harden, *J. Biol. Chem.* **266,** 14217 (1991).

[12] J. R. Hepler, T. Kozasa, A. V. Smrcka, *et al., J. Biol. Chem.* **268,** 14367 (1993).

[13] T. Kozasa, J. R. Hepler, A. V. Smrcka, *et al., Proc. Natl. Acad. Sci. U.S.A.* **90,** 9176 (1993).

[14] I.-H. Pang and P. C. Sternweis, *J. Biol. Chem.* **265,** 18707 (1990).

[15] J. L. Blank, A. H. Ross, and J. H. Exton, *J. Biol. Chem.* **266,** 18206 (1991).

[16] T. T. Amatruda III, D. A. Steele, V. Z. Slepak, and M. I. Simon, *Proc. Natl. Acad. Sci. U.S.A.* **88,** 5587 (1991).

[17] M. Strathmann and M. I. Simon, *Proc. Natl. Acad. Sci. U.S.A.* **87,** 9113 (1990).

[18] I.-H. Pang and P. C. Sternweis, *Proc. Natl. Acad. Sci. U.S.A.* **86,** 7814 (1989).

20–30 mg of $\beta\gamma$ subunit can be purified from bovine cortical membranes (500 g wet weight; 10–15 g total protein) following established procedures.[19] Purified $\beta\gamma$ subunit, which consists of a mixture of isoforms, is covalently coupled to ω-aminobutylagarose (Sigma, St. Louis, MO) using the cross-linking agent sulfo-SMPB [sulfosuccinimidyl 4-(p-maleimidophenyl)butyrate; Pierce, Rockford, IL], as described,[18] to generate an affinity resin that remains active and is stable at 4° for at least 6 months.

Baculoviruses Encoding G-Protein β and γ Subunits. Full-length cDNAs encoding G-protein β and γ subunits have been described.[20] Recombinant baculoviruses encoding G-protein β_2 and γ_2 subunits for expression in Sf9 cells (fall armyworm ovary, *Spodoptera frugiperda*) are generated as described.[21]

Phospholipase C-β_1. Recombinant (r) $G_q\alpha$, $G_{11}\alpha$, and $G_{16}\alpha$ are detected during purification, in part, by measuring their capacity to activate PLC-β_1 from bovine brain. PLC-β_1 is a relatively abundant protein in brain, and milligram quantities can be purified with relative ease from bovine cortical membranes as described.[22,23] If PLC-β_1 is to be used merely as a reagent to aid in the purification of G_q and G_q-like subunits, highly purified enzyme is not required, and protein enriched over ion-exchange and heparin-Sepharose columns serves adequately.[22]

Antisera. Rabbit polyclonal antisera raised against peptides specific for $G_q\alpha$, $G_{11}\alpha$, and $G_{16}\alpha$ are used to detect recombinant proteins and endogenous Sf9 $G_q\alpha$-like protein(s) during purification. Antisera that recognize internal and/or carboxyl-terminal sequences of $G_q\alpha$, $G_{11}\alpha$, and $G_{16}\alpha$ have been generated.[12,13,22,24] Rabbit anti-$G_q\alpha$ serum W082 was raised to a synthetic 19-amino acid peptide representing an internal sequence (amino acids 115–133) unique to $G_q\alpha$. Rabbit anti-$G_{11}\alpha$ serum B825 was raised to a synthetic 20-amino acid peptide representing an internal sequence (amino acids 114–133) unique to $G_{11}\alpha$. Rabbit anti-$G_{q/11}\alpha$ serum Z811 was raised to a synthetic 15-amino acid peptide representing the carboxyl terminus shared by $G_q\alpha$ and $G_{11}\alpha$. Rabbit anti-$G_{16}\alpha$, B861, serum was raised to an 11-amino acid synthetic peptide representing the carboxyl terminus of $G_{16}\alpha$.

---

[19] P. C. Sternweis and I.-H. Pang, in "Receptor–Effector Coupling: A Practical Approach" (E. J. Hulme, ed.), p. 1. Oxford Univ. Press, Oxford, 1990.

[20] J. Iñiguez-Lluhi, C. Kleuss, and A. G. Gilman, *Trends Cell Biol.* **3**, 230 (1993).

[21] J. Iñiguez-Lluhi, M. I. Simon, J. D. Robishaw, and A. G. Gilman, *J. Biol. Chem.* **267**, 23409 (1992).

[22] A. V. Smrcka and P. C. Sternweis, *Methods Neurosci.* **18**, 72 (1993).

[23] S. H. Ryu, K. S. Cho, K.-Y. Lee, P.-G. Suh, and S. G. Rhee, *J. Biol. Chem.* **262**, 12511 (1987).

[24] S. Gutowski, A. V. Smrcka, L. Nowak, D. Wu, M. I. Simon, and P. C. Sternweis, *J. Biol. Chem.* **266**, 20519 (1991).

Measurement of Phospholipase C-β

The PLC activity is measured by reconstituting the enzyme with activated G-protein α subunits and mixed phospholipid vesicles containing lipid substrate.[9,12] Substrate consists of a mixture of phosphatidylinositol 4,5-bisphosphate (PIP$_2$; Sigma) and phosphatidylethanolamine (PE; Sigma) in a ratio of 1 : 10 with 5000 to 10,000 counts/min (cpm) of [^{3}H]PIP$_2$ (New England Nuclear, Boston, MA), per assay. The final concentration of PIP$_2$ is usually 50 μM (3000 pmol). Assays are performed in a volume of 60 μl using a final buffer containing 50 mM sodium HEPES (pH 7.2), 3 mM EGTA, 0.2 mM EDTA, 0.83 mM MgCl$_2$, 20 mM NaCl, 30 mM KCl, 1 mM dithiothreitol (DTT), 0.1 mg/ml ultrapure bovine albumin (Calbiochem, La Jolla, CA), 0.16% sodium cholate, and 1.5 mM CaCl$_2$ (to yield ~150–200 nM free Ca^{2+}).

Prior to assay, solutions containing the reaction components are prepared separately. These include the following: (1) Gα subunit mix (10 μl/assay); (2) PE–PIP$_2$ substrate mix (20 μl/assay); (3) PLC mix (20 μl/assay), and (4) Ca^{2+} mix (10 μl/assay). The Gα subunits are first activated in incubation buffer 1 (50 mM sodium HEPES, pH 7.2, 1 mM EDTA, 3 mM EGTA, 5 mM MgCl$_2$, 2 mM DTT, 100 mM NaCl, and 1% sodium cholate) with either 1 mM GTPγS for 1 hr at 30° or with 10 mM NaF and 30 μM AlCl$_3$ for 15 min at room temperature. Following activation, samples are stored on ice until further use. The Gα subunits in this solution exist at a concentration six times higher than that desired in the final assay. Appropriate amounts of lipids (stored in chloroform at $-20°$) are dried under nitrogen gas at room temperature prior to sonication in incubation buffer 2 (50 mM sodium HEPES, pH 7.2, 3 mM EGTA, 1 mM DTT, 80 mM KCl). The PLC-β_1 reagent is also prepared in incubation buffer 2 containing bovine serum albumin (1 mg/ml). The amount of PLC-β required for each assay should be determined empirically. If the PLC-β_1 is very pure, 0.5–1 ng/assay will give a low basal activity and substantial activation by G$_q\alpha$; greater amounts (2–5 ng/assay) of less pure PLC-β_1 (e.g., enzyme enriched by anion-exchange and heparin-Sepharose chromatography) are required to give similar activities. The Ca^{2+} mix is prepared as a 9 mM solution of CaCl$_2$ in incubation buffer 2.

Before assay, 10 μl of the activated Gα subunit mix is added to each tube on ice, followed by 10 μl of the Ca^{2+} mix. The PE : PIP$_2$ substrate mix and PLC mix are then combined (20 μl of each/tube), and these components (40 μl assay) are added to each tube on ice. The assay mixture is vortexed, and the assay is started by transferring the tubes to a 30° water bath for the desired time. Assays are terminated by addition of 200 μl of 10% trichloroacetic acid. Tubes are then immediately transferred to an ice bath, followed by addition of 100 μl of bovine serum albumin (10 mg/ml). Centrifugation at 2000 g for 10 min separates unhydrolyzed [^{3}H]PIP$_2$ (pellet) from [^{3}H]Ins(1,4,5)P$_3$ (supernatant). Radioactivity in the supernatant is measured by liquid scintillation counting.

Under these conditions, the assay is linear for 10 min with 50 μM PIP$_2$ as substrate; assays are usually carried out for 3–5 min. Typically, only 50–60% of the substrate is accessible for hydrolysis. To quantify purification of Gα subunits, PLC activity is expressed as quantity of InsP$_3$ per minute per milligram of total protein in the crude or pure sample containing the Gα subunit.

Preparation of Transfer Vectors and Recombinant Baculoviruses for $G_q\alpha$, $G_{11}\alpha$, and $G_{16}\alpha$

For general cloning methods, see Sambrook et al.[25] To construct baculoviruses encoding Gα subunits, we have used the pVL1392/pVL1393 transfer vector containing the polyhedron promoter and an ampicillin resistance gene.[26] Full-length cDNAs encoding rG$_q$α, rG$_{11}$α, and rG$_{16}$α have been generated as described.[16,17] To construct the rG$_q$α expression vector, the entire coding region of G$_q$α [1.2-kilobase (kb) fragment] is excised by digestion with BamHI and SspI and is subcloned into the BamHI and SmaI sites of pVL1393. To construct the transfer vector for G$_{11}$α, the cDNA is digested first with XhoI and filled with the Klenow fragment of DNA polymerase to yield a linearized fragment with blunt ends. This DNA is cleaved with EcoRI to yield a 1.2-kb fragment, which is then subcloned into the EcoRI and SmaI sites of pVL1392 (identical to pVL1393 except the polycloning site is in an inverted orientation). To construct the G$_{16}$α transfer vector, the coding sequence is excised from pKSG16 by digestion with BalI and XbaI and is subcloned into the SmaI and XbaI sites of pVL1393.

The resulting transfer vectors are mixed individually with linearized AcPR-lacZ viral DNA (InVitrogen, San Diego, CA), and are transfected into Sf9 cell monolayers by lipofection (Lipofectin; BRL, Gaithersburg, MD). Recombinant virus is then plaque-purified by blue–white selection as described,[26] except that neutral red and X-Gal (5-bromo-4-chloro-3-indolyl-β-D-galactopyranoside) are added to the cells on the fourth day after infection in a second agarose overlay. Positive viral clones are identi-

[25] J. Sambrook, E. F. Fritsch, and T. Maniatis, "Molecular Cloning: A Laboratory Manual." Cold Spring Harbor Laboratory, Cold Spring Harbor, New York, 1989.
[26] M. D. Summers and G. E. Smith, "A Manual of Methods for Baculovirus Vectors and Insect Cell Culture Procedures." Texas Agricultural Experiment Station, Bulletin No. 1555, College Station, 1987.

fied by their capacity to direct the expression of the appropriate $G\alpha$ subunit as detected by immunoblotting (using antiserum W082 for $G_q\alpha$, B825 for $G_{11}\alpha$, and B861 for $G_{16}\alpha$).

Sf9 Cell Culture

All methods for the culture of Sf9 insect cells have been described.[26,27] Stock cultures of Sf9 cells (50 ml) are grown in suspension in IPL-41 medium (GIBCO, Grand Island, NY) containing 1% pluronic F68, 10% fetal calf serum (heat inactivated), Fungizone, and gentamicin. Large-scale cultures (8 to 12 1-liter cultures) are grown in IPL-41 medium containing 1% fetal calf serum, 1% lipid mix (GIBCO), gentamicin, and Fungizone. Cells are maintained in room air at 27° with constant shaking (125 rpm). Generally, cells are seeded at a density of 0.5×10^6 cells/ml and are allowed to multiply for 3 days to $4-6 \times 10^6$ cells/ml before subsequent passage.

Purification of $G_q\alpha$ and $G_{11}\alpha$

A number of factors complicate the purification of $G_q\alpha$ and $G_{11}\alpha$. First, $G_q\alpha$ and $G_{11}\alpha$ do not readily exchange guanine nucleotides and thus cannot be detected during purification with standard guanine nucleotide-binding assays. Second, Sf9 cells contain significant quantities of an endogenous $G_q\alpha$-like protein(s). This endogenous material shares with $G_q\alpha$ and $G_{11}\alpha$ amino acid sequences that are recognized by the common $G_q\alpha/G_{11}\alpha$ antiserum (Z811) and also a (reduced) capacity to activate brain PLC-β_1. This is important, as $G_q\alpha$ and $G_{11}\alpha$ are expressed at modest levels in Sf9 cells (~0.01% of detergent-extracted protein). It is thus essential to define conditions that resolve recombinant protein from the endogenous α subunit(s).

A third complication is the unusual propensity of $G_q\alpha$ and $G_{11}\alpha$ to aggregate. Conditions that previously favored synthesis of active recombinant $G\alpha$ subunits in prokaryotic or eukaryotic systems have proved unsuccessful for $G_q\alpha$ and $G_{11}\alpha$. When expressed as the free α subunit in bacteria, $G_q\alpha$ and $G_{11}\alpha$ aggregate, with loss of activity. When $G_q\alpha$ is expressed alone in Sf9 cells, the majority of the immunoreactive material is cytosolic; gel filtration reveals that most of the material is inactive and aggregated (Fig. 1). Although much less immunoreactivity is associated with the

[27] D. R. O'Reilly, L. K. Miller, and V. A. Luckow, "Baculovirus Expression Vectors: A Laboratory Manual." Freeman, New York, 1993.

membrane fraction, the activity (capacity to activate PLC-β_1) is roughly equal in the cytosol and membrane extract. Attempts to purify cytosolic $rG_q\alpha$ have been unsuccessful. In contrast, when $rG_q\alpha$ is expressed together with G-protein β and γ subunits (β_2 and γ_2), the majority of the active $rG_q\alpha$ is associated with membranes and is not aggregated following extraction with sodium cholate (Fig. 1); most of the $G_q\alpha$ that remains in

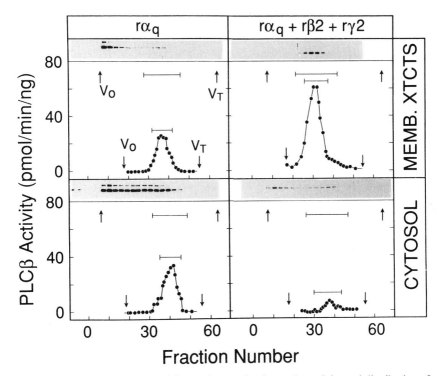

FIG. 1. Effect of expression of G-protein $\beta\gamma$ subunits on the activity and distribution of recombinant $G_q\alpha$ in Sf9 cells. Sf9 cells were infected either with virus encoding $rG_q\alpha$ alone (left, *top* and *bottom*) or with different viruses encoding $G_q\alpha$, β_2, and γ_2 subunits (right, *top* and *bottom*). Cytosol and cholate extracts of membranes were isolated as described. Membrane extracts (*top*, left and right) and cytosol (*bottom*, left and right) were chromatographed on an Ultrogel AcA 34 gel filtration column in the presence of 10 mM NaF, 10 mM MgCl$_2$, and 30 μM AlCl$_3$, and individual fractions were assayed for the capacity to stimulate PLC-β_1 (2.9 ng). Column fractions were also tested for immunoreactivity with anti-$G_q\alpha$ antiserum (W082) (see associated autoradiograms). V_O marks the void volume of the column, and V_T marks the total volume. The lines marking the peak fractions of PLC activity correspond to the underlined immunoreactive fractions. PLC-β_1 activity in this and all subsequent figures is expressed as picomoles InsP$_3$ formed per minute per nanogram PLC-β_1. (Reprinted from Ref. 12 with permission from The American Society for Biochemistry and Molecular Biology.)

the cytosol is aggregated and inactive. For these reasons, the starting material for purification of $G_q\alpha$ and $G_{11}\alpha$ is Sf9 cells that have been triply infected with viruses encoding G-protein α, β, and γ subunits.

Preparation of Extracted Membranes. All buffers used for purification of $rG_q\alpha$ and $rG_{11}\alpha$ from Sf9 cells are defined in Table I. In general, starting material for purification of $G_q\alpha$ and $G_{11}\alpha$ consists of 8–12 liters of Sf9 cells (1 × 10^6 cells/ml) infected for 48 hr with 2–5 plaque-forming units (pfu) of virus encoding either $G_q\alpha$ or $G_{11}\alpha$ and an equal amount of viruses encoding G-protein β_2 subunits and γ_2 subunits. Under these conditions, most of the $G_q\alpha$ and $G_{11}\alpha$ is associated with the plasma membrane, active, and nonaggregated following extraction (Fig. 1). Based on electrophoresis and immunostaining, $rG_{11}\alpha$ is expressed as a single protein with an apparent molecular mass of 42 kDa. Recombinant $G_q\alpha$ is expressed as a pair of proteins with apparent molecular masses of 42 and 43 kDa. This is apparently the result of unexpectedly efficient reading of the altered (ATG to ATT) polyhedron initiator codon upstream of the inserted $G_q\alpha$ sequence in the original pVL1393 expression vector. If a particular sequence is in frame with this altered start site, concurrent expression of a slightly longer polyhedron fusion protein and the normal protein is common.[28] This can be avoided by shifting the $G_q\alpha$ coding sequence out-of-frame with the altered polyhedron initiator codon.

To harvest cells, cultures are collected and gently centrifuged (750 g; 20 min at 4°). Cell pellets are then suspended in 400 ml of ice-cold buffer A (Table I) containing protease inhibitors [phenylmethylsulfonyl fluoride (PMSF), 0.02 mg/ml; leupeptin, 0.03 mg/ml; 1-chloro-3-tosylamido-7-amino-2-heptanone hydrochloride (TLCK), 0.02 mg/ml; L-1-p-tosylamino-2-phenylethyl chloromethyl ketone (TPCK), 0.02 mg/ml; and lima bean trypsin inhibitor, 0.03 mg/ml]. The suspended cells are lysed by nitrogen cavitation (Parr bomb) at 500 psi for 45 min at 4°. The lysate is centrifuged at 500 g for 10 min at 4° to remove a small number of intact cells and nuclei. The supernatant is centrifuged at 100,000 g for 30 min, and the pelleted membranes are suspended in 300 ml of buffer A containing protease inhibitors and homogenized (20 strokes with a Potter–Elvehjem homogenizer). The resulting membrane suspension (~5 mg/ml of protein) can be frozen in liquid nitrogen and stored at −80°.

When ready for purification, $G_q\alpha$- or $G_{11}\alpha$-containing membranes are thawed and extracted by addition of sodium cholate to a final concentration of 1%. Impure grades of sodium cholate should be purified as described.[19] After stirring for 90 min at 4°, the suspension is centrifuged at 100,000 g

---

[28] B. Beames, S. Braunagel, M. D. Summers, and R. E. Lanford, *BioTechniques* **11**, 378 (1991).

TABLE I
BUFFERS FOR PURIFICATION OF rG$_q$α AND rG$_{11}$α[a]

Buffer component	Lysis A	Phenyl-Sepharose				Mono Q		βγ-Agarose					
		B	C	D	E	F	G	H	I	J	K	L	M
Sodium HEPES, pH 7.2	50	50	50	50	50	50	50	50	50	50	50	50	50
EDTA	1	1	1	1	1	1	1	1	1	1	1	1	1
EGTA	3	3	3	3	3	3	3	3	3	3	3	3	3
MgCl$_2$	5	5	5	5	5	5	5	5	5	5	5	5	50
DTT	3	3	3	3	3	3	3	3	3	3	3	3	3
GDP	0.1	0.1	0.1	0.1	0.1	0.1	0.1	0.1	0.1	0.1	0.1	0.1	0.1
NaCl	25	400	575	400	—	—	1000	100	400	100	100	100	100
NaF	10	10	10	—	—	—	—	10	—	—	—	10	10
AlCl$_3$	0.03	0.03	0.03	—	—	—	—	0.03	—	—	—	0.03	0.03
Sodium cholate (w/v)	—	0.25%	—	0.25%	1.5%	—	—	0.2%	—	0.2%	1%	1%	1%
C$_{12}$E$_{10}$[b]	—	—	—	—	—	1%	1%	—	0.1%	—	—	—	—
Octylglucoside	—	—	—	—	—	—	—	—	—	—	—	—	—

[a] All concentrations are millimolar (mM) unless otherwise noted.
[b] C$_{12}$E$_{10}$ is polyoxyethylene 10 lauryl ether.

for 30 min at 4° to yield approximately 300 ml of supernatant containing 1.5–2 g of protein.

Phenyl-Sepharose Chromatography. $G_q\alpha$ can be resolved reasonably well from other proteins by hydrophobic chromatography using phenyl-Sepharose CL-4B (Pharmacia, Piscataway, NJ). The resin (200 ml) is first equilibrated with 1 liter of buffer B (Table I). The membrane extract (300 ml) is diluted 4-fold (slowly) with constant stirring at 4° with 900 ml of buffer C containing protease inhibitors and is then applied to the column. The resin is washed with 300 ml of buffer D; $AlCl_3$ and NaF are removed from the preparation at this point and are omitted in subsequent steps. Protein is finally eluted from the column with a linear gradient (750 ml each of buffers D and E) of ascending concentrations of cholate (0.25 to 1.5%) and descending concentrations of NaCl (400–0 mM). Fractions (60 total, 25 ml each) are collected and assayed (5 μl) for their capacity to activate PLC-β_1 and for immunoreactivity with specific antisera (W082 for $G_q\alpha$; B825 for $G_{11}\alpha$). Elution of $rG_q\alpha$ from the phenyl-Sepharose column is shown in Fig. 2. $G_q\alpha$, $G_{11}\alpha$, and the Sf9 $G_q\alpha$-like protein usually elute between fractions 20 and 30. In the example given for $G_q\alpha$, fractions 22 to 28 are pooled (175 ml). It is then necessary to reduce the total volume and switch the sample into a buffer containing octylglucoside (octyl-β-D-glucopyranoside; Calbiochem) prior to Mono Q chromatography. This is achieved by concentration to 10 ml by ultrafiltration (176 mm PM30 filter, Amicon, Danvers, MA), and dilution to 100 ml by the slow addition of 90 ml of buffer F with constant stirring.

Mono Q Anion-Exchange Chromatography. A 10-ml Mono Q column for FPLC (fast protein liquid chromatography, Pharmacia) is equilibrated with 50 ml of buffer F (Table I). The sample can then be loaded under pressure at a constant flow rate of 0.5 ml/min. The column is washed with an additional 15 ml of buffer F. Protein is then eluted as 50, 3-ml fractions with a 150-ml linear gradient of NaCl (25–300 mM using buffers F and G). All of the $G_q\alpha$, $G_{11}\alpha$, or Sf9 $G_q\alpha$ should elute from the column by 200–250 mM NaCl. The concentration of NaCl is increased to 1000 mM over an additional 10 fractions and held constant through fraction 65. The endogenous Sf9 $G_q\alpha$-like protein(s) elute after either $G_q\alpha$ or $G_{11}\alpha$ (Fig. 3). Figure 3 (*bottom*) shows the clear resolution of $rG_{11}\alpha$ from Sf9 $G_q\alpha$. In contrast to the situation with $rG_{11}\alpha$, $rG_q\alpha$ is expressed at sufficiently high levels to obscure the activity associated with Sf9 $G_q\alpha$ on the Mono Q column. Because of this, protein yield should be sacrificed for purity, and only the active fractions of $rG_q\alpha$ with the earliest elution times should be pooled to ensure complete resolution of $rG_q\alpha$ from Sf9 $G_q\alpha$.

Most of the free recombinant $\beta\gamma$ elutes from the Mono Q column before either $rG_q\alpha$ or $rG_{11}\alpha$ (fractions 15–20). Nevertheless, a roughly

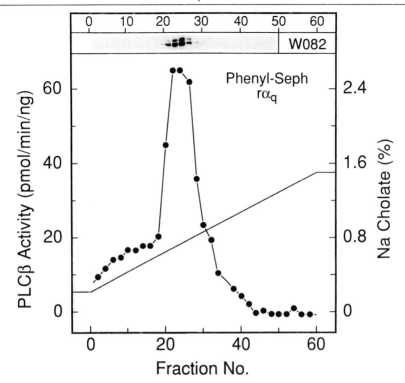

FIG. 2. Phenyl-Sepharose hydrophobic chromatography of recombinant $G_q\alpha$. Cholate extracts from a 12-liter preparation of Sf9 cells infected simultaneously with viruses encoding $G_q\alpha$, β_2, and γ_2 subunits were chromatographed on a phenyl-Sepharose column. Individual column fractions were assayed for the capacity to stimulate PLC-β_1 (2.9 ng) and for immunoreactivity with anti-$G_q\alpha$ antiserum (W082). (Reprinted from Ref. 12 with permission from The American Society for Biochemistry and Molecular Biology.)

stoichiometric amount of residual $\beta\gamma$ is retained in the pooled samples. Activating the sample with $AlCl_3$ and NaF prior to loading to the Mono Q column does not resolve $rG_q\alpha$ or $rG_{11}\alpha$ from $\beta\gamma$. Therefore, free α subunits are resolved from $\beta\gamma$ by subsequent affinity chromatography using $\beta\gamma$-agarose.

$\beta\gamma$-*Agarose Chromatography*. The $\beta\gamma$-agarose (5 ml) is equilibrated with 20 ml of buffer H (Table I) and then transferred to a 50-ml polypropylene centrifuge tube. To facilitate dissociation of residual $\beta\gamma$ from $rG_q\alpha$ or $rG_{11}\alpha$, $AlCl_3$ (30 μM final) and NaF (10 mM final) are added to the pooled sample, which is incubated at room temperature for 30 min. The sample is then added to the $\beta\gamma$-agarose and is incubated with constant mixing for 30 min at room temperature. EDTA in excess of the Mg^{2+} present (e.g., 20 mM final) is then added to the slurry to chelate Mg^{2+}

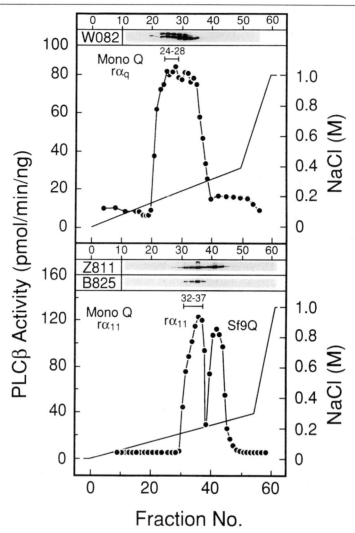

FIG. 3. Mono Q anion-exchange chromatography of recombinant $G_q\alpha$ and $G_{11}\alpha$. Fractions containing rGα protein from the phenyl-Sepharose column were chromatographed on a Mono Q FPLC column. *Top*, rG$_q\alpha$; *bottom*, rG$_{11}\alpha$. Individual fractions were assayed for the capacity to stimulate PLC-β_1 and for immunoreactivity with selective antisera. W082 is an anti-G$_q\alpha$ antiserum, B825 is an anti-G$_{11}\alpha$ antiserum, and Z811 is an anti-G$_q\alpha$/G$_{11}\alpha$ antiserum that recognizes the carboxy terminus of both α chains. (Reprinted from Ref. 12 with permission from The American Society for Biochemistry and Molecular Biology.)

and Al^{3+} and promote binding of α subunit to the affinity resin. The sample and resin are incubated overnight at 4° with constant mixing. The slurry is transferred to a 10-ml disposable column (Bio-Rad, Richmond, CA), and the flow-through is collected, passed a second time over the packed resin, and collected again. The column is washed at 4° with 15 volumes each of buffer H and buffer I to elute nonspecifically bound protein and with 3 volumes of buffer J to remove residual $C_{12}E_{10}$ (polyoxyethylene 10 lauryl ether), which inhibits PLC activity. Specifically bound $G_q\alpha$ or $G_{11}\alpha$ is eluted with 1% sodium cholate in 5 volumes of buffer K. To ensure that all α subunit has been eluted, the column is washed further with 5 volumes of buffer L, which contains $AlCl_3$ and NaF (AlF_4^-), and with 5 volumes of buffer M. Elution of $rG_q\alpha$ from βγ-agarose is shown in Fig. 4.

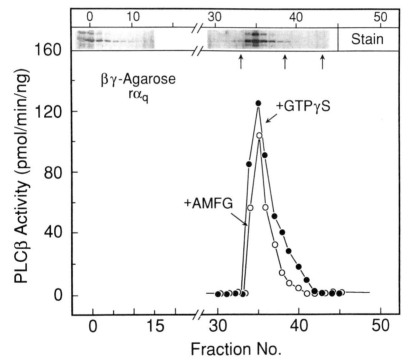

FIG. 4. Affinity chromatography of recombinant $G_q\alpha$ on βγ-agarose. Fractions containing $rG_q\alpha$ from the Mono Q column were pooled and chromatographed on βγ-agarose. Individual column fractions were assayed for the capacity to stimulate PLC-β_1 (2.9 ng), and proteins were visualized by staining with silver nitrate (see associated gel). Samples were activated with either 1 mM GTPγS or $AlCl_3$ (30 μM), NaF (10 mM), $MgCl_2$ (5 mM), and GDP (0.1 mM). (Reprinted from Ref. 12 with permission from The American Society for Biochemistry and Molecular Biology.)

The eluted protein is essentially homogeneous and, although somewhat dilute (10–30 μg/ml), is still sufficiently concentrated to be used directly for assay. This material may be divided into aliquots and frozen at −80° for months without apparent loss of activity. Attempts to concentrate the protein using ultrafiltration or centrifugal concentration devices (Centricon, Amicon) have resulted in losses and are not recommended. Protein may be further concentrated using a Mono Q column and a steep gradient of NaCl. If necessary, this buffer may be changed by gel filtration.[12,25]

A summary of the purification of $rG_q\alpha$ and $rG_{11}\alpha$ is presented in Table II. Final pools containing native bovine brain $G_{q/11}\alpha$, $rG_q\alpha$, $rG_{11}\alpha$, and Sf9 $G_q\alpha$ are shown in Fig. 5. Following treatment with N-ethylmaleimide, all proteins migrate on sodium dodecyl sulfate–polyacrylamide gel electrophoresis (SDS–PAGE) as 40- to 42-kDa proteins. $G_q\alpha$ is expressed as two proteins: the authentic $rG_q\alpha$ with a molecular mass of 42 kDa (75% of total protein) and the polyhedron $rG_q\alpha$ fusion protein at 43 kDa (25%). Endogenous Sf9 $G_q\alpha$ is also purified as two 42-kDa proteins. Antiserum W082, raised against an internal sequence specific for $G_q\alpha$, recognizes brain $G_{q/11}\alpha$, $rG_q\alpha$, and Sf9 $G_q\alpha$ but fails to recognize $rG_{11}\alpha$. Antiserum B825, raised against an internal sequence specific for $G_{11}\alpha$, recognizes both $rG_q\alpha$ and $G_{11}\alpha$, but fails to recognize Sf9 $G_q\alpha$. Antiserum Z811, raised against the 15 carboxyl-terminal amino acids shared by $G_q\alpha$ and $G_{11}\alpha$, recognizes all $G\alpha$ subunits in all four preparations. A 12-liter culture of infected cells yields approximately 125 μg of pure $rG_q\alpha$, whereas a similar preparation of $rG_{11}\alpha$ yields 15–30 μg of pure protein.

TABLE II
PURIFICATION OF RECOMBINANT $G_q\alpha$ AND $G_{11}\alpha$

Step	Purification of $rG_q\alpha$			Purification of $rG_{11}\alpha$		
	Volume (ml)	Protein (mg)	Stimulated PLC activity[a] (nmol InsP$_3$/min/mg protein)	Volume (ml)	Protein (mg)	Stimulated PLC activity[a] (nmol InsP$_3$/min/mg protein)
Cholate extract	1200	1630	9.8[b]	1200	1520	13[b]
Phenyl-Sepharose	175	119	24[b]	150	228	73[b]
Mono Q	18	11.2	104[b]	18	5.1	1070
βγ-Agarose	5	0.125	14,600	3	0.015	49,300

[a] PLC activity was measured with 2.9 ng of bovine brain PLC-β_1 per assay.
[b] PLC activity includes that stimulated by endogenous (Sf9 cell) PLC-activating G proteins.

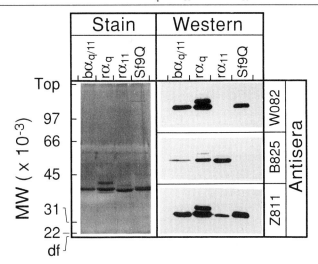

FIG. 5. Analysis by SDS–PAGE and immunoblotting of purified bovine brain $rG_{q/11}\alpha$ ($bG_{q/11}\alpha$), $rG_q\alpha$, $rG_{11}\alpha$, and endogenous Sf9 cell PLC-activating proteins (Sf9Q). Approximately 75 ng of each preparation was treated with N-ethylmaleimide and then resolved on 9.5% polyacrylamide gels. *Left:* Silver stain of the proteins; *right:* immunoblots using the indicated antisera. (Reprinted from Ref. 12 with permission from The American Society for Biochemistry and Molecular Biology.)

Purification of $G_{16}\alpha$

The biochemical properties of $G_{16}\alpha$ are distinct from those of G_q and G_{11}, and the unique strategies that proved successful for purification of $G_q\alpha$ and $G_{11}\alpha$ do not apply to $G_{16}\alpha$. Unlike $G_q\alpha$ and $G_{11}\alpha$, $G_{16}\alpha$ does not aggregate in the absence of $\beta\gamma$. When expressed alone, $G_{16}\alpha$ is detected in both the particulate and soluble fractions of Sf9 cells in roughly equal amounts (Fig. 6A). When $G_{16}\alpha$ is expressed together with G protein β_2 and γ_2 subunits, most of the associated immunoreactivity is found in the membranes (Fig. 6A). Although $G_{16}\alpha$ does not aggregate like $G_q\alpha$ and $G_{11}\alpha$, it is nevertheless expressed together with G protein β and γ subunits to generate greater quantities of membrane-associated protein, which is stable and active; cytosolic $G_{16}\alpha$ is not active (Fig. 6B).

The $G_{16}\alpha$ protein also differs from $G_q\alpha$ and $G_{11}\alpha$ in that it is very unstable unless activated with GTPγS. Surprisingly, neither AlF$_4^-$ nor GDP will stabilize $G_{16}\alpha$, nor will AlF$_4^-$ activate the protein (Fig. 7). Thus, $G_{16}\alpha$ must be purified in the presence of GTPγS. Unfortunately, generation of GTPγS-activated $rG_{16}\alpha$ precludes the use of $\beta\gamma$-agarose for purification.

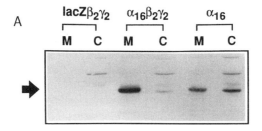
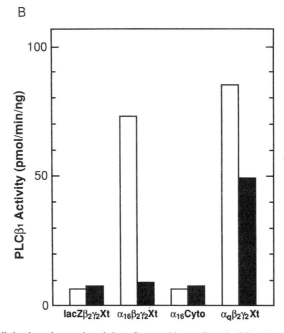

FIG. 6. Cellular location and activity of recombinant $G_{16}\alpha$ in Sf9 cells. Sf9 cells (50-ml culture) were infected with recombinant baculoviruses encoding *lacZ*, β_2, and γ_2; $G_{16}\alpha$, β_2, and γ_2; $G_q\alpha$, β_2, and γ_2; or $G_{16}\alpha$. Membrane and cytosolic fractions were prepared. (A) Membrane (M) and cytosolic (C) fractions (10 μg of each) of *lacZ*$\beta_2\gamma_2$-, $G_{16}\alpha\beta_2\gamma_2$-, and $G_{16}\alpha$-expressing cells were subjected to SDS–PAGE and analyzed by immunoblotting using anti-$G_{16}\alpha$ antiserum B861. The arrow indicates the position of recombinant $G_{16}\alpha$. (B) Membranes from cells expressing lacZ$\beta_2\gamma_2$, $G_{16}\alpha\beta_2\gamma_2$, and $G_q\alpha\beta_2\gamma_2$ were extracted (Xt) with 1% sodium cholate in the presence of AMF plus 200 μM GDP. The cytosolic fraction (Cyto) from $G_{16}\alpha$-expressing cells was similarly supplemented with sodium cholate, AMF, and GDP. Each fraction (5 μl) was assayed for the capacity to stimulate bovine brain PLC-β_1 (2.9 ng) in the presence of either 120 μM GTPγS (open bar) or AMF plus 200 μM GDP (filled bar). The fractions that were assayed with GTPγS were incubated with 1 mM GTPγS at 30° for 30 min prior to assay. (Reprinted from Ref. 13.)

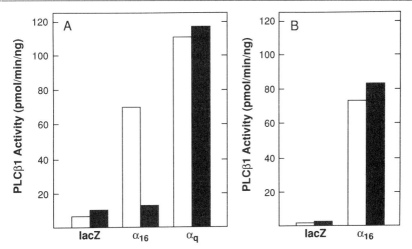

FIG. 7. Ability of AMF and GTPγS to stabilize recombinant $G_{16}\alpha$. (A) Membranes from cells expressing $lacZ\beta_2\gamma_2$ (lacZ), $G_{16}\alpha\beta_2\gamma_2$ (α_{16}), and $G_q\alpha\beta_2\gamma_2$ (α_q) were extracted with 1% sodium cholate in the presence of AMF plus 200 μM GDP. The extracts (5 μl) were assayed for the capacity to stimulate bovine brain PLC-β_1 (2.9 ng) immediately after extraction (open bar) or after 24 hr at 4° (filled bar). The extracts were incubated with 1 mM GTPγS at 30° for 30 min prior to assay. The final concentration of GTPγS in the assay was 120 μM. (B) Membranes from cells expressing $lacZ\beta_2\gamma_2$ (lacZ) and $G_{16}\alpha\beta_2\gamma_2$ (α_{16}) were extracted with 1% sodium cholate in the presence of 5 μM GTPγS. These fractions (5 μl) were assayed for the capacity to stimulate bovine brain PLC-β_1 (2.9 ng) immediately after extraction (open bar) or after 24 hr at 4° (filled bar). The final concentration of GTPγS in the assay was 120 μM. (Reprinted from Ref. 13.)

Preparation of $G_{16}\alpha$ Membranes and Extracts. All buffers used in the purification of r$G_{16}\alpha$ from Sf9 cells are defined in Table III. The starting material for purification of recombinant $G_{16}\alpha$ is 8–10 liters of Sf9 cells (1.5 × 10^6 cells/ml) infected for 48 hr with 2–5 pfu each of recombinant baculoviruses encoding $G_{16}\alpha$ and G-protein β_2 and γ_2 subunits. Cells are harvested and membranes are prepared as described above.

It is critical that 5 μM GTPγS be added to membranes, which are then incubated with the nucleotide for 30 min at 30°, prior to addition of detergent. To extract $G_{16}\alpha$ from the membranes, sodium cholate is added (1% final concentration) with constant stirring at 4° for 1 hr. The extracted membranes are then centrifuged at 100,000 g for 40 min at 4°, and the supernatant containing $G_{16}\alpha$ is collected.

Q-Sepharose Anion-Exchange Chromatography. Q-Sepharose (100 ml; Pharmacia) is equilibrated with 500 ml of buffer O (Table III). After application of the membrane extract, the column is washed with an additional 150 ml of buffer O. Protein is eluted from the column with a 500-

TABLE III
BUFFERS FOR PURIFICATION OF RECOMBINANT $G_{16}\alpha$[a]

Buffer component	Lysis N	Q-Sepharose		Phenyl-Sepharose				Mono Q I		Mono Q II			
		O	P	Q	R	S	T	U	V	W	X	Y	Z
Sodium HEPES, pH 7.2	50	50	50	50	50	50	50	50	50	50	50	50	50
EDTA	1	1	1	1	1	1	1	1	1	1	1	1	1
EGTA	3	3	3	3	3	3	3	3	3	3	3	3	3
MgCl$_2$	5	5	5	5	5	5	5	5	5	5	5	5	5
DTT	3	3	3	3	3	3	3	3	3	3	3	3	3
GTPγS	—	0.001	0.001	0.001	0.001	0.001	0.001	0.001	0.001	0.001	0.001	0.001	0.001
NaCl	50	50	600	50	—	—	—	—	500	—	—	50	600
Sodium cholate (w/v)	—	1%	1%	—	0.3%	1.5%	1.5%	—	1%	1%	2%	1%	2%
Octylglucoside (w/v)	—	—	—	—	—	—	—	1%	1%	—	—	—	—
Glycerol (w/v)	—	—	—	35%	35%	35%	—	—	—	—	—	—	—
(NH$_4$)$_2$SO$_4$	—	—	—	1250	1000	—	—	—	—	—	—	—	—

[a] All concentrations are millimolar (mM) unless otherwise indicated.

ml linear gradient (50–600 mM) of NaCl (250 ml each of buffers O and P). The column eluate is collected in 7-ml fractions and is assayed for the presence of $G_{16}\alpha$ by its capacity to activate bovine brain PLC-β_1 and by immunoblotting with a $G_{16}\alpha$-specific antiserum (B861). Elution of r$G_{16}\alpha$ from the Q-Sepharose column is shown in Fig. 8A; $G_{16}\alpha$ elutes rather broadly around 250 mM NaCl. In the example given, fractions 23–42 are pooled, yielding 120 ml of sample. The pool is slowly diluted 4-fold with constant stirring with buffer Q containing 1.25 M (NH$_4$)$_2$SO$_4$. Any precipitate is removed by centrifugation at 100,000 g for 30 min.

Phenyl-Sepharose Chromatography. Phenyl-Sepharose CL-4B (30 ml; Pharmacia) is equilibrated with 150 ml of buffer R (Table III). The diluted sample is applied to the column, which is washed with an additional 50 ml of buffer R. Protein is eluted from the column with a 120-ml gradient (60 ml each of buffers R and S) containing linearly decreasing concentrations of (NH$_4$)$_2$SO$_4$ (1–0 M) and linearly increasing concentrations of sodium cholate (0.3–1.5%). The column is finally washed with 120 ml of buffer T. Glycerol is included in the buffers to balance the increase in density resulting from the high concentrations of (NH$_4$)$_2$SO$_4$. However, glycerol also increases the rate of dissociation of guanine nucleotides from some G proteins. To ensure the stability of $G_{16}\alpha$, the column eluate is collected as 3-ml fractions in tubes containing 3 μl of 10 mM GTPγS. Fractions are then assayed for $G_{16}\alpha$ by immunoblotting with antiserum B861. Recombinant $G_{16}\alpha$ elutes at the end of the gradient and during the final wash; the endogenous $G_q\alpha$-like protein(s) in Sf9 cells elutes from this column in the flow-through and early gradient fractions. Phenyl-Sepharose column fractions cannot be assayed for their capacity to stimulate PLC-β_1 because of the inhibitory effect of (NH$_4$)$_2$SO$_4$. Fractions with peak immunoreactivity are pooled and concentrated to 20 ml by ultrafiltration using an Amicon PM30 membrane. It is necessary to change the detergent in the sample buffer from sodium cholate to octylglucoside before applying the sample to the next column. This is achieved by twice diluting the sample 5-fold with buffer U and concentrating it to 20 ml.

First Mono Q Chromatography (Octylglucoside). A Mono Q 5/5 HR anion-exchange column for FPLC is equilibrated with 5 volumes (25 ml) of buffer U (Table III). The sample containing $G_{16}\alpha$ is applied to the column, followed by a 5-ml wash with buffer U. Bound protein is eluted with a 40-ml linear gradient of NaCl (0–500 mM; 20 ml each of buffers U and V). In the presence of octylglucoside, $G_{16}\alpha$ fails to bind to the matrix, and the majority of $G_{16}\alpha$ activity (PLC-β_1 activation) and immunoreactivity is detected in the flow-through and subsequent wash. These fractions are pooled, diluted 10-fold with buffer W (containing sodium cholate), and concentrated to 20 ml.

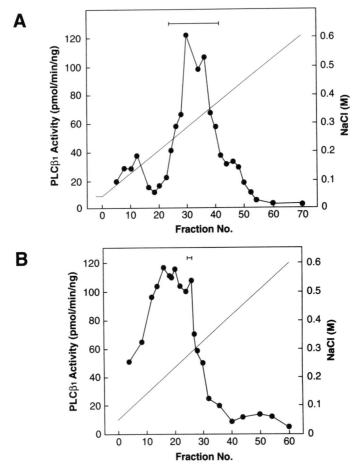

FIG. 8. Purification of recombinant $G_{16}\alpha$. (A) A cholate extract of membranes from Sf9 cells expressing $G_{16}\alpha\beta_2\gamma_2$ was chromatographed on Q-Sepharose. Each fraction (5 μl) was assayed for the capacity to stimulate bovine brain PLC-β_1 (2.9 ng). The bar indicates the peak fractions that were pooled for chromatography on phenyl-Sepharose. The peak fractions from the phenyl-Sepharose column were then chromatographed on a Mono Q column with octylglucoside as the detergent. (B) The peak fractions from the first Mono Q column were chromatographed again over a second Mono Q column using sodium cholate as the detergent. Fractions (5 μl) were assayed for the capacity to stimulate bovine brain PLC-β_1 (2.9 ng). The bar indicates the fractions that contained nearly homogeneous $G_{16}\alpha$. (Reprinted from Ref. 13.)

TABLE IV
PURIFICATION OF RECOMBINANT $G_{16}\alpha$

Purification step	Volume (ml)	Protein (mg)	Specific activity[a] (nmol InsP$_3$/min/mg protein)
Cholate extract	700	1470	3.8
Q-Sepharose	160	121	10
Phenyl-Sepharose	110	4.5	n.d.[b]
Mono Q (octylglucoside)	20	0.44	1200
Mono Q (sodium cholate)	1	0.018	5000

[a] PLC activity was measured with 2.9 ng of bovine brain PLC-β_1 per assay.
[b] Not determined.

Second Mono Q Chromatography (Sodium Cholate). A Mono Q HR 5/5 anion-exchange column is equilibrated with 10 volumes of buffer X (Table III) to saturate the binding sites with cholate. The column is then washed with 5 volumes of buffer Y. The sample containing $G_{16}\alpha$ is applied to the column, which is then washed with 5 ml of the same buffer. The majority of the $G_{16}\alpha$ binds to the matrix in the presence of sodium cholate. Bound protein is eluted from the column by washing with a 30-ml linear

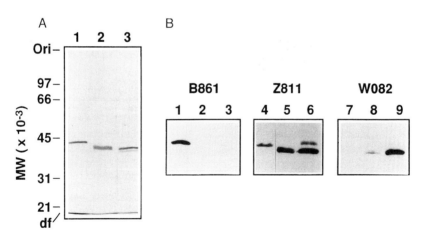

FIG. 9. Analysis by SDS–PAGE and immunoblotting of purified $G_{16}\alpha$, $G_q\alpha$, and endogenous Sf9 cell $G_q\alpha$-like protein(s). Purified r$G_{16}\alpha$, Sf9 cell $G_q\alpha$, and r$G_q\alpha$ (50 ng of each) were subjected to SDS–PAGE (11% gel). (A) Silver stain of the gel. Lane 1, $G_{16}\alpha$; lane 2, Sf9 $G_q\alpha$; lane 3, $G_q\alpha$. (B) Immunoblots of $G_{16}\alpha$ (lanes 1, 4, 7), Sf9 cell $G_q\alpha$ (lanes 2, 5, 8), and $G_q\alpha$ (lanes 3, 6, 9) with antisera B861 (lanes 1, 2, 3), Z811 (lanes 4, 5, 6), and W082 (lanes 7, 8, 9). Recombinant $G_q\alpha$ is expressed as a doublet because of unexpectedly efficient reading of the altered polyhedron initiator codon upstream of the inserted $G_q\alpha$ sequence in pVL1393.[12] (Reprinted from Ref. 13.)

gradient of NaCl (50–600 mM; 15 ml each of buffers Y and Z). Elution of r$G_{16}\alpha$ from the second Mono Q column is shown in Fig. 8B. Analysis of the fractions (PLC-β_1 activation and immunoblotting) indicates that $G_{16}\alpha$ elutes between 150 and 250 mM NaCl. Further analysis of fractions by silver staining of SDS–polyacrylamide gels indicates that the latter half of the activity peak consists of essentially pure $G_{16}\alpha$. These fractions are pooled, aliquoted, and frozen at $-80°$. The protein at this stage, although rather dilute (18 μg/ml), can be assayed directly or switched to another buffer of choice by gel filtration.

A summary of the purification of r$G_{16}\alpha$ is shown in Table IV. Final pools comparing $G_{16}\alpha$ with $G_q\alpha$ and endogenous Sf9 $G_q\alpha$-like proteins are shown in Fig. 9. Fractions pooled from the second Mono Q column contain only $G_{16}\alpha$, which is visualized as a single 44-kDa protein. Antiserum B861 recognizes r$G_{16}\alpha$ specifically and does not react with either r$G_q\alpha$ or Sf9 cell $G_q\alpha$. Antiserum Z811 recognizes $G_{16}\alpha$ only weakly, whereas antiserum W082 does not recognize $G_{16}\alpha$. Because affinity chromatography using $\beta\gamma$-agarose cannot be employed as a final purification step, yields are relatively low, for example, 18 μg of pure protein from an 8-liter culture.

Acknowledgments

Work from the authors' laboratory was supported by National Institutes of Health Grant GM34497, American Cancer Society Grant BE30-O, The Perot Family Foundation, The Lucille P. Markey Charitable Trust, The Raymond Willie Chair of Molecular Neuropharmacology, National Research Service Award GM13569 to J.R.H., and a grant from the International Human Frontier Science Program to T.K.

[17] Expression and Purification of G-Protein α Subunits Using Baculovirus Expression System

By STEPHEN G. GRABER, ROBERT A. FIGLER, and JAMES C. GARRISON

Introduction

The heterotrimeric G proteins are presently classified by the nature of the α subunit. Molecular cloning techniques have identified approximately 21 distinct mammalian α subunits that may be grouped into four general classes based on amino acid identities.[1] Unfortunately, the func-

[1] M. I. Simon, M. P. Strathmann, and N. Gautam, *Science* **252**, 804 (1991).

tional role of specific α subunits is not defined by this classification. Different α subunits are capable of interacting with the same effectors, and individual receptors are capable of activating several distinct α subunits.[2,3] A similar level of complexity is becoming evident with the β and γ subunits. Four distinct β subunits have been defined in mammals, and multiple γ subunits have been identified.[1] Perhaps the large number of potential α, β, and γ combinations explains the specificity of hormone signaling evident in intact cells.

To date, the most convincing assignments of function to specific G-protein α subunits have involved reconstitution of signaling functions using purified components. Whereas purification of G proteins with careful attention to the existence of multiple subtypes has yielded purified α subunits of defined type,[4-9] The structural similarities of these proteins, the likely existence of additional species, and the low abundance in most tissues limits the usefulness of this approach. An alternative approach has been the expression of recombinant α subunits in *Escherichia coli*.[10-13] Although these recombinant proteins have reproduced the known interactions with various effectors observed with native proteins in a qualitative sense, the affinities of the recombinant G proteins for the effectors appear to be reduced compared with the native proteins.[11,13,14]

In an attempt to combine the advantages of these approaches, recombinant G-protein α subunits have been produced using the baculovirus expression system originally developed by Smith *et al.*[15,16] Among the major

---

[2] G. L. Johnson and N. Dhanasekaran, *Endocr. Rev.* **10**, 317 (1989).
[3] L. Birnbaumer, J. Abramowitz, and A. M. Brown, *Biochim. Biophys. Acta* **1031**, 163 (1990).
[4] I. Pang and P. C. Sternweis, *J. Biol. Chem.* **265**, 18707 (1990).
[5] S. J. Taylor, J. A. Smith, and J. H. Exton, *J. Biol. Chem.* **265**, 17150 (1990).
[6] I. Kobayashi, H. Shibasaki, K. Takahashi, K. Tohyama, Y. Kurachi, H. Ito, M. Ui, and T. Katada, *Eur. J. Biochem.* **191**, 499 (1990).
[7] S. E. Senogles, A. M. Spiegel, E. Padrell, R. Iyengar, and M. G. Caron, *J. Biol. Chem.* **265**, 4507 (1990).
[8] D. J. Carty, E. Padrell, J. Codina, L. Birnbaumer, J. D. Hildebrandt, and R. Iyengar, *J. Biol. Chem.* **265**, 6268 (1990).
[9] A. V. Smrcka, J. R. Hepler, K. O. Brown, and P. C. Sternweis, *Science* **251**, 804 (1991).
[10] M. P. Graziano, P. J. Casey, and A. G. Gilman, *J. Biol. Chem.* **262**, 11375 (1987).
[11] A. Yatani, R. Mattera, J. Codina, R. Graf, K. Okabe, E. Padrell, R. Iyengar, A. M. Brown, and L. Birnbaumer, *Nature (London)* **336**, 680 (1988).
[12] R. Mattera, M. P. Graziano, A. Yatani, Z. Zhou, R. Graf, J. Codina, L. Birnbaumer, A. G. Gilman, and A. M. Brown, *Science* **243**, 804 (1989).
[13] M. E. Linder, D. A. Ewald, R. J. Miller, and A. G. Gilman, *J. Biol. Chem.* **265**, 8243 (1990).
[14] M. P. Graziano, M. Freissmuth, and A. G. Gilman, *J. Biol. Chem.* **264**, 409 (1989).
[15] G. E. Smith, M. J. Fraser, and M. D. Summers, *J. Virol.* **46**, 584 (1983).
[16] G. E. Smith, M. D. Summers, and M. J. Fraser, *Mol. Cell. Biol.* **3**, 2156 (1983).

advantages of this expression system are the very abundant production of recombinant proteins combined with the protein modification, processing, and transport systems of a higher eukaryotic cell.[17,18] The methods below describe the construction of recombinant baculovirus vectors encoding four G-protein α subunits (G_{i1}, G_{i2}, G_{i3}, and G_o), infection of *Spodoptera frugiperda* (Sf9) cells with the viruses to generate large amounts of the recombinant protein, and the chromatographic procedures used to purify the α subunits.

Construction of Baculovirus Transfer Vectors

General Considerations

A major reason for the utility of the baculovirus/Sf9 cell expression system is that the system uses the powerful promoter of the polyhedrin gene in the *Autographica californica* nuclear polyhedrosis virus (AcNPV). This promoter drives the gene encoding the polyhedrin protein that packages the virus.[19] If the cDNA for another protein is spliced into a virus in place of the gene encoding the polyhedrin protein, a high level of expression of the foreign protein will occur in insect cells infected with the virus. Because the large size of the baculovirus genome (130 kilobase pairs) makes direct manipulations of the DNA impractical, recombinant virus is usually produced by the cotransfection of wild-type viral DNA with a transfer vector containing the foreign cDNA sequence. Transfer vectors are modified bacterial plasmids which contain the two flanking portions of the polyhedrin gene bounding a multiple cloning site. Homologous recombination *in vivo* substitutes the foreign cDNA sequence in the polylinker of the transfer vector for the polyhedrin gene product, producing a recombinant virus. The recombinant virus can be isolated by visual screening or other methods (see below).

Selection of a transfer vector is dependent on individual requirements such as whether a fusion or nonfusion protein product is desired and the specific cloning sites available in the cDNA to be inserted into the vector. A variety of vectors are available with and without markers to facilitate selection of the recombinant virus.[17,20,21] Some commercially available vectors incorporate sequences coding for affinity tags to facilitate purifica-

[17] V. A. Luckow and M. D. Summers, *Bio/Technology* **6**, 47 (1988).
[18] M. J. Fraser, *In Vitro Cell. Dev. Biol.* **25**, 225 (1989).
[19] L. K. Miller, *Annu. Rev. Microbiol.* **42**, 177 (1988).
[20] J. Vialard, M. Lalumiere, T. Vernet, D. Briedis, G. Alkhatib, D. Henning, D. Levin, and C. Richardson, *J. Virol.* **64**, 37 (1990).
[21] N. R. Webb and M. D. Summers, *Technique* **2**, 173 (1990).

tion of the protein product. The simplest subcloning strategies are those which involve directional ligation into existing sites in the polylinker, but constructs may also be produced through the use of oligonucleotide adaptors and linkers or polymerase chain reaction (PCR)-effected mutagenesis of the specific cDNA clone to be inserted. Subcloning strategies are of necessity specific to each individual cDNA insert and the transfer vector selected, but certain general rules apply. No modifications should be made which alter the sequences downstream from the polyhedrin promoter and upstream from the multiple cloning site, since this region possesses elements which affect transcription.[22,23] Furthermore, the 5' noncoding sequence upstream of the insert start codon should be minimized or eliminated if possible as the level of protein expression is decreased by increasing distance between the polyhedrin promoter sequences and the start codon of the inserted gene. Noncoding sequences and modifications (i.e., affinity tags) placed at the 3' end of the construct do not appear to affect gene expression. The identity of the constructs should always be confirmed prior to transfection either by restriction mapping or, better, by DNA sequencing using universal primers made for the baculovirus DNA. An excellent description of these techniques can be found in the manual provided by Summers and Smith.[24]

Subcloning Strategy for G-Protein α Subunits

The cDNAs for $G_{i1}\alpha$, $G_{i2}\alpha$, $G_{i3}\alpha$ and $G_o\alpha$ were obtained from Dr. R. Reed, Johns Hopkins University (Baltimore, MD). To take advantage of the *Nco*I site in the ATG initiation codons of $G_{i1}\alpha$ and $G_o\alpha$, an *Nco*I site was inserted between the *Bam*HI and *Xba*I sites of the baculovirus transfer vector pVL1393 (kindly provided by Dr. M. Summers, Texas A&M University, College Station, TX) using a synthetic linker containing an *Nco*I and a *Sma*I site between *Bam*HI and *Xba*I cohesive ends. Figure 1 presents the sequence of the region surrounding the insert added to pVL1393 to generate the *Nco*I site. This vector has been designated pVLSG and was used to subclone the appropriate *Nco*I/*Eco*RI fragments of $G_{i1}\alpha$ and $G_o\alpha$ [1728 and 1363 base pairs (bp) respectively]. The 1618-bp *Bgl*I/*Eco*RI fragment of the $G_{i2}\alpha$ cDNA was subcloned into the *Bam*HI/*Eco*RI sites of the pVL1393 using a single-stranded adaptor. The 2998-bp *Xma*III/*Eco*RI fragment of the $G_{i3}\alpha$ cDNA was subcloned into the same sites in pVL1392. As no changes were made to the coding region of the cDNAs

---

[22] V. A. Luckow and M. D. Summers, *Virology* **167**, 56 (1988).
[23] B. G. Ooi, C. Rankin, and L. K. Miller, *J. Mol. Biol.* **210**, 721 (1989).
[24] M. D. Summers and G. E. Smith, *Texas Agric. Exp. Stn. Bull.* **1555**, 1 (1987).

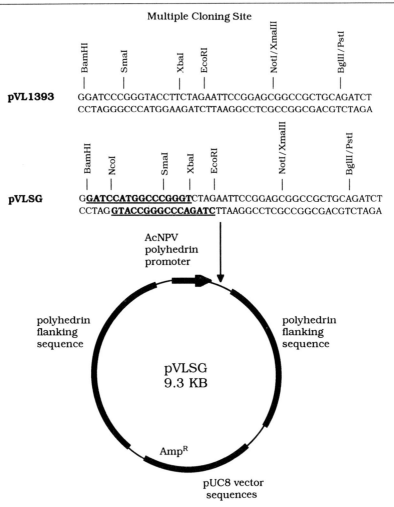

FIG. 1. Changes made in pVL1393 to generate pVLSG. A *Nco*I site was inserted between the *Bam*HI and the *Sma*I sites in pVL1393. The oligonucleotides used to construct the *Nco*I site are underlined and in boldface type.

during insertion into the transfer vector, the identity of each construct was confirmed by restriction mapping.

Sf9 Cell Culture

Spodoptera frugiperda (Sf9) cells are obtained from the American Type Culture Collection (Rockville, MD, ATCC CRL 1711). Cells are maintained in logarithmic growth in Grace's medium[25] supplemented with yeastolate and lactalbumin hydrolyzate (TNM–FH medium)[24] containing 10% (v/v) fetal bovine serum. Normally, cells are maintained in 75-ml suspension cultures in spinner flasks at 27° in an atmosphere of 50% O_2/50% N_2, and stirred at a rate of 60 rpm. For normal passage, spinner flasks are seeded at a density of 1.2×10^6 cells/ml. The doubling time for the cells is typically 18–22 hr, and they are therefore subcultured every third day. Aliquots from each spinner are diluted 1:1 with 0.4% (w/v) trypan blue in 0.85% (w/v) normal saline (GIBCO, Grand Island, NY) and counted using a Neubauer hemocytometer. Routinely, subculturing is done by removing cell suspension and replacing it with fresh medium. Periodically, to prevent the accumulation of toxic by-products, the cells are pelleted by gentle centrifugation (600 g for 6 min) and resuspended in fresh medium. In our experience subculturing is needed every 3 days. Because the health of the cells is essential to high levels of protein expression, the schedules should be arranged so that cell viability remains greater than 95% and the cell borders are cleanly rounded.

Production and Purification of Recombinant Baculovirus

Recombinant baculoviruses are obtained by transfecting Sf9 cells with a 5:1 (mass/mass) mixture of transfer vector and circular wild-type viral DNA using a calcium phosphate precipitation technique modified for insect cells.[24] Plasmid and viral DNAs are mixed in 950 μl HEBS/CT [137 mM NaCl, 6 mM D(+)-glucose, 5 mM KCl, 0.7 mM $Na_2HPO_4 \cdot 7H_2O$, 20 mM HEPES with 15 μg/ml calf thymus DNA, pH 7.05] and precipitated with 125 mM $CaCl_2$ for 30 min at room temperature. The mixture is then pipetted onto a preattached Sf9 cell monolayer (2.5×10^6 cells in a 25-cm^2 tissue culture flask covered with 2 ml of 1× Grace's medium) and absorbed for 4 hr. The anticipated frequency of recombinants utilizing this technique is 0.1–3.0%. Recombinants are selected by visually screening for the occ^- phenotype. Typically, the occ^- phenotype is distinguished from the normal phenotype by a gap in the monolayer associated with swollen cells having a grainy cytoplasm, trypan-stained cells and cell debris, and the absence of inclusion bodies. Recombinant viruses have also been produced in Sf9 cells by transfection using linear viral DNA, and it has been demonstrated that the frequency of recombinants can be improved to greater than 30%.[26]

Medium from transfection is harvested by centrifugation at 2000 g and

---

[25] T. D. C. Grace, *Nature (London)* **195**, 788 (1962).
[26] P. A. Kitts, M. D. Ayres, and R. D. Possee, *Nucleic Acids Res.* **18**, 5667 (1990).

titered by plaque assay. Plaque assays are performed according to the method of Summers and Smith.[24] Briefly, the medium containing the virus is diluted 10^4-fold, 10^5-fold, and 10^6-fold in TNM–FH medium supplemented with 10% fetal bovine serum, 50 µg/ml gentamicin (GIBCO), and 2.5 µg/ml amphotericin B (GIBCO/BRL, Gaithersburg, MD), and 1 ml is placed on cells plated at 2.0×10^6 cells/plate in 60-mm polystyrene tissue culture plates (Corning, Corning, NY). The plates are incubated for 1 hr at 27°, after which the medium is aspirated and the plates covered with 4 ml of 50% low melt agarose (3%, w/v)/50% 2× Grace's medium supplemented with 20% fetal bovine serum and the above antibiotics. After 2–3 days, an additional 2 ml of low melt agarose overlay containing 10% (w/v) trypan blue is added to assist in the identification of recombinant plaques. Following titer of the transfection medium, 25–50 plates are set up at density of 125–150 plaques/plate to screen for recombinants. Plaques are identified based on the observation of the occ^- phenotype.

Recombinant viruses identified are purified through four rounds of plaque purification. Owing to the possibility of false recombinant viruses and the fact that different recombinant clones from a single transfection may express protein at different levels, several recombinants are picked and followed through the initial screening process. Analysis of recombinant proteins is performed by solubilization of infected Sf9 cells with Laemmli sample buffer,[27] resolution of the proteins on a sodium dodecyl sulfate (SDS)–polyacrylamide gel, transfer of the proteins to nitrocellulose, and immunoblotting with peptide antisera specific for the G-protein α subunit under study.[28] If an antibody is not available, screening can also be done by dot hybridization using the cDNA clone inserted in the transfer vector as a probe.[24] The utility of the polymerase chain reaction for screening recombinant clones has been demonstrated.[29]

Protein Expression

For large-scale expression of G-protein α subunits, cells are infected with the appropriate virus at a 2- to 10-fold multiplicity of infection for 1 hr at a density of 10×10^6 cells/ml. After infection, the cells are seeded in 75- or 300-ml spinner cultures at a density of 3.0×10^6 cells/ml in TNM–FH medium supplemented with 10% fetal bovine serum and antibiotics. At these volumes, surface diffusion will provide sufficient oxygen-

---

[27] U. K. Laemmli, *Nature (London)* **227**, 680 (1970).
[28] S. M. Mumby and A. G. Gilman, this series, Vol. 195, p. 215.
[29] A. C. Webb, M. K. Bradley, S. A. Phelan, J. Q. Wu, and L. Gehrke, *BioTechniques* **11**, 512 (1991).

ation. At larger volumes or in bioreactors, gas sparging is necessary to provide sufficient oxygen to the culture, and surfactants may be added to prevent foaming.[30] Infected cells are harvested between 40 and 72 hr depending on their appearance and density, normally when trypan blue staining reaches 20–25%. Because infectivity and expression levels vary depending on the recombinant virus used, an immunological or functional time course study should be done on a case by case basis to determine accurately the infection period yielding the highest level of protein expression. This will also allow determination of the optimum multiplicity of infection and will assess the degree of proteolysis which may occur as the infection progresses. For harvest, cells are washed 3 times in ice-cold phosphate-buffered saline (PBS: 7.3 mM NaH_2PO_4, 55 mM KCl, 47 mM NaCl, 6.8 mM $CaCl_2$, pH 6.2), resuspended as 1.0–5.0 g wet weight/5.0 ml in homogenization buffer consisting of 10 mM Tris-Cl, 25 mM NaCl, 10 mM $MgCl_2$, 1 mM EGTA, 1 mM dithiothreitol (DTT), 0.1 mM phenylmethylsulfonyl fluoride (PMSF), 10 mM NaF, 10 μM GDP, pH 8.0, at 4°, snap frozen in liquid nitrogen, and stored at $-70°$.

Frozen harvested cells are thawed in 15 times the wet weight of ice-cold homogenization buffer and burst by N_2 cavitation (600 psi, 20 min on ice). After disruption by decompression, cells are centrifuged for 10 min at 1000 g at 4° to remove unbroken nuclei and cell debris. The supernatant from the low-speed spin is centrifuged for 1 hr at 100,000 g at 4°. Based on estimations from Coomassie blue-stained gels run using the cell lysates, the recombinant proteins represent from 2 to 10% of the total cell protein at the time of harvest. After disruption of the cells, roughly 30–50% of the $G_i\alpha_s$ and $G_o\alpha$ remain soluble through the 100,000 g centrifugation step. The stained gel presented in Fig. 2 shows the amount of $G_{i2}\alpha$ subunit retained during the centrifugation steps and purified by the chromatographic procedures outlined below.

Purification of G-Protein α Subunits

All chromatographic steps are carried out at 4° and are similar to procedures used by numerous laboratories to purify G proteins. A Waters (Milford, MA) Model 650 Advanced Protein Purification System has proved to be a convenient and reproducible means of purifying G-protein α subunits from a 100,000 g supernatant of infected Sf9 cells. The chromatography scheme presented below could certainly be reproduced using conventional, low-pressure systems, as identical or similar resins are available in bulk from various suppliers; however, the Waters system provides

---

[30] D. W. Murhammer and C. F. Goochee, *Bio/Technology* **6**, 1411 (1988).

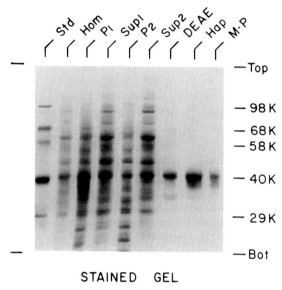

FIG. 2. Analysis by SDS–polyacrylamide gel electrophoresis of fractions obtained during the purification of recombinant $G_{i2}\alpha$. Aliquots from each step of a purification of recombinant $G_{i2}\alpha$ were electrophoresed on a 10% SDS–polyacrylamide "minigel" (Hoefer, San Francisco, CA), and the gel was stained with Coomassie blue. The apparent molecular weights are indicated at right. The lanes contained the following samples: lane Std, 0.5 μg protein of each molecular weight standard, except 40K which was 1 μg; Hom, 11 μg N_2 cavitated cells; P1, 14 μg 1000 g pellet; Sup1, 13 μg 1000 g supernatant; P2, 12 μg 100,000 g pellet; Sup2, 12 μg 100,000 g supernatant; DEAE, 11 μg DEAE pool; Hap, 6 μg hydroxyapatite pool; M-P, 1 μg Mono P pool.

excellent reproducibility. All elution buffers and samples (i.e., the 100,000 g supernatant) are filtered through a 0.45-μm filter prior to use to avoid potential damage to the pumps and columns by particulates. This precaution could probably be omitted if conventional chromatography systems are employed. Initial purifications[31] include 40 μM GDP or GTP and 1 mM $MgCl_2$ in all elution buffers; however, this does not seem to be necessary, and these additions are not routinely employed.

DEAE Chromatography. We have had good success packing a Waters AP-5 column (5 × 10 cm) with Waters DEAE 15HR bulk packing following the manufacturer's instructions. The Waters DEAE 15HR column (5 × 10 cm) is equilibrated with TED (50 mM Tris-Cl, 0.02 mM EDTA, 1 mM DTT, pH 8.0 at 4°) and the 0.45-μm filtered 100,000 g supernatant is loaded at a flow rate of 10 ml/min. We have loaded up to 250 mg of protein

[31] S. G. Graber, R. A. Figler, and J. C. Garrison, *J. Biol. Chem.* **267**, 1271 (1992).

without loss of resolution. The flow is maintained at 10 ml/min, and the column is washed with 170 ml (~1 column volume) of TED and eluted with a 48-min gradient ranging from 0 to 210 mM NaCl in TED. Fractions of 10 ml are collected, and 10 μl of each fraction is assayed for GTP binding activity (see below).

A representative DEAE elution profile is shown in Fig. 3. The GTP binding activity typically elutes between 90 and 150 mM NaCl. Figure 3 shows three distinct peaks of GTP binding activity eluting in this region. Although this is a "typical" result, we have also observed two or even one GTP binding peak in this region in other preparations. When the three peaks are analyzed individually on Coomassie blue-stained SDS–polyacrylamide gels, they all contain the same prominent band in the 40- to 42-kDa region (see Fig. 2). We have some preliminary evidence which suggests that the third peak represents the myristoylated form of the α subunit, whereas the middle peak is nonmyristoylated. It is also possible that the various peaks may represent α subunits with or without bound GDP. However, the individual peaks have not been extensively characterized, and we routinely pool all of the GTP binding activity.

Hydroxyapatite Chromatography. The GTP binding activity pool from the DEAE column is diluted with 4 volumes of 10 mM potassium phosphate, 10 mM Tris-Cl, pH 8.0 at 4°, and loaded onto a hydroxyapatite column (3 × 15 cm, Mitsui Toatsu Chemicals, Tokyo, Japan) at 10 ml/min using the Waters 650 system. After loading, the column is eluted at 10 ml/min with a 40-min linear gradient to 200 mM potassium

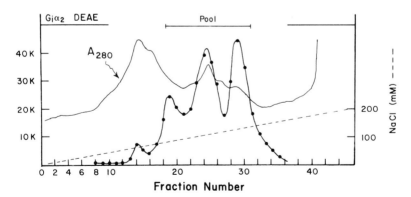

FIG. 3. DEAE chromatography of recombinant $G_{i2}\alpha$. Sf9 cells were infected as described in the text and distributed to six 75-ml spinner cultures. The size of the cell pellet at harvest was 7.82 g (wet weight). After initial processing as described in the text, the 100,000 g supernatant was applied to a Waters AP5 DEAE HR15 column and eluted with NaCl as shown.

phosphate 10 mM Tris-Cl, pH 8.0. Fractions of 10 ml are collected, and 10 μl of each fraction is assayed for GTP binding activity (see below). The GTP binding activity typically elutes as a broad peak between 80 and 150 mM KPO$_4$. However, as shown in Fig. 4, two distinct GTP binding peaks are often seen on the hydroxyapatite column. As was the case with the DEAE peaks, differences between the proteins in these peaks are not apparent on Coomassie blue-stained SDS–polyacrylamide gels. If the peaks are resolved separately on the chromatofocusing column (see below) they differ in elution pH by approximately 0.1 pH unit, suggesting that they may differ by the presence or absence of abound GDP. They do not differ in any of the functional properties that have been examined to date.

Chromatofocusing. The GTP binding activity from the hydroxyapatite column is diluted with an equal volume of 50 mM Bis–Tris, adjusted to pH 6.3 at 4°, and loaded onto a Mono P HR 5/20 chromatofocusing column (Pharmacia, Piscataway, NJ) at a flow rate of 1 ml/min. The column is washed with 10 ml of 25 mM Bis–Tris, pH 6.3, and eluted with 40 ml (1 ml/min) of 1:10 Polybuffer (Pharmacia), pH 4.0 at 4°. The GTP binding activity elutes between pH 4.9 and 4.6, and a typical chromatogram is shown in Fig. 5. The last three lanes of Fig. 2 present the purification of recombinant $G_{i2}\alpha$ via the three column steps. Table I presents the data from a typical purification of recombinant $G_{i2}\alpha$ using the scheme described. Figure 6 depicts a silver-stained SDS–polya-

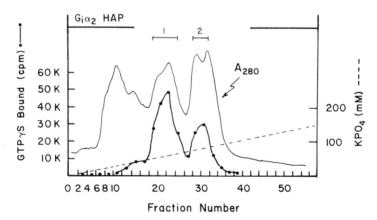

FIG. 4. Hydroxyapatite chromatography of recombinant $G_{i2}\alpha$. The GTP binding pool from the DEAE column shown in Fig. 3 was diluted with 4 volumes of 10 mM potassium phosphate, 10 mM Tris-Cl, pH 8.0 at 4°, applied to a Mitsui Toatsu HCA column (3 × 15 cm), and eluted with potassium phosphate as shown.

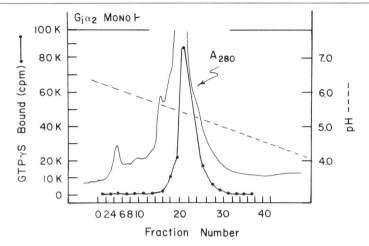

FIG. 5. Chromatofocusing of recombinant $G_{i2}\alpha$. Pool 1 of the hydroxyapatite column shown in Fig. 4 was mixed with an equal volume of 50 mM Bis–Tris, pH 6.3. After being adjusted to pH 6.3, the sample was applied to a Mono P HR 5/20 chromatofocusing column (Pharmacia) and eluted as shown.

crylamide gel of the three G_i and G_o α subunits obtained at this stage of purification.

To prepare the α subunits for storage, the 1 : 10 Polybuffer is exchanged with 50 mM HEPES, 1 mM EDTA, 1 mM DTT, pH 8.0 (usually supple-

TABLE I
PURIFICATION OF RECOMBINANT $G_{i2}\alpha$[a]

Fraction	Volume (ml)	Protein (mg)	[³H]GTP Binding		Yield[b] (%)
			nmol	nmol/mg	
N_2 cavitated cells	95	213	80.6[c]	0.38	100
100,000 g supernatant	86	103	75.1[c]	0.73	93
DEAE	100	47	57.0[c]	1.21[d]	71
Hydroxyapatite	90	9.9	19.8	2.00[d]	25
Mono P	4	1.5	18.0	12.00	22

[a] Approximately 1.14×10^6 Sf9 cells/ml were infected and cultured in six 75-ml spinner cultures. Cells were harvested at 72 hr postinfection and yielded 6.1 g wet weight. The column chromatography was as shown in Figs. 3–5.

[b] Yield may not be accurate due to the presence of other GTP-binding proteins in the first three fractions.

[c] Other GTP-binding proteins are likely to be present in these fractions.

[d] Specific activities in these fractions varied widely among preparations.

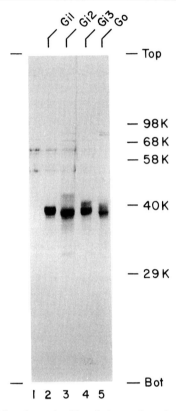

FIG. 6. Analysis by SDS-polyacrylamide gel electrophoresis of purified recombinant Gα subunits. Purified recombinant Gα subunits were electrophoresed on a 12% SDS-polyacrylamide gel. Electrophoresis was continued for 1 hr after the dye front had migrated off the bottom of the gel, and the gel was stained with silver. The apparent molecular weights are indicated at right. Lanes 2, 3, 4, and 5 contained 500 ng of recombinant $G_{i1}\alpha$, $G_{i2}\alpha$, $G_{i3}\alpha$, and $G_o\alpha$, respectively. Lane 1 was not loaded with protein, and the faint bands visible at 65K and 55K in all lanes are artifacts of the silver staining method.

mented with 10 μM GDP and 1 mM $MgCl_2$), using an HR 10/10 Fast Desalting column (Pharmacia) or by repeated dilution and concentration using Centricon 30 microconcentrators (Amicon, Danvers, MA). Proteins are concentrated to a final concentration of 0.5–4.0 mg/ml, divided into aliquots, and stored frozen at −70° until use. Proteins have been stored for up to 18 months at −70° in the HEPES, EDTA, DTT buffer without loss of activity; however, significant loss of activity has been observed in samples stored for longer than 2 years. Similarly, repeated freeze–thaw

TABLE II
[^{35}S]GTPγS BINDING^a

G protein	k_{on}app (min^{-1})	K_D (nM)	Activity (nmol/mg)
rG$_{i1}$α	0.04 ± 0.005	6.8 ± 0.61	23.5 ± 1.26
rG$_{i2}$α	0.06 ± 0.010	25.1 ± 2.63	19.5 ± 1.11
rG$_{i3}$α	0.07 ± 0.003	not done	21.3 ± 2.85
rG$_o$α	0.14 ± 0.053	not done	24.5 ± 3.27

^{a} The values for k_{on}app and the specific binding activity for GTPγS were taken from GTPγS binding time course experiments using 2 μM [^{35}S]GTPγS continued for 120 min. Samples were taken in triplicate every 5 min for the first 20 min and at 20-min intervals thereafter. The values shown are means ± S.E. as determined by a nonlinear least-squares fit to the equation $B = B_{eq}(1 - e^{-kt})$. The values for K_D were taken from GTPγS equilibrium binding experiments using concentrations of [^{35}S]GTPγS ranging from 2.8 to 200 nM and recombinant Gα subunits at 10 nM. Incubations were for 1 hr at 30°. The values shown are means ± S.E. as determined by a nonlinear least-squares fit to the equation $B = K_D[L]/(B_{max} + [L])$.

cycles result in a loss of activity, although usually 3–4 freeze thaw episodes may be tolerated.

GTP Binding Assay

Column fractions are assayed for GTP binding activity using [^{3}H]GTP or [^{35}S]GTPγS (New England Nuclear, Boston, MA) by a protocol similar to that developed by Sternweis and Robishaw.[32] Each assay tube contains 75 mM sodium HEPES, pH 8.0, 50 mM MgCl$_2$, 100 mM NaCl, 1 mM EDTA, 0.5 mM ATP, 1 μM GTP or GTPγS [including 400,000–800,000 disintegrations/min (dpm) of the radioligand used], and 5 or 10 μl of the fraction to be assayed in a total volume of 150 μl. Tubes are incubated for 20 min at 30°, and binding is terminated by the addition of 3 ml of ice-cold wash buffer (50 mM Tris-Cl, 100 mM NaCl, 25 mM MgCl$_2$, pH 8.0 at 4°) followed by rapid filtration through nitrocellulose filters (Millipore, Bedford, MA, HAWP). Filters are rinsed three times with 3 ml of ice-cold wash buffer and counted in 4 ml of Beckman Ready Protein Scintillant (Beckman, Inc., Palo Alto, CA). Specific binding activities are obtained from incubations continued to steady state, which is reached in 1 hr under these conditions, using 2 μM GTPγS with 1 μCi [^{35}S]GTPγS/ml and 30 to 80 ng of Mono P-purified α subunits in 150 μl of the same buffer used to assay the column fractions. Table II summarizes the GTP binding properties of representative preparations of each α subunit.

[32] P. C. Sternweis and J. D. Robishaw, *J. Biol. Chem.* **259**, 13806 (1984).

Conclusion

In summary, using Sf9 cells infected with recombinant baculoviruses expressing three G_i and the G_o α subunits, we have been able to obtain milligram amounts of pure α subunits after three column steps. These α subunits are myristoylated[31] as are native α subunits, and they are functional as measured by the ability to interact with angiotensin II receptors[31] and adenosine receptors. The baculovirus/Sf9 insect cell protein expression system appears to offer a number of advantages for the expression of these G-protein α subunits.

[18] Analysis of G-Protein α and $\beta\gamma$ Subunits by *in Vitro* Translation

By Eva J. Neer, Bradley M. Denker, Thomas C. Thomas, and Carl J. Schmidt

Introduction

The analysis of G-protein subunits that have been modified by point mutation, deletion, or formation of chimeras requires that the altered protein be expressed. Cell-free transcription and translation of modified cDNAs offer certain advantages and disadvantages compared to various cellular systems. An important disadvantage is that cell-free translation yields only a small amount of synthesized protein that is generated in a concentrated mixture of many cellular proteins. It would, therefore, be very difficult to purify the small amount of specific protein from the very large amount of cellular components. This disadvantage is counterbalanced by the ability to radiolabel the specific protein selectively. Because only specific mRNA is added to the *in vitro* translation mixture, the desired protein usually represents the only radioactive band produced. Thus, if methods exist to evaluate the structure and function of the radioactive protein, without purifying it from the nonradioactive proteins present, then *in vitro* translation offers a rapid and attractive way of screening the consequences of modifications introduced into proteins. In this brief chapter, we focus on the techniques that can be used to assay the function of G-protein α, β, and γ subunits that have been synthesized *in vitro*. We also review modifications that have been made to standard *in vitro* transcription–translation systems in order to enhance the yield of G-protein subunits; in general, however, the methods of transcription and trans-

Transcription and Translation

The efficiency of translation is strongly dependent on structural features of the mRNA, including the structure of the 5' and 3' ends.[1] The effect of 5' untranslated sequences on the efficiency of translation varies among the α subunits. For example, the 5' untranslated region of rat α_o inhibits translation. We routinely transcribe α_o mRNA from a plasmid-containing α_o cDNA from which all but 14 base pairs (bp) upstream of the ATG have been removed.[2] Removal of the 5' untranslated region of α_o increases the efficiency of translation about 5-fold. Sanford et al.[3] found that the 5' untranslated region of α_o inhibited translation of the γ subunit when they transferred this region in front of the translational start site of the γ subunit.

Sanford et al.[3] developed another method to increase the yield of translated product, as well as the fidelity of initiation. The authors modified the pGEM plasmid by inserting 36 bp from the 5' untranslated region of the alfalfa mosaic virus RNA-4. The composition of this sequence is given by Herson et al.[4] Transcription and translation from the plasmid containing the alfalfa mosaic virus sequence routinely gave between 0.05 and 0.1 pmol/μl and diminished the amount of incorrectly initiated translation products.

Plasmids

Most of the experiments in our laboratory have been done using cDNAs cloned into the Bluescript plasmid (Strategene, La Jolla, CA). This plasmid has the advantage that mRNA can be transcribed from two promoters in opposite orientations (T7 RNA polymerase and T3 RNA polymerase). Either polymerase can be used depending on the orientation of the insert. We have also used the pSP64t plasmid and the SP6 RNA polymerase to direct the synthesis of γ_1. Other laboratories have used pGEM[5,6] and pIXI.[7]

Plasmids to be used for *in vitro* transcription are prepared from 5-ml

---

[1] M. Kozak, *J. Cell Biol.* **115**, 887 (1991).
[2] B. M. Denker, E. J. Neer, and C. J. Schmidt, *J. Biol. Chem.* **267**, 6272 (1992).
[3] J. Sanford, J. Codina, and L. Birnbaumer, *J. Biol. Chem.* **266**, 9570 (1991).
[4] D. Herson, A. Schmidt, S. Seal, A. Marcus, and L. van Vloten-Doting, *J. Biol. Chem.* **254**, 8245 (1979).
[5] J. Olate, R. Mattera, J. Codina, and L. Birnbaumer, *J. Biol. Chem.* **263**, 10394 (1988).
[6] W. A. Maltese and J. D. Robishaw, *J. Biol. Chem.* **265**, 18071 (1990).
[7] L. Journot, C. Pantaloni, J. Bockaert, and Y. Audigier, *J. Biol. Chem.* **266**, 9009 (1991).

overnight cultures by the alkaline lysis method.[8] The plasmid is further purified by 13% (w/v) polyethylene glycol (PEG) precipitation for 1 hr at 4°.

mRNA Transcription

One microgram of linearized plasmid is transcribed using the appropriate RNA polymerase and according to the directions contained in the mCAP RNA capping kit (Stratagene). The reaction mixture contains 40 mM Tris, pH 7.5, 0.4 mM ATP, 0.4 mM GTP, 0.4 mM CTP, 0.4 mM UTP, 0.5 mM 4'5^{me}Gppp5'G, 50 mM NaCl, 8 mM MgCl$_2$, 2 mM spermidine, 1 mM dithiothreitol (DTT), and 5 units of T7, T3, or SP6 RNA polymerase. Transcription is carried out at 37° for 30 min. Following transcription, the reaction is treated for 10 min at 37° with 10 units of DNase. Subsequently, 150 μl of TNE (10 mM Tris, pH 7.4, 250 mM NaCl, 0.1 mM Na$_2$EDTA) is added, and the reaction mixture is extracted with an equal volume of phenol/chloroform (1:1, v/v). RNA is precipitated with 3 volumes of ethanol. Precipitated RNA is resuspended in 10 mM Tris, pH 7.4, at a concentration of 1 mg/ml.

In Vitro Translation

Translations are carried out using the rabbit reticulocyte lysate system (Promega, Madison, WI) at 30° for 60 min. Between 5 and 50 μCi of [^{35}S]methionine (>1000 Ci/mmol; Du Pont/New England Nuclear, Boston, MA) is added to monitor synthesis. For translation of individual subunits, 0.5–1 μg of α_o mRNA, 1 μg of β mRNA, or 0.5 μg of γ mRNA (γ_2 or γ_t) are used to direct synthesis. For cotranslations, 1 μg of β mRNA and 0.5 μg of γ mRNA are added to the same translation mixture. Following translation, aliquots of the translation mixture are analyzed on an 11 or 13% sodium dodecyl sulfate (SDS)–polyacrylamide gel and the products identified by autoradiography. Total incorporation of [^{35}S]methionine is monitored by trichloroacetic acid precipitation followed by scintillation counting.

Sodium Dodecyl Sulfate–Polyacrylamide Gel Electrophoresis

Gels containing 11 or 13% polyacrylamide are prepared according to Laemmli[9] or as modified by Christy with 0.1 M sodium acetate in the lower chamber.[10] It is helpful to load as little lysate as possible in each

---

[8] J. Sambrook, E. F. Fritsch, and T. Maniatis, "Molecular Cloning: A Laboratory Manual." Cold Spring Harbor Laboratory, Cold Spring Harbor, New York, 1989.
[9] U. K. Laemmli, *Nature (London)* **227**, 680 (1970).
[10] K. G. Christy, Jr., D. B. Latart, and H. W. Osterhoudt, *BioTechniques* **7**, 692 (1989).

lane of the gel as the large amount of hemoglobin in the lysate frequently distorts the bands of γ (8 kDa) or of the smaller tryptic products. Diluting the lysate also helps reduce the viscosity. Following electrophoresis, gels are soaked in En^{3}Hance (Du Pont/New England Nuclear), dried, and then exposed to film at $-70°$.

Analysis of Structure and Function of *in Vitro* Translated G-Protein Subunits

Analysis of Patterns of Tryptic Proteolysis

A powerful way to probe the conformation of the *in vitro* synthesized subunits is to compare the tryptic cleavage products with those derived from native bovine brain α or $\beta\gamma$ subunits. The α and β subunits each have more than 30 potential tryptic cleavage sites. However, in the native molecules, the number of sites available is extremely limited. The specificity of the native tryptic cleavage patterns gives a simple and reliable way to assess the formation of $\beta\gamma$ subunits from cotranslates of β and γ or from mixtures of the two subunits, and to evaluate the ability of α subunits to bind GTPγS, GTP, or GDP.[2,11-13]

Dimerization of $\beta\gamma$ Subunits. The β subunit has only one tryptic cleavage site accessible in the native molecule that splits it into two fragments of 27 and 14 kDa.[14,15] The β subunit synthesized without γ subunits does not give the characteristic 27- and 14-kDa tryptic fragments seen with $\beta\gamma$ purified from animal tissues. Instead, β_1 gives a 19-kDa product. An example of the tryptic cleavage patterns of *in vitro* translated $\beta\gamma$ is shown in Fig. 1. This product is not seen with β_2 and β_3. When β and γ are cotranslated or when separate translations of β and γ are mixed, approximately 25–30% of the β subunit forms a $\beta\gamma$ dimer that can be separated from aggregated or incorrectly folded protein by gel filtration (see Fig. 2). This dimer gives a tryptic cleavage pattern like that of purified $\beta\gamma$. A mixture of undimerized β and $\beta\gamma$ gives all three bands (19, 27, 14 kDa) on tryptic cleavage. Thus, the appearance of the 27-kDa band is a convenient index of the formation of $\beta\gamma$ dimers folded into a coformation that makes 32 lysines and arginines inaccessible to trypsin. We have found the 14-kDa band less easy to detect than the 27-kDa product, especially in unpurified lysates because of the

[11] B. M. Denker, C. J. Schmidt, and E. J. Neer, *J. Biol. Chem.* **267**, 9998 (1992).
[12] C. J. Schmidt and E. J. Neer, *J. Biol. Chem.* **266**, 4538 (1991).
[13] C. J. Schmidt, T. C. Thomas, M. A. Levine, and E. J. Neer, *J. Biol. Chem.* **267**, 13807 (1992).
[14] B. K.-K. Fung and C. R. Nash, *J. Biol. Chem.* **258**, 10503 (1983).
[15] J. W. Winslow, J. R. Van Amsterdam, and E. J. Neer, *J. Biol. Chem.* **261**, 7571 (1986).

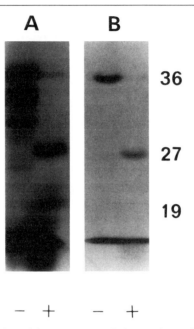

FIG. 1. Trypsin digestion of the *in vitro* translation products. Samples were incubated for 1 hr at 37° in the absence (−) or presence (+) of 20 pmol of trypsin. Samples were (A) crude cotranslates of β and γ$_2$ mRNA and (B) *in vitro* translated βγ$_2$ dimers purified by gel filtration and sucrose density gradient centrifugation. Following tryptic digestion, samples were analyzed on 11% SDS–polyacrylamide gels. The gels were soaked in En^{3}Hance, dried, and exposed to film at −70° for 3 days (A) or 2 weeks (B).

distortion of the SDS–polyacrylamide gel by hemoglobin (subunit molecular weight 16,400). It is also harder to detect because it tends to be broad (even in native purified βγ) and because there are fewer methionines than in the 27-kDa fragment, so it is less labeled. By this assay, and by other assays described below, we found that βγ could be formed either by cotranslation of the subunits or by mixing separately translated β and γ. Because the yield of individually translated subunits was higher than in cotranslations, mixing after translation became the standard method.

Guanine Nucleotide Binding to α Subunits. The tryptic cleavage pattern of α subunits depends on whether a guanine nucleoside diphosphate or triphosphate is bound at the active site. When a nonhydrolyzable guanine nucleoside triphosphate analog, such as GTPγS, is bound to α$_o$ or other members of the α$_i$ family, a 2-kDa amino-terminal peptide is cleaved,

[16] J. R. Hurley, M. I. Simon, D. B. Teplow, J. D. Robishaw, and A. G. Gilman, *Science* **226**, 860 (1984).

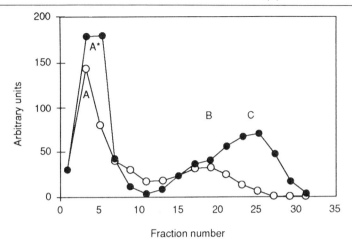

FIG. 2. Ultrogel AcA 34 column chromatography of translation products. Translation products were applied to a 7-ml Ultrogel AcA 34 column equilibrated with TMED plus 0.3% Lubrol and eluted at a flow rate of 5 ml/hr at 4°. Blue dextran was added to the samples to indicate the void volume. Fractions of 200 μl were collected, and, beginning at the void volume, aliquots were analyzed by SDS-PAGE followed by densitometric scanning of the autoradiogram. Open circles show the elution profile of a translate of β_1 alone. Filled circles show the β_1 products of a cotranslation of β_1 and γ_2. A and A* indicate the β_1 subunits that eluted with the void volume of this column. B indicates the elution position of the β subunits from the translation of β_1 mRNA alone that were included in the column volume. C indicates the elution position of the β subunits of a cotranslation of β_1 and γ_2 mRNA that were included in the column volume.

leaving a stable 37- to 39-kDa fragment depending on the α subunit.[14-16] Similarly, tryptic cleavage of GTPγS-liganded α_s gives stable fragments 1 kDa smaller than the native molecule.[17] In contrast, when the α subunit is liganded with GDP, the 37- to 39-kDa fragment is cleaved further to relatively stable 25- to 26-kDa and 17- to 18-kDa polypeptides. The latter contain the carboxy terminus.[15] Figure 3 shows an example of tryptic cleavage of rat α_o. Because the formation of the 37-kDa fragment from GTPγS-liganded α subunits requires the sequestration of a large number of potential tryptic cleavage sites, it is a very sensitive method for assessing the folding of α subunits. All α subunits are tested to determine the pattern of tryptic fragments generated in the presence of GTPγS or GDP before any further studies are done.

The tryptic assay can also be used to evaluate the relative affinities of different mutant α subunits for guanine nucleotides. For such experiments, the *in vitro* translated proteins must first be passed over a desalting

[17] T. H. Hudson, J. F. Roeber, and G. L. Johnson, *J. Biol. Chem.* **256**, 1459 (1981).

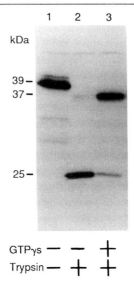

FIG. 3. Tryptic digestion of α_o. *In vitro* translated subunits were incubated for 10 min at 30° with or without 25 μM GTPγS and digested with 20 pmol of trypsin for 30 min at 30°. The rabbit reticulocyte lysate contains approximately 50 μM GTP (Promega).[18] The control without trypsin is shown in the first lane. The proteolysis was stopped with 10 mM benzamidine, and the sample was analyzed by SDS–PAGE and autoradiography.

column to remove creatine phosphate. Otherwise, the nucleotide-regenerating system, which is present in the reticulocyte lysate, will convert added GDP back to GTP. To remove the creatine phosphate, 50- to 100-μl samples are passed over 1- to 2-ml columns of Sephadex G-50, equilibrated with 50 mM HEPES, 6 mM MgCl$_2$, 75 mM sucrose, 1 mM EDTA, 1 μM GDP, pH 7.8, and either 0.3% Lubrol PX or 0.9% cholate. The hemoglobin in the lysate makes a convenient marker so that fractions of 150–200 μl can be taken in the region of peak color. Different concentrations of guanine nucleotides can then be added back to the samples and tryptic proteolysis carried out. The amount of 27-kDa product is quantitated by densitometry of the autoradiograph. The EC$_{50}$ for GTPγS or GDP can be determined in such an experiment. Comparing the EC$_{50}$ for wild-type and mutated α subunits gives an indication of their relative affinities for guanine nucleotides and their ability to change conformation on binding of nucleotide.

Experimental Procedures. Proteins are cleaved with tosylphenylalanyl chloromethyl ketone-treated trypsin (Cooper Biomedical, Malvern, PA) made up in 1 mM HCl at a concentration of 20 μM. Samples of 2–10 μl

[18] R. B. Pennel, in "The Red Blood Cell" (D. M. Surgenor, ed.), Academic Press, New York, 1974.

of translation mix are diluted to 10–15 µl with 50 mM Tris, pH 7.6, 6 mM Mg^{2+}, 1 mM EDTA, and 1 mM DTT (TMED), and 1 µl of the trypsin solution (20 pmol of trypsin) is added. The digestion is carried out for 10 to 60 min at 30°. In an alternative protocol (used for column fractions), the final trypsin concentration is 0.7 mM, and incubation is for 15 min at 30°. Proteolysis is terminated by making the reaction 10 mM in benzamidine, followed by addition of Laemmli sample buffer and immediate boiling for 3 min. Products were analyzed by SDS–polyacrylamide gel electrophoresis (SDS–PAGE).

Determination of Native Molecular Weights in Vitro Synthesized α and βγ Subunits

The most direct way to document dimerization of β and γ and interactions between α and βγ subunits is by showing an increase in molecular weight when they form βγ dimers or an αβγ heterotrimer. The physical properties of particular proteins in crude mixtures can be determined provided that the proteins do not interact with other components of the solution. To determine explicitly the molecular weight of *in vitro* synthesized subunits, it is necessary to measure the Stokes radius (a), sedimentation coefficient ($s_{20,w}$), and partial specific volume ($\bar{v}$) of the particle in order to solve the following equation[19,20]:

$$MW = \frac{6\pi\eta_{20,w}N}{(1 - \bar{v}\rho_{20,w})}(s_{20,w})(a)$$

where $\eta_{20,w}$ is the viscosity of water at 20° (1.002 centipoise), N is Avogadro's number, $\rho_{20,w}$ is the density of water at 20° (0.998 ml/g), $s_{20,w}$ is the sedimentation coefficient (S), a is the Stokes radius (Å), and $\bar{v}$ is the partial specific volume (g/ml). To compare the size of *in vitro* translated α and βγ subunits with purified, detergent-solubilized proteins, all measurements of physical properties are made in detergent solutions.

Stokes Radius. The Stokes radius can be measured by gel filtration. The elution position of the subunits is compared to the elution of protein of known Stokes radius included in the sample being analyzed. This procedure is valid provided that the *in vitro* synthesized subunits form peaks of the same width as the monodisperse marker proteins analyzed on the same column. If the peak of the test protein is as sharp as the markers, it is reasonable to conclude that the *in vitro* synthesized protein is also monodisperse.

Analysis of *in vitro* synthesized α subunits shows that the α subunits

---

[19] C. Tanford, "Physical Chemistry of Macromolecules," p. 364. Wiley, New York, 1961.
[20] E. J. Neer, *J. Biol. Chem.* **249**, 6527 (1974).

have the same profile as the marker proteins and have the same Stokes radius as purified α subunits.[2] Analysis of the Stokes radius of $\beta\gamma$ is less straightforward than evaluation of the α subunits because a substantial fraction of *in vitro* translated β subunits aggregate either together or with other proteins in the lysate and elute at the void volume of gel-filtration columns (see Fig. 2). The complexes are quite large and are even excluded from AcA 22 which has an exclusion limit of 3×10^6 kDa. However, some free β and $\beta\gamma$ form an included peak. The Stokes radius of β is larger than $\beta\gamma$, presumably because the molecule is in an extended, partly denatured conformation.

Stokes radii can be determined by gel-filtration chromatography over Ultrogel AcA 34 (IBF Inc., Columbia, MD) equilibrated with TMED (50 mM Tris-HCl, pH 7.5, 6 mM MgCl$_2$, 1 mM EDTA, 1 mM DTT) and 0.3% Lubrol PX. Between 50 and 100 μl of the *in vitro* translates are applied to a 0.7×18 cm column along with 50 μg each of catalase (Stokes radius 52 Å) and bovine serum albumin (BSA) (Stokes radius 37 Å) and carbonic anhydrase (Stokes radius 23 Å) as markers. Samples are eluted at 5 ml/hr. Fractions containing the *in vitro* translated products are identified by SDS–PAGE followed by autoradiography. Elution profiles are analyzed by scanning densitometry of the autoradiogram, and Stokes radii are determined by comparison with the elution positions of the added markers of known Stokes radius.

Sedimentation Coefficient and Partial Specific Volume. The sedimentation coefficient of α subunits can be determined by loading the translation mixture onto 5–20% sucrose density gradients. However, before the sedimentation coefficient of $\beta\gamma$ can be determined, the large complexes must be removed since they tend to sediment as a broad smear across sucrose density gradients. The complexes can be resolved from the $\beta\gamma$ dimers by gel filtration as described above. The included peak of $\beta\gamma$ dimers separated from the high molecular weight material can then be further analyzed by sucrose density gradient centrifugation. These $\beta\gamma$ dimers behave as a monodisperse population on sucrose density gradient centrifugation. The peaks are sharp and have the same width at half-height as do the marker enzymes. The sharpness of the peaks is further evidence that the *in vitro* translated $\beta\gamma$ subunits are not interacting with the other proteins present in the reticulocyte lysate.

To determine the molecular weight of hydrophobic proteins, such as $\beta\gamma$,[21] the partial specific volume ($\bar{v}$) must be determined. The $\bar{v}$ for most hydrophilic proteins is 0.73–0.74 ml/g, but the $\bar{v}$ of hydrophobic proteins in detergent solution may be quite different from that value. Some commonly used nonionic detergents (such as Lubrol PX or Triton X-100) have

---

[21] R. M. Huff, J. M. Axton, and E. J. Neer, *J. Biol. Chem.* **260**, 10864 (1985).

values for $\bar{v}$ much higher than the average protein. A complex of hydrophobic protein and detergent will have a $\bar{v}$ intermediate between the protein and the detergent and proportional to the fraction of each. To determine the partial specific volume, the sedimentation in sucrose density gradients made up in H_2O or D_2O must be compared. The calculation of $\bar{v}$ from such experiments is described in Refs. 20–22.

Linear sucrose density gradients (5–20%) are prepared in TMED plus 0.3% Lubrol PX in either H_2O or D_2O. When appropriate, 1 μM GDPβS or 10 μM GTPγS is added to all solutions. Samples of the *in vitro* translate are mixed with markers: bovine serum albumin ($s_{20,w}$ 4.35), carbonic anhydrase ($s_{20,w}$ 2.95), and cytochrome c ($s_{20,w}$ 1.8 S). It is essential that the markers be included in the sample to get an accurate measure of the relative sedimentation rate. Purified α_o and $\beta\gamma$[23] from bovine brain are also added to each gradient either for internal comparison with *in vitro* translated subunits or to determine whether the *in vitro* translated subunits could associate with their partners. To measure association of *in vitro* translated $\beta\gamma$ with α_o, 0.5 μg of purified $\beta\gamma$ and 3 μg of purified α_o are added to the 150-ml sample loaded on the gradient. The proportions are reversed when binding of *in vitro* translated α_o to pure $\beta\gamma$ was measured. The 4-ml gradients are centrifuged at 59,000 rpm for 6 hr (H_2O) or 18 hr (D_2O) in a Beckman SW 60 rotor at 4°. Gradients are collected in 25–35 fractions from the bottom of the tube. Aliquots of the fractions are analyzed by SDS–PAGE. Following electrophoresis, gels are soaked for 1 hr in En³Hance (Du Pont/New England Nuclear), dried overnight, and then exposed to Kodak (Rochester, NY) XAR film for varying lengths of time at $-70°$. Elution profiles are analyzed by scanning densitometry and apparent $s_{20,w}$ values determined based on comparison with sedimentation rates of marker proteins of known $s_{20,w}$. The correct values for $s_{20,w}$ are calculated taking into account the partial specific volume.[20,22]

Analysis of Association Using Cross-Linking Reactions

Formation of $\beta\gamma$ Dimers. Deducing the formation of $\beta\gamma$ dimers from the native tryptic cleavage pattern becomes problematic when pairs of $\beta\gamma$ dimers are analyzed that have not yet been purified from cells or tissues. For example, the tryptic cleavage pattern of purified $\beta_3\gamma$ has not been determined. An extra accessible tryptic site could lead to loss of the characteristic pattern despite formation of $\beta\gamma$ dimers. An alternative method that does not depend on the distribution of lysines and arginines makes use of the ability of β and γ subunits to be covalently cross-linked

---

[22] J. E. Sadler, J. L. Rearick, J. C. Paulson, and R. L. Hill, *J. Biol. Chem.* **254**, 4434 (1979).
[23] E. J. Neer, J. M. Lok, and L. G. Wolf, *J. Biol. Chem.* **259**, 14222 (1984).

with 1,6-bismaleimidohexane (BMH) which links cysteine residues.[24] This reagent cross-links β_1 to γ_1 and γ_2, cross-links β_2 to γ_2 but not to γ_1, and does not cross-link β_3 to either of these forms of γ. Clearly, negative experiments must be verified by determination of the native molecular weight, which is the explicit test of $\beta\gamma$ formation. Nevertheless, cross-linking forms another convenient screening procedure for $\beta\gamma$ formation.

Prior to cross-linking, the *in vitro* translated β and γ monomers and the β plus γ mixtures are incubated for 1 hr at 37° (Fig. 4). Subsequently, the reactions are applied to a 1-ml Sephadex G-50 column (Pharmacia LKB Biotechnology, Inc., Piscataway, NJ) equilibrated with 50 mM HEPES (pH 8), 6 mM MgCl$_2$, 1 mM EDTA, and 0.4% Lubrol PX to remove the DTT. The hemoglobin in the reticulocyte lysate serves as a marker for the void volume, so the entire colored peak can be collected. For cross-linking, BMH prepared in dimethyl sulfoxide is added to a final concentration of 2 mM, and the samples are incubated at 4° for 20 min. The products are analyzed by SDS-PAGE.

Cross-Linking of α to $\beta\gamma$. The BMH reagent also cross-links $\beta\gamma$ to α subunits to give two characteristic bands at 140 and 122 kDa.[25] These products form specifically even in the presence of the large excess of other proteins in the reticulocyte lysate, and they are an index of the formation of $\alpha\beta\gamma$ heterotrimers.[25a] The method used is the same as above, except that 0.5 μg purified α_o or α_i subunits or 0.2–0.5 μg of purified $\beta\gamma$ subunits are added to the translation mixtures of $\beta\gamma$ or α, respectively. The yield of cross-linked product in a translation mix is less than the yield with purified proteins.[25] The reason for this has not yet been determined. However, in a translation mixture, the concentration of available sulfhydryl groups (e.g., from hemoglobin) is high enough to diminish significantly the effective BMH concentration. The insolubility of BMH, together with the need to keep the concentration of dimethyl sulfoxide in the final reaction at 5% or less, makes it difficult to raise the concentration further.

Pertussis Toxin-Catalyzed ADP-Ribosylation

In some cases, mutant α subunits are able to interact with $\beta\gamma$, but not stably enough to remain associated over several hours of centrifugation. Such weaker interactions may be detected by pertussis toxin-catalyzed, $\beta\gamma$-dependent ADP-ribosylation of the α subunits. Although the ADP-ribose group is added to the carboxyl terminus of α_i/α_o group of proteins, the toxin appears to recognize the $\alpha\beta\gamma$ heterotrimer and not the free α

---

[24] T. C. Thomas, T. Sladek, F. Yi, T. Smith, and E. J. Neer, *Biochemistry* **32**, 8628 (1993).
[25] F. Yi, B. M. Denker, and E. J. Neer, *J. Biol. Chem.* **266**, 3900 (1991).
[25a] T. C. Thomas, C. J. Schmidt, and E. J. Neer, *Proc. Natl. Acad. Sci. U.S.A.* **90**, 10295 (1993).

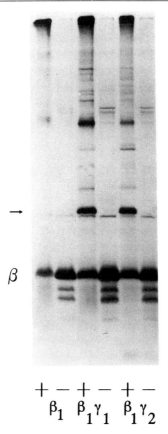

FIG. 4. Cross-linking of $\beta\gamma$ dimers mediated by BMH. Samples of β alone or mixtures of β plus γ were incubated for 1 hr at 37°, and DTT was removed by gel filtration. Subsequently, samples were cross-linked with 2 mM BMH at 4° for 20 min. Products were analyzed by SDS–PAGE, followed by autoradiography. The non-cross-linked γ subunits were intentionally run off the bottom of the gel to increase resolution of the 46-kDa cross-linked product (arrow) from an endogenous product of the rabbit reticulocyte lysate. For lanes marked −, DTT was added before BMH to prevent cross-linking; lanes marked + received DTT following the 20-min incubation with BMH.

subunit.[23,26] Thus, ADP-ribosylation of α subunits reflects their ability to interact with $\beta\gamma$. As the transient interaction is monitored by an irreversible covalent modification, ADP-ribosylation is a very sensitive way to detect weak interactions between α and $\beta\gamma$ subunits.

[26] S. C. Tsai, R. Adamik, Y. Kanaho, E. L. Hewlett, and J. Moss, J. Biol. Chem. **259**, 15320 (1984).

In vitro translated proteins are made as described above, with nonradioactive methionine and ADP-ribosylated with pertussis toxin (List Biological, Campbell, CA). Twenty microliters of lysate in a 40-μl total reaction volume are ADP-ribosylated in duplicate. Each reaction contains final concentrations of 5 μM NAD$^+$, 3 mM ATP, 1.25 mM isoniazid, 10 mM thymidine, 0.1 mM GTP, and 0.1% Lubrol in 50 mM Tris, pH 7.6. Pertussis toxin (100 μg/ml) is activated with DTT (final concentration of 20 mM) for 15 min at 30° prior to addition to the reaction mixture. The final concentration of toxin in the reaction mix is 10 μg/ml. [^{32}P]NAD$^+$ [1–5 μCi, 1000 Ci/mmol (Amersham Corp., Arlington Heights, IL)] is added, and the reaction is carried out for 30 min at 37°. Reactions are performed with or without 0.2–0.5 μg of purified bovine brain βγ subunit. Control ADP-ribosylations of *in vitro* translation with no added mRNA are included with each experiment to account for the pertussis toxin labeling of the endogenous substrates in the rabbit reticulocyte lysate. Thirteen percent SDS–PAGE gels are run for 16–20 hr at 50–100 V, allowing the dye front to migrate off the gel.

Direct Assay of α_s Function

When the yield of a G-protein subunit is high enough, it is possible to use the *in vitro* translated protein to regulate an effector such as adenylyl cyclase (adenylate cyclase). Because activation of the enzyme by α_s is very specific, the assay can be carried out without purifying α_s from the reticulocyte lysate.[5,27]

Lipid Modification of G-Protein Subunits

Rabbit reticulocyte lysates contain the enzymes necessary for myristoylation of α subunits and for prenylation of γ subunits. Different lysates appear to vary in how efficiently these modifications are carried out, perhaps because the amount of substrates for the enzymes is different.

Myristoylation. *In vitro* translation is performed as described above, except that nonradioactive methionine is added to the translation mixture, and 1 μCi of [9,10-^{3}H]myristic acid (53 Ci/mmol, Amersham Corp.) is included with each translation. Samples are analyzed by SDS–PAGE, stained with Coomassie blue, destained, soaked in En^{3}Hance (Du Pont/New England Nuclear), dried, and analyzed by autoradiography. Parallel translations should be carried out with [^{35}S]methionine to monitor the level of protein synthesis.

Polyisoprenylation. Translation of γ subunits in rabbit reticulocyte lysates supplemented with 13 μM [^{3}H]mevalonic acid lactone (and without

---
[27] Y. Audigier, this volume [19].

[^{35}S]methionine) showed that the γ subunits could be polyisoprenylated.[3,6] Polyisoprenylation of γ_t (but not of γ_2) was reported to increase its susceptibility to proteolysis after 30 min of incubation.[3] Unprenylated γ_2 is capable of forming βγ dimers, but these dimers have diminished ability to interact with α subunits.[28] To avoid potential artifacts due to incomplete polyisoprenylation, we now routinely add 10 μM mevalonic acid lactone to all translations of γ subunits.

Conclusion

In vitro translation of G-protein α and βγ subunits provides a rapid, flexible method to screen the consequences of site-directed mutations. The results of such *in vitro* studies can give new information about the structure and function of G-protein subunits and act as a convenient guide in determining which modified proteins merit more detailed analysis, either by large-scale synthesis in other expression systems or by transfection into appropriate cell lines.

Acknowledgments

This work was supported by National Institutes of Health Grant GM36295 to E.J.N. and an American Cancer Society Grant CD496 to C.J.S. T.C.T. was supported by Public Health Service Training Grant T32 DK07737. B.M.D. was supported by a Clinical Investigator Award, K08 DK02110. The authors are grateful to Mrs. Paula McColgan for expertly typing the manuscript.

[28] J. A. Iniguez-Lluhi, M. I. Simon, J. D. Robishaw, and A. G. Gilman, *J. Biol. Chem.* **267**, 23409 (1992).

[19] Assays for Studying Functional Properties of *in Vitro* Translated $G_s\alpha$ Subunit

By YVES AUDIGIER

Introduction

The relationship between the structure and function of a protein has been widely explored by *in vitro* mutagenesis of the cDNA encoding the protein and expression of the mutated protein in an eukaryotic cell.[1,2] If the synthesis of the endogenous protein can be abolished, this approach

[1] R. C. Mulligan, B. H. Howard, and P. Berg, *Nature (London)* **277**, 108 (1979).
[2] M.-J. Gething and J. Sambrook, *Nature (London)* **293**, 620 (1981).

becomes more fruitful because the functional properties of the mutated protein can be directly assayed on the transfected cell.

Isolation of the cyc^- variant of the mouse lymphoma S49 cell line,[3] which does not synthesize the α subunit of the GTP-binding protein G_s,[4,5] has provided this ideal situation for studying the functional domains of $G_s\alpha$ subunits. Unfortunately, transfection of S49 cells by current procedures is not easy, and time-consuming approaches, such as those using retroviral vectors, have been developed.[6] An alternative approach, based on the synthesis of large amounts of α subunit in bacteria and its use for reconstitution of cyc^- membranes, is more attractive, but it faces the problem of posttranslational modifications which are lacking in the bacterially expressed proteins.[7]

In vitro translation of $G_s\alpha$ subunits in eukaryotic cell-free systems generates a fully active protein which restores the coupling between the β-adrenergic receptor and adenylyl cyclase (adenylate cyclase) on cyc^- membranes.[8] We have therefore developed a strategy which circumvents the two disadvantages of the above-mentioned approaches: the translation of $G_s\alpha$ subunits is performed in reticulocyte lysates and the functional properties of the *in vitro* translated protein are characterized either in the soluble state directly in translation medium or in the membrane-associated form after reconstitution of cyc^- membranes.[9,10] Because the messenger RNA used for *in vitro* translation is transcribed from the cDNA encoding $G_s\alpha$ subunits, genetic modifications can be introduced at the nucleotide level and their consequences immediately analyzed at the aminoacid level by studying the phenotype of the mutated protein.[10] We report here the different procedures that we adapted or designed in order to study the functional properties of mutated *in vitro* translated $G_s\alpha$ subunits.

In Vitro Translation of $G_s\alpha$ Subunit

In all these experiments, the messenger RNA is obtained by *in vitro* transcription of the cDNA coding for the long form of $G_s\alpha$ subunit (a generous gift from Dr. Birnbaumer and Dr. Codina, Baylor College of

---

[3] H. R. Bourne, P. Coffino, and G. M. Tomkins, *Science* **187**, 750 (1975).
[4] G. L. Johnson, H. R. Kaslow, and H. R. Bourne, *J. Biol. Chem.* **253**, 7120 (1978).
[5] B. A. Harris, J. D. Robishaw, S. M. Mumby, and A. G. Gilman, *Science* **229**, 1274 (1985).
[6] S. B. Masters, K. A. Sullivan, R. T. Miller, B. Beiderman, N. G. Lopez, J. Ramachandran, and H. R. Bourne, *Science* **241**, 448 (1988).
[7] M. P. Graziano, P. J. Casey, and A. G. Gilman, *J. Biol. Chem.* **262**, 11375 (1987).
[8] J. Olate, R. Mattera, J. Codina, and L. Birnbaumer, *J. Biol. Chem.* **263**, 10394 (1988).
[9] L. Journot, J. Bockaert, and Y. Audigier, *FEBS Lett.* **251**, 230 (1989).
[10] L. Journot, C. Pantaloni, J. Bockaert, and Y. Audigier, *J. Biol. Chem.* **266**, 9009 (1991).

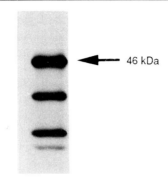

FIG. 1. *In vitro* translation of $G_s\alpha$ subunit. The plasmid pIBI-$G_s\alpha$-2 was transcribed *in vitro* using T7 polymerase, and the transcription product was then translated in a reticulocyte lysate system with [^{35}S]methionine. The ^{35}S-labeled polypeptides were resolved in a 12% sodium dodecyl sulfate (SDS)–polyacrylamide gel and visualized by autoradiography.

Medicine, Houston, TX) that we subcloned into the pIBI vector under the control of the promoter for T7 RNA polymerase.[9] As previously described,[11] a typical translation medium contains 7.5 µl of nuclease-treated reticulocyte lysate (Promega, Madison, WI), 0.5 µl of each of 19 amino acids lacking methionine at 1 mM, 1 µl of 10 µM [^{35}S]methionine (1000 Ci/mmol), and 1 µl of the messenger RNA encoding $G_s\alpha$ subunit (2–6 ng/µl). The translation is carried out at 30° for 45 min. It is important to stress that reticulocyte lysate contains a high concentration of protein (50–60 mg/ml), and, therefore, for subsequent electrophoretic analysis, no more than 2 µl of translation medium should be loaded per slot about 10 mm wide and 1.5 mm thick.

Figure 1 shows the translation of $G_s\alpha$ mRNA which generates a major 46-kDa ^{35}S-labeled band and other bands of lower molecular mass. As previously demonstrated by immunoprecipitation,[9] the low molecular mass bands correspond to initiation at internal AUG codons (methionine-60, -110, or -135). Interestingly, these translation products represent amino-terminal deleted mutants which should be useful for studying the functional role of the amino-terminal domain of $G_s\alpha$ subunits. This translation medium can then be used directly for studying the properties of soluble $G_s\alpha$ subunits or those of the membrane-bound $G_s\alpha$ subunits after reconstitution of cyc^- membranes.

[11] Y. Audigier, L. Journot, C. Pantaloni, and J. Bockaert, *J. Cell Biol.* **111,** 1427 (1990).

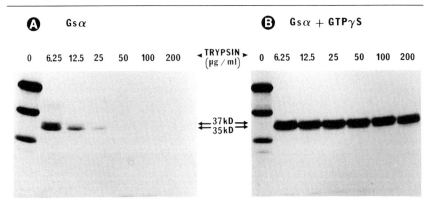

FIG. 2. Trypsin proteolysis of translation products. The translation medium was preincubated in the absence (A) or presence of 100 μM GTPγS and 10 mM MgCl$_2$ (B) for 10 min at 37°. The ^{35}S-labeled translation products were then digested for 60 min at 30° with the indicated concentrations of trypsin. The ^{35}S-labeled proteins were separated on a 12% SDS–polyacrylamide gel and visualized by autoradiography.

Assays for Studying Properties of Soluble $G_s\alpha$ Subunit

GTPγS-Induced Activation

During the activation of the α subunit, the replacement of GDP by GTP is associated with a conformational change which can be visualized by a differential sensitivity to trypsin degradation.[12–14] As far as *in vitro* translated $G_s\alpha$ subunit is concerned, similar tryptic fragments ranging between 35 and 37 kDa can be generated in the absence or presence of GTPγS; however, at high concentrations of trypsin, the GDP form is fully degraded, whereas digestion of the GTPγS-liganded form leads to the same protected 35–37 kDa fragments even when the ratio of enzyme to substrate is more than 1000 : 1 (Fig. 2). Consequently, the differential pattern of digestion in the presence of a high concentration of trypsin (100 μg/ml or more) clearly reveals whether a mutation has affected the property of an α_s chain to bind GTPγS and to undergo a conformational change.[10]

[12] B. K.-K. Fung and C. R. Nash, *J. Biol. Chem.* **258**, 10503 (1983).
[13] J. W. Winslow, J. R. van Amsterdam, and E. J. Neer, *J. Biol. Chem.* **261**, 7571 (1986).
[14] R. T. Miller, S. B. Masters, K. A. Sullivan, B. Beiderman, and H. R. Bourne, *Nature (London)* **334**, 712 (1988).

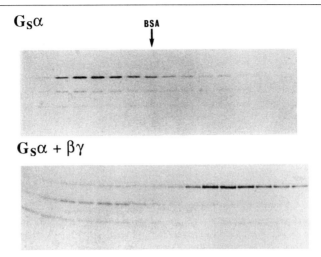

FIG. 3. Effect of $\beta\gamma$ subunits on the sedimentation rate of $G_s\alpha$ subunit. The translation medium was incubated in the absence (*top*) or presence of 5 ng purified $\beta\gamma$ subunits (*bottom*) for 18 hr at 0° and then applied to linear 5–20% sucrose gradients as described in Ref. 10. The fractions were collected and precipitated by 10% trichloroacetic acid. The ^{35}S-labeled polypeptides were resolved in a 12% SDS–polyacrylamide gel and visualized by autoradiography. The position of one of the marker proteins, bovine serum albumin (BSA), is indicated by the arrow.

Interaction with $\beta\gamma$ Subunits

The shift in the sedimentation rate on sucrose density gradients of $G_s\alpha$ subunits in the presence of $\beta\gamma$ subunits is a good index of subunit association. Although it is the subject of another chapter [18] of this volume, I would like to point out that the association between *in vitro* translated $G_s\alpha$ subunit and purified $\beta\gamma$ subunits occurs with great difficulty and requires unique conditions such as the absence of detergent in the sucrose buffers, high EDTA concentrations (10 mM), and low temperature (4°).[10] Besides its importance for the association, the absence of detergent allows the protein to sediment at the expected value corresponding to the molecular weight of the monomer or the heterotrimer (see Fig. 3).

Cholera Toxin-Catalyzed ADP-Ribosylation

Cholera toxin specifically catalyzes ADP-ribosylation of $G_s\alpha$ subunits.[15] Using [^{32}P]NAD$^+$, it is possible to radiolabel *in vitro* translated

[15] D. Cassel and Z. Selinger, *Proc. Natl. Acad. Sci. U.S.A.* **74**, 3307 (1977).

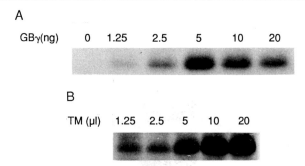

FIG. 4. Cholera toxin-catalyzed ADP-ribosylation of the translation products. $G_s\alpha$ mRNA was translated in the presence of unlabeled methionine. Various amounts of the translation medium (TM) were preincubated in the presence of the indicated amounts of purified $\beta\gamma$ subunits and then ADP-ribosylated by cholera toxin in the presence of $[^{32}P]NAD^+$. The ^{32}P-labeled bands were resolved in a 12% SDS–polyacrylamide gel and visualized by autoradiography. (A) Labeling efficiency as a function of increasing amounts of $\beta\gamma$ subunits. (B) Effect of increasing amounts of $G_s\alpha$ subunit on the rate of ADP-ribosylation performed in the presence of 5 ng of $\beta\gamma$ subunits.

$G_s\alpha$ subunits. Preliminary experiments revealed that ADP-ribosylation was not very efficient in the translation medium (Fig. 4A), but addition of $\beta\gamma$ subunits produced a dose-dependent increase of a ^{32}P-labeled 46-kDa protein (Fig. 4A). However, large amounts of $\beta\gamma$ subunits (5 ng) decrease the efficiency of ADP-ribosylation.

Translation of $G_s\alpha$ mRNA is performed with unlabeled methionine. Two microliters of translation medium is incubated at 0° for 18 hr in the presence or absence of 5 ng $\beta\gamma$ subunits purified from bovine brain (a generous gift from Dr. Sternweis, University of Texas Southwestern Medical Center, Dallas, TX). If we assume that 1 μl of mRNA gives rise to 2 fmol of $G_s\alpha$ subunit and that the molecular weight of $\beta\gamma$ subunits is 50,000, it follows that the interaction takes place between 4 fmol of α subunit and 100 fmol of $\beta\gamma$ subunits. However, given the respective amounts of G_s and other GTP-binding proteins in bovine brain, less than 10% of the $\beta\gamma$ subunits would be available to interact with the α subunit of G_s, and therefore the $\beta\gamma$ subunits are probably only in apparent excess.

The mixture (10 μl) is then incubated in a total volume of 20 μl for 30 min at 37° with 50 μg/ml of the activated toxin (activation of 0.5 mg/ml cholera toxin is carried out with 25 mM dithiothreitol at 37° for 30 min) and 2 HCi $[^{32}P]NAD^+$ (800–1000 Ci/mmol; Du Pont–New England Nuclear, Boston, MA) in 200 mM potassium phosphate buffer (pH 6.8), 10 mM MgCl$_2$, 1 mM ATP, 0.1 mM GTP, and 10 mM thymidine. The reaction is terminated by addition of 20 μl of 20% (w/v) trichloroacetic acid (TCA).

The medium is left on ice for 15 min and then centrifuged at 10,000 g for 5 min. The resulting pellet is washed twice with ether and resuspended in 25 μl of sample buffer. Samples are loaded on 12% sodium dodecyl sulfate (SDS)–polyacrylamide gels.

For a constant amount of $\beta\gamma$ subunits, the intensity of ^{32}P-labeling is proportional to the amount of translation medium until it reaches saturation (Fig. 4B). Consequently, the relationship between the amount of *in vitro* translated $G_s\alpha$ subunit and the labeling allows one to determine whether a change in the ADP-ribosylation rate of a mutated α chain results from a lower affinity or is due to a decreased efficiency of labeling.

All these assays clearly reveal that the *in vitro* translated $G_s\alpha$ subunit displays the intrinsic properties required for the function of G_s protein, namely, association with $\beta\gamma$ subunits and activation by either GTPγS or cholera toxin-catalyzed ADP-ribosylation. Because G_s protein exerts its transducing role at the level of plasma membrane, it is essential to demonstrate that the *in vitro* translated $G_s\alpha$ subunit is able to interact with the membrane.

Assays for Studying Membrane Association of $G_s\alpha$ Subunit

In Vitro Reconstitution of cyc$^-$ Membranes

cyc^- Membranes do not contain the α subunit of G_s protein, although $\beta\gamma$ subunits are present. Consequently, cyc^- membranes represent an ideal system for a reconstitution assay, since the stoichiometry of the different components is altered only at the level of the α subunit, the amount of which can be easily adjusted by varying the quantity of *in vitro* translated protein. When the translation medium is incubated with cyc^- membranes, most translation products cosediment with the membrane fraction,[9] and consequently the amount of membrane-bound ^{35}S-labeled 46-kDa protein is a good index of binding of $G_s\alpha$. Furthermore, these experiments clearly demonstrate that association of *in vitro* translated $G_s\alpha$ subunits can occur posttranslationally.

We first showed that only a fraction of *in vitro* translated $G_s\alpha$ subunit was able to associate with the membrane.[16] We thus decided to eliminate the unbound fraction by discarding the supernatant fraction and resuspend the pellet in order to obtain reconstituted cyc^- membranes at the previous protein concentration of 2 mg/ml.[16] Centrifugation of the reconstituted

---

[16] Y. Audigier, *in* "Signal Transduction" (G. Milligan, ed.), p. 57. IRL Press, New York, 1992.

membranes clearly revealed that 100% of ^{35}S-labeled $G_s\alpha$ subunit was then recovered in the pellet fraction.[16]

To define the optimal conditions, we analyzed membrane association as a function of the amount of *in vitro* translated $G_s\alpha$ subunit or cyc^- membranes (Fig. 5A). Densitometric analysis of the dose–response curves for the two different membrane concentrations showed that, although the membrane association saturated at high amounts of *in vitro* translated $G_s\alpha$ subunit, the lower quantity of membranes (20 μg) bound more $G_s\alpha$ subunit per milligram of membrane protein (Fig. 5B).

The standard assay for the membrane association is therefore carried out as follows: 10 μl of translation medium is incubated at 37° for 30 min with an equal volume of cyc^- membranes (2 mg protein/ml). The reconstitution medium is centrifuged at 10,000 g for 5 min. The supernatant is discarded, and the reconstituted cyc^- membranes are resuspended in 10 μl of a homogenization buffer containing 2 mM MgCl$_2$, 1 mM EDTA, 1 mM dithiothreitol, and sodium HEPES at pH 8. By using this very simple assay, we showed that proteolytic removal by V8 protease of carboxyl-terminal residues abolished membrane association[11] and that addition of amino acids 367–376 of $G_s\alpha$ was sufficient to promote membrane association of the soluble amino-terminal deleted $G_{i1}\alpha$.[17]

In Vivo Targeting to Plasma Membrane

In vivo targeting to the plasma membrane is interesting for determining how deletions or mutations can affect the subcellular location of a mutated α chain. Because of the difficulties in transfecting S49 cells, simian COS cells have been mainly used for such studies.[18,19] The membrane localization of the transiently expressed α subunit is analyzed by immunoblotting[18] or immunoprecipitation.[19] We used a similar approach with $G_s\alpha$ subunits, and, for this purpose, the cDNA coding for the long form of $G_s\alpha$ subunit was subcloned in the CDM8 vector between *Hin*dIII and *Xho*I sites.

Calcium phosphate-mediated transfection of the COS-7 cells is performed according to the method of Chen and Okayama,[20] and we analyze the transient expression of the transfected DNA by characterization of the $G_s\alpha$ subunit protein in a plasma membrane fraction using antibodies

---

[17] L. Journot, C. Pantaloni, M.-A. Poul, H. Mazarguil, J. Bockaert, and Y. Audigier, *Proc. Natl. Acad. Sci. U.S.A.* **88**, 10054 (1991).

[18] S. M. Mumby, R. O. Heukeroth, J. I. Gordon, and A. G. Gilman, *Proc. Natl. Acad. Sci. U.S.A.* **87**, 728 (1990).

[19] T. L. Z. Jones, W. F. Simonds, J. J. Merendino, M. R. Brann, and A. M. Spiegel, *Proc. Natl. Acad. Sci. U.S.A.* **87**, 568 (1990).

[20] C. Chen and H. Okayama, *Mol. Cell. Biol.* **7**, 2745 (1987).

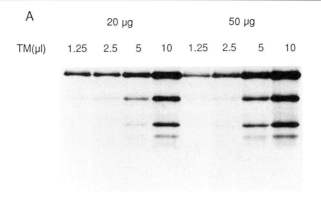

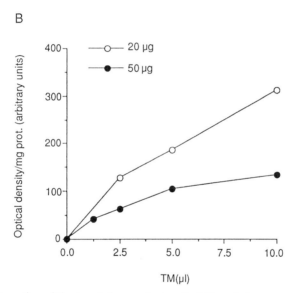

FIG. 5. Interaction of the translation products with S49 cyc^- plasma membranes. The $G_s\alpha$ mRNA was translated in the presence of [^{35}S]methionine. Various amounts of translation medium were mixed with 20 or 50 μg of S49 cyc^- plasma membranes and incubated at 37° for 30 min. The medium was then centrifuged and the supernatant discarded. The membrane pellet was resuspended and centrifuged again. The pellet fraction was analyzed by 12% SDS–polyacrylamide gel electrophoresis and visualized by autoradiography. (A) Autoradiogram of ^{35}S-labeled proteins that were present in the pellet fraction. (B) Densitometric analysis of the autoradiogram.

directed against the carboxyl-terminal decapeptide of $G_s\alpha$ subunit. In this protocol, the calcium phosphate–DNA complex is formed gradually in the medium during incubation with cells and precipitates on them. The crucial factors for high transformation efficiency are the pH of the buffer (BES; pH 6.95) used for the precipitation, a low CO_2 atmosphere (3%) during incubation of DNA with the cells, and a large amount (20 µg) of the circular form of DNA.

After an overnight incubation with calcium phosphate–DNA complex under 3% CO_2, cells on a 10-cm plate are rinsed twice with 10 ml growth medium (Dulbecco's modified Eagle's medium with sodium pyruvate and 1 g/liter glucose containing 10% fetal calf serum), refed, and incubated for 48 hr at 37° under 5% CO_2. Then we prepare from the transfected cells a fraction enriched in plasma membranes by the following procedure. The cells are rinsed twice with 10 ml of phosphate-buffered saline (PBS) without Ca^{2+} and Mg^{2+}. The washed cells are scraped, resuspended in 1.6 ml PBS, and transferred into 2-ml Eppendorf tubes. The tubes are centrifuged at 200 g for 3 min at 4°. The cell pellet is lysed for 10 min on ice with 0.5 ml of homogenization buffer containing 2 µM leupeptin, 0.3 µM aprotinin, 1 µM pepstatin, and 100 µg/ml soybean trypsin inhibitor. The cells are homogenized by trituration through a 20-gauge needle and centrifuged at 1000 g for 3 min at 4° to remove the nuclear pellet. The supernatant (~400 µl) is then centrifuged at 30 psi for 1 hr in a Beckman airfuge. The supernatant fraction is separated from the pellet fraction, which is resuspended in 100 µl of homogenization buffer containing protease inhibitors in a Dounce homogenizer. The protein concentration in the pellet and supernatant fractions is determined by the Bradford method[21] with bovine serum albumin (BSA) as a standard. From a 10-cm plate, we obtain approximately 300 µg of proteins in the supernatant fraction and 200 µg of proteins in the pellet fraction.

Fifty micrograms of the pellet fraction and an equivalent amount of supernatant (~75 µg) are precipitated with 10% (w/v) TCA on ice for 15 min. After centrifugation at 10,000 g for 5 min, the pellet is resuspended in 20 µl of sample loading buffer and incubated for 30 min at 50°. The solubilized pellet is then heated at 100° for 5 min, and the proteins are resolved by 12% (w/v) SDS–polyacrylamide gel electrophoresis (SDS–PAGE) by the method of Laemmli.[22]

Proteins are transferred onto nitrocellulose sheets, as previously described.[23] The nitrocellulose sheets are dried and stained with red

---

[21] M. M. Bradford, *Anal. Biochem.* **72**, 248 (1976).
[22] U. K. Laemmli, *Nature (London)* **227**, 680 (1970).
[23] Y. Audigier, C. Pantaloni, J. Bigay, P. Deterre, J. Bockaert, and V. Homburger, *FEBS Lett.* **189**, 1 (1985).

Ponceau S in order to assess the quality of the transfer. After blocking nonspecific binding with 3% gelatin, blots are incubated overnight at room temperature in 10 mM Tris-HCl, pH 7.5, 500 mM NaCl containing 0.3% gelatin, and 1:200 antiserum raised against the carboxyl-terminal decapeptide of $G_s\alpha$ subunit. The blots are then washed and incubated at room temperature with ^{125}I-labeled protein A [100,000 counts/min (cpm)/ml] for 1 hr. After extensive washing, the blots are dried and exposed to Kodak (Rochester, NY) XAR-5 with image-intensifying screens at $-80°$.

Figure 6 clearly reveals that the $G_s\alpha$ subunit transfected in COS cells is essentially distributed in the plasma membrane (P) and not in the cytoplasm (S). Interestingly, the polypeptides corresponding to initiations at internal AUG codons are also expressed *in vivo*, and they too are associated with the plasma membrane, suggesting that the determinant for $G_s\alpha$ targeting to the plasma membrane is located in the carboxyl-terminal half of the protein.

The same approach was previously used for the localization of the crucial residues involved in membrane association and *in vivo* targeting of $G_i\alpha$ and $G_o\alpha$ subunits.[18,19] The results demonstrated the importance of the myristoylated amino-terminal glycine in both membrane anchorage and targeting to the plasma membrane.

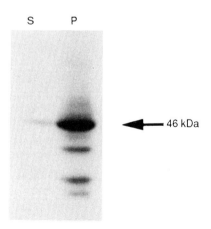

FIG. 6. Membrane association of transiently expressed $G_s\alpha$ subunit after transfection in COS cells. COS cells were transfected with DNA encoding α_s subcloned in the pCDM8 vector. Membrane and cytoplasmic fractions of the cells were prepared 2 days after transfection. The proteins from each fraction were resolved by SDS-PAGE and transferred to nitrocellulose. A 1:200 dilution of crude α_s antiserum and ^{125}I-labeled protein A was used to detect $G_s\alpha$ subunits on the blot. The blot was exposed to film for 24 hr at $-70°$ with an intensifying screen.

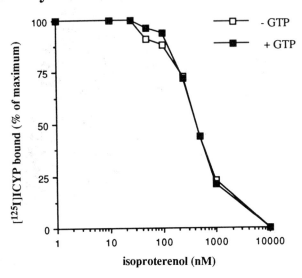

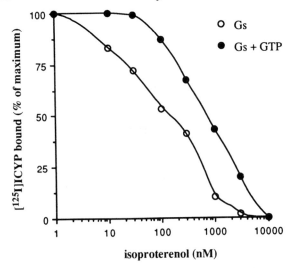

FIG. 7. Effect of GTP on the affinity of agonist–receptor interactions measured on cyc^- membranes and reconstituted cyc^- membranes. The cyc^- membranes were incubated with reticulocyte lysate alone (A) or reconstituted with *in vitro* translated $G_s\alpha$ subunit (B). Binding of 75 pM [^{125}I]iodocyanopindolol ([^{125}I]ICYP) to the membranes in competition with increas-

Assays for Studying Properties of Membrane-Bound $G_s\alpha$ Subunit

The properties of reconstituted cyc^- membranes are expected to be identical to those of the wild-type membranes if the membrane-bound $G_s\alpha$ subunit fulfills its function in transmembrane signaling: the coupling of the β-adrenergic receptor to adenylyl cyclase should be restored by posttranslational addition of $G_s\alpha$ subunit.

GTP Effect and Receptor Coupling

Because it represents the functional consequence of the interaction between the receptor and the GTP-binding protein, restoration of the GTP effect on agonist affinity for the β-adrenergic receptor is a corollary to the effective reconstitution of cyc^- membranes by *in vitro* translated $G_s\alpha$ subunit. The GTP effect is not observed with antagonists and therefore can be measured by determining the agonist-induced displacement of a radiolabeled antagonist such as [^{125}I]iodocyanopindolol, a specific antagonist of β-adrenergic receptors. For example, the agonist isoproterenol displays a high affinity for the receptor in the absence of GTP, whereas it has a lower affinity in the presence of GTP.[24] When [^{125}I]iodocyanopindolol binding is performed on cyc^- membranes, the curves of displacement by isoproterenol are identical in the presence or absence of GTP (Fig. 7A). Reconstitution of cyc^- membranes by *in vitro* translated $G_s\alpha$ subunit shifts to the left the agonist-induced displacement determined in the absence of GTP (Fig. 7B).

To reduce the nonspecific sticking of the iodinated ligand, we prepare a dilution buffer which is made up with 50 mM sodium HEPES (pH 8), 10 mM MgSO$_4$, and 1 mg/ml BSA. This dilution buffer is used for resuspending the reconstituted cyc^- membranes at a final protein concentration of 0.5 mg/ml and for diluting the radioligand. Then the resuspended membranes are divided into two batches, one to which no GTP is added and the other containing 300 μM GTP. Nonspecific sticking can also be decreased by adding first the membranes (30 μl) and the effector (10 μl) at 0°. The assay is then started by adding 10 μl of 350 pM [^{125}I]iodocyano-

---

[24] H. R. Bourne, D. Kaslow, H. R. Kaslow, M. R. Salomon, and V. Licko, *Mol. Pharmacol.* **20**, 435 (1981).

---

ing concentrations of isoproterenol was then measured in the absence (open symbols) or presence of 300 μM GTP (filled symbols). Maximum binding represents the specific binding of radioligand in the absence of competing ligands. Each data point is the mean of triplicate determinations in a single representative experiment.

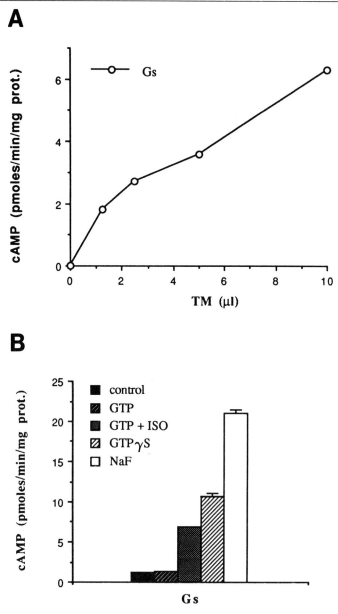

FIG. 8. Adenylyl cyclase activity of reconstituted cyc^- membranes. The cyc^- membranes were reconstituted with *in vitro* translated $G_s\alpha$ subunit. cAMP levels were measured as described in Ref. 25 after incubation of 20 µg of reconstituted cyc^- membranes with the indicated effectors at the following concentrations: 0.1 mM GTP, 10 µM isoproterenol (ISO),

pindolol. After 90 min at 30°, the reaction is stopped by the addition of 3 ml of ice-cold Tris-HCl, 50 mM, pH 8, and the medium is filtered over Whatman (Clifton, NJ) 2.4 cm GF/C filters. The tube is washed twice with the same volume of Tris buffer, and the filters are again washed twice with 3 ml of Tris buffer. We find that it is necessary to soak the Whatman 2.4 cm GF/C filters for 2 hr in 10 mM $MgSO_4$ and 1 mg/ml BSA in order to improve the reproducibility of binding assays.

Adenylyl Cyclase Activation

Validation of the reconstitution is also provided by restoration of the basal activity of adenylyl cyclase as well as by restoration of effector responsiveness to various agents known to act at different levels of the G_s-mediated transduction pathway. As far as reconstitution is concerned, we have found that lowering the reconstitution temperature to 30° does not alter membrane association, whereas it strongly decreases inactivation of adenylyl cyclase. The methodology used for measuring cAMP levels has been reported previously[25] and is described in Chapter 3 of this volume.

As expected, activation of adenylyl cyclase by GTPγS (Fig. 8A) or NaF (data not shown) is dependent on the amount of $G_s\alpha$ subunit that has been incorporated in cyc^- membranes. However, it reaches a plateau because increasing the amount of translation medium above 10 μl does not further stimulate adenylyl cyclase activity. The results parallel those obtained for the membrane association when the ratio of $G_s\alpha$ subunit to membranes is changed. Furthermore, not only is the basal activity of adenylyl cyclase restored, but adenylyl cyclase also becomes stimulated by isoproterenol, an agonist of the β-adrenergic receptor, and GTPγS, a hydrolysis-resistant guanine nucleotide, or a mixture of NaF and $AlCl_3$, which directly activate $G_s\alpha$ subunit (Fig. 8B).

In conclusion, these assays can be used for the characterization of the phenotype of mutated α_s chains in order to locate the functional domains

---

[25] J. Bockaert, P. Deterre, C. Pfister, G. Guillon, and M. Chabre, *EMBO J.* **4**, 1413 (1985).

---

10 μM GTPγS, and 10 mM NaF plus 200 μM $AlCl_3$. The values represent the means ± S.E. of triplicate determinations from one representative experiment. (A) GTPγS-induced stimulation of adenylyl cyclase as a function of increasing amounts of translation medium (TM). (B) Adenylyl cyclase activity in the presence of various effectors.

of $G_s\alpha$ subunits and to define the crucial residues required for each functional property.

Acknowledgments

I would like to thank R. van der Neut, L. Journot, C. Pantaloni, and J. Bockaert for various contributions to the information contained in this chapter.

[20] Myristoylation of G-Protein α Subunits

By SUSANNE M. MUMBY and MAURINE E. LINDER

Introduction

Many G-protein α subunits are modified by the 14-carbon saturated fatty acid myristic acid. As in other myristoylated proteins, the site of the amide-linked myristate in α subunits is apparently an amino-terminal glycine.[1-3] Although not required, another common feature of myristoylated proteins is the presence of a threonine or serine four residues from the N-terminal glycine.[4] The known myristoylated α subunits, including $G_o\alpha$, $G_i\alpha$, and $G_z\alpha$, all have an N-terminal glycine (following cleavage of the initiator methionine) at position 2 and a serine at position 6, features that are underlined in Table I. The α subunit of transducin ($G_t\alpha$) also has these features but is now known to be modified heterogeneously by a variety of saturated and unsaturated 12- and 14-carbon fatty acids.[5,6] It is possible that other α subunits are modified by more than one fatty acid as well.

The $G_s\alpha$ subunit is not myristoylated.[1,2] It has an N-terminal glycine but no serine at position 6. We have substituted a serine for the asparagine at position 6 of $G_s\alpha$ by site-directed mutagenesis and expressed the mutant

---

[1] J. E. Buss, S. M. Mumby, P. J. Casey, A. G. Gilman, and B. M. Sefton, *Proc. Natl. Acad. Sci. U.S.A.* **84**, 7493 (1987).
[2] S. M. Mumby, R. O. Heuckeroth, J. I. Gordon, and A. G. Gilman, *Proc. Natl. Acad. Sci. U.S.A.* **87**, 728 (1990).
[3] T. L. Z. Jones, W. F. Simonds, J. J. Merendino, Jr., M. R. Brann, and A. M. Spiegel, *Proc. Natl. Acad. Sci. U.S.A.* **87**, 568 (1990).
[4] D. A. Towler, J. I. Gordon, S. P. Adams, and L. Glaser, *Annu. Rev. Biochem.* **57**, 69 (1988).
[5] T. A. Neubert, R. S. Johnson, J. B. Hurley, and K. A. Walsh, *J. Biol. Chem.* **267**, 18274 (1992).
[6] K. Kokame, Y. Fukada, T. Yoshizawa, T. Takao, and Y. Shimonishi, *Nature (London)* **359**, 749 (1992).

TABLE I
α Subunit Amino-Terminal Amino Acid Alignment and Myristoylation

		Criteria for myristoylation			
α Subunit	Amino-terminal sequence	Sequence[a]	[^{3}H]Myr incorporation[b]	Chemical analysis[c]	Mass spectrometry[d,e]
o	MGCTLSAE	+	+	+	
i1	MGCTLSAE	+	+	+	
i2	MGCTVSAE	+	+		
i3	MGCTLSAE	+	+		
z	MGCRQSSE	+	+		
t1	MGAGASAE	+	+	−[f]	+
s	MGCLGNSK	−	−	−	
olf	MGCLGNSS	−			

[a] Positive (+) sequence criteria for myristoylation are G at position 2 and S (or T) at position 6. Position 2 is not a G in α subunits q, 11, 12, 13, 14, 15, and 16; therefore, the sequences of these proteins are not shown.

[b] S. M. Mumby, R. O. Heuckeroth, J. I. Gordon, and A. G. Gilman, *Proc. Natl. Acad. Sci. U.S.A.* **87,** 728 (1990).

[c] J. E. Buss, S. M. Mumby, P. J. Casey, A. G. Gilman, and B. M. Sefton, *Proc. Natl. Acad. Sci. U.S.A.* **84,** 7493 (1987).

[d] T. A. Neubert, R. S. Johnson, J. B. Hurley, and K. A. Walsh, *J. Biol. Chem.* **267,** 18274 (1992).

[e] K. Kokame, Y. Fukada, T. Yoshizawa, T. Takao, and Y. Shimonishi, *Nature (London)* **359,** 749 (1992).

[f] Myristate was not found by chemical analysis of $α_t$ purified from bovine retina.[1] However, tritium from [^{3}H]myristic acid was incorporated into $α_t$ expressed in COS cells by transfection.[2] Subsequently, it was determined by mass spectral analysis that $α_t$ from bovine retina is heterogeneously acylated by saturated and unsaturated 12- and 14-carbon fatty acids.[5,6] The amount of myristate found by mass spectrometry was below the detection limit in previous chemical analyses.[3]

protein in simian COS cells. However, tritiated myristate was not incorporated into the protein.[7] The signal for myristoylation of α subunits therefore involves more than the N-terminal glycine and a serine at position 6. The deduced amino acid sequence of a number of α subunits (q, 11, 12, 13, 14, 15, and 16) does not include an amino-terminal glycine. Therefore, these proteins are not anticipated to be myristoylated (and the sequences are not shown in Table I).

We have explored the function of α-subunit acylation by myristate in two systems: (1) by site-directed mutagenesis and expression of mutant α subunits in COS cells and (2) by biochemical characterization of myris-

[7] M. E. Linder and S. M. Mumby, unpublished (1991).

toylated and nonmyristoylated forms of recombinant $G_o\alpha$ purified from *Escherichia coli*. Myristoylation of α occurs in *E. coli* when the α subunit is coexpressed with *N*-myristoyltransferase (NMT).[8,9] Wild-type $G_o\alpha$ or $G_i\alpha$, when expressed in COS cells, incorporates tritium from [^{3}H]myristate and is associated with the membranes of fractionated cells. If the N-terminal glycine is substituted with alanine (by site-directed mutagenesis), tritium from [^{3}H]myristate is not incorporated, and the protein is found in the soluble fraction of the cells.[2,3] Thus, it was concluded that myristoylation facilitates membrane association of α subunits. Sternweis proposed that $\beta\gamma$ subunits act as membrane anchors for α subunits.[10] Although nonmyristoylated recombinant $G_o\alpha$ purified from *E. coli* has a reduced affinity for $\beta\gamma$, myristoylated recombinant $G_o\alpha$ was indistinguishable from bovine brain $G_o\alpha$ in its subunit interactions.[11] Thus, myristoylation of the α subunit apparently increases its affinity for $\beta\gamma$. Although a reduced affinity of nonmyristoylated α for $\beta\gamma$ may contribute to the cytoplasmic distribution of the protein in COS cells, $\beta\gamma$ does not appear to be entirely responsible for membrane targeting of α. Myristoylated α subunits can be overexpressed by transfection to greatly exceed the available $\beta\gamma$, yet are still membrane-associated.[2,12] The myristate of α subunits may therefore interact with other components of the cell membrane in addition to $\beta\gamma$.

In this chapter we describe how to reconstitute N-myristoylation of α subunits in *E. coli* by concurrent expression of α and NMT. The advantage of the bacterial expression system is the ability to purify large quantities of an α subunit without contamination by other forms of α, a task that is virtually impossible from mammalian tissues. In addition, we give protocols for the incorporation of [^{3}H]myristate into α subunits expressed in *E. coli* and in mammalian cells and for isolation of radiolabeled α by immunoprecipitation. Some of the information given here has been reviewed in previous publications.[13,14]

---

[8] The systematic name for NMT is myristoyl-CoA: protein *N*-myristoyltransferase (EC 2.3.1.97).

[9] R. J. Duronio, E. Jackson-Machelski, R. O. Heuckeroth, P. O. Olins, C. S. Devine, W. Yonemoto, L. W. Slice, S. S. Taylor, and J. I. Gordon, *Proc. Natl. Acad. Sci. U.S.A.* **87,** 1506 (1990).

[10] P. C. Sternweis, *J. Biol. Chem.* **261,** 631 (1986).

[11] M. E. Linder, I.-H. Pang, R. J. Duronio, J. I. Gordon, P. C. Sternweis, and A. G. Gilman, *J. Biol. Chem.* **266,** 4654 (1991).

[12] W. F. Simonds, R. M. Collins, A. M. Spiegel, and M. R. Brann, *Biochem. Biophys. Res. Commun.* **164,** 46 (1989).

[13] R. J. Duronio, D. A. Rudnick, R. L. Johnson, M. E. Linder, and J. I. Gordon, *Methods (San Diego)* **1,** 253 (1990).

[14] S. M. Mumby and J. E. Buss, *Methods (San Diego)* **1,** 216 (1990).

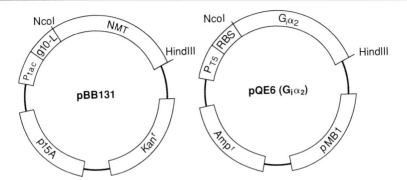

FIG. 1. Plasmid constructs for coexpression of $G_i\alpha_2$ and NMT. Plasmids designated pBB131 and pQE6($G_i\alpha_2$) were used to express yeast NMT and rat $G_i\alpha_2$, respectively, in *E. coli*. See text for explanation.

Coexpression of G-Protein α Subunits and *N*-Myristoyltransferase in *Escherichia coli*

The dual plasmid system developed by Duronio *et al.*[9] was used initially to synthesize myristoylated recombinant α subunits.[11,13,15] We describe here modifications of that system that result in increased yields of recombinant protein.

Construction of Plasmids

As shown in Fig. 1, the cDNAs for NMT and a protein substrate ($G_i\alpha_2$) are cloned into separate plasmids, each under the regulation of a promoter inducible by isopropyl-1-thio-β-D-galactopyranoside (IPTG). The plasmids carry either ampicillin (Amp^r) or kanamycin (Kan^r) resistance markers and different (but compatible) origins of replication (p15A, pMB1). The *Saccharomyces cerevisiae NMT1* gene is subcloned into a plasmid designated pBB131. The promoter for NMT (P_{tac}) is fused to a translational "enhancer" derived from the gene 10 leader region (g10-L) of bacteriophage T7.[16] The cDNAs for $G_i\alpha_2$ (or $G_i\alpha_1$, $G_i\alpha_3$, or $G_o\alpha$) are cloned into pQE6, a plasmid available from Qiagen (Chatsworth, CA). pQE6 has an *E. coli* bacteriophage T5 promoter containing two *lac* operator sequences (P_{T5}) and a synthetic ribosomal binding site (RBS). cDNAs encoding rat $G_o\alpha$, $G_i\alpha_1$, $G_i\alpha_2$, and $G_i\alpha_3$[17] were provided by R. Reed (Johns Hopkins

---

[15] R. J. Duronio, D. A. Rudnick, S. P. Adams, D. A. Towler, and J. I. Gordon, *J. Biol. Chem.* **266**, 10498 (1991).
[16] P. O. Olins and S. H. Rangwala, *J. Biol. Chem.* **264**, 16973 (1989).
[17] D. T. Jones and R. R. Reed, *J. Biol. Chem.* **262**, 14241 (1987).

University, Baltimore, MD). NcoI sites had been introduced into the $G_i\alpha_2$ and $G_i\alpha_3$ cDNAs at a position corresponding to the initiator methionine by site-directed mutagenesis.[13] An NcoI site at nucleotide 411 in the $G_i\alpha_2$ cDNA had been removed by site-directed mutagenesis.[13] The DNA encoding $G_i\alpha_2$ in vector NpT7-5[18] is digested with HindIII to generate the 3' end of the fragment and then partially digested with NcoI. Restriction fragments (NcoI–BglII and BglII–HindIII) from vector pQE6 and the NcoI–HindIII fragment containing the coding sequence of $G_i\alpha_2$ are ligated and transformed into E. coli strain JM109. Positive clones are identified by restriction enzyme analysis of plasmid DNA. The resulting plasmid is designated pQE6($G_i\alpha_2$). pBB131 is transformed into JM109 cells carrying pQE6($G_i\alpha_2$), and transformants are selected for resistance to both kanamycin and ampicillin. The same strategy is employed to generate pQE6 plasmids containing $G_o\alpha$, $G_i\alpha_1$, and $G_i\alpha_3$.

Expression of G-Protein α Subunits

To assay for expression of $G_i\alpha_2$, cells harboring pBB131 and pQE6($G_i\alpha_2$) are grown at 30° to an OD_{600} of 0.5–0.7 in enriched medium (2% tryptone, 1% yeast extract, 0.5% NaCl, 0.2% glycerol, and 50 mM KH_2PO_4, pH 7.2) supplemented with 50 µg/ml ampicillin and 50 µg/ml kanamycin. IPTG is added to a final concentration of 100 µM to induce synthesis of NMT and Gα, and the culture is incubated for 90 min. Cultures (0.5 ml) are harvested by centrifugation at 13,000 g for 5 min at 4° in a microcentrifuge. The cell pellets are washed once with phosphate-buffered saline (PBS) and lysed by boiling for 5 min in 25 µl sodium dodecyl sulfate–polyacrylamide gel electrophoresis (SDS–PAGE) sample buffer. Cellular debris is removed by centrifugation at 13,000 g for 5 min at room temperature. For immunoblot analysis, a 10-µl aliquot is subjected to SDS–PAGE and transferred to nitrocellulose. The immunoblot is processed as previously described[19] using antiserum P-960 which cross-reacts with most G-protein α subunits.[20] Similar and other useful antisera are commercially available from Calbiochem (La Jolla, CA), Du Pont NEN (Boston, MA), Upstate Biotechnology Inc. (Lake Placid, NY) and Gramsch Laboratories (Schwabhausen, Germany). Lysates from cells labeled with [^{3}H]myristate are subjected to SDS–PAGE and fluorography. Immunoblot analysis reveals that $G_i\alpha_2$ is expressed (Fig. 2, lanes 1 and 2).

[18] M. E. Linder, D. A. Ewald, R. J. Miller, and A. G. Gilman, *J. Biol. Chem.* **264,** 8243 (1990).
[19] B. A. Harris, J. D. Robishaw, S. M. Mumby, and A. G. Gilman, *Science* **229,** 1274 (1985).
[20] P. J. Casey, H. K. W. Fong, M. I. Simon, and A. G. Gilman, *J. Biol. Chem.* **265,** 2383 (1990).

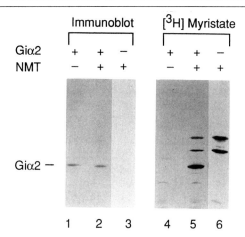

FIG. 2. Coexpression of $G_i\alpha_2$ and NMT in *E. coli*. *Escherichia coli* cells expressing $G_i\alpha_2$ (lanes 1, 2, 4, and 5) and/or NMT (lanes 2, 3, 5, and 6) were incubated with [^{3}H]myristate for 45 min following induction and processed as described in the text. The cell pellet from an 0.5-ml culture was resuspended in 25 μl SDS-PAGE sample buffer, and 10 μl was loaded on the gel (lanes 4-6). A second culture without radioactive label was processed in parallel. Samples (10 μl) from the cultures grown without [^{3}H]myristate (lanes 1-3) were resolved on an 11% polyacrylamide gel and transferred to nicrocellulose. The immunoblots were probed with antiserum P-960 at a 1 : 1000 dilution (lanes 1-3). Samples labeled with [^{3}H]myristate (lanes 4-6) were subjected to SDS-PAGE and fluorography. The blot and fluorogram were exposed to Kodak XAR-5 film overnight.

Purification of Myristoylated Recombinant $G_i\alpha_2$

The following protocol is for 10 1-liter cultures. Inoculate a 3-ml culture with bacteria from a glycerol stock or a single colony from a plate. Grow the cells for approximately 8 hr. Add this culture to 150 ml of enriched medium containing antibiotics and grow overnight. Add 10 ml overnight culture to each of 10 liters of medium. Grow the cells to an OD_{600} of 0.5-0.7 and subsequently add IPTG to a final concentration of 100 μM and chloramphenicol to a final concentration of 1 μg/ml. The addition of chloramphenicol appears to enhance the production of some α subunits. Grow the cells overnight (16 hr) at 30° with gentle shaking (200 rpm). The optimal time to harvest cells after induction varies with the individual α subunits. For $G_i\alpha_1$, $G_i\alpha_2$, and $G_o\alpha$, 16 to 18 hr yielded the best expression. Expression of $G_i\alpha_3$ was optimal at 6 hr postinduction.

The cells are harvested in a Beckman JA10 rotor at 7000 rpm (9000 g) for 10 min at 4°. The medium is discarded, and the cell pellets are scraped from the bottles, flash-frozen in liquid N_2, and stored at $-80°$. Even if the cells will be processed immediately, the cell pellets should undergo a freeze-thaw cycle as this probably facilitates cell lysis.

The following steps are all performed at 4°. The frozen cell paste is added to a beaker containing 1.8 liters of TEDP [50 mM Tris, pH 8, 1 mM ethylenediaminetetraacetic acid (EDTA), 1 mM dithiothreitol (DTT), and 0.1 mM phenylmethylsulfonyl fluoride (PMSF)] and is resuspended with gentle stirring. Resuspend any clumps with a syringe and 18-gauge cannula. Add lysozyme to a final concentration of 0.2 mg/ml and incubate for 30 min on ice. The lysate should become viscous. Add $MgSO_4$ to a final concentration of 5 mM and 20 mg DNase I (in powder form) and incubate for 30 min. Centrifuge the lysate in a Beckman JA14 rotor at 14,000 rpm (30,000 g) for 1 hr at 4°. Collect the supernatant fraction. The $G_i\alpha_2$ protein is purified from the soluble fraction as we have not been able to extract significant amounts of active protein from the particulate fraction with detergent.

Myristoylated recombinant subunits are purified according to a modification of the protocol used to purify nonmyristoylated recombinant proteins.[18] The soluble fraction is mixed with 200 ml of DEAE-Sephacel (Pharmacia, Piscataway, NJ) that has been equilibrated in TEDP. The resin is incubated with the extract for 20 min with occasional stirring and is then collected on a Whatman (Clifton, NJ) No. 4 filter in a Büchner funnel. The resin is washed with 1.5 liters of TEDP. Protein is eluted from the resin with three 200-ml volumes of TEDP containing 300 mM NaCl.

The DEAE eluate is adjusted to 1.2 M in $(NH_4)_2SO_4$ by the addition of 0.5 volume (300 ml) of 3.6 M $(NH_4)_2SO_4$. GDP is added to a concentration of 25 μM. (GDP is included in the buffers during this stage of purification since high ionic strength facilitates dissociation of the nucleotide from G-protein α subunits.[21] The protein is more sensitive to denaturation when in the nucleotide-free form.) The mixture is incubated on ice for 10 min and is centrifuged at 8500 rpm for 10 min in a Beckman JA14 rotor (11,000 g) to remove any precipitated protein. The supernatant fraction is applied to a 100-ml phenyl-Sepharose (Pharmacia) column (2.6 × 20 cm) that is equilibrated in TEDP containing 25 μM GDP and 1.2 M $(NH_4)_2SO_4$. Protein is eluted with a 1-liter descending gradient of $(NH_4)_2SO_4$ (1.2 to 0 M) in TEDP, and 25 μM GDP. Glycerol is added (35%, v/v, final concentration) to increase the density of the dilution buffer (stabilizing gradient formation) and to slow the rate of dissociation of GDP from the α subunit.[21] The column is then washed with an additional 250 ml of TEDP containing 25 μM GDP. Fractions (15 ml) are collected across the gradient and the final wash step. Aliquots (1–2 μl) of the fractions from the phenyl-Sepharose column are assayed by pertussis toxin-catalyzed ADP-ribosylation.[18]

---

[21] K. M. Ferguson, T. Higashijima, M. D. Smigel, and A. G. Gilman, *J. Biol. Chem.* **261**, 7393 (1986).

The elution profile of $G_i\alpha_2$ from the phenyl-Sepharose column is shown in Fig. 3. Myristoylated protein is separated from nonmyristoylated protein on this column. The first peak of activity elutes at a position where the nonmyristoylated form of the protein is normally found[18]; the second elutes at low salt concentration, consistent with the behavior of a more hydrophobic protein. Analysis of $G_o\alpha$[11] expressed in this system demonstrated that $G_o\alpha$ purified from the first peak of activity was not myristoylated. $G_o\alpha$ purified from the second peak had a blocked amino terminus, and chemical analysis of the hydrolyzed protein revealed the presence of myristic acid.

The second peak of activity from the phenyl-Sepharose column is pooled (100–125 ml), desalted over a PM30 membrane in an Amicon (Danvers, MA) ultrafiltration device, and diluted into Q buffer (50 mM Tris-HCl, pH 8, 0.02 mM EDTA, and 1 mM DTT). The protein is taken through successive concentration and dilution cycles until the $(NH_4)_2SO_4$ concentration is reduced below 20 mM. The protein is applied to a 100-ml column of Q-Sepharose (Pharmacia) that has been equilibrated in Q buffer. Protein is eluted with a 240-ml gradient of NaCl (0–300 mM) in Q buffer. Fractions of 8 ml are collected, and aliquots (2.5 μl) of the fractions are assayed by guanosine 5'-O-(3-thio)triphosphate (GTPγS) binding.[18] Other GTPγS-binding proteins were resolved from the α subunit in the previous

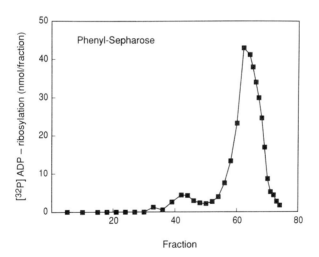

FIG. 3. Phenyl-Sepharose chromatography of $G_i\alpha_2$ from *E. coli* that coexpressed NMT. The DEAE eluate was chromatographed on a phenyl-Sepharose column, as described in the text. Aliquots (1.5 μl) of the fractions indicated were assayed by pertussis toxin-catalyzed ADP-ribosylation in the presence of 5 pmol βγ subunits purified from bovine brain. The first peak represents nonacylated $G_i\alpha_2$, and the second peak represents myristoylated $G_i\alpha_2$.

step. The peak fractions are pooled (usually 25 ml) and adjusted to a phosphate concentration of 10 mM by the addition of 1/100 volume of 1 M potassium phosphate (pH 8). The protein solution is then diluted with an equal volume of Hap buffer (10 mM Tris-HCl, pH 8, 10 mM potassium phosphate, pH 8, and 1 mM DTT) and applied to a 20-ml column (1 × 20 cm) of hydroxylapatite (Bio-Rad HTP, San Francisco, CA) that has been equilibrated in Hap buffer. The column is washed with 25 ml Hap buffer, and protein is eluted by a 200-ml gradient of phosphate (10–300 mM) in 10 mM Tris-HCl (pH 8) and 1 mM DTT. The gradient is collected in 50 4-ml fractions. Fractions containing $G_i\alpha_2$ are identified by SDS–PAGE and GTPγS binding, pooled, and concentrated to 4 ml using an Amicon ultrafiltration device (PM30 membrane).

The remaining contaminating proteins are removed by gel filtration on an Ultrogel AcA 44 (Pharmacia-LKB, Piscataway, NJ) column (175 ml resin, 2.6 × 34 cm) equilibrated in 50 mM Tris (pH 8), 1 mM EDTA, 1 mM DTT, 100 mM NaCl, and 25 μM GTP. Fractions of 4 ml are collected, assayed by GTPγS binding, and checked for purity by SDS–PAGE. The final pool of purified recombinant myristoylated $G_i\alpha_2$ is shown in Fig. 4 (fourth lane). Note that the myristoylated protein migrates with a faster electrophoretic mobility than nonmyristoylated $G_i\alpha_2$. The yield of $G_i\alpha_2$ from this preparation was 8 mg. The amount of active protein in the final preparation can be determined by measurement of GTPγS binding[18] and total protein. We observe GTPγS-binding stoichiometries in the range of 40–100%.

Recombinant myristoylated $G_i\alpha_1$ (r-myr$G_i\alpha_1$) has been purified using similar protocols (Fig. 4, second lane). A yield of 60 mg was obtained from 11 liters of culture. Because the expression of $G_i\alpha_1$ was much higher than that of $G_i\alpha_2$, the protein was purified to homogeneity after hydroxylapatite chromatography. Expression of r-myr$G_i\alpha_3$ was lower; a yield of 4 mg was obtained from 11 liters of culture using the same protocol described for the purification of r-myr$G_i\alpha_2$. For purification of r-myr$G_o\alpha$, the early chromatography steps are the same as those used to purify $G_i\alpha$. The final step in the purification protocol is chromatography over phenyl-Superose using a Pharmacia fast protein liquid chromatography system (FPLC) instead of gel filtration over AcA 44. The pool containing $G_o\alpha$ after hydroxylapatite chromatography is diluted into TEDP containing 1.2 M (NH$_4$)$_2$SO$_4$ and 50 μM GDP and applied to a phenyl-Superose HR 10/10 column that has been equilibrated in the same buffer. Protein is eluted at 1 ml/min with a 240-ml gradient of decreasing (NH$_4$)$_2$SO$_4$ from 1.2 to 0 M. Fractions of 4 ml are collected, and 5-μl aliquots are assayed by SDS–PAGE for purity. The yield of r-myr$G_o\alpha$ from 11 liters of culture was 15 mg.

FIG. 4. Analysis by SDS–urea gel electrophoresis of purified recombinant myristoylated and nonmyristoylated $G_i\alpha_1$ and $G_i\alpha_2$. Each protein (2 µg) was subjected to electrophoresis on a 9% SDS–polyacrylamide gel containing 4 M urea. The gel was stained with Coomassie blue. The myristoylated proteins (second and fourth lanes, +) migrate ahead of the nonmyristoylated proteins (first and third lanes, −).

Protocols for Metabolic Labeling with Radioactive Myristic Acid

Volume and Concentration of Radioactive Medium

Cells are incubated with 9,10-tritiated myristic acid because of its high specific activity and to reduce reincorporation of radioactivity into other metabolic precursors. ^{14}C-labeled fatty acid (particularly that labeled in the 1 position) should not be used because of susceptibility to degradation by β-oxidation.[22] Myristic acid is commercially available, but the concentration of radioactivity is usually too low or the solvent is too toxic to add directly to cells. The solvent must therefore be evaporated and the radioactive fatty acid redissolved in ethanol, dimethyl sulfoxide, or culture medium as described below.

The volume of radiolabeling medium required and the concentration

---

[22] A. I. Magee, *in* "Posttranslational Modification of Proteins by Lipids" (U. Brodbeck and C. Bordier, eds.), p. 59. Springer-Verlag, New York, 1988.

of radioactivity desired for efficient incorporation are determined prior to exchange of the solvent for the radioactive precursor. When considering the concentration of radioactivity to detect tritium-labeled proteins by fluorography, the investigator should be influenced by the pool size of the lipid precursor, the financial expense of the radioactive precursor, and the patience required for long fluorographic exposures. The cellular pool size of myristate is relatively small compared to other fatty acids such as palmitate. A suitable range of concentration of [^{3}H]myristic acid is 0.05–1.0 mCi/ml. The volume should be minimal to conserve the radioactive precursor yet great enough to maintain cells during the incubation period. Mammalian cells grown on tissue culture plates require a minimum of 0.4 or 0.8 ml on a 35- or 60-mm diameter plate, respectively, for an overnight incubation in a well-humidified incubator.

Exchange of Solvent for Radioactive Myristic Acid

It is good practice to check the radioactivity of the precursor by liquid scintillation counting of diluted samples before and after solvent exchange.

Exchange of Solvent for Ethanol or Dimethyl Sulfoxide

1. Evaporate the solvent from the radioactive solution with a stream of nitrogen or under reduced pressure in a concentrator such as a Savant Speed-Vac (Farmingdale, NY).
2. Dissolve the [^{3}H]myristic acid in a volume of ethanol or dimethyl sulfoxide equivalent to 1% of the total volume of medium desired.
3a. For bacterial cells add the redissolved [^{3}H]myristic acid directly to a culture (0.5 ml) of *E. coli* in enriched medium supplemented with antibiotics.
3b. For mammalian cells add serum followed by culture medium (supplemented with nonessential amino acids and sodium pyruvate) to the redissolved [^{3}H]myristic acid. Determine that the radioactivity has been dissolved (by liquid scintillation counting). Replace the culture medium on cells with the radioactive medium.

Exchange of Solvent Directly for Aqueous Culture Medium

1. Transfer radioactive precursor in ethanol (*not* toluene if using plastic vessels) to a sterile culture tube or to a well of a multiwell plate (without cells). The size of the well depends on the volume of ethanol to be evaporated; it will not be used for culturing cells.
2. Evaporate the ethanol passively from the uncovered plate in a tissue culture hood (without the ultraviolet light illuminated) or from the tube under a stream of nitrogen.

3a. For bacterial cells, add a culture (in enriched medium supplemented with antibiotics) directly to the tube of dried [^{3}H]myristic acid.

3b. For mammalian cells, dissolve the radioactive precursor by two sequential washings of the well or tube with half of the total desired volume of medium supplemented with serum, nonessential amino acids, and sodium pyruvate. Determine that the radioactivity has been dissolved (by liquid scintillation counting). Replace the medium on a culture of mammalian cells with the radioactive medium.

Time of Incubation

The length of time cells are incubated with the radioactive medium can be critical. Longer periods are desirable to maximize incorporation of radioactive precursor into protein. This is particularly true for myristate-modified proteins for which incorporation of myristate is closely linked with protein synthesis. Shorter periods of time minimize the occurrence of metabolic reincorporation of radiolabel. However, this can vary with cell type, with some mammalian cells exhibiting specific labeling even after 24 hr of incubation.[22] We have found convenient times for specific labeling of myristoylated proteins to be 45 min and overnight (15–17 hr) for bacterial and mammalian cells, respectively. Mammalian cell culture medium is supplemented with sodium pyruvate (1 mM) to supply the acetyl-CoA requirements of the cell and thus reduce reincorporation of label. Nonessential amino acids also serve to supplement the medium and thereby reduce reincorporation of the label. Incubation of mammalian cells with radioactive medium should be restricted to a single incubator, well marked with radioactivity caution signs. The incubator, particularly the water reservoir used to humidify the chamber, should be monitored for radioactivity. Charcoal may be used to attempt to trap radioactive compounds released.[23]

Results of Metabolic Radiolabeling

Following an overnight incubation, we find 70–90% of the radioactivity from [^{3}H]myristic acid associated with mammalian cells and 10–30% remaining in the medium. Owing to the relatively few number of proteins modified by myristate, most myristoylated proteins can be resolved by one-dimensional SDS–PAGE. A useful control is to compare samples prepared from whole cells incubated with [^{3}H]myristic acid and radioactive amino acid.[22] The labeling patterns of fluorograms should be substantially

[23] J. Meisenhelder and T. Hunter, *Nature (London)* **335**, 120 (1988).

different, providing evidence that the radioactive myristic acid is not being metabolized and the label incorporated into amino acids.

Recombinant α subunits expressed in mammalian cells or in bacteria and labeled by [^{3}H]myristic acid are easily identified when compared to control cells (that do not express the recombinant protein) and to purified α subunit standards (Fig. 2, lane 5 versus lane 6, and Fig. 5,

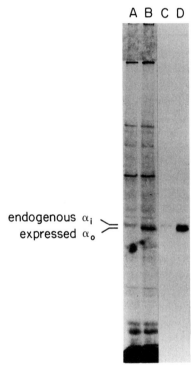

FIG. 5. Comparison of [^{3}H]myristic acid-labeled proteins in total cell lysates and immunoprecipitates of G-protein α subunits. Plates of COS-M6 cells (35 mm) were transfected with vector DNA (pCMV, lanes A and C) or DNA encoding the α subunit of G_o (lanes B and D).[2] The cells were incubated with 1 mCi/ml [9,10-^{3}H]myristic acid (Du Pont NEN) overnight and were lysed in 1 ml of RIPA buffer (see text). The lysate was cleared by centrifugation at 200,000 g for 15 min in a Beckman TL 100.3 rotor at 3°. Aliquots (15 μl) of the cleared lysates were loaded in lanes A and B. A 150-μl sample of the cleared lysates was immunoprecipitated with affinity-purified antibodies from G-protein α-subunit antiserum A569 [S. M. Mumby, R. A. Kahn, D. R. Manning, and A. G. Gilman, *Proc. Natl. Acad. Sci. U.S.A.* **83**, 265 (1986)]. The antibodies immunoprecipitated endogenous COS cell $α_i$ and the $α_o$ expressed as a result of the transfection. The proteins were resolved by SDS–PAGE utilizing a 9% acrylamide gel and visualized by fluorographic exposure of Kodak XAR-5 film for 36 hr. These data have been published elsewhere.[14]

lane A versus B). Note the endogenous *E. coli* proteins of 45 and 55 kDa that incorporate [^{3}H]myristate in the presence of NMT (Fig. 2, lanes 5 and 6). Myristoylation of $G_o\alpha$ and $G_i\alpha$, but not $G_z\alpha$, are recapitulated in *E. coli* when coexpressed with *Saccharomyces cerevisiae NMT1*.[15] However, $G_z\alpha$ is myristoylated when it is expressed with the human NMT enzyme.[24]

The identity of a tritium-labeled protein as a G-protein α subunit (either endogenous or overexpressed) can be verified by immunoprecipitation (Fig. 5, lanes C and D). Cells prepared for immunoprecipitation are extracted with RIPA buffer comprised of 100 mM NaCl, 50 mM sodium phosphate buffer (pH 7.2), 1% (w/v) deoxycholate, 1% (v/v) Triton X-100, 0.5% (w/v) SDS, 1 mM DTT, and 0.05% (v/v) aprotinin (Sigma, St. Louis, MO). The DNA is removed by centrifugation at 200,000 g for 20 min at 2° (Beckman TLA-100.3 rotor). Antibodies are incubated with a portion of the cleared cell extract overnight. The incubation and following processing steps are performed at 4°. The amount of extract and antibodies necessary for efficient immunoprecipitation and for a detectable signal are determined empirically. Antibodies are commercially available from Calbiochem, Du Pont NEN, Upstate Biotechnology Inc., and Gramsch Laboratories. Following the overnight incubation, a 10% suspension of fixed *Staphylococcus aureus* (Pansorbin from Calbiochem) is added to precipitate the immunoglobulin by incubation for 30 min. The precipitates are sedimented by centrifugation (13,000 g for 1 min) and resuspended in 0.1 ml RIPA. The suspension is layered over 0.9 ml of RIPA containing 20% sucrose (w/v) and sedimented by centrifugation for 6 min. The pellet is washed with 0.5 ml of phosphate-buffered saline and extracted by heating for 1–3 min at 90–100° in SDS-PAGE sample buffer containing 20 mM DTT and 1% 2-mercaptoethanol. The extracts from the immunoprecipitates are resolved by SDS–PAGE.[25]

Detection of tritium in gels requires fluorography. If staining of the gel is necessary to locate standards, it should be performed with Coomassie blue (do not silver stain) prior to impregnation of the gel with fluor. Gels may be prepared for fluorography with a solution of 2,5-diphenyloxazole in dimethyl sulfoxide[26] or with commercial products such as En^{3}Hance (Du Pont New England Nuclear). Kodak (Rochester, NY) XAR-5 film is suitable and sensitive for detection of fluorographic signals. Preflashing of the film is considered to increase sensitivity and is required to obtain a linear correlation between film exposure and radioactivity on the gel.[26]

---

[24] R. J. Duronio, S. I. Reed, and J. I. Gordon, *Proc. Natl. Acad. Sci. U.S.A.* **89,** 4129 (1992).
[25] U. K. Laemmli, *Nature (London)* **227,** 680 (1970).
[26] R. A. Laskey, this series, Vol. 65, p. 363.

Once an α subunit is found to incorporate tritium from myristate, further work is necessary to verify the nature of the modification. Myristate linked to protein through an amide bond is stable to base but can be hydrolyzed by acid. The hydrolyzed lipid can be extracted by organic solvent and analyzed chromatographically together with known standards.[1,11,13,27] A more definitive identification can be made if a lipid-containing peptide is isolated and analyzed by mass spectrometry.[5,6]

Acknowledgments

We acknowledge with gratitude the contributions of our collaborators, Alfred G. Gilman (University of Texas Southwestern Medical Center at Dallas), Jeffrey Gordon (Washington University, St. Louis, MO), and Janice E. Buss (Iowa State University, Ames, IA). Our work has been funded by grants from the National Institute of Health (GM34497 to A.G.G.), the American Cancer Society (BE30N to A.G.G.), and the American Heart Association, Texas Affiliate (91R-077 to S.M.M.).

[27] P. J. Casey and J. E. Buss, in "Posttranslational Modification of Proteins by Lipids" (U. Brodbeck and C. Bordier, eds.), p. 64. Springer-Verlag, New York, 1988.

[21] Specificity and Functional Applications of Antipeptide Antisera Which Identify G-Protein α Subunits

By GRAEME MILLIGAN

Introduction

Early attempts to define the specificity of interactions between receptors, G proteins, and effectors made considerable use of exotoxins produced by the bacteria *Vibrio cholerae* and *Bordetella pertussis*. These toxins respectively modulated stimulatory and inhibitory regulation of cyclic AMP production by causing mono-ADP-ribosylation of the G proteins which allow communication between receptors and adenylyl cyclase (adenylate cyclase, EC 4.6.1.1). Furthermore, both toxins interfered with the functional interactions between receptors and G proteins. In the case of cholera toxin-catalyzed ADP-ribosylation of G_s, the G protein was converted to an essentially irreversibly activated form which no longer required agonist occupation of a receptor to produce regulation of adenylyl cyclase, whereas in the case of pertussis toxin-catalyzed ADP-ribosylation of the "G_i-like" G proteins, the covalent modification prevented productive coupling between receptors and these G proteins. However, with the realization that there could be multiple G-protein gene products which acted as substrates for these toxins expressed in a single cell or tissue, it became clear that more selective tools would be required to discriminate between receptor functions transmitted to effector moieties by the individual G proteins.

Immunological probes were initially generated against purified G-protein fractions.[1,2] Although these have been highly useful in a wide range of studies to identify and locate a variety of G proteins, difficulties in the purification of individual G proteins to homogeneity and the now appreciated sequence similarities of the α subunits of many G proteins meant that the absolute specificity of a number of the antisera was unknown. The isolation of cDNA species corresponding to G-protein α subunits revolutionized the field by allowing the generation of antisera against short peptides which represent sections of the primary amino acid sequence unique to specific G proteins.[3,4] A number of antisera generated in this manner have been demonstrated to be useful in interfering with receptor–G-protein interactions and hence with agonist control of effector function. To date, antisera of this type have been used to examine stimulatory regulation of adenylyl cyclase by both "seven transmembrane element"[5,6] and other[6,7] receptors, inhibitory regulation of adenylyl cyclase,[8–11] photon-driven regulation of cyclic GMP hydrolysis,[12] G-protein regulation of classes of Ca^{2+} channels,[13] receptor regulation of phosphoinositidase C action,[14] and the interactions of somatostatin receptors with

---

[1] A. M. Spiegel, in "G-Proteins" (R. Iyengar and L. Birnbaumer, eds.), p. 115. Academic Press, San Diego, 1990.

[2] G. Milligan, in "G-Proteins as Mediators of Cellular Signalling Processes" (M. D. Houslay and G. Milligan, eds.), p. 31. Wiley, Chichester, 1990.

[3] S. M. Mumby, R. A. Kahn, D. R. Manning, and A. G. Gilman, Proc. Natl. Acad. Sci. U.S.A. **83**, 265 (1986).

[4] P. Goldsmith, P. Gierschik, G. Milligan, C. G. Unson, R. Vinitsky, H. L. Malech, and A. M. Spiegel, J. Biol. Chem. **262**, 14683 (1987).

[5] W. F. Simonds, P. K. Goldsmith, C. J. Woodward, C. G. Unson, and A. M. Spiegel, FEBS Lett. **249**, 189 (1989).

[6] B. G. Nair, B. Parikh, G. Milligan, and T. B. Patel, J. Biol. Chem. **265**, 21317 (1990).

[7] M. del C. Vila, G. Milligan, M. L. Standaert, and R. V. Farese, Biochemistry **29**, 8735 (1990).

[8] W. F. Simonds, P. K. Goldsmith, J. Codina, C. G. Unson, and A. M. Spiegel, Proc. Natl. Acad. Sci. U.S.A. **86**, 7809 (1989).

[9] F. R. McKenzie and G. Milligan, Biochem. J. **267**, 391 (1990).

[10] S. J. McClue, E. Selzer, M. Freissmuth, and G. Milligan, Biochem. J. **284**, 565 (1992).

[11] M. Tallent and T. Reisine, Mol. Pharmacol. **41**, 452 (1992).

[12] R. A. Cerione, S. Kroll, R. Rajaram, C. Unson, P. Goldsmith, and A. M. Spiegel, J. Biol. Chem. **263**, 9345 (1988).

[13] I. McFadzean, I. Mullaney, D. A. Brown, and G. Milligan, Neuron **3**, 177 (1989).

[14] S. Gutowski, A. Smrcka, L. Nowak, D. Wu, M. Simon, and P. C. Sternweis, J. Biol. Chem. **266**, 20519 (1991).

G proteins.[15] Other potentially selective "knockout" techniques such as the application of antisense strategies[16,17] are likely to provide complimentary information in the future, but to date antipeptide antibody approaches have had the greatest success and most widespread application. This chapter discusses strategies for the generation of effective antipeptide G-protein α-subunit antisera, characterization of the specificities of these probes, and their use in functional analyses of the specificity of G-protein function.

Choice of Synthetic Peptide Sequence

Antisera which are likely to be useful in the analysis of receptor–G-protein interactions must be highly selective for individual G-protein α subunits, and this limits the choice of peptides to regions of relative dissimilarity between the individual G proteins. Moreover, if antisera which might be useful for studying the functional interactions of G proteins with receptors and effectors are required, the choice is more restricted still. However, if it is accepted a priori that differences in sequence and hence conformation would be required to allow selectivity or specificity of contacts between receptors and the range of G proteins coexpressed in a particular cell, then the above restrictions are not as draconian as they might seem.

It is clear that sections of G-protein subunits which can be deduced to play key roles in interactions with receptors and effectors are likely to be most appropriate for development of antisera. Both biochemical evidence from knowledge of the site of action and functional effects of pertussis toxin[18] and genetic analysis of the locus of the mutation in S49 lymphoma *unc* cells[19] had identified the extreme C-terminal region of G-protein α subunits as a key site for functional interactions between receptors and G proteins, and on this basis antibodies generated against peptides corresponding to the last 10–12 amino acids of individual G-protein α subunits have been the most widely used to study functional interactions between receptors and G proteins. Other antisera have been useful, particularly when the approach has been to attempt to coimmunoprecipitate receptors

[15] S. F. Law, D. Manning, and T. Reisine, *J. Biol. Chem.* **266**, 17885 (1991).
[16] C. Kleuss, J. Hescheler, C. Ewel, W. Rosenthal, G. Schultz, and B. Wittig, *Nature (London)* **353**, 43 (1991).
[17] C. Kleuss, H. Scherubl, J. Hescheler, G. Schultz, and B. Wittig, *Nature (London)* **358**, 424 (1992).
[18] G. Milligan, *Biochem. J.* **255**, 1 (1988).
[19] K. A. Sullivan, R. T. Miller, S. B. Masters, B. Beiderman, H. Heideman, and H. R. Bourne, *Nature (London)* **330**, 758 (1987).

and G proteins using antipeptide anti-G protein antisera.[15,20] However, as sites of interaction between proteins may often be determined by multiple contact points or be dependent on spatial conformation rather than a linear peptide sequence, inherent limitations may exist for antipeptide antisera, particularly as they must recognize the native conformation of the G protein to be useful for such studies (see later).

Specificity of Anti-G Protein Antisera

Examination of the primary amino acid sequences of individual G proteins as defined from the nucleotide sequences of corresponding cDNA species can indicate likely peptides to allow the generation of specific antisera, but following generation of the antisera (using approaches similar or identical to those described in Refs. 21 and 22) specificity must be carefully defined by a variety of approaches.

Enzyme-Linked Immunosorbent Assays

Initial information on specificity may be obtained by enzyme-linked immunosorbent assays (ELISA). Antigen [100 μl of a 1 μg/ml solution in phosphate-buffered saline (PBS, 8.5 g NaCl, 1.28 g Na_2HPO_4, 0.156 g NaH_2PO_4 per liter, pH 7.4)] or peptides corresponding to the equivalent regions of other G-protein α subunits are immobilized in wells of a multi-well polystyrene plate by incubation overnight at 4° and the plates subsequently washed twice with PBS. Then 100 μl blocker (1% dried milk powder in PBS) is added to each well and left for 2 hr. The blocker is removed and each well washed 3 times with PBS contining 0.05% Tween 20 (PBS–Tween). Antiserum (dilutions between 1:200 and 1:200,000 in PBS) is incubated in triplicate wells overnight at 30° and then removed. The plate is washed three times with PBS–Tween and then twice with PBS, and a 1:1000 dilution in blocker plus 0.1% Tween 20 of a horseradish peroxidase-linked donkey anti-rabbit immunoglobulin G (IgG) antiserum is added and allowed to incubate at 30° for 2 hr. The secondary antiserum is removed and the plate rinsed as before with PBS–Tween and PBS. Then 100 μl of substrate [o-phenylenediamine (0.4 mg/ml) in citrate–phosphate buffer (17.9 ml of 0.1 M citric acid, 32.1 ml of 0.2 M Na_2HPO_4, and 50 ml water) plus 0.01% hydrogen peroxide] is added and the plate left for 20 min in the dark. Next, 2 M H_2SO_4 (50 μl) is added to stop the reaction, and the absorbance of each well is determined at 492 nm with a plate

[20] Y. Okuma and T. Reisine, *J. Biol. Chem.* **267,** 14826 (1992).
[21] S. M. Mumby and A. G. Gilman, this series, Vol. 195, p. 215.
[22] M. Reichlin, this series, Vol. 70, p. 159.

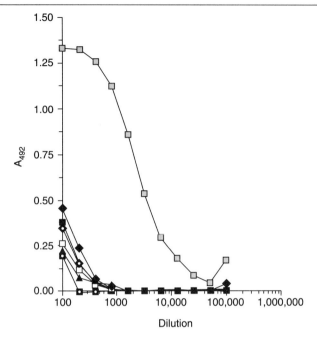

FIG. 1. Use of ELISA to indicate the likely specificity of an antipeptide antiserum raised against a synthetic peptide representing the C-terminal decapeptide of the α subunit of G_{13} for this polypeptide. The ELISA reactivity of varying dilutions of antiserum 13CB against peptides which correspond to the C-terminal decapeptide of different G proteins was determined. The ELISA procedures were performed using 100 ng of peptides corresponding to G_{13} (▣), G_{12} (■), G_q/G_{11} (▲), G_z (◆), G_{i1}/G_{i2} (□), G_{i3} (◇), or G_{o1} (▨). [Data from G. Milligan, I. Mullaney, and F. M. Mitchell, FEBS Lett. **297**, 186 (1992), with permission.]

scanner. Lack of cross-reactivity with peptides which correspond to the equivalent region of other G-protein α subunits[23,24] suggests (Fig. 1), but is insufficient evidence in isolation, that the antiserum will not cross-react either with the immobilized G proteins (for Western blotting studies) or with the native proteins (for functional studies).

Western Blotting to Determine Specificity of Recognition of G-Protein α Subunits

The most useful standards for assessment of specificity of anti-G-protein antipeptide antisera are purified mammalian G proteins or recombi-

[23] F. M. Mitchell, I. Mullaney, P. P. Godfrey, S. J. Arkinstall, M. J. O. Wakelam, and G. Milligan, *FEBS Lett.* **287**, 171 (1991).
[24] G. Milligan, I. Mullaney, and F. M. Mitchell, *FEBS Lett.* **297**, 186 (1992).

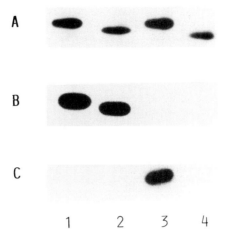

FIG. 2. Specificity of antiserum AS7 (raised against a peptide which represents the C-terminal decapeptide of transducin 1α) and I3B (raised against a peptide which represents the C-terminal decapeptide of $G_{i3}\alpha$) to identify recombinant pertussis toxin-sensitive G proteins expressed and purified from *E. coli*. Recombinant $G_{i1}\alpha$ (lane 1), $G_{i2}\alpha$ (2), $G_{i3}\alpha$ (3), and $G_o\alpha$ (4) (50 ng/lane) were resolved by SDS–PAGE and stained with silver (A) or immunoblotted with either antiserum AS7 (B) or I3B (C). [Data from S. J. McClue, E. Selzer, M. Freissmuth, and G. Milligan, *Biochem. J.* **284**, 565 (1992), with permission.]

nant proteins produced in either *Escherichia coli*[25] or baculovirus[26] expression systems. Clear and unequivocal data on both qualitative and quantitative aspects of G-protein α-subunit recognition can be achieved when such standards are available (Fig. 2). In many situations a full panel of recombinant proteins is not available to an investigator, and other approaches which involve immunoblotting of cells and tissues with previously highly characterized G-protein profiles can be usefully employed to define adequately the G-protein reactivity profile of a particular antiserum. It is worth noting that the detailed profile of cross-reactivity of antisera produced by injecting the same conjugate into separate rabbits can be noticeably different and even that this can change in different bleeds of the same rabbit.

Owing to the similarity in molecular size of α subunits of many of the G proteins, in the absence of adequate recombinant standards, sodium dodecyl sulfate–polyacrylamide gel electrophoresis (SDS–PAGE) conditions often have to be tailored to the resolution of particular G proteins when examination of the sequences might suggest cross-reactivity of a

[25] M. E. Linder, D. A. Ewald, R. J. Miller, and A. G. Gilman, *J. Biol. Chem.* **265**, 8243 (1990).
[26] S. G. Graber, R. A. Figler, and J. C. Garrison, *J. Biol. Chem.* **267**, 1271 (1992).

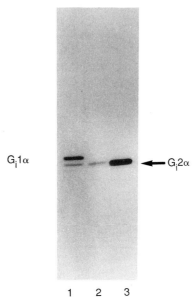

FIG. 3. Resolution of the α subunits of G_{i1} and G_{i2} in SDS–polyacrylamide gels containing reduced concentrations of bisacrylamide. For SDS–PAGE in which the bisacrylamide concentration is 0.26%, the α subunits of G_{i1} and G_{i2} essentially comigrate. Reduction of the bisacrylamide concentration (in this case to 0.15%, but often 0.0625% is used) results in substantial resolution of the two polypeptides. Rat brain cortex expresses both G proteins (lane 1), whereas NIH 3T3 (lane 2) and glioma C_6 cells (lane 3) express detectable levels of only G_{i2}. [Data from G. Milligan, S.-A. Davies, M. D. Houslay, and M. J. O. Wakelam, *Oncogene* **4**, 659 (1989), with permission.]

particular antiserum with more than one G protein. For example, in 10% acrylamide–0.26% bisacrylamide gels, the α subunits of G_{i1}, G_{i2}, and G_{i3} essentially comigrate. Considerable resolution of G_{i2} from the two other G_i-like G proteins can be achieved by increasing the acrylamide concentration to 12.5% and decreasing the concentration of bisacrylamide to 0.0625%[27] (Fig. 3). Under these conditions G_{i2} migrates more rapidly through the gel than the other two polypeptides.

Antisera generated against the C-terminal decapeptide of either rod and cone transducins or G_{i1} and G_{i2} cross-react with each of these G proteins (Fig. 3) because of the sequence similarities (G_{i1} and G_{i2} are identical in this region, and the transducins differ from this sequence by a single conservative substitution). Thus, antibodies of this type such as

[27] F. M. Mitchell, S. L. Griffiths, E. D. Saggerson, M. D. Houslay, J. T. Knowler, and G. Milligan, *Biochem. J.* **262**, 403 (1989).

AS7 (produced by the laboratory of A. Spiegel) and SG1 (produced by the laboratory of G. Milligan) [antisera of this type are sold commercially by Du Pont New England Nuclear (Boston, MA; Cat. No. NEI 801), Calbiochem (La Jolla, CA; Cat. No. 371723), Gramsch Laboratories (Schwabhausen, Germany, Cat. No. 3A-130), and Biomac, among others] are useful for unequivocal definition of receptor–G-protein interactions only if only one of these four individual G proteins are expressed in the cell or tissue under examination. As the transducins are restricted to photoreceptor-containing tissues and G_{i1} is expressed in detectable levels by a limited repertoire of cells,[28] these antisera can often be used as a specific probe to examine functions of G_{i2}.[8,9] However, as G_{i2} appears to be expressed universally,[28] these antisera cannot be used as a specific probe to examine the function of G_{i1}.

The G_{i1} and G_{i3} proteins are 94% identical at the primary sequence level, and we have been unable to develop one-dimensional SDS–PAGE conditions able to resolve these polypeptides reproducibly. Thus, assessments of cross-reactivity between antisera directed against peptide segments of these two G proteins must be performed either with recombinant protein or in tissues which do not express G_{i1} (such as neutrophils and related cells[4,29]). The SDS–PAGE systems which incorporate deionized urea[30] (usually either 4 M or a 4–8 M urea gradient) have also been shown to be useful for the resolution of highly similar G-protein α subunits to allow assessments of antiserum specificity. For example, the inclusion of 4 M urea in a 10% acrylamide gel can allow the resolution of multiple splice forms of G_o.[31]

Surprising cross-reactivity can also occasionally be noted with antipeptide antisera. For example, we have produced an antiserum (CQ1) directed against the C-terminal decapeptide common to the α subunits of G_q and G_{11}[23] which can be noted to have weak but distinct cross-reactivity with the α subunit of G_s (Fig. 4A), although a cursory examination of the relevant sequences would not obviously suggest this. Antisera produced by three other rabbits (CQ2–4) to the same antigen conjugate all identified G_q and G_{11} proteins but did not cross-react with G_s. Another example of unexpected cross-reactivity which we have noted came from an attempt to produce an immunological probe which would be specific for $G_q\alpha$ versus $G_{11}\alpha$. We generated antiserum (IQ1) against amino acids 119–134 of $G_q\alpha$.

[28] G. Milligan, *Cell. Signalling* **1**, 411 (1989).
[29] P. M. Murphy, B. Edie, P. Goldsmith, M. Brann, P. Gierschik, A. Spiegel, and H. L. Malech, *FEBS Lett.* **221**, 81 (1987).
[30] N. M. Scherer, M. J. Toro, M. L. Entman, and L. Birnbaumer, *Arch. Biochem. Biophys.* **259**, 431 (1987).
[31] I. Mullaney and G. Milligan, *J. Neurochem.* **55**, 1890 (1990).

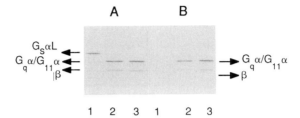

FIG. 4. Cross-reactivity of an antiserum directed toward the C-terminal decapeptide common to the α subunits of G_q and G_{11} with $G_s\alpha$. Ten nanograms of *E. coli* expressed $G_s\alpha$ (long form) (lanes 1), 10 ng of a mixed purified preparation of liver G_q and G_{11} (a gift from Dr. J. H. Exton, Nashville, TN) (2), and 10 ng purified turkey erythrocyte G_{11} (a gift from Dr. T. K. Harden, Chapel Hill, NC) (3) were resolved on a Bio-Rad (Richmond, CA) minigel [10% (w/v) acrylamide] and immunoblotted (A) with a mixture of antipeptide antiserum (CQ1) directed toward the C-terminal decapeptide of $G_q\alpha/G_{11}\alpha$ and an antipeptide antiserum (BN1) directed against the N-terminal decapeptide of β_1 subunit (to check the loadings of the mammalian and avian G_q/G_{11} preparations) or (B) with an antipeptide antiserum (IQ1) directed against amino acids 119-134 of mammalian $G_q\alpha$ plus antiserum BN1. Cross-reaction of antiserum CQ1 but not antiserum IQ1 with $G_s\alpha$ as well as with G_q/G_{11} is shown. Furthermore, antiserum IQ1 clearly identifies the turkey G_{11} (shown to be a G_{11} rather than a G_q species[34]) (see also Fig. 5), indicating that although designed to be a specific probe for $G_q\alpha$ it is not. Such studies highlight the importance of adequate quality control for the specificity of antisera.

This region differs in six positions from the corresponding sequence of $G_{11}\alpha$:

G_q (119-134)   EKVSAFENPYVDAIKS
G_{11}(119-134)   EKVTTFEHQYVNAIKT

and, furthermore, the sequence of $G_q\alpha$ contains a proline residue at position 127 which is not conserved in $G_{11}\alpha$. Owing to the physicochemical properties of proline, it was anticipated that the presence or absence of this amino acid might provide a distinct conformational difference in this region in the two peptides which could be exploited by the immune system for the generation of a discriminatory antiserum.

Immunoblotting of purified preparations of a mixture of liver G_q/G_{11}[32] under SDS-PAGE conditions which can resolve the two polypeptides[32] demonstrated that the antiserum was able to identify the two polypeptides equally well and further that the predicted, putative G_q-selective antiserum was able to immunoidentify the purified G_q/G_{11}-like polypeptide from turkey erythrocytes,[33] which both migrates on SDS-PAGE in a manner

[32] J. L. Blank, A. H. Ross, and J. H. Exton, *J. Biol. Chem.* **266**, 18206 (1991).
[33] G. L. Waldo, J. L. Boyer, A. J. Morris, and T. K. Harden, *J. Biol. Chem.* **266**, 14217 (1991).
[34] D. H. Maurice, G. L. Waldo, A. J. Morris, R. A. Nicholas, and T. K. Harden, *Biochem. J.* **290**, 765 (1993).

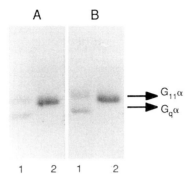

FIG. 5. Turkey erythrocyte G_p is a G_{11} rather than a G_q protein. Ten nanograms of either a purified mixture of mammalian liver G_q and G_{11} (lanes 1) or turkey erythrocyte G_p (2) were resolved by SDS–PAGE [13% (w/v) acrylamide] as described by Blank et al.[32] The samples were immunoblotted with antisera CQ2 (A) or IQ1 (B) (see legend to Fig. 4 for details). The mammalian mixture of G_q and G_{11} was clearly resolved into two polypeptides ($G_q\alpha$ migrates more rapidly through the gel), whereas the avian G_p migrated as a single band in a position equivalent to mammalian G_{11}. Cloning and sequencing of the turkey protein indicate it to be closely related to mammalian G_{11}.[34] Antiserum IQ1 (as well as CQ2) identified both G_q and G_{11} from mammalian liver, demonstrating that it is not a specific probe for G_q although it was designed for this purpose.

similar to mammalian G_{11} rather than G_q (Fig. 5) and has a sequence consistent with its being a G_{11} rather than a G_q type.[34] Such studies indicate the importance of definitive quality control.

Immunoprecipitation

The most appropriate test of specificity for antisera to be used in functional uncoupling studies is that they can immunoprecipitate the relevant G proteins, but not others. Such an analysis defines that the antiserum can identify selectively the native conformation of the G-protein α subunit. For such immunoprecipitation assays membrane pellets are resuspended in 50 μl of 1% (w/v) SDS and boiled for 3 min to solubilize the membranes. After cooling, 950 μl of solubilization buffer [1% Triton X-100, 10 mM EDTA, 100 mM NaH$_2$PO$_4$, 10 mM NaF, 100 μM Na$_3$VO$_4$, 50 mM HEPES, pH 7.2 at 4°, containing 1 mM phenylmethylsulfonyl fluoride (PMSF), 3 mM benzamidine, 2 μg/ml soybean trypsin inhibitor, 10 μM leupeptin, 10 μg/ml aprotinin, 1 μg/ml antipain, and 1 μg/ml pepstatin A] (buffer B) is added and the samples left on ice for at least 1 hr. The samples are centrifuged in a microcentrifuge for 10 min at 4° to pellet nonsolubilized material. The supernatant is removed and antiserum added at appropriate dilution (usually 1:100). Samples are then incubated overnight at 4° with rotary mixing. A 1:1 suspension of protein A agarose (50 µl) is then added to each sample and the solution mixed gently for 2 hr at 4°. The immune complex is pelleted by centrifugation in a microcentrifuge for 5 min and washed 3 times with 1% Triton X-100, 100 mM NaCl, 100 mM NaF, 50 mM NaH$_2$PO$_4$, 50 mM HEPES, pH 7.2, at 4° (this buffer can be supplemented with between 0.1 and 0.5% SDS depending on the stringency of the wash conditions required).

The final immune complex is resuspended with 30 µl Laemmli sample buffer and the samples boiled for 5 min. The samples are then resolved by SDS-PAGE and the presence of individual G proteins in the immunoprecipitate assessed by immunoblotting with antisera known to identify individual G proteins. In the example shown in Fig. 6, antiserum CQ2 generated against the C-terminal decapeptide shared by the α subunits of G_q and G_{11} successfully immunoprecipitated G_q/G_{11} from NG108-15 neuroblastoma × glioma hybrid cells but did not immunoprecipitate G_{i2},

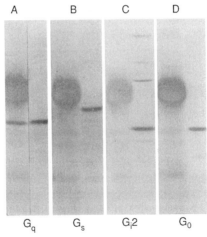

FIG. 6. Selective immunoprecipitation of G_q/G_{11} with antiserum CQ2. Membranes of NG108-15 cells were immunoprecipitated with antiserum CQ2 (raised against a peptide corresponding to the C-terminal decapeptide of G_q/G_{11}). The immunoprecipitates (left-hand lane of each pair) along with NG108-15 cell membranes (right-hand lane of each pair) were subsequently resolved by SDS-PAGE and immunoblotted with antisera directed against the C-terminal decapeptides of (A) G_q/G_{11}, (B) G_s, (C) G_{i2}, and (D) G_o. The data indicate that membranes of NG108-15 cells express all of these G proteins but that only G_q/G_{11} is present in the CQ2 immunoprecipitate. Such studies indicate that the antiserum displays the same specificity for native G_q/G_{11} as it does in immunoblots.

G_o, or G_s, although immunoblotting of native membranes of these cells indicated the expression of high levels of this polypeptide.

Immunoblotting to detect the presence in the immunoprecipitate of the G protein which the primary antiserum is designed to recognize should ideally be performed with a second antiserum directed against a separate segment of the G-protein α-subunit primary sequence to limit potential concerns relating to immunorecognition of a spurious, irrelevant polypeptide.

Quantitative information from the immunoprecipitation controls are also vital if information is required to assess the likely usefulness of a particular G-protein antiserum in functional uncoupling studies. For an antiserum to be useful in such an approach it is necessary that the antiserum can be shown to immunoprecipitate a substantial fraction of the G protein of interest from the tissue source being analyzed. The antiserum is unlikely to be a useful functional probe if, while immunoprecipitation occurs, only a small proportion of the available antigen is immunoneutralized.

We have examined this question for a variety of antipeptide antisera directed toward distinct epitopes of forms of $G_o\alpha$.[31] Immunoprecipitation of $G_o\alpha$ from rat brain cortical membranes with antiserum OC2 (directed toward amino acids 345–354 of $G_o\alpha$) or with antiserum ON1 (directed toward amino acids 2–17 of $G_o\alpha$) (these sequences are identical in the α subunits of mammalian G_{o1} and G_{o2}) resulted in both antisera immunoprecipitating greater than 70% of total $G_o\alpha$ as assessed by immunoblotting both rat cortical membranes and the immunoprecipitates with a third antipeptide antiserum (IM1, directed toward amino acids 22–35, a further region conserved between G_{o1} and G_{o2} (Fig. 7). In contrast, an antiserum (O1B, raised against amino acids 308–317 or G_{o1}), which is specific for forms of G_{o1} and does not cross-react with G_{o2}, while able to identify G_{o1} in immunoblotting procedures was able to immunoprecipitate only a small fraction of rat cortical membrane G_{o1} (data not shown). Such data indicate that antisera OC2 and ON1 are potentially useful probes for functional studies of receptor interactions with G_o but that antiserum O1B is unlikely to be useful for such studies, although it is a highly useful probe for the selective detection of forms of $G_{o1}\alpha$ in immunoblotting studies.

Antibody Uncoupling Experiments on Membrane Preparations

Preparation of Immunoglobulin G Fractions from Crude Anti-G Protein Antipeptide Antisera

Crude antisera (5 ml in glycine buffer containing 1.5 M glycine, 3 M NaCl, pH 8.9) are added to a column of protein A-Sepharose (1.5 × 2

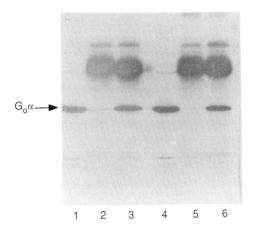

FIG. 7. Anti-$G_o\alpha$ antisera immunoprecipitate $G_o\alpha$ from rat cortical membranes. Rat cortical membranes (100 μg) were either solubilized as described in the text (lanes 2, 3, 5, 6) and immunoprecipitated (15 μg IgG) with either antiserum OC2 (lane 3), antiserum ON1 (lane 6), or normal rabbit serum (lanes 2, 5) or subjected directly to SDS–PAGE [10% (w/v) acrylamide] (lanes 1, 4). The resolved samples were then immunoblotted using antiserum IM1 as primary antiserum. The results indicate that both antisera were able to immunoprecipitate at least 70% of the $G_o\alpha$ in these experiments.

cm) and allowed to equilibrate overnight with gentle rotary mixing. The column is drained and washed with glycine buffer until the eluate has an A_{280} of 0.0. The column is then eluted with 100 mM citric acid, pH 4.0, into 2 M Tris/HCl, pH 7.5, and the IgG-containing fractions are dialyzed overnight against 1000 volumes of 10 mM Tris-HCl, 0.1 mM EDTA, pH 7.5 (buffer A) and subsequently lyophilized. Samples are stored at $-80°$ and reconstituted with buffer A to the required working dilution just prior to use.[9,35]

Use of Anti-G Protein Antipeptide Antisera to Prevent Interactions between Receptors and Guanine Nucleotide-Binding Proteins

Assuming that the generated antisera meet all of the criteria for specificity in identification and immunoprecipitation protocols as detailed above, it should be possible to use the antisera in functional uncoupling studies. Although we routinely use protein A-Sepharose purified IgG fractions for such studies, certain other laboratories (e.g., Spiegel and co-

---

[35] S. J. McClue and G. Milligan, *Mol. Pharmacol.* **40**, 627 (1991).

workers[5,8]) use specific immune fractions from the antisera which are generated by chromatography on columns of immobilized antigenic peptide. With the availability of recombinant protein which can be generated in large amounts, it should be equally appropriate to use columns of immobilized protein for such affinity purifications. One attraction of such purifications is that estimates based on amounts of true immune immunoglobulin can be used to calculate amounts of G protein present in a preparation, whereas with preparation of a total IgG fraction no information is available to define what proportion of the IgG is targeted against the antigen.

In experiments to assess the specificity of functional interactions between receptors and guanine nucleotide-binding proteins, regulation of each of agonist binding affinity to the receptor,[9,36] agonist stimulation of high-affinity GTPase activity,[9,10] and agonist control of effector activity[8-10] have been measured. In each of these protocols we routinely preincubate a cell membrane preparation with protein A-Sepharose-purified fractions of anti-G protein antisera for 1 hr at 30°–37° prior to performing the functional assay. A dilution of the purified IgG fraction equivalent to a 1 : 100 dilution of crude antiserum is usually the amount (~ 10 μg protein) taken for initial analyses. Effects of the antibody are controlled for by inclusion of equivalent amounts of IgG fractions isolated from normal rabbit serum in parallel assays. In experiments which examine the effects of the antisera on receptor regulation of high-affinity GTPase and of adenylyl cyclase activity, preincubation of the membranes with the immune fractions are performed in the presence of all chemical components required for the assay except receptor ligands and detection radiolabels, which are added after the preincubation period to then measure the effect of the antibodies. We generally note that the preincubation phase reduces measured agonist regulation of the signaling system, but although other workers have performed the antibody preincubations on ice to limit this decay, in our hands the balance between decay of the agonist signal and time of preincubation is best achieved as noted above.

We have used C-terminal antisera able to discriminate selectively between G_{i2} and G_{i3}, the two G_i-like G proteins which appear to be expressed in all cell types, to demonstrate that agonist activation of the α_2C10 adrenoceptor following transfection and expression in rat 1 fibroblasts can produce activation of both of these G proteins but that, at least in this system, only G_{i2} and not G_{i3} can subsequently cause inhibition of adenylyl cyclase[10] (Table I). Activation of both G proteins by agonist activation of the receptor was demonstrated by the ability of both the anti-G_{i2} and anti-G_{i3} antisera

---

[36] S. J. McClue and G. Milligan, *FEBS Lett.* **269**, 430 (1990).

TABLE I
ANTISERA INHIBITION OF AGONIST ACTIVITY[a]

Antiserum	Decrease (%) of	
	UK14304 stimulation of high-affinity GTPase activity	UK14304 inhibition of forskolin-amplified adenylyl cyclase activity
AS7	41	85
I3B	25	0

[a] Protein A-Sepharose-purified IgG fractions from antisera AS7 (anti-$G_{i1}\alpha/G_{i2}\alpha$) and I3B (anti-$G_{i3}\alpha$) (see Fig. 2) from normal rabbit serum were incubated with membranes of rat 1 fibroblasts expressing the α_2C10 adrenoceptor. Subsequently the ability of the α_2-adrenoceptor agonist UK14304 to stimulate high-affinity GTPase activity or to inhibit forskolin-amplified adenylyl cyclase was assessed. Both antisera were able to inhibit partially the agonist stimulation of high-affinity GTPase activity, indicating that both G_{i2} and G_{i3} are activated by agonist occupation of the receptor, but the G_{i3} antibody was completely unable to attenuate agonist inhibition of adenylyl cyclase, which indicates that G_{i3} does not act to mediate inhibition of adenylyl cyclase in these cells. Data adapted from E. McClue, M. Selzer, M. Freissmuth, and G. Milligan, *Biochem. J.* **284,** 565 (1992), with permission.

to inhibit partially the agonist-stimulated high-affinity GTPase activity (Table I). However, the anti-G_{i3} antiserum was unable to modify agonist-mediated inhibition of adenylyl cyclase, which was essentially attenuated by the anti-G_{i2} antiserum.

Conclusions

Antipeptide antisera directed against sections of the primary amino acid sequence of the α subunits of a variety of G proteins have provided, and are likely to continue to provide, a wide range of information on G-protein distribution, function, and regulation. The specificities of such antisera must be rigorously proved for the application in which they are being used, and for studies on the functional interactions between receptors and G proteins the ability of the antisera to immunoprecipitate the relevant G protein selectively and quantitatively must be demonstrated prior to use. Assuming that such quality controls are performed, such antisera are likely to continue to provide novel information on the interactions between G proteins and other polypeptides in native membrane systems and in whole cells.

Acknowledgments

Work in the author's laboratory is supported by the Medical Research Council, the Agriculture and Food Research Council, the Wellcome Trust, and the British Heart Foundation.

[22] Identification of Receptor-Activated G Proteins: Selective Immunoprecipitation of Photolabeled G-Protein α Subunits

By KARL-LUDWIG LAUGWITZ, KARSTEN SPICHER, GÜNTER SCHULTZ, and STEFAN OFFERMANNS

Introduction

In recent years, it has become obvious that receptor–G protein interaction is neither totally specific nor absolutely nonspecific. Receptor–effector coupling by G proteins appears to be a fine-tuned process which occurs in a tissue-specific manner and depends on the expression pattern of different signal transduction components in the particular tissue.

Reconstitution and cotransfection experiments are powerful tools to study the interaction of receptors, G proteins, and effectors. To examine receptor–G-protein coupling at natural concentrations of the individual components, methods for *in situ* determination of receptor–G-protein interaction have been developed. These approaches are based on the action of specific antisense oligonucleotides[1] or on the functional effects of subtype-specific antibodies[2,3]. In some cases, G proteins coupled to receptors can be identified with specific antibodies in isolated receptor–G-protein complexes[4,5] or by receptor-stimulated ADP-ribosylation of α subunits by cholera toxin.[6,7] We have described the application of $[\alpha\text{-}^{32}P]GTP$ azidoanilide as a tool to label α subunits of receptor-activated G proteins.[8] Because

[1] C. Kleuss, G. Schultz, and B. Wittig, this volume [27].
[2] F. R. McKenzie, E. C. Kelly, C. G. Unson, A. M. Spiegel, and G. Milligan, *Biochem. J.* **249,** 653 (1988).
[3] W. F. Simonds, P. K. Goldsmith, J. Codina, C. G. Unson, and A. M. Spiegel, *Proc. Natl. Acad. Sci. U.S.A.* **86,** 7809 (1989).
[4] S. F. Law, D. Manning, and T. Reisine, *J. Biol. Chem.* **266,** 17885 (1991).
[5] R. Munshi, I.-H. Pang, P. C. Sternweis, and J. Linden, *J. Biol. Chem.* **266,** 22285 (1991).
[6] P. Gierschik, D. Sidiropoulos, and K. H. Jakobs, *J. Biol. Chem.* **264,** 21470 (1989).
[7] G. Milligan, C. Carr, G. W. Gould, I. Mullaney, and B. E. Lavan, *J. Biol. Chem.* **266,** 6447 (1991).
[8] S. Offermanns, G. Schultz, and W. Rosenthal, this series, Vol. 195, p. 286.

the number of known G proteins, which are defined by their α subunits, has constantly increased,[9] it was necessary to improve the selective identification of G-protein α subunits photolabeled in response to receptor activation. This was achieved by combining photolabeling of receptor-activated G proteins by [α-^{32}P]GTP azidoanilide in membranes with immunoprecipitation of the photolabeled G-protein α subunits by subtype-specific antisera. Whereas this experimental approach is not suitable for identification of $\beta\gamma$ complexes involved in receptor–G-protein coupling, it allows the exact identification of receptor-activated G proteins provided that specific precipitating antisera are available[10–12] and, moreover, allows a relative quantification of the different G proteins coupling to a receptor in a given tissue.[11]

Materials and Equipment

For preparation of [α-^{32}P]GTP azidoanilide used in immunoprecipitation experiments, [α-^{32}P]GTP of the highest specific activity available (3000 Ci/mmol) is routinely employed to increase the sensitivity of the method. Other materials and equipment for preparation of [α-^{32}P]GTP azidoanilide and photolabeling of G-protein α subunits have been described.[8] Detergents and protein A-Sepharose for immunoprecipitation are from Sigma (St. Louis, MO). A rotating incubator suitable for 4° operation is needed for incubation of samples with antisera.

Antisera

Antisera directed against G-protein α subunits are obtained after injection into rabbits of synthetic peptides corresponding to specific regions of α subunits coupled to keyhole limpet hemocyanin (KLH). Depending on the coupling reagent used, the peptides contain an additional cysteine residue to facilitate coupling to KLH. Antisera are not necessarily suitable for immunoprecpitation even if they specifically recognize G-protein α subunits in immunoblotting experiments. Table I shows sequences of peptides used for the generation of selective antisera which have been described to be precipitating. Some antisera are commercially available (e.g., from Du Pont–New England Nuclear, Boston, MA, and Upstate

---

[9] M. I. Simon, M. P. Strathmann, and N. Gautam, *Science* **252**, 802 (1991).
[10] R. L. Wange, A. V. Smrcka, P. C. Sternweis, and J. H. Exton, *J. Biol. Chem.* **266**, 11409 (1991).
[11] K.-L. Laugwitz, S. Offermanns, K. Spicher, and G. Schultz, *Neuron* **10**, 233 (1993).
[12] S. Offermanns, K.-L. Laugwitz, K. Spicher, and G. Schultz, *Proc. Natl. Acad. Sci. U.S.A.* **90**, in press (1994).

Biotechnology, Inc., Lake Placid, NY). Antisera are employed as crude sera for immunoprecipitation experiments.

General Procedure

Preparation of [α-^{32}P]GTP Azidoanilide and Agonist-Dependent Photolabeling of G-Protein α Subunits in Membranes

Preparation of and photolabeling with [α-^{32}P]GTP azidoanilide are principally performed as described.[8] To increase the amount of photolabeled G-protein α subunits, the assay volume is increased to 120 μl containing 200 μg of membranes and 4–8 μCi of [α-^{32}P]GTP azidoanilide. Depending on the G-protein subtype examined, assays are performed in the presence or absence of 10 μM GDP, and the incubation time varies between 3 and 30 min. (see below). Glutathione (2 mM) is used as a scavenger (instead of dithiothreitol used previously[8]).

Immunoprecipitation of Photolabeled G-Protein α Subunits

Precipitation is performed in 1.5-ml Eppendorf-type test tubes at a temperature of 4°.

1. Photolabeled membranes are centrifuged for 10 min at 12,000 g, solubilized in 40 μl of 4% (w/v) sodium dodecyl sulfate (SDS), and kept for 10 min at room temperature in order to denature membrane proteins. This predenaturation step increases the efficiency of immunoprecipitation with certain antisera (see below) but is optional.

2. Membrane proteins are solubilized or diluted in 280 μl of precipitation buffer containing 1% (w/v) Nonidet P-40, 1% (w/v) deoxycholate, 0.5% (w/v) SDS, 150 mM NaCl, 1 mM dithiothreitol (DTT), 1 mM EDTA, 0.2 mM phenylmethylsulfonyl fluoride (PMSF), 10 μg/ml aprotinin, and 10 mM Tris-HCl (pH 7.4). When membrane proteins are predenatured, SDS is diluted to a final concentration of 0.5% (w/v) by addition of precipitation buffer. In some cases it is advantageous to use nonionic detergents other than Nonidet P-40, for example, CHAPS or Lubrol (see below).

3. Solubilized membranes are centrifuged at 4° for 10 min at 12,000 g to remove insoluble material.

4. (*Optional*) To reduce the amount of protein precipitated nonspecifically, solubilized membranes are precleared. Fifty microliters of 10% (w/v) protein A-Sepharose beads (prepared as described under Step 7) or 10–30 μl of preimmune serum together with protein A-Sepharose is added to the supernatant. After 1–2 hr, Sepharose beads are removed by centrifugation.

TABLE I
Peptide Antibodies Immunoprecipitating G-Protein α Subunits

Designation	Code number	Peptide sequence	Amino acids	Precipitation shown for	Ref.
α_{common}	AS8	(C)GAGESGKSTIVKQMK	40–54[a]	α_i, α_o	b
	A-569	(C)GAGESGKSTIVKQMK	40–54[a]	α_i, α_s	c
	1398	(C)GAGESGKSTIVKQMK	40–54[a]	α_i	d
$\alpha_{s,Arg}$	A-572	(C)KQLQRDRQVYRATHR	28–42[a]	α_s	c
α_s	RM	RMHLRQYELL	385–394[a]	α_s	e, f
	AS348	RMHLRQYELL	385–394[a]	α_s	o
	CS1	RMHLRQYELL	385–394[a]	α_s	g
	1190	RMHLRQYELL	385–394[a]	α_s	h
α_o common	AS6	(C)NLKEDGISAAKDVK	22–35[i]	α_{o1}, α_{o2}	b
	2353	(C)EYGDKERKADSK	94–105[a]	α_o	j
α_{o1}	AS248	SKNRSPNKEIYCHM	310–323[i]	α_{o1}	b
α_i common	AS266	(C)NLREDGEKAAREV	22–34[a]	α_{i1}, α_{i2}, α_{i3}	b
	8730	KNNLKDCGLF	345–354[a]	α_{i1}, α_{i2}, α_{i3}	j, d
	8645	(C)FDVGGQRSERKK	200–211[a]	α_i, α_z	d
α_t/α_i	AS/7	KENLKDCGLF	341–350[k]	α_{i2}, α_{i3}	g, l
	AS	KENLKDCGLF	341–350[k]	α_{i1}	e
α_{i1}	AS190	(C)LDRIAQPNYI	159–168[a]	α_{i1}	b
	3646	(C)LDRIAQPNYI	159–168[a]	α_{i1}	j
α_{i2}	AS269	(C)TGANKYDEAAS	295–305[m]	α_{i2}	b
	1521	(C)LERIAQSDYI	160–169[a]	α_{i2}, α_z	j, d
α_{i3}	AS105	(C)LDRISQSNYI	159–168[m]	α_{i3}	b
	SQ	(C)LDRISQSNYI	159–168[m]	α_{i3}	j
	I3B	KNNLKECGLY	345–354[a]	α_{i2}, α_{i3}	g
	1518	(C)IDFGEAARADDAR	93–105[a]	α_{i3}	j
α_z	AS227	(C)HLRSESQRQRREI	22–34[n]	α_z	o
	2919	(C)TGPAESKGEITPELL	111–125[n]	α_z	d
	2921	QNNLKYIGLC	346–355[n]	α_z	d
$\alpha_{q/11}$	X-384/Z811	(C)ILQLNLKEYNLV	348–359[p]	$\alpha_{q/11}$	q, r, o
	AS368	(C)LQLNLKEYNLV	349–359[p]	$\alpha_{q/11}$	o
α_{12}	AS233	(C)QENLKDIMLQ	370–379[s]	α_{12}	t
α_{13}	AS243	(C)LHDNLKQLMLQ	367–377[s]	α_{13}	t

[a] D. T. Jones and R. R. Reed, *J. Biol. Chem.* **262**, 14241 (1987).
[b] K.-L. Laugwitz, S. Offermanns, K. Spicher, and G. Schultz, *Neuron* **10**, 233 (1993).
[c] J. E. Buss, S. M. Mumby, P. J. Casey, A. G. Gilman, and B. M. Sefton, *Proc. Natl. Acad. Sci. U.S.A.* **84**, 7493 (1987).
[d] K. E. Carlson, L. F. Brass, and D. R. Manning, *J. Biol. Chem.* **264**, 13298 (1989).
[e] T. L. Jones, W. F. Simonds, J. J. Merendino, Jr., M. R. Brann, and A. M. Spiegel, *Proc. Natl. Acad. Sci. U.S.A.* **87**, 568 (1990).
[f] W. F. Simonds, P. K. Goldsmith, C. J. Woodard, C. G. Unson, and A. M. Spiegel, *FEBS Lett.* **249**, 189 (1989).
[g] M. Bushfield, G. J. Murphy, B. E. Lavan, P. J. Parker, V. J. Hruby, G. Milligan, and M. D. Houslay, *Biochem. J.* **268**, 449 (1990).
[h] Y. Okuma and T. Reisine, *J. Biol. Chem.* **267**, 14826 (1992).
[i] W. H. Hsu, U. Rudolph, J. Sanford, P. Bertrand, J. Olate, C. Nelson, L. G. Moss, A. E. Boyd III, J. Codina, and L. Birnbaumer, *J. Biol. Chem.* **265**, 11220 (1990).

5. Antiserum (10–30 µl) is added to the supernatant.
6. Samples are incubated for at least 2 hr at 4° under constant rotation.
7. Protein A-Sepharose beads [60 µl of 12.5% (w/v)] are added. Prior to addition, the Sepharose beads are allowed to swell for 30 min in precipitation buffer and are washed three times with 1 ml of precipitation buffer.
8. Samples are further incubated overnight (incubation should be performed for at least 2 hr).
9. Sepharose beads are pelleted (12,000 g for 5 min) and washed twice with 1 ml of washing buffer A containing 1% (w/v) Nonidet P-40, 0.5% (w/v) SDS, 600 mM NaCl, and 50 mM Tris-HCl (pH 7.4) and once with 1 ml of washing buffer B containing 300 mM NaCl, 10 mM EDTA, and 100 mM Tris-HCl (pH 7.4).
10. In preparations for SDS–polyacrylamide gel electrophoresis (SDS–PAGE), 100 µl of sample buffer[13] is added to Sepharose beads, the mixture is incubated for 10 min at 100°, then centrifuged, and the supernatant is subjected to SDS–PAGE.

Determination of [α-^{32}P]GTP Azidoanilide Incorporated into Immunoprecipitated G-Protein α Subunits

Washed immunoprecipitates are routinely analyzed by SDS–PAGE. Gels are dried and exposed to X-ray films. A semiquantitative evaluation is obtained by laser densitometry of the autoradiograms.

[13] U. K. Laemmli, *Nature* (*London*) **227**, 680 (1970).

---

[j] S. F. Law, D. Manning, and T. Reisine, *J. Biol. Chem.* **266**, 17885 (1991).
[k] T. Tanabe, T. Nukada, Y. Nishikawa, K. Sugimoto, H. Suzuki, H. Takahashi, M. Noda, T. Haga, A. Ichiyama, K. Kangawa, N. Minamino, H. Matsuo, and S. Numa, *Nature* (*London*) **315**, 242 (1985).
[l] P. L. Rothenberg and C. R. Kahn, *J. Biol. Chem.* **263**, 15546 (1988).
[m] W. N. Suki, J. Abramowitz, R. Mattera, J. Codina, and L. Birnbaumer, *FEBS Lett.* **220**, 187 (1987).
[n] H. K. Fong, K. K. Yoshimoto, P. Eversole-Cire, and M. I. Simon, *Proc. Natl. Acad. Sci. U.S.A.* **85**, 3066 (1988).
[o] K.-L. Laugwitz, K. Spicher, G. Schultz, and S. Offermanns, unpublished. (See Figs 1–3.)
[p] M. Strathmann and M. I. Simon, *Proc. Natl. Acad. Sci. U.S.A.* **87**, 9113 (1990).
[q] A. V. Smrcka, J. R. Hepler, K. O. Brown, and P. C. Sternweis, *Science* **251**, 804 (1991).
[r] R. L. Wange, A. V. Smrcka, P. C. Sternweis, and J. H. Exton, *J.Biol. Chem.* **266**, 11409 (1991).
[s] M. Strathmann and M. I. Simon, *Proc. Natl. Acad. Sci. U.S.A.* **88**, 5582 (1991).
[t] S. Offermanns, K.-L. Laugwitz, K. Spicher, and G. Schultz, *Proc. Natl. Acad. Sci. U.S.A.*, **90**, in press (1994).

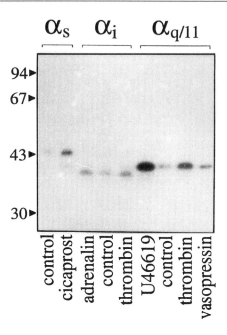

FIG. 1. Selective immunoprecipitation of α_s, α_i, and $\alpha_{q/11}$ from human platelet membranes photolabeled in the presence of different receptor agonists. Membranes were photolabeled in the absence of agonists (control) or in the presence of prostacyclin receptor agonist cicaprost (1 μM), adrenalin (100 μM), thrombin (1 U/ml), the thromboxane A_2 receptor agonist U46619 (5 μM), or vasopressin (1 μM) as indicated. Incubation was performed for 10 min in a buffer containing 10 mM $MgCl_2$, 0.1 mM EDTA, 30 mM NaCl, and 50 mM HEPES (pH 7.4). Samples for precipitation with the $\alpha_{i\ common}$ antiserum (α_i) additionally contained 10 μM GDP. After photolysis, solubilized membranes were incubated with the indicated antisera, and immunocomplexes were precipitated and separated by SDS–PAGE. Shown is an autoradiogram of a dried gel with the molecular masses (kDa) of marker proteins on the left. α_s, α_i, and $\alpha_{q/11}$ indicate the use of antisera AS348, AS266, and AS368, respectively.

Exact quantification of specifically immunoprecipitated radioactivity is achieved by direct counting. For counting of radioactivity in the immunoprecipitate, 1 ml of 2% (w/v) SDS is added to the precipitated beads, which are then boiled for 10 min and pelleted. Supernatants are diluted with water and counted for radioactivity. Nonspecific radioactivity precipitated by preimmune sera is subtracted from the radioactivity precipitated by the corresponding specific antisera.

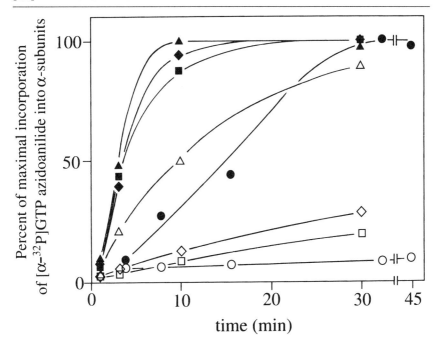

FIG. 2. Effect of incubation time on photolabeling of α_i, α_s, $\alpha_{q/11}$, α_{12} in the absence and presence of receptor agonists. Membranes of human platelets were photolabeled in the absence (open symbols) or presence (closed symbols) of 100 μM adrenalin ($\triangle$/▲), 1 μM cicaprost ($\diamond$/♦), and 5 μM U46619 ($\square$/■/○/●), respectively. Platelet membranes were incubated for the indicated time periods (abscissa) in a buffer containing 10 mM MgCl$_2$, 0.1 mM EDTA, 30 mM NaCl, and 50 mM HEPES (pH 7.4). Samples for precipitation with the $\alpha_{i\,common}$ antiserum (α_i) additionally contained 10 μM GDP. Platelet membranes photolabeled in the absence or presence of receptor agonists were solubilized and α_i, α_s, $\alpha_{q/11}$, and α_{12} were immunoprecipitated with antisera AS266, AS348, AS368, and AS233 ($\triangle$/▲, $\diamond$/♦, $\square$/■, and ○/●, respectively). Immunoprecipitates were separated by SDS–PAGE and incorporation of [α-^{32}P]GTP azidoanilide into α subunits was quantitated by laser densitometry of autoradiograms from dried gels. Shown is the incorporation of [α-^{32}P]GTP azidoanilide into α subunits as percent of the maximal incorporation.

Receptor-Dependent Stimulation of Photolabeling of Individual G-Protein α Subunits

Because α subunits of G_i and G_o proteins rapidly accumulate the nonhydrolyzable GTP analog [α-^{32}P]GTP azidoanilide even in the absence of receptor agonists, a considerable receptor-dependent stimulation of photolabeling of the α subunits can only be observed at relatively short incubation times (3–10 min, Figs. 1, 2, and 4). The α subunits of G_s,

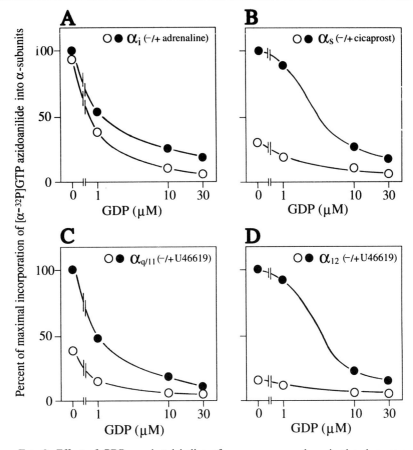

FIG. 3. Effect of GDP on photolabeling of α_i, α_s, $\alpha_{q/11}$, and α_{12} in the absence and presence of receptor agonists. Membranes of human platelets were photolabeled in the absence (open symbols) or presence (closed symbols) of 100 μM adrenalin (A), 1 μM cicaprost (B), and 5 μM U46619 (C, D), respectively. Platelet membranes were incubated with the indicated GDP concentrations (abscissa) in a buffer containing 10 mM MgCl$_2$, 0.1 mM EDTA, 30 mM NaCl, and 50 mM HEPES (pH 7.4). Incubation was performed for 10 min (A, B, C) or for 30 min (D). Platelet membranes photolabeled in the absence or presence of receptor agonists were solubilized, and α_i, α_s, $\alpha_{q/11}$, and α_{12} were immunoprecipitated with antisera AS266, AS348, AS368, and AS233 (A, B, C, D, respectively). Immunoprecipitates were separated by SDS–PAGE and incorporation of [α-^{32}P]GTP azidoanilide into α subunits was quantitated by laser densitometry of autoradiograms from dried gels. Shown is the incorporation of [α-^{32}P]GTP azidoanilide into α-subunits as percent of the maximal incorporation.

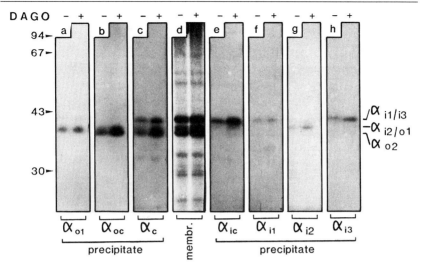

FIG. 4. Selective immunoprecipitation of G-protein α subunits from SH-SY5Y cell membranes after photolabeling in the absence and presence of DAGO. Membranes from SH-SY5Y cells were incubated with [α-^{32}P]GTP azidoanilide in the absence (−) or presence (+) of the μ-opioid receptor agonist DAGO (1 μM). Incubation was performed for 3 min in a buffer containing 10 mM MgCl$_2$, 10 μM GDP, 0.1 mM EDTA, 30 mM NaCl, and 50 mM HEPES (pH 7.4). After photolysis, solubilized membranes either were subjected directly to SDS–PAGE (d) on gels containing 9% (w/v) acrylamide and 6 M urea or were incubated with different antisera as indicated (a–c, e–h). After addition of protein A-Sepharose, immunocomplexes were precipitated and separated by SDS–PAGE on gels as described above. Shown are autoradiograms of dried gels with the molecular masses (kDa) of marker proteins on the left-hand side. α_{o1}, α_{oc}, α_c, α_{ic}, α_{i1}, α_{i2}, and α_{i3}, indicate the use of antisera AS248, AS6, AS8, AS266, AS190, AS269, and AS105, respectively. "Membr." indicates that photolabeled membranes were directly subjected to SDS–PAGE.

$G_{q/11}$, and $G_{12/13}$, in contrast, exhibit a slow basal nucleotide exchange rate,[10,12,14] (Fig. 2) and stimulation of [α-^{32}P]GTP azidoanilide binding to $\alpha_{q/11}$ and $\alpha_{12/13}$ can also be determined at longer incubation times (see Figs. 1, 2, and 5). The different basal nucleotide exchange rates obviously lead to different effects of GDP when present in the incubation mixture. Whereas the presence of GDP is not necessary for an effective stimulation of photolabeling of α_s, $\alpha_{q/11}$, and $\alpha_{12/13}$[10,12] (see Figs. 1, 3, and 5), receptor-mediated activation of G_i and G_o proteins, when measured as accumulation of hydrolysis-resistant GTP analogs, is significantly enhanced in the presence of GDP at micromolar concentrations[15,16] (see Fig. 3).

[14] I. H. Pang and P. C. Sternweis, *J. Biol. Chem.* **265**, 18707 (1990).
[15] V. A. Florio and P. C. Sternweis, *J. Biol. Chem.* **264**, 3909 (1989).
[16] S. Offermanns, G. Schultz, and W. Rosenthal, *J. Biol. Chem.* **266**, 3365 (1991).

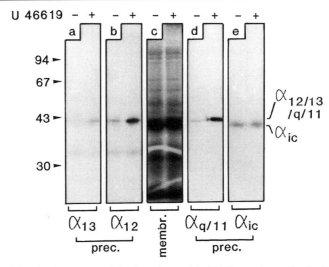

FIG. 5. Selective immunoprecipitation of photolabeled G-protein α subunits from human platelets. Membranes from platelets were photolabeled in the absence (−) or presence (+) of 3 μM of the thromboxane A$_2$ receptor agonist U46619. Incubation was performed for 30 min in a buffer containing 10 mM MgCl$_2$, 0.1 mM EDTA, 30 mM NaCl, and 50 mM HEPES (pH 7.4). After photolysis, solubilized membranes either were subjected directly to SDS–PAGE (c) on gels containing 10% (w/v) acrylamide or were incubated with different antisera as indicated (a, b, d, e). After addition of protein A-Sepharose, immunocomplexes were precipitated and separated by SDS–PAGE on gels as described above. Shown are autoradiograms of dried gels with the molecular masses (kDa) of marker proteins on the left-hand side. $α_{13}$, $α_{12}$, $α_{q/11}$, and $α_{ic}$ indicate the use of antisera AS343, AS233, Z811, and AS 266, respectively. "Membr." indicates that photolabeled membranes were directly subjected to SDS–PAGE.

Incorporation of [α-^{32}P]GTP azidoanilide into G-protein α subunits is strictly dependent on the presence of Mg^{2+} and increases up to Mg^{2+} concentrations of about 10 mM. The ratio of agonist-stimulated to basal photolabeling of $G_{i/o}$, G_s, $G_{q/11}$, and $G_{12/13}$ α subunits, however, shows no considerable dependence on different Mg^{2+} concentrations. Concentrations of 3–10 mM Mg^{2+} are recommended in order to achieve an optimal sensitivity of the method.

Specificity of Immunoprecipitation

The specificity of immunoprecipitation of G-protein α subunits by selective antisera can be controlled by parallel precipitation with the corresponding preimmune sera or by blocking the precipitation with the peptides against which the respective antisera were raised. For the latter control experiments, antisera are preincubated with peptides (10–14 amino acids) employed at concentration of 1–10 μg/ml for 6 hr at 4° under constant agitation.

In some cases, a specific immunoprecipitation of G-protein α subunits is difficult to achieve. The specificity of immunoprecipitation can be improved by preabsorption of nonspecifically binding proteins to preimmunesera and/or protein A-Sepharose (see Step 4 of the general procedure and Fig. 6). An additional alternate washing of the precipitate with washing buffers A and B also is a suitable measure to increase the specificity of immunoprecipitation.

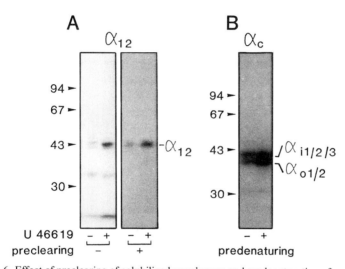

FIG. 6. Effect of preclearing of solubilized membranes and predenaturation of membrane proteins on immunoprecipitation with α_{12} and α_{common} antiserum, respectively. (A) Platelet membranes were photolabeled in the absence (−) or presence (+) of 3 μM U46619 as described in Fig. 3. Prior to immunoprecipitation with AS233, preclearing of solubilized membranes with protein A-Sepharose (Step 4 of the general procedure) was performed (preclearing +) or omitted (preclearing −). Immunoprecipitates were subjected to SDS–PAGE on gels containing 10% (w/v) acrylamide. Shown are autoradiograms of dried gels with molecular masses (kDa) of marker proteins on the left-hand side. (B) Membranes from SH-SY5Y cells were incubated for 10 min with [α-^{32}P]GTP azidoanilide and photolyzed. Membrane proteins were either predenatured (Step 1 of the general procedure) prior to addition of precipitation buffer (predenaturation +) or were directly solubilized in precipitation buffer (predenaturation −). Immunoprecipitation was performed with antiserum AS8, and immunoprecipitates were separated by SDS–PAGE on gels containing 9% (w/v) acrylamide without urea. Shown is an autoradiogram of a dried gel with the molecular masses (kDa) of marker proteins on the left-hand side and the positions of α_i and α_o proteins on the right-hand side.

Efficiency of Immunoprecipitation

The efficiency of immunoprecipitation is mainly dependent on the accessibility of the protein region against which the antiserum is directed. This factor can be influenced by the choice of ionic and nonionic detergents used during the precipitation procedure. Predenaturation of the photolabeled proteins with SDS prior to addition of nonionic detergents and antisera (Step 1 of the general procedure) in some cases leads to a profound increase in the amount of protein precipitated by antisera. The amount of photolabeled G_o α subunits precipitated by antisera AS 6 and AS 8 (see Fig. 6) can be increased severalfold by predenaturation of membrane proteins, whereas the amount of precipitated G_i α subunits remains the same. We also observe striking differences in the efficiency of immunoprecipitation when comparing the effect of several nonionic detergents. Immunoprecipitation with the antisera AS266, AS190, and AS269, for example, was much more efficient when using 1% (w/v) CHAPS instead of 1% (w/v) Nonidet P-40 in the precipitation buffer. Finding the optimal detergent is empirical, and the most useful detergent has to be determined for each individual antiserum/α subunit.

Finally, it is advisable for optimal efficiency of immunoprecipitation to keep the precipitation volume low (e.g., 200–300 μl for 200 μg of solubilized membranes and 10–30 μl of antiserum). In addition, this saves antiserum.

Acknowledgments

The authors wish to thank Evelyn Bombien for helpful technical assistance, Dr. Klaus-Dieter Hinsch for providing antisera AS6 and AS8, and Dr. Paul C. Sternweis for the kind donation of antiserum Z811. Our studies reported herein were supported by the Deutsche Forschungsgemeinschaft.

[23] Identification of Mutant Forms of G-Protein α Subunits in Human Neoplasia by Polymerase Chain Reaction-Based Techniques

By JOHN LYONS

Introduction

G-protein α subunits bind GTP and GDP and cycle between an inactive GDP-bound form and an active GTP-bound form.[1,2] The role of α subunits in signal transduction may be defined genetically and biochemically, when a G-protein α subunit is expressed in a given cell type, it may interact with an appropriate set of activated receptors (G-protein receptors), bind GTP for GDP, and transduce its signal to second messengers via a relatively slow intrinsic hydrolysis of GTP to GDP, thereby instigating a complex sorting and amplification cascade resulting, finally, in either mitogenic effects, hormone secretion, or cell differentiation.

Members of the α-subunit families may be divided into four different groups based on amino acid sequences. In the first group the α_s and the olfactory subunits activate adenylyl cyclase (adenylate cyclase, EC 4.6.1.1) and elevate levels of cAMP within the cell.[3] The α subunits of the inhibitory class are largely ADP-ribosylated by pertussis toxin and include G_{i1}, G_{i2}, G_{i3}, G_o, and transducin. ADP-ribosylation at the carboxyl terminus by pertussis toxin results in the modification of amino acid sequences in the α chain and uncouples the α subunit from its receptor and, consequently, from its signal transduction pathway. A further group of α subunits has been cloned, the G_q class. This group of α subunits activates phospholipase C and is pertussis toxin insensitive.[4-7] A fourth group, consisting of $G\alpha_{12}$ and $G\alpha_{13}$, has been cloned, and their sequences have been shown to be unique from the other three groups.[8]

---

[1] H. R. Bourne, D. A. Sanders, and F. McCormick, *Nature* (*London*) **348**, 117 (1991).
[2] H. R. Bourne, D. A. Sanders, and F. McCormick, *Nature* (*London*) **348**, 125 (1990).
[3] J. E. Dumont, J. C. Jauniaux, and P. P. Roger, *Trends Biochem. Sci.* **14**, 67 (1989).
[4] M. Strathmann, T. M. Wilkie, and M. I. Simon, *Proc. Natl. Acad. Sci. U.S.A.* **86**, 7407 (1989).
[5] S. G. Rhee and K. D. Choi, *J. Biol. Chem.* **267**, 12393 (1992).
[6] A. Schenker, P. Goldsmith, C. G. Unson, and A. M. Spiegel, *J. Biol. Chem.* **266**, 9309 (1991).
[7] D. Wu, C. H. Lee, S. G. Rhee, and M. I. Simon, *J. Biol. Chem.* **267**, 1811 (1992).
[8] M. P. Strathmann and M. I. Simon, *Proc. Natl. Acad. Sci. U.S.A.* **88**, 5582 (1991).

A number of trophic hormones (such as thyroid-stimulating hormone) are capable of stimulating cAMP and causing mitogenic effects in cells. A subset of patients with pituitary tumors exhibited elevated levels of growth hormone, cAMP, and constitutive activation of adenylate cyclase. When the gene encoding the α subunit of G_s was sequenced in these patients, two codons were found to be mutated; one at R201 was the site of ADP-ribosylation by cholera toxin, and the other was Q227, cognate to Q61 of the protooncogene p21*ras*. When expressed these mutated proteins showed low GTPase activity and caused constitutive activation of adenylate cyclase[9]: these results prompted us to search for mutations in $G_s\alpha$ in other tumor types at these two codons and in other G proteins at equivalent codons. There was, however, no direct evidence for activating mutations in other G proteins in human tumors, as happened for the $G_s\alpha$ mutations and their accompanying biochemical properties, so our broad search relied on the biochemical criteria for each G protein.[10-14]

At the outset of the study we made two inferences. On the one hand, highly conserved amino acids such as codon 201 of G_s, the site of ADP-ribosylation by cholera toxin, and codon 227, the cognate amino acid to codon 61 of the protooncogene *ras*, should have the same functional role in any other G protein. Moreover, we also inferred that activation of either of these amino acids in a G-protein α subunit would exert a mitogenic and, hence, a growth advantage to tumor cells of that tissue type harboring the activated G protein. The study was initiated by screening a large number of over 250 human neoplasia for $G_s\alpha$ subunits using oligonucleotide probes specific for point mutations at the codons 201 and 227. Subsequently, we chose G_{i2}, since we were able to amplify exon and intron sequences from formalin fixed paraffin-embedded tumor tissue more easily than the other targets, initially.

Assays for Detection of Point Mutations in G-Protein α Subunits

Sequences of amplifying primers and conditions for the polymerase chain reaction (PCR) optimized for paraffin-embedded and formalin-fixed

---

[9] C. A. Landis, S. B. Masters, A. Spada, A. M. Pace, H. R. Bourne, and L. Vallar, *Nature (London)* **340**, 692 (1989).
[10] D. T. Jones and R. R. Reed, *J. Biol. Chem.* **262**, 14241 (1987).
[11] T. Asano, R. Semba, N. Ogasawara, and K. Kato, *J. Neurochem.* **48**, 1617 (1987).
[12] W. W. Y. Lo and J. Hughes, *FEBS Lett.* **220**, 327 (1987).
[13] S. B. Masters, M. W. Martin, T. K. Harden, and J. H. Brown, *Biochem. J.* **227**, 933 (1985).
[14] P. F. Worley, J. M. Baraban, E. Van Dop, E. J. Neer, and S. H. Snyder, *Proc. Natl. Acad. Sci. U.S.A.* **83**, 4561 (1986).

tumor DNA are given in Table I. We chose at the outset to study five different subunits based on their biochemical, mitogenic, and oncogenic (for G_s) properties, as noted above, although more α subunits have been cloned since. The oligonucleotide sequences are derived from published sequences,[15-19] each amplifier set presented in Table I is optimized for paraffin-embedded tissue as template, is generally derived from exon sequences in genomic DNA, and results in a single product on an agarose gel stained with ethidium bromide. In Table II all possible single-base changes resulting in new amino acids at codons 201 and 227 of the $G_s\alpha$ subunit are listed; likewise, for each other α subunit 20-mer oligonucleotides are synthesized with all possible amino acid changes at the cognate 201 and 227 amino acid positions from single-point mutations. Each set of oligonucleotides is hybridized against the amplified material from over 250 tumors, including 42 pituitary adenomas, 11 adrenal cortical tumors, 25 thyroid tumors, 16 melanomas, 29 glioblastomas, 10 ovarian adenocarcinomas, 14 gastric adenomacarcinomas, 17 renal cell carcinomas, 25 breast adenomacarcinomas, 9 bladder transitional cell carcinomas, 12 pancreatic adenocarcinomas, 8 acute myelogenous leukemias, 33 squamous cell carcinomas, 20 colonic adenocarcinomas, 12 hepatomas, 12 prostatic tumors, and 10 ovarian sex cord stromal tumors.

Dot-Blot Procedure

We chose the dot-blot procedure initially, because mutations originally described for pituitary tumors were found only at either of two codons. Screening by oligonucleotide hybridization is extremely rapid, with a turnaround of analysis of 2–4 days. We expected from our inferences to find clonal expansions of *gsp* mutations in the tumor samples being investigated, thereby simplifying the detection of mutations.

After the PCR, an aliquot (1/20) is analyzed on an agarose gel and visualized by ethidium bromide staining. Approximately 5 μl/dot of each successful PCR amplification (as judged by agarose gel electrophoresis)

[15] S. Lavu, J. Clark, R. Swarup, K. Matsushima, K. Paturu, J. Moss, and H.-F. Kung, *Biochem. Biophys. Res. Commun.* **150**, 811 (1988).
[16] H. K. W. Fong, K. K. Yoshimoto, P. Eversole-Cire, and M. I. Simon, *Proc. Natl. Acad. Sci. U.S.A.* **85**, 3066 (1988).
[17] M. Matsuoka, H. Itoh, T. Kozasa, and Y. Kaziro, *Proc. Natl. Acad. Sci. U.S.A.* **85**, 5384 (1988).
[18] T. Kozasa, H. Itoh, T. Tsukamoto, and Y. Kaziro, *Proc. Natl. Acad. Sci. U.S.A.* **85**, 2081 (1988).
[19] H. Itoh, R. Toyama, T. Kozasa, T. Tsukamoto, M. Matsuoka, and Y. Kaziro, *J. Biol. Chem.* **263**, 6656 (1988).

TABLE I
POLYMERASE CHAIN REACTION FOR G-PROTEIN α SUBUNITS[a]

Gene product	Amino acid position	Amplimer pair	Cycling parameters			Number of cycles	Product size (bp)
G_{i3}	178	GATCTGGATAGAATATCCCAG GGTGAAATGTGTTTCTACAAT	95° 30 sec	55° 30 sec	72° 30 sec	40	99
G_{i3}	204	TTCCCCTTGCGCAGGATGTTT CAGAACAAGGTCATAATCACT	95° 30 sec	60° 30 sec	72° 30 sec	40	120
$G_s\alpha$	201	GTGATCAAGCAGGCTGACTATGTG TAACAGTTGGCTTACTGGAA	95° 30 sec	55° 30 sec	72° 30 sec	40	222
$G_s\alpha$	227	CCCAGTCCCTCTGGAATAACCAG GCTGCTGGCCACCACGAAGATGAT	95° 30 sec	55° 30 sec	72° 30 sec	50	165
G_{i2}	179	ATTGCACAGAGTGACTACATCCCC GCTCACTTGAAGTGTAGG	95° 60 sec	57° 30 sec	72° 30 sec	50	118
G_{i2}	201	CTGCAGGATGTTTGATGTGG GGCGCTCAAGGCTACGCAGAA	95° 30 sec	57° 30 sec	72° 30 sec	50	~123
G_o	201	CTGGACAGCCTGGATCGGATTGGG GAGGTTCTTGAATGTGAAGTGGGT	95° 60 sec	57° 60 sec	72° 60 sec	35	87
G_o	227	AACCTCCACTTCAGGCTGTTT GTGGAGCACCTGGTCATAGCCGCT	95° 30 sec	48° 90 sec	72° 30 sec	50	129
G_z	201 + 227	AACGACCTGGAGCGCATCGCC TGTGACGCCCTCGAAGCAGT	95° 60 sec	57° 60 sec	72° 60 sec	40	~200

[a] Primers were synthesized for the codons corresponding to R201 and Q227 of G_s, G_{i2}, G_{i3}, G_o, and G_z. Each set of primer pairs were optimized for PCR with human genomic DNA samples as template. The annealing times and number of cycles in columns 4 and 5 correspond to a 2- to 5-μl aliquot of formalin-fixed paraffin-embedded tissue. The size of the PCR product corresponds to the distance between the primers according to the published sequences.

TABLE II
POSSIBLE SINGLE-BASE CHANGE COMBINATIONS IN AMINO ACIDS R201
AND Q227 OF $G_s\alpha$ GENE[a]

Amino acid	Type of change	Single-base changes				
Arginine-201 (CGT)	Transitions	TGT Cys	CAT His			
	Transversions	CCT Pro	AGT Ser	GGT Gly	CTT Leu	
Glutamine-227 (CAG)	Transitions	CGG Arg	TAG Stop			
	Transversions	CCG Pro	CAT/C His	CTG Leu	AAG Lys	GAG Glu

[a] Encoding new amino acids or a stop codon in the G_s protein sequence.

is denatured in 50 μl of 0.4 M NaOH and 25 mM EDTA for 5 min in a water bath at 50°. Usually, 40 μl of PCR product (for 8 duplicate filters) is added to 400 μl of denaturing solution. Using a multipipette, 12 samples are applied to duplicate nylon filters (Pall Biodyne), presoaked in water and fixed on a Bio-Rad (Richmond, CA) dot-blot apparatus. Amplified DNA is then covalently cross-linked to the filters with a Stratalinker (Stratagene, La Jolla, CA) at automatic setting. The filters are then air dried and prehybridized in a solution of 5× SSPE (10 mM sodium phosphate, 0.1 M NaCl, and 1 mM EDTA) and 0.5% sodium dodecyl sulfate (SDS) for 5–10 min at 50° in a shaking water bath. Two picomoles of each mutant oligonucleotide and the wild-type oligonucleotide are kinased in a 10-μl reaction in the presence of 1 μl of [γ-^{32}P]ATP, 1 μl of 10× kinase buffer (NEB kinase buffer), and 1–2 units of polynucleotide kinase for 30 min at 37°. The samples are diluted into 600 μl of water, and 200 μl (~300–650 fmol) of each labeling reaction is incubated with a separate filter in 5–7 ml of 5× SSPE, 0.5% SDS at 50° for 2 hr in a shaking water bath.

After hybridization the filters are rinsed twice in 2× SSPE, 0.1% SDS at ambient temperature and subsequently washed in 5× SSPE, 0.5% SDS for 20 min at 50°. A stringent wash in 3 M TMACl (tetramethylammonium chloride), 0.1% SDS at a temperature previously determined empirically is then performed. The temperature allowing optimum discrimination between full matched hybrids and mismatched species is obtained empirically for each set of oligonucleotide probes as follows: triplicate filters are hybridized in parallel with a labeled wild-type and a mutant oligonucleotide and subsequently washed at increasing stringent temperatures beginning at 58°, since an oligonucleotide containing 20 nucleotides should have no or only minimal GC or AT preferences in 3 M tetramethylammonium chloride and should be discriminated only by length. The temperature is increased in increments of 2° until only the wild-type signals remain and the mutant oligonucleotide signals are washed off, as determined by autoradiography. Because this experiment may be carried out by hybridizing both oligonucleotides in parallel, optimal discriminating temperatures for probes for each α subunit can be obtained in a single experiment.

Results from Dot-Blot Procedure

Of the 42 pituitary tumors tested, 18 proved positive for *gsp* mutations at either codon 201 or 227. All of the positive samples were from patients exhibiting growth hormone-secreting pituitary tumors (see Fig. 1) and elevated basal levels of cAMP.[20] When compared to patients with normal cAMP levels, it was found that the tumors harboring *gsp* mutations were smaller at presentation, had lower growth hormone levels, and were also more susceptible to glucose suppression.[21] None of the other hormone-secreting pituitary tumors tested positive for mutations in *gsp* or any of the other G proteins tested at the cognate codons. A single thyroid tumor from the 25 tested proved likewise positive for an activating mutation in the *gsp* gene. In the screen of tumors for mutations in G_{i2}, 3 of 10 ovarian sex cord tumors and 3 of 11 adrenal cortical tumors scored positive for mutations at codon R179, changing the amino acid either to histidine or cysteine. We were unable to detect any other mutations in any of the other G-protein α chains we amplified (Table I) using the dot-blot hybridization procedure, suggesting that (a) we missed the tumor phenotype in our screen (i.e., the screen was not broad enough), (b) other G proteins of either the q family or the fourth family of α subunits play a more general role in human neoplasia, (c) α-subunit activation is a late event in tumor progression in some of the tumor types we investigated, and (d) G protein involvement in human neoplasia is restricted to endocrine tumors only.

Alternative Assays for Activating Mutations in Gα Subunits

A number of other PCR-based methodologies, besides the use of oligonucleotide probes, have been applied to the detection of genetic alterations, such as mutation, insertion, or deletion of nucleic acids in a given

---

[20] J. Lyons, C. A. Landis, G. Harsh, L. Vallar, K. Grünewald, H. Feichtinger, Q. Y. Duh, O. H. Clark, E. Kawasaki, H. R. Bourne, and F. McCormick, *Science* **249**, 655 (1990).

[21] C. A. Landis, G. Harsh, J. Lyons, R. L. Davis, F. McCormick, and H. R. Bourne, *J. Clin. Endocrinol. Metab.* **71**, 1416 (1990).

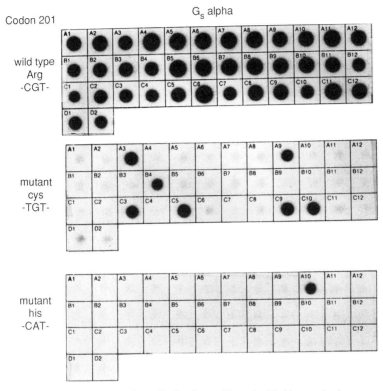

FIG. 1. Thirty eight different formalin-fixed, paraffin-embedded human brain tumor samples were amplified around the amino acids R201 and Q227, spotted onto nylon filters, and hybridized with radioactively labeled oligonucleotide probes for wild-type 201 sequence and the mutant specific oligonucleotides R201C and Q227H. Eight growth hormone-secreting pituitary samples, A3, A9, A10, B4, C3, C5, C9, and C10, scored positive for *gsp* mutations after stringent washing and autoradiography.

region of amplified DNA. One of these methods, denaturing gradient gel electrophoresis (DGGE),[22] is described in [24] of this volume. Generally, there are two different approaches to the detection of alterations; one is based on the near certainty that the genetic alteration of interest will lead to a clonal expansion of the aberration in all the tumor and that the tumor sample will be homogeneous for the aberration being investigated. Therefore, amplification of the region of interest on both alleles and subsequent detection either by dot blotting and oligonucleotide hybridization,

[22] L. S. Weinstein, A. Shenker, P. V. Gejman, M. J. Merino, E. Friedman, and A. Spiegel, *N. Engl. J. Med.* **325**, 1688 (1991).

sequencing the fragment of DNA directly, or separating heterozygous strands because of altered mobility in electrophoretic systems will result in a strong and positive mutant signal.[23] However, this presumes that a high percentage of the sample contains the mutant sequence. This is not always the case, since it is possible that tumors might be heterogeneous for a given aberration. In some cases, multiple alterations can occur in the same tumor, resulting in a decrease of relative detection of each of the mutant species.

To overcome these difficulties, a number of allele-specific PCR-based techniques have been developed. Allele-specific amplification of mutant *ras* sequences has been achieved by decreasing the total deoxynucleoside triphosphate concentration in the PCR, thereby increasing the stringency of amplification of the mutant specific primer. In an alternative technique that has also found application in the detection of *ras* mutations in human neoplasia, a novel restriction site is introduced into the amplified product through the PCR, including the codon of interest.[24] Both wild-type and mutant sequences will be amplified. A deletion, insertion, or point mutation at the codon of interest, however, will result in loss of the novel restriction site. Fragments with a mutation in the restriction site in the sequence only will remain intact if one digests this PCR product. If this product is subject to an additional round of amplification, the resulting PCR fragment will be enormously enriched for specific mutant sequences. Because heterozygous fragments are present as template in the second PCR amplification, wild-type sequences will always be present. Dilution experiments based on Kirsten *ras* codon 12 mutations predict a detection limit of one mutant allele in 1000 alleles, or one heterozygous cell in 500.[24]

A method for specific amplification of mutant *gsp* sequences after incorporation of a restriction site at codon 201 in a first-round PCR is presented here. The method is based on the incorporation of an *Eag*I restriction site by means of a mismatched oligonucleotide (T to G) in a PCR amplification at amino acid positions 200 and 201 of the $G_s\alpha$ subunit gene (see Scheme I and Fig. 2a). Amplification with this mismatched oligonucleotide introduces a new base at the first position of codon 200 into all amplified DNA fragments. This modified sequence constitutes an *Eag*I restriction site that spans and includes the first two bases of the wild-type arginine in amino acid 201, namely, CG (the restriction sequence for *Eag*I is CGGCCG). Thus, a wild-type allele in a tumor sample will be digested by the enzyme *Eag*I after the first round of PCR amplification.

[23] J. Lyons, *CANCER Suppl.* **69,** 1527 (1992).
[24] S. Levi, A. Urbano-Ispizua, R. Gill, D. M. Thomas, J. Gilbertson, C. Foster, and C. J. Marshall, *Cancer Res.* **51,** 3497 (1991).

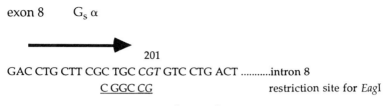

SCHEME I

However, if either of the two bases C or G at position 201 is mutated, the restriction enzyme will not recognize these templates and will be unable to restrict them. Now, if a new round of PCR is initiated with templates that are refractory to restriction by EagI, there will be a specific amplification of only mutant strands (note that some fragments may be heterozygous for the fragment and, hence, refractory to digestion, since late cycles encourage reannealing of different, i.e., wild-type and mutant PCR strands to one another). Direct sequencing with a single biotinylated primer on solid-phase magnetic beads allows facile detection of any mutated sequences at codon 201 (see Fig. 2).

The product from the initial round of PCR is readily digested, as seen from Fig. 3, after overnight incubation with EagI. Subsequently, the digested PCR product is coupled to a solid phase to remove the restricted fragments without biotin; this step enhances PCR amplification of the template in a second round in our hands. Finally, direct sequencing of the second-round PCR product on magnetic beads results in high-quality and readily reproducible sequencing gels from tissue samples. The presence of aberrant bands from mutated sequences can be readily detected from autoradiograms. The method can also be applied to amino acid 227 of $G_s\alpha$ and may be extended to a number of different α subunits to include the equivalent activating amino acids.

Alternative Assay Procedure

In the first-round PCR amplification, 50 pmol of the primers SG07B TTGTTTCAGGACCTGCTTCGCGGC (sense and biotinylated on the 5' end) and SG08 CCACTTGCGGCGTTCATCGCGCCG (antisense) and 5 µl of paraffin-embedded, formalin-fixed DNA solution, previously digested with proteinase K to release the DNA into solution, are cycled 35 times at 95° for 1 min, 60° for 1 min, and 72° for 1 min with 2 units of *Taq* polymerase (Perkin-Elmer, Norwalk, CT) with buffer and deoxynucleoside triphosphates. After PCR, the product is precipitated in the presence of glycogen and dried. All of the PCR product is digested at 37° with 10

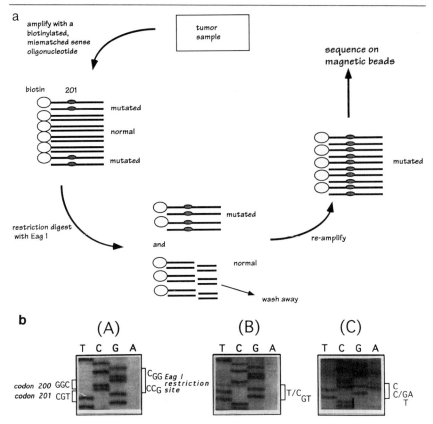

FIG. 2. (a) Schematic representation of PCR-mediated restriction analysis to detect mutations in the $G_s\alpha$ gene at codon 201. One mismatched oligonucleotide is used in a standard PCR to amplify sequences around codon 201. This round of PCR introduces a novel EagI restriction site spanning codon 201. If the PCR product is refractory to digestion, it can be used to initiate a second round of amplification with the same sense oligonucleotide but an inner antisense oligonucleotide. Direct sequencing of single-stranded PCR product after separation on a solid phase by biotin and streptavidin results in reproducible sequencing gels. (b) Results of sequencing reactions from three patients with thyroid cancer (A) without a gsp mutation at R201 (CGT), (B) with an R201C (TGT) mutation, and (C) with a double mutation, R201P (CCT) and R201H (CAT). The EagI restriction site (CGGCCG) spans codons 200 and 201 and is incorporated into both amplified sequences, as described in the text. Sequencing was performed as described in the text.

units of EagI (New England Biolabs, Beverly, MA) and 0.5 μl bovine serum albumin (BSA) (10 mg/ml) overnight in the appropriate buffer (NEB buffer 3). An additional purification step is added at this stage to increase the stringency of the second-round PCR. Then 20 μl of magnetic beads

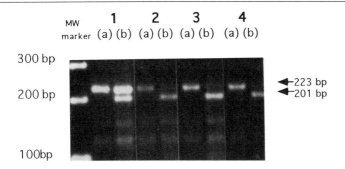

FIG. 3. Four thyroid samples were amplified with the oligonucleotides SG07B and SG09 as described in the text. Each sample was digested with 10 units of *Eag*I at 37° overnight, and aliquots were electrophoresed on a 2% TBE-agarose gel with a 100-bp molecular weight marker. (a) Twelve microliters of each PCR mixture was loaded prior to digestion and (b) 12 μl of the restriction digest from each sample, after overnight incubation with the enzyme.

(Dynal) and 50 μl of water are added directly to the digestion mixture and incubated with agitation for 1 hr. Next, 1/10th volume of the beads are used after washing, for the second-round PCR. In the second round of PCR SG07B TTGTTTCAGGACCTGCTTCGCGGC and SG09 AAGAAACCATGATCTCTGTTATAT are cycled under the same conditions as in the first round.

The PCR product is finally "genecleaned" (Bio 101, La Jolla, CA) and sequenced on magnetic beads, as described by the manufacturer, using standard PCR product dideoxy-sequencing protocols (United States Biochemicals, Cleveland, OH, Sequenase kit) with the sequencing primer SG11, AAGAAACCATGATCTCTG, at 6 pmol per reaction and [^{35}S]dATP [NEN Du Pont, Boston, MA (~4000 Ci/ml)] for radioactive labeling of the strands during the dideoxy-sequencing protocol. Briefly, after strand denaturation in 100 μl of 0.15 N NaOH, the beads are washed once in water and resuspended in 8 μl of water. To the 8 μl of suspension are added 2 μl of 5× reaction buffer (U.S. Biochemicals) and 6 pmol of oligonucleotide to 10 μl total. The reaction mixture is denatured for 3 min at 95° and placed on ice. To the denatured bead suspension are added 1 μl dithiothreitol (DTT) solution, 1.75 μl enzyme dilution buffer, 2 μl of labeling mixture diluted 1:10 with water, 0.5 μl of [^{35}S]dATP, and 0.25 μl of Sequenase. The reaction mixture is incubated at room temperature for 5 min. Subsequently, 3.5 μl of the labeling reaction is added to each of the termination mixes. The incubation is extended for a further 5 min at 45°. Four microliters of stop solution is added to each well, and the reaction mixtures are denatured at 75° for 20 min before loading on a

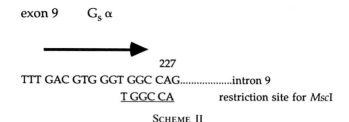

SCHEME II

6% sequencing gel. The gels are fixed in acetic acid/methanol and dried. Autoradiography is overnight at room temperature without screens.

Similarly, for codon 227, restriction by MscI can discern mutant from wild-type PCR strands. However, in this case no new nucleotides need be incorporated into the design of a restriction digest assay around the 227 codon, as seen in Scheme II; normal PCR conditions allow amplification of the gene, and subsequent restriction and direct sequencing may be carried out as in the case of codon 201.

Implications for G-Protein Research

In what way has our knowledge about activating mutations of G proteins in human neoplasia taught us anything about G-protein biochemistry? In a first experiment Zachary et al.[25] were able to show that transfection of activating $G_s\alpha$ subunits at either position 201 or 227 in NIH 3T3 cells caused constitutive activation of adenylate cyclase and accumulation of cAMP (albeit in the presence of a phosphodiesterase inhibitor). Detection of gsp mutations in a restricted tumor population, namely, growth hormone-secreting pituitary tumors and thyroid nodules, led others to search for syndromes displaying endocrine dysfunction. At codon 201 gsp mutations were detected in endocrine dysplasia in the McCune–Albright syndrome,[22,26–28] clearly pointing to the gsp oncogene as the long sought autosomal dominant gene responsible for the mosaicism in this syndrome. These mutations must occur early in embryogenesis, since the mutation is randomly distributed throughout the endocrine glands. An important observation from these results is that gsp mutations are detected in many

[25] I. Zachary, S. B. Masters, and H. R. Bourne, *Biochem. Biophys. Res. Commun.* **168,** 1184 (1990).
[26] R. Happle, *Clin. Genet.* **29,** 321 (1986).
[27] W. F. Schwindinger, C. A. Francomano, M. A. Levine, and V. A. McCusick, *Lancet* **338,** 1454 (1991).
[28] L. S. Weinstein, P. V. Gejman, E. Friedman, T. Kadowaki, R. M. Collins, E. S. Gershon, and A. Spiegel, *Proc. Natl. Acad. Sci. U.S.A.* **87,** 8287 (1990).

tissues from these patients, including testis, ovary, adrenal, lung, liver, kidney, and thymus; this points to the probable involvement of *gsp* mutations in tumor development in these different endocrine tissues, because each mutation was clonal by virtue of the method of detection used.

The novel restriction method described here may be of importance in trying to explain some questions left unanswered by the simple dot-blot hybridization methodology which detects clonal evolution with a detection limit of approximately 15% of absolute mutant signal. One question is why do we find *gsp* mutations in 42% of growth hormone-secreting pituitary tumors and in only 10% of thyroid tumors[29] and only in these tumors? Why is the frequency of *gsp* mutations so low in thyroid tumors? It is well established that thyroid tumors are heterogeneous, evolving from multinodular goiters (a nonneoplastic nodular hyperplasia), and can give rise to either papillary or follicular tumors. Might this difference in the frequency of mutation in thyroid cancer be due to the appearance of *gsp* mutations at a later stage in the clonal progression of disease, or, indeed, might the *gsp* mutation be selected against by more malignant subclones harboring other activating oncogenes[30] giving rise to a multiclonal disease? The role of thyrotropin (thyroid-stimulating hormone or TSH) as a positive mitogen for thyrocytes, which results in the stimulation of adenylate cyclase, has been well documented.[3] In fact, overexpression of the TSH receptor which is coupled to G_s in thyrocytes induces a transformed phenotype in rat thyroid cells,[31] suggesting that a higher preponderance of *gsp* mutations will be detected by this novel technique.

Biochemical studies have in the first line confirmed the original inference, namely, that the α subunit of G_{i2} is an oncogene, inducing neoplastic transformation of rat-1 cells.[32] In addition, *gip2*, when transfected into either NIH 3T3 cells[33] or Chinese hamster ovary (CHO) cells, can inhibit the stimulation of adenylate cyclase constitutively and can attenuate the stimulation of phospholipase A_2 by the thrombin receptor in the CHO cell system.[34] In rat-1a cells activated *gip2* has also been shown to constitu-

---

[29] G. Suarez, J. A. du Villard, B. Caillou, M. Schlumberger, C. Parmentier, and R. Monier, *Oncogene* **6**, 677 (1991).

[30] S. M. Jhiang, D. R. Caruso, E. Gilmore, Y. Ishizaka, T. Tahira, M. Nagao, I.-M. Chiu, and E. L. Mazzaferri, *Oncogene* **7**, 1331 (1992).

[31] M. Derwahl, M. Broecker, S. Aeschmann, H. Schatz, and H. Studer, *Biochem. Biophys. Res. Commun.* **183**, 220 (1992).

[32] A. M. Pace, Y. H. Wong, and H. R. Bourne, *Proc. Natl. Acad. Sci. U.S.A.* **88**, 7031 (1991).

[33] Y. H. Wong, A. Federman, A. M. Pace, I. Zachary, T. Evans, J. Pouysségur, and H. R. Bourne, *Nature* (London) **351**, 63 (1991).

[34] J. M. Lowndes, S. K. Gupta, S. Osawa, and G. L. Johnson, *J. Biol. Chem.* **266**, 14193 (1991).

tively activate MAP kinase, a protein rapidly activated by growth factor stimulation of tyrosine kinase and G-protein receptors.[35]

Activating mutations of G_{i2} were detected by us in human neoplasia before its biochemical role was elucidated.[20] Others have exploited the reverse situation by transfecting activating members of G_q and G_{11} at codon 205, cognate to G_s codon 227, into Cos-7 cells.[7] This resulted in persistent activation of phospholipase C-β_1 and high basal levels of inositol phosphate. Transfection with activating members of G_o and G_z had no effect on inositol phosphate formation.

In conclusion, perturbation of G-protein signaling pathways as a result of oncogenic G-protein α subunits can lead to the development of human endocrine neoplasia. This knowledge will be of help in the design of more rational therapeutics which might interfere with and block downstream effectors of the affected pathways.

Acknowledgments

Dr. Peter Goretzki (University of Düsseldorf) is acknowledged for thyroid tumor samples. I wish also to thank Sandrina Phipps (Roche Molecular Systems, Alameda, CA) and Stefan Gras [Centre de Research Pierre Fabre (CRPF), Castres, France] for excellent technical assistance and Prof. Henry Bourne (University of California, San Francisco) for helpful comments on the manuscript. Dr. Robert Kiss (CRPF) is also thanked for continuous support on the novel restriction strategy.

[35] S. K. Gupta, C. Gallego, G. L. Johnson, and L. E. Heasley, *J. Biol. Chem.* **267,** 7987 (1992).

[24] Detection of Mutations and Polymorphisms of $G_s\alpha$ Subunit Gene by Denaturing Gradient Gel Electrophoresis

By PABLO V. GEJMAN and LEE S. WEINSTEIN

Introduction

The genomic structure of the α subunit of the guanine nucleotide-binding protein G_s ($G_s\alpha$) is composed of 13 exons and 12 introns spanning more than 20 kilobases (kb),[1] whereas the transcribed message of the gene is approximately 1.9 kb in length.[2] Denaturing gradient gel electrophoresis

[1] T. Kozasa, H. Itoh, T. Tsukamoto, and Y. Kaziro, *Proc. Natl. Acad. Sci. U.S.A.* **85,** 2081 (1988).
[2] P. Bray, A. Carter, C. Simons, V. Guo, C. Puckett, J. Kamholz, A. Spiegel, and M. Nirenberg, *Proc. Natl. Acad. Sci. U.S.A.* **83,** 8893 (1986).

(DGGE) analysis detected several disease-related mutations in individuals affected with Albright hereditary osteodystrophy (AHO)[3] and with the McCune–Albright syndrome (MAS),[4] and it also uncovered a polymorphism that was instrumental to the genetic linkage mapping of the $G_s\alpha$ gene.[5] For the reader unacquainted with DGGE we review some basic principles of DNA denaturation and DGGE applications before discussing the specific experimental protocols applied to the analysis of $G_s\alpha$. Many aspects of DGGE have been reviewed.[6–10]

Principles of DNA Melting and Applications to Denaturing Gradient Gel Electrophoresis

The stability of the double helical structure of DNA is the result of a balance of enthalpic and entropic forces. Cooperative noncovalent interactions of the nucleotide sequence including stacking energies, hydrogen bonding, backbone conformational tendencies, electrostatic forces, and solvation effects preserve the DNA helical structure.[7] Contrary to a common misconception, the hydrogen bonds contribute little to the helical stability. The prime support of DNA helix stability comes from short-range electrostatic attractions among stacked bases.[11]

The study of DNA melting by spectrophotometry and electron microscopy indicates that double-stranded DNA is composed of a succession of regions with different energies known as domains which denature successively as a response to changes in temperature. DNA denaturation is correlated with a rise in spectrophotometric absorbance; the temperature associated with the midpoint of the absorbance increase is known as T_m. The T_m is affected by changes in nucleotide composition, pairing, and sequence. Nucleotide base composition refers to the mole percent of guanine plus cytosine (normally expressed as GC content); a greater GC

---

[3] L. S. Weinstein, P. V. Gejman, E. Friedman, T. Kadowaki, R. M. Collins, E. S. Gershon, and A. M. Spiegel, *Proc. Natl. Acad. Sci. U.S.A.* **87,** 8287 (1990).
[4] L. S. Weinstein, A. Shenker, P. V. Gejman, M. J. Merino, E. Friedman, and A. M. Spiegel, *N. Engl. J. Med.* **325,** 1688 (1991).
[5] P. V. Gejman, L. S. Weinstein, M. Martinez, A. M. Spiegel, Q. Cao, W.-T. Shieh, M. R. Hoehe, and E. S. Gershon, *Genomics* **9,** 782 (1991).
[6] S. G. Fischer and L. S. Lerman, this series, Vol. 68, p. 183.
[7] A. Wada, S. Yabuki, and Y. Husimi, *Crit. Rev. in Biochem.* **9,** 87 (1980).
[8] L. S. Lerman, S. G. Fischer, I. Hurley, K. Silverstein, and N. Lumelsky, *Annu. Rev. Biophys. Bioeng.* **13,** 399 (1984).
[9] L. S. Lerman, K. Silverstein, and E. Grinfeld, *Cold Spring Harbor Symp. Quant. Biol.* **51,** 285 (1986).
[10] R. M. Myers, T. Maniatis, and L. S. Lerman, this series, Vol. 155, p. 501.
[11] S. G. Delcourt and R. D. Blake, *J. Biol. Chem.* **266,** 15160 (1991).

content corresponds to a greater thermal stability.[12] Pairing refers to the existence of perfectly (complementary) or imperfectly (mismatched) paired bases and is a major component of DNA stability.[13,14] Variations in the DNA sequence can also affect the T_m as is demonstrated by the effect of diverse single nucleotide substitutions on melting profiles.[15] However, the effect of substitutions on T_m is less marked than the effect of mismatches.

Partial Denaturation of DNA and Electrophoretic Mobility

The observation that the thermal denaturation properties of DNA as well as the DNA electrophoretic mobility are both affected by the nucleotide sequence made possible the development of DGGE as a tool for detecting mutations.[16] In DGGE experiments a temperature gradient is supplanted by a chemical gradient made of a linearly increasing concentration of a urea–formamide mixture in a polyacrylamide gel. Because the denaturation properties of the urea–formamide gradient are reasonably equivalent to those of a thermal gradient at temperatures near the melting temperatures, electrophoresis is performed at 60°.[17,18]

Partial DNA strand dissociation in a gel results in a dramatic reduction in electrophoretic mobility.[16] The melting domain with the lowest T_m determines the point in a denaturing gradient where the first mobility change (retardation) occurs, which in turn substantially affects the final position of the molecule in the gel. Fisher and Lerman showed that distinct DNA fragments containing the same least-stable sequence reach the same concentration of denaturants in the gel.[15] After partial denaturation, DNA mobility is mainly determined by the length of the melted DNA sequence. Branching due to perturbations in the helical geometry of DNA explains the retardation effect of partial melting on DNA electrophoretic mobility.[19,20]

---

[12] J. Marmur and P. Doty, *J. Mol. Biol.* **5**, 109 (1962).
[13] R. M. Myers, N. Lumelsky, L. S. Lerman, and T. Maniatis, *Nature (London)* **313**, 495 (1985).
[14] R. M. Myers and T. Maniatis, *Cold Spring Harbor Symp. Quant. Biol.* **51**, 275 (1986).
[15] S. G. Fischer and L. S. Lerman, *Proc. Natl. Acad. Sci. U.S.A.* **77**, 4492 (1980).
[16] S. G. Fischer and L. S. Lerman, *Cell (Cambridge, Mass.)* **16**, 191 (1979).
[17] H. Klump and W. Burkart, *Biochim. Biophys. Acta* **475**, 601 (1977).
[18] K. Nishigaki, Y. Husimi, M. Masuda, K. Kaneko, and T. Tanaka, *J. Biochem. (Tokyo)* **95**, 627 (1984).
[19] E. S. Abrams and V. P. Stanton, Jr., this series, Vol. 212, p. 71.
[20] L. S. Lerman and K. Silverstein, this series, Vol. 155, p. 482.

DNA Stability, Modeling, and Manipulation

Fisher and Lerman showed that when DNA molecules with more than one melting domain are examined by DGGE, only substitutions in the domain with the lowest T_m are distinguishable from the wild-type sequences, and the removal of the lowest melting domain allows detection of mutations in the domain with the next lowest T_m.[21] Myers et al. subsequently demonstrated that the attachment of a GC-rich DNA sequence (GC clamp) to a DNA fragment can alter the melting contours of a DNA fragment in a predicted manner while also preventing complete DNA denaturation during DGGE.[22,23] This strategy made possible the detection of DNA sequence differences situated in higher melting domains. The use of GC clamps became increasingly practical when Sheffield et al. demonstrated that a GC clamp can be attached onto a DNA fragment by including in the polymerase chain reaction (PCR) a primer with a 40-base pair (bp) GC clamp sequence onto its 5' end.[24]

A computer algorithm written by Lerman and Silverstein predicts the thermal stability and mobility in DGGE of DNA molecules based on sequence.[20,25,26] The mid-melting point temperature, the temperature at which there is a 50% probability of double-helical conformation at each nucleotide position, is a widely used output of this program. A bidimensional plot of the mid-melting point temperature versus DNA sequence position is known as a melting map (e.g., see Fig. 1).[20]

To search for disease mutations and polymorphisms in genomic DNA, melting maps of genomic fragments encompassing the targeted exons and corresponding donor and acceptor splicing sites are generated. On the basis of this analysis a decision is made on the precise DNA sequence to amplify and whether to attach a GC clamp to the 5' end of either the upstream or downstream PCR primer. Ideally a PCR-amplified DNA sequence should display only two melting domains: a uniform low melting domain spanning the genomic sequence where DNA sequence differences should be detectable and a high melting domain spanning only the GC clamp. This is not always possible in practice. If amplified DNA fragments are either too long (more than 500 nucleotides) or if the melting maps

---

[21] S. G. Fischer and L. S. Lerman, *Proc. Natl. Acad. Sci. U.S.A.* **80**, 1579 (1983).
[22] R. M. Myers, S. G. Fischer, T. Maniatis, and L. S. Lerman, *Nucleic Acids Res.* **13**, 3111 (1985).
[23] R. M. Myers, S. G. Fischer, L. S. Lerman, and T. Maniatis, *Nucleic Acids Res.* **13**, 3131 (1985).
[24] V. C. Sheffield, D. R. Cox, L. S. Lerman, and R. M. Myers, *Proc. Natl. Acad. Sci. U.S.A.* **86**, 232 (1989).
[25] M. Fixman and J. J. Friere, *Biopolymers* **16**, 2693 (1977).
[26] D. Poland, *Biopolymers* **13**, 1859 (1974).

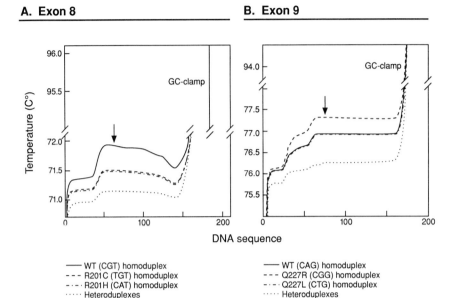

Fig. 1. Melting maps of exons 8 and 9 of $G_s\alpha$. Melting maps were generated for genomic DNA fragments exon 8 (A) and exon 9 (B) using the computer algorithm of Lerman and Silverstein.[20] The mid-melting point temperature is plotted as a function of the DNA sequence position. For all curves the input sequence for the program has a GC clamp attached to the 3' end. The position of the mutations within the sequence is shown by an arrow. In each case the melting map for the wild-type (WT) homoduplex is shown as a solid line, homoduplexes containing specific activating mutations as broken lines, and heteroduplexes as dotted lines. Both the R201C and R201H homoduplexes in exon 8 have a similar melting map with a lower melting temperature than the wild type, consistent with the similar position of both mutant homoduplexes above the wild type in parallel DGGE (see Fig. 3). In exon 9, the Q227R homoduplex has a higher melting temperature than wild type whereas Q227L is similar to wild type, consistent with the results for parallel DGGE (see Fig. 4). To generate these melting maps, a destabilization value of $-50°$ was assumed, regardless of the kind of mismatch or sequence position.[20]

show several distinct melting domains, the effect of a GC clamp may not be sufficient to achieve optimal results. Amplifying the genomic fragment in segments may avert this problem.

Patterns of Mutations in Denaturing Gradient Gel Electrophoresis

The pattern displayed by mutations in DGGE depends on the presence of DNA molecules in the form of homoduplexes and/or heteroduplexes and on the orientation of the gradient of denaturants to the electric field (perpendicular versus parallel). Heteroduplexes are DNA molecules containing mismatches between strands. Homoduplexes are DNA molecules which do not show alterations in pairing but may bear one or more sequence differences with respect to a wild-type molecule. Amplification by PCR from a heterozygous individual at the examined locus will produce heteroduplex and homoduplex DNA molecules, whereas PCR amplification of a locus at which an individual is homozygous or hemizygous will only produce homoduplexes.

In parallel DGGE, heteroduplexes produce one or more extra bands with lighter intensity and decreased mobility, compared to the wild-type homoduplex sequence. Homoduplexes that bear a single nucleotide substitution in the lowest melting domains frequently show mobility retardation at slightly different concentration of denaturant than the wild-type sequence, reflecting changes in T_m due to modifications in stacking forces. Mobility differences between homoduplexes are occasionally marginal and sometimes undetectable. Nevertheless, if the mutation is present in a heterozygous state, the presence of heteroduplexes will be detected.

In perpendicular DGGE, samples are applied to a slot that spans the width of the gel. This slot is parallel to the gradient of denaturants and perpendicular to the electrical axis (Fig. 2). The DNA has greater mobility on the side with lower denaturants since it is not melted. The mobility will begin to decrease at the point in the gradient where melting of the lowest melting domain starts and continues to decrease up to a point in the gradient where the DNA electrophoretic mobility is not significantly affected by the concentration of denaturants. If heteroduplexes were formed during the PCR or if homoduplexes with different T_m are present,

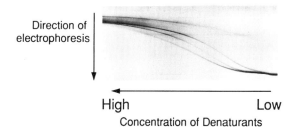

FIG. 2. Perpendicular DGGE of a genomic DNA fragment encompassing exons 10–11 of $G_s\alpha$ from a patient with a missense mutation in codon Ser^{250} (AGC → AGG). The DNA sample was applied to a slot that spans the width of the gel and then electrophoresed through the polyacrylamide gel with a 40–80% gradient of denaturants. The loading slot is parallel to the gradient of denaturants and perpendicular to the electrical axis. The gel was silver stained in this experiment. Two heteroduplexes formed during the PCR emerge above the wild-type band below. The parallel DGGE for this mutation is shown in Fig. 3 as mutation C.

one or more additional electrophoretic bands diverge from the wild-type band at a certain point of the gradient (Fig. 2). In the absence of heteroduplexes or detectable homoduplexes, there should only be one band along the entire gradient.

Methods

The PCR and DGGE conditions for $G_s\alpha$ gene sequences in which mutations have been detected are summarized on Table I. The PCR amplifications are typically performed using 30 cycles for blood DNA and 35–45 cycles for DNA from paraffin-embedded tissues with an annealing temperature of 58°.[4] Using more than 40 PCR cycles or reamplifying from PCR aliquots should be avoided when possible because cumulative *Taq* polymerase errors may make the interpretation of data difficult. Conditions of amplification of a fragment containing exon 1 were reported by another group,[27] but we could not successfully amplify it by using this protocol or others.

Electrophoresis

The equipment and procedures for DGGE have been previously reviewed by others.[10,19] We briefly describe the DGGE conditions employed by us. The PCR products are analyzed in a Hoefer SE600 apparatus (San Francisco, CA). Gradient gels (24 cm length) containing 6–8% acrylamide (37.5:1, acrylamide:bisacrylamide) in 1× TAE buffer (40 mM Tris–acetate/1 mM EDTA, pH 8.3) are prepared from two stock solutions: 0% containing no denaturants and 100% containing 7 M urea and 40% (v/v) formamide and mixed by a gravitational gradient former. The gradient of denaturants can be chosen based on the calculated melting maps and the expression $T_e = T_b + x(\%$ denaturant), where T_e is the effective temperature, T_b is the bath temperature, and x is a constant whose value is approximately 0.3°/1% denaturant[8] or is determined from preliminary perpendicular DGGE experiments.

In parallel DGGE, if focusing is achieved, the volume of the DNA aliquot loaded in each well will not significantly affect the quality of the results. However, the volume of loaded samples should be minimized in perpendicular DGGE experiments. For perpendicular gradients we prefer to load aliquots smaller than 150 μl. Gels are electrophoresed at 60–100 V in a 60° circulating bath for 15–20 hr in parallel DGGE and for 2.5–5

[27] J. L. Patten, D. R. Johns, D. Valle, C. Eil, P. A. Gruppuso, G. Steele, P. M. Smallwood, and M. A. Levine, *N. Engl. J. Med.* **322,** 1412 (1990).

TABLE I
PCR AND DGGE CONDITIONS FOR $G_s\alpha$ GENE SEQUENCES

Parameter	Exon 4–5	Exon 8	Exon 9	Exon 10–11
Fragment length with GC clamp (bp)[a]	405	204	218	532
GC-clamp position	Intron 3 (upstream)	Intron 8 (downstream)	Intron 9 (downstream)	Intron 11 (downstream)
$MgCl_2$ (mM) in PCR mixture	1.5	2.5	2.5	1.5
DGGE gradient	5–75%	35–65%	35–90%	5–75%
Findings[b]	Two silent polymorphisms[c]: (A) Val^{128} (GTG to GTA), rare; (B) Ile^{131} (ATT to ATC); Alters FokI restriction site: FokI$^+$ frequency, 0.47; FokI$^-$ frequency, 0.53	Missense mutations resulting in constitutive activation: (A) Arg^{201} to Cys (CGT to TGT); (B) Arg^{201} to His (CGT to CAT)	Missense mutations resulting in constitutive activation: (A) Gln^{227} to Arg (CAG to CGG); (B) Gln^{227} to Leu (CAG to CTG)	Mutations resulting in disruption of gene expression: (A) donor splice junction substitution in first base of intron 10 (G to C); (B) deletion of 1 bp within exon 10 (first C in codon 272) resulting in frameshift; (C) missense mutation of Ser^{250} to Arg (AGG to AGC)

[a] Primer sequences are as follows: for exon 4–5, 5' GC clamp-<u>CAGTACTCCTAACTGACATG</u> 3' and 5' TTTGGATCC<u>ATGTTCCTATATGGACACTGT</u> 3'; for exon 8, 5' <u>TCGGTTGGCTTTGGTGAGATCCA</u> 3' and 5' GC clamp-<u>AGAAACCATGATCTCTGTTATA</u> 3'; for exon 9, 5' AACTGCAGCC<u>AGTCCCTCTGGAATAACCAG</u> 3' and 5' GC clamp-<u>AGCGACCCTGATCCCTAACAAC</u> 3'; for exon 10–11, 5' AAGAATTC<u>TTAGGGATCAGGGTCGCTGCTC</u> 3' and 5' GC clamp-<u>ATGAACAGCCAGCAAGAGTGGA</u> 3'. The GC-clamp sequence is CGCCCGCCGCGCCCCGCGCCCGTCCCGCCGCCCCCGCCCC. The underlined sequence is complimentary to $G_s\alpha$ gene sequence.
[b] Codon positions are based on Ref. 1.
[c] Double heterozygotes but not individual polymorphisms are detected by parallel DGGE.

hr in perpendicular DGGE. The combination of voltage and time chosen for electrophoresis should be large enough to allow the DNA fragments to reach the position in the parallel gradient where they will partially melt; DNA molecules will not migrate significantly after DGGE focusing is accomplished.

Gel Staining

Photochemical silver staining of double-stranded DNA using silver nitrate is more sensitive than ethidium bromide staining of double-stranded DNA and shows a better signal-to-noise (S/N) ratio.[28,29] It has been calculated that silver stains can detect 0.03 ng/mm^2 of DNA.[28] Sensitivity of silver staining for DNA is greater for fragments smaller than 300 bp. Silver staining is not affected by DNA denaturation, whereas ethidium bromide staining is greatly inhibited by DNA strand separation.

Silver images are generated by a difference in the oxidation–reduction potential between areas in a gel with and without nucleic acid.[30] Silver nitrate reacts with biopolymers under acidic conditions and subsequently is selectively reduced by formaldehyde in alkaline solution. It has been suggested that the purines are the active subunits in the silver staining reaction.[31] We perform silver staining on 1 mm thick gels with a commercial kit (Bio-Rad, Richmond, CA), diluting the kit components in deionized water. Overdeveloped gels or gels showing high background can be destained by using a photographic reducer and stained anew.

Gels can be stained first with ethidium bromide (2 μg/ml) for rapid visualization of DGGE results. Ethidium bromide staining does not adversely affect silver staining results. Alternatively, gels may be fixed in 10% ethanol and stored for long periods before performing silver staining.

Results

Albright Hereditary Osteodystrophy

Albright hereditary osteodystrophy (AHO) is characterized by short stature, focal skeletal abnormalities, obesity, subcutaneous calcifications, and mental retardation. This phenotype may occur alone or in association with multiple hormone resistance. In most cases a partial deficiency of functional G_s is present. Heterozygous mutations in the $G_s\alpha$ gene which presumably disrupt function were found in four kindreds by parallel DGGE (Fig. 3).[3] These include a G to C substitution at the donor splice junction of intron 10,[3] a coding frameshift created by a single-base deletion within exon 10,[3] and a missense mutation within exon 10 at codon Ser^{250}.[32] A 4-bp deletion was also found in exon 7 by aberrant migration of the hetero-

---

[28] D. Goldman and C R. Merril, *Electrophoresis* **3**, 24 (1982).
[29] J. L. Beidler, P. R. Hilliard, and R. L. Rill, *Anal. Biochem.* **126**, 374 (1982).
[30] C. R. Merril, *Nature (London)* **343**, 779 (1990).
[31] C. R. Merril and M. E. Pratt, *Anal. Biochem.* **156**, 96 (1986).
[32] L. S. Weinstein and P. V. Gejman, unpublished results (1989).

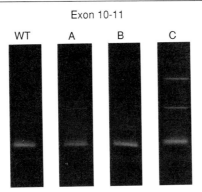

FIG. 3. Analysis by DGGE of PCR-amplified genomic fragments encompassing exons 10–11 from a normal subject (WT) and three unrelated patients with AHO (labeled A, B, and C). Conditions for PCR and DGGE are described in the text and summarized in Table I. The heterozygous mutations determined by sequencing and/or restriction analysis are as follows: (A) substitution of G to C in the first base position of intron 10, (B) single base deletion within exon 10 (first C in codon Leu^{272}) resulting in a coding frameshift, and (C) single base substitution resulting in a missense mutation, Ser^{250} (AGC) to Arg (AGG). Mutations A and B have been previously characterized.[3] The gel in this experiment was stained with ethidium bromide.

duplexes in both DGGE and nondenaturing polyacrylamide gel electrophoresis.[33] Miric and Levine have demonstrated an AHO mutation within exon 6 by DGGE.[34]

Activating $G_s\alpha$ Mutations

Missense mutations of the $G_s\alpha$ gene that encode substitutions of either residue Arg^{201} in exon 8 or Gln^{227} in exon 9 produce constitutively active forms of the $G_s\alpha$ protein with reduced guanosine triphosphatase activity.[35] These mutations have been previously detected in sporadic pituitary somatotrope tumors.[35–37] We have detected identical mutations in tissues from patients with the McCune–Albright syndrome (MAS),[4] a disease characterized by cafe-au-lait spots, polyostotic fibrous dysplasia, and hyperplasia

---

[33] L. S. Weinstein, P. V. Gejman, P. de Mazancourt, N. American, and A. M. Spiegel, *Genomics* **4**, 1319 (1992).

[34] A. Miric and M. A. Levine, in "G Proteins: Signal Transduction and Disease" (G. Milligan and M. Wakelam, eds.), p. 29. Academic Press, London, 1992.

[35] C. A. Landis, S. B. Masters, A. Spada, A. M. Pace, H. R. Bourne, and L. Vallar, *Nature (London)* **340**, 692 (1989).

[36] J. Lyons, C. A. Landis, G. Harsh, L. Vallar, K. Grunewald, H. Feichtinger, Q.-Y. Duh, O. H. Clark, E. Kawasaki, and H. R. Bourne, *Science* **249**, 655 (1990).

[37] E. Clementi, N. Malgaretti, J. Meldolesi, and R. Taramelli, *Oncogene* **5**, 1059 (1990).

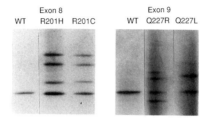

FIG. 4. Analysis by DGGE of DNA samples with activating $G_s\alpha$ mutations. The DGGE analysis of PCR-amplified genomic fragments (A) encompassing exon 8 (which contains codon Arg^{201}) and (B) those encompassing exon 9 (which contains codon Gln^{227}) is shown. Conditions for PCR and DGGE are described in the text and summarized in Table I. Paraffin-embedded normal adrenal tissue was used as the wild-type (WT) sample for both exons. DNA samples from pituitary tumors with known heterozygous activating $G_s\alpha$ mutations (provided by J. Lyons) were analyzed. In exon 8 the mutations are Arg^{201} (codon sequence, CGT) to His (CAT) or to Cys (TGT), labeled R201H and R201C, respectively. For both mutations, the mutant homoduplexes behave similarly and melt just above the wild-type band. The heteroduplexes are within the upper two bands for both mutations. In exon 9, the mutations were Gln^{227} (CAG) to Arg (CGG) or to Leu (CTG), labeled Q227R and Q227L, respectively. The mutant homoduplexes melt below the wild-type homoduplexes for Q227R and at the same position as the wild type for Q227L. Two upper heteroduplex bands are present in Q227R and one in Q227L. All these results are consistent with the differences in melting predicted by the computer algorithm of Lerman and Silverstein (see Fig. 1). The gel was silver stained in this experiment.

and hyperfunction of multiple endocrine glands. In MAS these mutations are somatic and have a widespread distribution.

Analysis by DGGE is sensitive for the detection of activating $G_s\alpha$ mutations at both codon Arg^{201} in exon 8 and Gln^{227} in exon 9 (Fig. 4). For MAS samples, in which the proportion of mutant to wild-type alleles varies between tissues, the ratio of the intensity of mutant bands to the wild-type band in DGGE varies similarly to the ratio of mutant to wild-type signal in allele-specific oligonucleotide hybridization.[4] Moreover, in samples with a small proportion of mutant alleles, the mutant homoduplex band is barely detectable or undetectable, presumably because the larger proportion of wild-type alleles favors the formation of heteroduplexes over homoduplexes. For all four activating $G_s\alpha$ mutations studied by DGGE, the effect of each specific base substitution on the melting map of the genomic fragment correlated well with the position of the mutant homoduplex in the parallel gradient (Figs. 3 and 4). The DGGE findings for one of the Arg^{201} mutations have been confirmed.[38] Evidence that mutations are present in the analogous codons of other G-protein α sub-

[38] W. F. Schwindinger, C. A. Francomano, and M. A. Levine, *Proc. Natl. Acad. Sci. U.S.A.* **89**, 5152 (1992).

units *in vivo*[36] and that they lead to constitutive activation[39-41] suggest that DGGE may be useful in searching for activating mutations within genes encoding other G proteins in other clinical states.

Genetic Mapping of $G_s\alpha$ Subunit Gene

Analysis of amplified fragments spanning exons 4 and 5 of $G_s\alpha$ of one individual displayed a pattern consistent with a double mutation.[5] Direct DNA sequencing detected two silent nucleotide sequence differences in the exon 5 sequence of this individual, one of which is recognized by the restriction enzyme *Fok*I. The observed heterozygosity of the *Fok*I biallelic system was 0.53, making it possible to map $G_s\alpha$ genetically.[5] The best estimate location of the $G_s\alpha$ locus relative to the map of fixed loci is between D20S15 (at 9 and 4 cM for males and females, respectively) and D20S25 (at 16 and 7 cM for males and females, respectively).[5,42]

Strategies of Denaturing Gradient Gel Electrophoresis Analysis

Analysis by DGGE reduces the work load of sequencing by directing sequencing efforts toward DNA samples which are likely to contain mutations or polymorphisms. It is therefore most useful when a large-scale screening of mutations is performed.

The degree of certainty that DNA samples excepted from sequencing through rigorous DGGE analysis do not contain a mutation is very high although not absolute.[43-46] The effectiveness of DGGE in detecting mutations depends on multiple factors, including the nature of the mutations (heterozygous or homozygous), the length of the analyzed DNA fragments, the profile of the melting map, and the orientation of the DGGE (parallel versus perpendicular).

---

[39] S. Hermouet, J. J. Merendino, Jr., J. S. Gutkind, and A. M. Spiegel, *Proc. Natl. Acad. Sci. U.S.A.* **88,** 10455 (1991).

[40] Y. H. Wong, A. Federman, A. M. Pace, I. Zachary, T. Evans, J. Pouyssegur, and H. R. Bourne, *Nature (London)* **351,** 63 (1991).

[41] J. M. Lowndes, S. K. Gupta, S. Osawa, and G. L. Johnson, *J. Biol. Chem.* **266,** 14193 (1991).

[42] T. P. Keith, K. Falls, D. W. Bowden, J. L. Weber, P. V. Gejman, J. Hazan, P. Phipps, and K. Serino, *Science* **258,** 67 (1992).

[43] M. D. Traystman, M. Higuchi, C. K. Kasper, S. E. Antonarakis, and H. H. Kazazian, Jr., *Genomics* **6,** 293 (1990).

[44] S.-P. Cai and Y. W. Kan, *J. Clin. Invest.* **85,** 550 (1990).

[45] M. Higuchi, S. E. Antonarakis, L. Kasch, J. Oldenburg, E. Economou-Petersen, K. Olek, M. Arai, H. Inaba, and H. H. Kazazian, Jr., *Proc. Natl. Acad. Sci. U.S.A.* **88,** 8307 (1991).

[46] F. Barbetti, P. V. Gejman, S. I. Taylor, N. Raben, A. Cama, E. Bonora, P. Pizzo, P. Moghetti, M. Muggeo, and J. Roth, *Diabetes* **41,** 408 (1992).

Differences in migration on DGGE are regularly detected between heteroduplexes and homoduplexes, but they are less prominent between homoduplexes with a sequence difference. Homoduplexes can be converted to heteroduplexes by mixing DNA from two or more individuals with different alleles at the tested locus. Two methods can be used to accomplish this: either coamplifying DNA fragments from two or more such individuals or mixing and boiling the DNA aliquots after PCR amplification, followed by rapid cooling in ice-cooled water and reannealing at room temperature.

When the melting map of the GC-clamped amplified sequence shows three or more melting domains, parallel DGGE experiments may fail to detect both homozygous and heterozygous mutations. For example, parallel DGGE analysis of the human insulin receptor gene failed to detect a heterozygous missense mutation in exon 2 (Lep/Ver-1) in an individual affected with severe insulin resistance.[46] The mutation was located in the second lowest domain (T_m about 5° higher than the least stable domain). Subsequent analysis with perpendicular DGGE detected this mutation.

A mutation located at the end of a DNA fragment may escape detection since the ends are less thermally stable than the central portion. Screening of the mutations may require either very short running times in parallel DGGE[19] or, alternatively, perpendicular DGGE analysis.

The problem of excessive length of amplified DNA can be obviated by amplifying smaller fragments. This approach will increase the number of experiments required unless multiplexing DGGE techniques are also employed. Furthermore, genomic DNA fragments which span coding sequences or splicing sites will have to be analyzed for the presence of mutations even if they show undesirable melting properties. The combination of perpendicular and parallel DGGE geometries will increase the sensitivity of detection of mutations in all these cases.

Acknowledgments

Research was supported in part by a grant from Mrs. Anita Kaskel Row to the Foundation for Advanced Education in the Sciences. We thank E. S. Gershon and A. M. Spiegel for support, Q. Cao for technical help, J. Lyons for DNA samples containing activating $G_s\alpha$ mutations, A. Shenker for assistance in analyzing activating mutations, and L. Lerman for the use of the MELT87 computer program.

[25] Construction of Mutant and Chimeric G-Protein α Subunits

By SIM WINITZ, MARIJANE RUSSELL, and GARY L. JOHNSON

Introduction

Among the G-protein α subunits there are both common and unique functions encoded in the structure of the α-chain polypeptide. Common features include the functions involved in regulation of the α subunit itself: (1) GDP/GTP binding, (2) intrinsic GTPase activity, and (3) binding sites for association with βγ subunits. Unique functions for each α chain include (1) selectivity for regulating specific effectors and (2) selectivity for coupling to specific receptors. The unique functions for each α subunit allow selectivity in cellular responses to the environment (hormones and neurotransmitters) and regulation of specific physiological signal functions (enzymes and ion channels).

The α-subunit sequences involved in two of the common functions for all G proteins (GDP/GTP binding and GTPase activity) are highly conserved at the amino acid level. The sequences involved in contact and regulation by βγ complexes are less apparent in the primary sequence of G-protein α subunits. The sequences involved in the unique functions of receptor and effector interaction and regulation are also predictably diverse and nonconserved.

The G proteins G_s and G_i regulate the common effector adenylyl cyclase (adenylate cyclase, EC 4.6.1.1). G_s activates and G_i inhibits adenylyl cyclase activity. G_s and G_i also preferentially couple to different receptors. For this reason, the α_s and α_i polypeptides have proved extremely valuable for characterizing the common and unique functions of G-protein α subunits by chimera and mutation analysis.

Construction of α_i/α_s Chimeras Using Polymerase Chain Reaction

Figure 1 illustrates the general procedure for using polymerase chain reactions (PCR)[1] in constructing rat G-protein α-subunit chimeras. Although G-protein α-subunit sequences are highly conserved among mammalian species, DNA sequences other than rat may vary and should be compared to those given here. Four oligonucleotides and the full-length

---

[1] M. A. Innis, D. H. Gelfand, J. J. Sninsky, and T. J. White, "PCR Protocols: A Guide to Methods and Applications." Academic Press, San Diego, 1990.

A. Oligonucleotides required for PCR generation of α_i/α_s chimeras

i) 5'-α_i sense oligonucleotide

```
5'-CCAAGCTTGAGAGCTTCCCGCAGAG-3'
   HindIII        α_i sense  →
```

ii) Chimeric oligonucleotides (see Panel B & Figure 2)

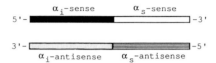

iii) α_s antisense oligonucleotide containing EcoRI restriction site

```
5'-ATAGAATTCAGGTGGGAA-3'
   EcoRI  α_s-antisense  →
```

B. Strategy for PCR generated α_i/α_s chimeras

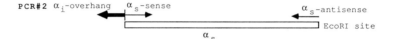

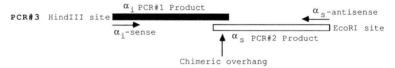

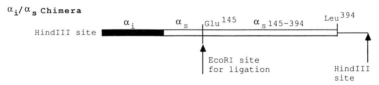

FIG. 1. Polymerase chain reaction for constructing G-protein α-subunit chimeras.

cDNAs of the respective rat α subunits[2] are required for the synthesis of a chimeric α-subunit cDNA. Two of the oligonucleotides may be used for the construction of many different chimeras, and two are unique for each chimera. The strategy is defined here for the construction of chimeras for the $G_{i2}\alpha$ subunit (α_i) and $G_s\alpha$ subunit (α_s), but the procedure is applicable for all known G-protein α subunits.

[2] D. T. Jones and R. R. Reed, *J. Biol. Chem.* **262**, 1424 (1987).

For α_i/α_s chimeras having α_i sequences at the amino terminus of the chimeric polypeptide, a sense strand oligonucleotide encoding α_i 5′-DNA sequence is used. This oligonucleotide may encode α_i sequence beginning at the translation start site methionine, but it is generally recommended to utilize 5′-noncoding sequence information. Excellent success has been achieved using a 25-base oligonucleotide (Fig. 1A, i). This oligonucleotide encodes 17 bases of α_{i2} noncoding sequence beginning at base number −105 relative to the translation start site methionine (A of ATG is base 1). A *Hin*dIII restriction site (AAGCTT) is incorporated at the 5′ end of the oligonucleotide for subsequent cloning with two additional bases at the 5′ end to increase the efficiency of *Hin*dIII restriction enzyme digestion at this site. The second oligonucleotide is a 28-base oligonucleotide in the antisense orientation and encodes 14 bases of α_i and 14 bases of α_s sequence information at the desired junction between α_i and α_s (Fig. 1A, ii; Fig. 2). These α_i 5′-sense and chimeric α_i/α_s antisense oligonucleotides are used in a polymerase chain reaction in the presence of the full-length α_i cDNA. The double-stranded PCR product (Fig. 1B, PCR #1) will encode a 5′ *Hin*dIII restriction site, 105 base pairs (bp) of α_i noncoding sequence and the desired amount of α_i coding sequences beginning at the methionine

$\alpha_i(7)/s$
5′-GCACCGTGAGCGCCGAGGAGAAGGCGCA-3′
3′-CGTGGCACTCGCGGCTCCTCTTCCGCGT-5′
$\alpha_i(1-7)$ $\alpha_s(15-394)$
-CysThrValSerAla$^{7-15}$GluGluLysAlaGln-

$\alpha_i(17)/s$
5′-CCGAGCGCTCTAAGAAGATCGAGAAGCA-3′
3′-GGCTCGCGAGATTCTTCTAGCTCTTGGT-5′
$\alpha_i(1-17)$ $\alpha_s(25-144)$
-AlaGluArgSerLys$^{17-25}$LysIleGluLysGln-

$\alpha_i(27)s$
5′-TGCGGGAGGACGGCCAGGTCTACCGGGC-3′
3′-ACGCCCTCCTGCCGGTCCAGATGGCCCG-5′
$\alpha_i(1-27)$ $\alpha_s(35-394)$
-LeuArgGluAspGly$^{27-35}$GlnValTyrArgAla-

$\alpha_i(34)/s$
5′-CGGCACGGGAGGTGCGCCTGCTGCTGCT-3′
3′-GCCGTGCCCTCCACGCGGACGACGACGA-5′
$\alpha_i(1-34)$ $\alpha_s(42-394)$
-AlaAlaArgGluVal$^{34-42}$ArgLeuLeuLeuLeu-

$\alpha_i(54)/s$
5′-TCAAGCAGATGAAGATCCTACATGTTAA-3′
3′-AGTTCGTCTACTTCTAGGATGTACAATT-5′
$\alpha_i(1-54)$ $\alpha_s(62-394)$
-ValLysGlnMetLys$^{54-62}$IleLeuHisValAsn-

$\alpha_i(64)/s$
5′-GCTACTCAGAGGAGCGCCGCGAAGAGGA-3′
3′-CGATGAGTCTCCTCGCGGCGCTTCTCCT-5′
$\alpha_i(1-64)$ $\alpha_s(72-394)$
-GlyTyrSerGluGlu$^{64-72}$GlyGlyGluGluAsp-

$\alpha_i(79)/s$
5′-GCAACACCATCCAGGCCATTGAAACCAT-3′
3′-CGTTGTGGTAGGTCCGGTAACTTTGGTA-5′
$\alpha_i(1-79)$ $\alpha_s(102-394)$
-TyrAsnThrIleGln$^{79-102}$AlaIleGluThrIle-

$\alpha_i(94)/s$
5′-ACCTGCAGATCGACGTGGAGCTGGCCAA-3′
3′-TGGACGTCTAGCTGCACCTCCAGCGGTT-5′
$\alpha_i(1-94)$ $\alpha_s(117-144)$
-AsnLeuAsnIleAsp$^{94-117}$ValGluLeuAlaAsn-

$\alpha_i(122)/s$
α_i-antisense
3′-TA<u>GAATTC</u>TTCCGGAAGCATGCGG-5′
  EcoRI site
   -^{122}GluProLeuMetGly^{118}-

$\alpha_{s/i}(36)$
5′-ATGGACGTCACTACATCTACACGCACTT-3′
3′-TACCTGCAGTGATGTAGATGTGCGTGAA-5′
$\alpha_s(1-358)$ $\alpha_i(320-355)$
-AspGlyArgHisTyr$^{358-320}$IleTyrThrHisPhe-

FIG. 2. α_i/α_s Chimeric oligonucleotides for PCR-generated chimeras.

translation start site, and a chimeric α_i/α_s junction encoding 14 bp of α_s sequence.

The second PCR reaction (Fig. 1B, PCR #2) uses a 28-base chimeric α_i/α_s oligonucleotide that is in the sense orientation encoding 14 bases of α_i and 14 bases of α_s sequence (Fig. 1A, ii; Fig. 2). This oligonucleotide is complimentary to the antisense 28-base chimeric α_i/α_s antisense oligonucleotide used in PCR #1 (see Fig. 2 for examples). The second oligonucleotide used in PCR #2 takes advantage of a single EcoRI restriction enzyme site (GAATTC) in the α_s sequence beginning at codon 145 of the rat α_s DNA sequence. An 18-base α_s antisense oligonucleotide (Fig. 1A, iii) is used having the EcoRI site near the 5' end of the oligonucleotide. The α_i/α_s chimeric sense and α_s antisense oligonucleotides are used in a polymerase chain reaction in the presence of full-length α_s cDNA. The double-stranded PCR product (Fig. 1B, PCR #2) will encode a 14-bp α_i sequence, the appropriate amount of α_s coding sequence, and the α_s EcoRI restriction site.

The products of PCR #1 and PCR #2 are then used as templates in a final polymerase chain reaction (PCR #3, Fig. 1B). This reaction utilizes the 14-base overhangs for α_s and α_i incorporated in the PCR products from reactions 1 and 2. The 5'-α_i sense and EcoRI α_s antisense oligonucleotides are used as primers in PCR #3. The final PCR product has a 5' HindIII restriction site, 105 bp of α_i noncoding sequence, the desired α_i and α_s coding sequences, and a 3' EcoRI restriction site. This PCR-generated DNA product is purified, then restricted with the HindIII and EcoRI enzymes. The restricted product from PCR #3 is then ligated into a plasmid such as pUC18 and sequenced to verify the fidelity of the PCR amplifications and coding sequence for the regions of α_i and α_s. For ligation of the chimeric α_i/α_s sequence the HindIII–EcoRI restriction fragment is isolated from the pUC18 vector used for sequencing. In a separate restriction enzyme digestion the EcoRI–HindIII α_s fragment encoding codons 145–395 and the 3'-untranslated region of the α_s cDNA is isolated. The HindIII–EcoRI chimeric restriction cDNA fragment and the EcoRI–HindIII α_s cDNA fragments are purified on horizontal 1.2% agarose gels (see below for details). Similarly pUC18 or an equivalent plasmid is linearized by restriction digestion using the HindIII restriction enzyme. A three-piece ligation[3] is then performed using the HindIII–EcoRI α_i/α_s chimeric cDNA fragment, the EcoRI–HindIII α_s fragment, and the HindIII-linearized pUC18. The final α_i/α_s chimeric ligation product is diagrammed in Fig. 1B.

---

[3] F. M. Ausubel, R. Brent, R. E. Kingston, J. A. Smith, J. G. Seidman, and K. Struhl, "Current Protocols in Molecular Biology." Wiley (Interscience), New York, 1987.

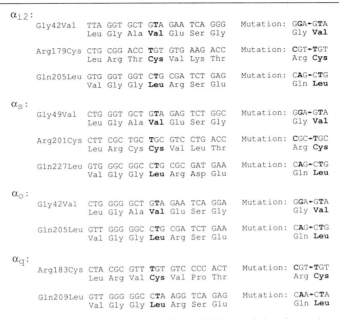

FIG. 3. Single amino acid mutations resulting in GTPase-deficient G-protein α subunits.

Figure 2 gives the sequences for the α_i/α_s sense (5'–3') and antisense (3'–5') oligonucleotides that have been successfully used for the construction of several α_i/α_s chimeras. These chimeras substitute a region of α_s with the corresponding region of α_i. The amino acid sequence at the chimera junction for α_i and α_s sequences is also given. A similar strategy for α_s/α_i carboxy-terminal sequences can also be used, but different 5'- and 3'-primer oligonucleotides would have to be designed.

Polymerase chain reaction mixtures generally consist of primer oligonucleotides (0.15 μM final concentration of each), PCR buffer (10 mM Tris-HCl, pH 8.3, 50 mM KCl, 1 mM MgCl$_2$), deoxynucleotides (dATP, dGTP, dCTP, dTTP, 200 μM each), 100 ng template (α_i or α_s cDNAs), 1–5 units Amplitaq DNA polymerase (Perkin-Elmer Cetus, Norwalk, CT) or equivalent thermal stable DNA polymerase, and sterile water in a final reaction volume of 100 μl. For the described oligonucleotides, the PCR is carried out for 30 cycles with an annealing temperature of 54°, a melting temperature of 94°, and an extension temperature of 72°. The final PCR product is gel purified using 1.2% agarose. The horizontal agarose gel is electrophoresed for 1–2 hr at 70 V and constant current. The PCR-generated cDNA fragment is detected by staining with ethidium bromide and visualized using a long-wavelength UV hand-held lamp. The DNA band is excised from the agarose gel using a razor blade. The DNA is extracted using NaI to dissolve the agarose (Geneclean Kit, Bio 101, La Jolla, CA) or equivalent procedure.[3]

Applications

In addition to the generation of G-protein α-subunit chimeras, it is possible to construct mutant α subunits that encode single amino acid mutations.[4] Single amino acid mutations that inhibit the intrinsic GTPase activity of any G-protein α subunit can be constructed. The conservation in sequence in the different G-protein α-subunit regions forming the GDP/GTP-binding domain allows construction of common mutations to inhibit the GTPase activity. The resulting phenotype when the mutationally GTPase-deficient α-subunit polypeptides are expressed in cells is a constitutive activation of the effector regulated by that G protein.

Figure 3 defines the mutations in α_{i2}, α_s, α_o, and α_q that inhibit intrinsic GTPase activity. For α_{i2} and α_s three different mutations resulting in single amino acid substitutions have been defined. For α_q only two mutations (Arg183Cys and Gln209Leu) have been characterized. For α_o the Gly42Val and Gln205Leu mutations have been characterized in regard to GTPase inhibition and functional expression.

To introduce single base mutations resulting in amino acid substitutions, 21-base oligonucleotides are synthesized. The appropriate base substitutions are incorporated in the middle of the oligonucleotide to allow efficient annealing with the wild-type α-subunit cDNA. Numerous mutagenesis kits are commercially available for the use of synthetic oligonucleotides to generate mutations in cDNAs. Generally, all of the kits rely on annealing of the oligonucleotide harboring the mutation to a single-stranded α-subunit cDNA. Double-stranded DNA is then generated using the mutant oligonucleotide as a primer. Double-stranded DNA encoding the mutation in both strands is eventually isolated and verified by DNA sequencing. The mutagenesis kits from Promega (Madison, WI), Bio-Rad (Richmond, CA), Amersham (Arlington Heights, IL), and other companies give detailed experimental protocols for these procedures and use straightforward molecular biology techniques for mutagenesis. Both the chimeric and single amino acid mutant α subunits can be shuttled into appropriate plasmids or retroviruses for expression in bacteria, insect cells, or mammalian cells using standard molecular biology techniques.

---

[4] M. P. Graziano and A. G. Gilman, *J. Biol. Chem.* **264**, 15475 (1989).

[26] Design of Degenerate Oligonucleotide Primers for Cloning of G-Protein α Subunits

By THOMAS M. WILKIE, ANNA M. ARAGAY, A. JOHN WATSON, and MELVIN I. SIMON

Introduction

Heterotrimeric G proteins couple the seven-transmembrane domain receptors to various intracellular effectors, such as adenylyl cyclase (adenylate cyclase, EC 4.6.1.1), phospholipase C-β, and ion channels.[1,2] Discrete signal transduction pathways are mediated through G proteins distinguished by the particular α, β, and γ subunit composition.[3,4] The G-protein α-subunit genes are encoded by a large and evolutionarily conserved multigene family,[5] as are the β and γ subunits[6] (see [37] in this volume). Sequence alignment of α-subunit genes from fungi, plants, and animals reveals several highly conserved amino acid motifs that are thought to contribute to GTP binding. Degenerate polymerase chain reaction (PCR) primers which target these motifs have been used to amplify G-protein α subunits selectively from mice[7] *Drosophila*,[8] *Caenorhabditis elegans*,[9] *Dictyostelium*,[10] *Arabidopsis*,[11] and *Neurospora*.[11a] The PCR products can be easily cloned, screened, and sequenced over a precise region to identify rapidly novel genes. The PCR cloning technique is a powerful means of revealing possible novel G-protein α subunits that play central roles in the biological processes of a variety of organisms.

The detailed descriptions of the methods that were used to identify G-protein α subunits are generally applicable to any multigene family that

---

[1] A. Gilman, *Annu. Rev. Biochem.* **56**, 615 (1987).
[2] A. Brown and L. Birnbaumer, *Annu. Rev. Physiol.* **52**, 197 (1990).
[3] C. Kluess, J. Hescheller, C. Ewel, W. Rosenthal, and G. Schultz, *Nature (London)* **353**, 43 (1991).
[4] C. Kluess, H. Scherubl, J. Hescheller, G. Schultz, and B. Wittig, *Nature (London)* **358**, 424 (1992).
[5] H. R. Bourne, D. A. Sanders, and F. McCormick, *Nature (London)* **349**, 117 (1991).
[6] M. I. Simon, M. P. Strathmann, and N. Gautam, *Science* **252**, 802 (1991).
[7] M. P. Strathmann and M. I. Simon, *Proc. Natl. Acad. Sci. U.S.A.* **87**, 9113 (1990).
[8] M. A. Lochrie, J. E. Mendel, P. W. Sternberg, and M. I. Simon, *Cell Regul.* **2**, 135 (1991).
[9] J. A. Hadwiger, T. M. Wilkie, M. P. Strathmann, and R. A. Firtel, *Proc. Natl. Acad. Sci. U.S.A.* **88**, 8213 (1991).
[10] H. Ma, M. Yanofsky, and E. Meyerowitz, *Proc. Natl. Acad. Sci. U.S.A.* **87**, 3821 (1990).
[11] K. B. Mullis and F. A. Faloona, this series, Vol. 155, p. 335.
[11a] G. E. Turner and K. A. Borkovich, *J. Biol. Chem.* **268**, 14805 (1993).

exhibits blocks of conserved amino acid motifs, such as G-protein-coupled receptors and protein kinases.[12-14] We discuss how to choose a target sequence for PCR amplification, how to design degenerate PCR primers, and how to bias the PCR primers toward particular subfamilies.

Design of Degenerate Polymerase Chain Reaction Primers

The PCR requires two oligonucleotide primers that are complementary to opposite strands of the target sequences.[11,15,16] To identify new genes among the G-protein α subunits, or any other multigene family, at least two regions of highly conserved amino acid sequence should be common to most members within the family. The amino acid sequence of the conserved motifs will dictate the DNA sequence of degenerate oligonucleotides that are to serve as PCR primers. The proper orientation of the primers is critical to the PCR; the sense primer should target the upstream conserved motif, and the opposing antisense primer should complement the coding strand of the downstream motif. These conserved motifs should not be less than 5 amino acid residues in length and should be separated from one another by at least 20 amino acids. When cDNA is used as the starting template in the PCR, it is advantageous for the primers to flank an intron, thus reducing the possibility of amplifying and cloning genomic DNA that may have contaminated the original RNA sample. To distinguish the targeted sequences from nonspecific PCR products that might be cloned, the amino acid sequence between the conserved motifs must be diagnostic for G-protein α subunits.

The PCR requires that the target sequence within the degenerate primers be at least 15 nucleotides long, but it may contain up to 27 or 30 nucleotides. This translates to a conserved motif of at least 5 amino acid residues, but the best primers are generally derived from 6 to 8 amino acids of uninterrupted identity. Longer primers are useful when two blocks of 2 to 4 highly conserved amino acids are separated by 2 or 3 residues that are less well conserved. The DNA sequence of degenerate primers can be made to take into account all possible nucleotide sequences coding for the amino acids at the primer binding site. This approach is recom-

---

[12] A. Kamb, M. Weir, B. Rudy, H. Varmus, and C. Kenyon, *Proc. Natl. Acad. Sci. U.S.A.* **86,** 4372 (1989).

[13] A. F. Wilks, *Proc. Natl. Acad. Sci. U.S.A.* **86,** 1603 (1989).

[14] F. Libert, M. Parmentier, A. Lefort, C. Dinsart, J. Van Sande, C. Maenhaut, M.-J. Simons, J. Dumont, and G. Vassart, *Science* **244,** 569 (1989).

[15] R. K. Saiki, S. Scharf, F. Faloona, K. B. Mullis, G. Horn, H. A. Ehrlich, and N. Arnheim, *Science* **230,** 1350 (1985).

[16] S. J. Scharf, G. T. Horn, and H. A. Erlich, *Science* **233,** 1076 (1986).

mended for amino acid motifs that are identical in every known member of the G-protein α subunits, for example, DVGGQR (Table I). If the motifs are not highly conserved, the degeneracy of the primers can become too great to support the PCR. When many members within a multigene family are known, degeneracy can be reduced if sets of primers are synthesized to target specific classes within the family (Table I). Another approach is to synthesize primers containing a consensus sequence derived from the nucleotide sequence of the known genes across the conserved region.[14,17] Degeneracy can be further reduced if inosine is used at any nucleotide position that has 3- or 4-fold degeneracy.[14] We have analyzed the cDNA sequence of several genes at the PCR primer binding site and have not yet detected a bias in inosine pairing to the starting template. We found that guanine specifically replaces inosine in the DNA sequence of the cloned PCR fragments, suggesting that *Taq* polymerase incorporates cytosine opposite inosine during the PCR. In at least one study, only the degenerate primers which contained inosine at degenerate positions amplified the target sequence.[18]

Degeneracy at the 3' ends of PCR primers is best held to a minimum.[19] If possible, the last one or two amino acids should be conserved among all family members in the targeted region. In addition, the PCR primers usually do not include degeneracy at the position of the 3' nucleotide. This is partly for expediency because oligonucleotides are synthesized from 3' to 5', with the 3' nucleotide coupled to a column. One or two inosine residues can be used at the 3' end if this position must be degenerate; however, inosine-coupled columns are not yet commercially available, and degenerate 3' ends greatly reduce the efficiency of PCR amplification.

Cloning Polymerase Chain Reaction Products

Degenerate primers may be synthesized to include restriction enzyme sites on the 5' ends. This provides the obvious advantage that the PCR products may be directionally cloned. In our experience, *Bam*HI and *Eco*RI have always performed well, but *Hin*dIII, *Sal*I, and *Pst*I have also been used.[14,16] Addition of two nucleotides (G or C) beyond the restriction sites at the 5' end of the primers facilitates enzyme recognition.

One disadvantage of using restriction enzyme sites to clone the PCR products directionally may be encountered if the recognition sequences are also present within the target DNA between the primers. In this case,

---

[17] A. F. Wilks, R. R. Kurban, C. M. Hovens, and S. J. Ralph, *Gene* **85,** 67 (1989).
[18] J. E. Johnson, S. Birren, and D. Andersen, *Nature (London)* **346,** 858 (1990).
[19] R. Sommer and D. Tautz, *Nucleic Acids Res.* **17,** 6749 (1989).

TABLE I
SEQUENCE OF Gα DEGENERATE OLIGONUCLEOTIDE PRIMERS

Oligonucleotide[a]	Amino acid[b]	Nucleotide[c]	Restriction site	Bias[d]
CT14 →	ESGKST	GTCTAGAGARTCIGGIAARTCIAC	XbaI	s, i, q, 12
CT15 →	ESGKST	GTCTAGAGARTCIGGIAARAGYAC	XbaI	s, i, q, 12
CT16 →	ESGKST	GTCTAGAGARAGYGGIAARTCIAC	XbaI	s, i, q, 12
CT17 →	ESGKST	GTCTAGAGARAGYGGIAARAGYAC	XbaI	s, i, q, 12
iMP41 →	DVGGQR	CGGATCCGAYGTIGGIGGICARHG	BamHI	s, i, q, 12
iMP19 →	KWIHCF	CGGATCCAARTGGATICAYTGYTT	BamHI	i, q
CT48 →	KWIQCF	CGGATCCAARTGGATICARTGYTT	BamHI	s
CT56 →	(RK)W(MILF)(QE)CF	CGGATCCARRTGGITILARTGYTT	BamHI	12
CT57 →	(RK)W(MILF)(CWSR)CF	CGGATCCARRTGGITIJGITGYTT	BamHI	Not s, i, q, 12
iMP20 ←	FLNKKD	GGAATTCRTCYTYTYTRTTIAGRAA	EcoRI	i, q
MP21 ←	FLNKKD	GGAATTCRTCYTYTYTRTTYAARAA	EcoRI	i, q
CT49 ←	FLNKQD	GGAATTCRTCYTGYTTRTTIARRAA	EcoRI	s
Ta29 ←	TCATDT	GGGATCCIGTRTCIGTIGCRCAIGT	BamHI	i, q
CT50 ←	TCAVDT	GGAATTCIGTRTCIACIGCRCAIGT	EcoRI	s
CT45 ←	NL(RK)(DEG)CG	CGAATTCCRCAIYCYYTIAGRTT	EcoRI	i
CT46 ←	NY(RK)(DEG)CG	CGAATTCCRCAIYCYYTYAARTT	EcoRI	i

[a] Oligonucleotides are listed as in Fig. 1 from the amino to carboxy terminus of a G-protein α subunit. Sense and antisense oligonucleotides are indicated by right- and left-facing arrows, respectively.
[b] Amino acid sequence targeted by each oligonucleotide is written from the amino to carboxy end of the coding sequence.
[c] Antisense oligonucleotides are written as the reverse complement of the amino acid sequence shown; restriction sites are underlined. I, Inosine; R, A, G; Y, C, T; H, A, C; L, C, G; J, A, T.
[d] The sequence of the degenerate oligonucleotide PCR primers may have identity with every nucleotide position in all known mammalian G-protein α subunits or may have complete identity with only certain classes, thus introducing a bias in the PCR toward amplification of α subunits in those classes. Four classes of mammalian G-protein α subunits are recognized: G_s (s), G_i (i), G_q (q), and G_{12} (12) [M. I. Simon, M. P. Strathmann, and N. Gautam, Science **252**, 802 (1991); T. M. Wilkie, D. J. Gilbert, A. S. Olsen, X. N. Chen, T. T. Amatruda, J. R. Korenberg, B. J. Trask, P. de Jong, R. R. Reed, M. I. Simon, N. A. Jenkins, and N. G. Copeland, Nat. Genet. **1**, 85 (1992); S. Yokoyama and W. T. Starmer, J. Mol. Evol. **35**, 230 (1992).]

two approaches are available. First, the PCR products can be cloned into a linearized vector with a single overhanging thymidine at the 5' end of either DNA strand. *Taq* polymerase frequently incorporates an A residue at 3' ends[20] which are then complimentary to the free ends of this PCR cloning vector, thus facilitating cloning of the PCR products without using restriction enzyme digestion prior to ligating the PCR products to the cloning vector. Promega (Madison, WI), In Vitrogen (San Diego, CA), and Novagen (Madison, WI) all sell commercial kits to perform TA cloning. The method is simple and rapid, although it apparently cannot be used with ligations in low melting point agarose gels. Therefore, if the PCR yields more than a single band or significant amounts of primer–dimer amplification products, the desired band must be purified from the agarose. Small DNA fragments less than 200 base pairs (bp) are preferentially ligated by this method, increasing the probability of cloning primer–dimer fragments. The method also suffers from occasional high background transformation by religated vector; unfortunately, these colonies are not always blue on X-Gal (5-bromo-4-chloro-3-indolyl-β-D-galactopyranoside)/IPTG (isopropyl-1-thio-β-D-galactopyranoside) plates and seem to arise from small deletions in the polylinker cloning region of the vectors. In our hands, the Promega kit performs best.

Another approach to obtaining the entire sequence between the PCR primers is to ligate the PCR products directly into a blunt-end cloning site (*Sma*I and *Eco*RV) of any common vector. If this approach is taken, it is recommended that the cloning site in the vector be dephosphorylated with alkaline phosphatase to reduce vector reclosure. For ligation into a dephosphorylated vector it is essential to phosphorylate either the degenerate primers or the PCR products with polynucleotide kinase.[21] As mentioned above, *Taq* polymerase frequently adds an adenine nucleotide to the 3' end of duplex DNA fragments.[20] Therefore, prior incubation of the PCR products with the Klenow fragment of DNA polymerase I will increase the efficiency of blunt-end ligations. Alternatively, a thermostable polymerase that possesses 3' to 5' proofreading activity, such as New England Biolabs (Beverly, MA) Vent polymerase, which does not modify the 3' end of the amplification product, could be used to perform the PCR. The PCR products can also be prepared for blunt-end ligation with T4 DNA polymerase.[22]

---

[20] J. M. Clark, *Nucleic Acids Res.* **16,** 9677 (1988).
[21] J. Sambrook, E. F. Fritsch, and T. Maniatis, "Molecular Cloning: A Laboratory Manual." Cold Spring Harbor Laboratory, Cold Spring Harbor, 1989.
[22] R. Kumar, *Technique* **1,** 133 (1989).

Design of Degenerate Polymerase Chain Reaction Primers Targeting G-Protein α Subunits

Most of the suggestions given above for designing degenerate oligonucleotide primers were applied to cloning members of the G-protein α-subunit multigene family. The search for PCR primer binding sites focused on the GTP-binding domains because they are highly conserved among G-protein α subunits. The motifs most well suited to PCR primer sites occur within or near the regions designated G1, G3, G4, and G5 (Fig. 1A). We first chose the amino acid sequences KWIHCF, just downstream of the G3 box, and FLNKKD, within the G4 box, to synthesize the sense and antisense oligonucleotide primers, respectively (iMP19 and MP21, Fig. 1C). These amino acid sequences were chosen for three reasons. (1) They were among the most highly conserved motifs found in G-protein α subunits, and thus they were not expected to amplify other genes in the superfamily of GTP-binding proteins, such as *ras* and EF-2. Furthermore, these sequences were found in all known members of the G_i class of α subunits, whereas $G\alpha_s$ had one amino acid substitution which resulted in a single nucleotide mismatch in each primer. Thus, MP19 and MP21 were expected to amplify preferentially members of the G_i class over the G_s

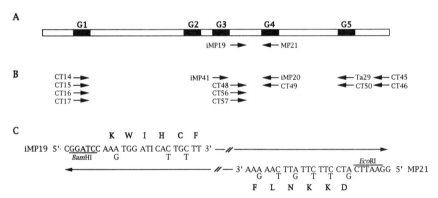

FIG. 1. Design of degenerate PCR primers. (A) The approximate location of the most highly conserved amino acid motifs, labeled G1 through G5,[5] and the degenerate oligonucleotides iMP19 and MP21 are shown in a linear depiction of the G-protein α subunit. (B) Additional degenerate oligonucleotides listed in Table I that were used to search for novel G-protein α subunits are shown in their approximate locations. (C) iMP19 and MP21 were synthesized in the sense and antisense orieintation, respectively. The DNA sequence of the PCR primers was derived from the consensus amino acids in boldface type above iMP19 and below MP21. Degenerate positions in the nucleotide sequence are indicated by either two nucleotides or an inosine (I). The restriction endonuclease sites *Bam*HI and *Eco*RI used for cloning the PCR products are identified. DNA polymerization during the PCR extends from the 3' end of either primer in the direction of the arrows.

TABLE II
CLONE FREQUENCY ANALYSIS OF G-PROTEIN α SUBUNITS AMPLIFIED WITH MP19 × MP20,21[a]

Class	Gα subunit	Oligonucleotide	Sequence[b]	Number[c]
G_s	$G\alpha_s/G\alpha_{olf}$	SG1	GCAGCAGCTACAACATGGT	11
G_i	$G\alpha_{i1}$	CT35	CTTCAGCAAGAACCAG	121
	$G\alpha_{i2}$	CT34	CCTCAGCCAGCACCAA	63
	$G\alpha_{i3}$	OP1	CAATTTCATGCTTTCA	11
	$G\alpha_{oA}$	OP3	GAGCATGAGAGACTCG	8
	$G\alpha_{oB}$	OP4	GAGCTTCAGGGATTCG	20
G_q	$G\alpha_q$	GQ4	ATTCGCTAAGCGCTACTAGA	2
	$G\alpha_{11}$	CT106	CTCGCTTAGTGCCACC	124
	$G\alpha_{14}$	CT107	TTCACTCAGAGCAACCAAGAAA	1
	$G\alpha_{15}$	CT58	GAGAACGTGATTGCCCTCATC	1
G_{12}	$G\alpha_{12}$	CT108	CTCGCTCGAGGACACCATGAAC	0
	$G\alpha_{13}$	CT109	TTCACTTGAAGACACAAGGAAA	0
Total[d]				362

[a] Random primed cDNA from mouse embryos was PCR amplified by iMP19 in combination with iMP20 and MP21. The PCR products were fractionated, isolated from agarose gels, and cloned using the Promega pGEM-T Vector System I. Individual clones were transferred to Hybond nylon filters and identified by hybridization to gene-specific oligonucleotide probes. Note the strong bias of the PCR primers against α subunits of the G_s and G_{12} classes.

[b] Oligonucleotide sequences are written 5' to 3' of the antisense strand, except for CT58, which is of the sense strand.

[c] Number of individual clones obtained from a total of 384 clones.

[d] In addition to the G-protein α subunit clones obtained, we found 12 clones that contained primer–dimers, formed by self-priming of the oligonucleotides during the PCR, 5 clones that contained ligation products whose DNA sequence had no similarity to GTP-binding proteins, and 5 clones generated by vector reclosure, for a total of 384 clones that were screened.

class, as was observed[23] (Table II). (2) The amino acid sequence between the primers was well conserved and diagnostic for the G-protein α subunits. (3) The primers were expected to generate a 203-nucleotide DNA fragment amenable to PCR amplification and sequence analysis.

Although it is important to include diagnostic sequence between the PCR primers, their presence could enhance template crossover during the PCR. Template crossover is a PCR artifact that joins two different sequences, ABC and αβγ, at a region of local sequence identity (B and β) to produce a novel PCR fragment of sequence AB/βγ. Template crossover is stimulated under conditions of incomplete extension from the

[23] M. P. Strathmann, T. M. Wilkie, and M. I. Simon, *Proc. Natl. Acad. Sci. U.S.A.* **86**, 7407 (1989).

primer, as occurs during the plateau phase of the PCR. Template crossover occurred in about 2% of the Gα subunit PCR products.[23] Sequence identity at 15 of 16 nucleotides between $G\alpha_z$ and $G\alpha_o$, in one case, and identity at 16 of 17 nucleotides between $G\alpha_{i1}$ and $G\alpha_{oA}$ in another case, was sufficient to achieve template crossover. It cannot be assumed that the presence of any two sequences which are identical through a portion of the PCR fragment, but then diverge, indicates a PCR artifact. One set of clones with this characteristic proved to be the product of alternate splicing in the $G\alpha_o$ subunit gene.[24] These cDNAs were cloned because the PCR primers flanked a highly conserved intron in the Gα multigene family that participated in alternative splicing in the $G\alpha_o$ gene.

The design of oligonucleotide primers influences the quality of the amplification products and their bias toward a class of G-protein α subunits. The degenerate oligonucleotide primer MP41 was used in combination with MP20 and MP21 to clone the broadest possible range of G-protein α subunits[23] and yielded clones form the four known classes of mammalian Gα subunit genes.[6,25-27] Further attempts to clone additional mammalian Gα genes usually yielded only sequences that had been previously obtained. For example, the primers CT56 and CT57, in combination with iMP20 and MP21, were designed to bias the PCR amplification toward the G_{12} class of α subunits. Although numerous $G\alpha_{12}$ and $G\alpha_{13}$ clones were obtained in excess over all other α subunits, no novel sequences were identified (Table III and data not shown). In contrast, PCR amplification with iMP19 in combination with iMP20 and MP21 strongly biased against $G\alpha_{12}$ and $G\alpha_{13}$ (Table II). Other degenerate oligonucleotides (listed in Table I) that were biased toward different classes of α subunits yielded similar results following PCR amplification of cDNA from mouse embryonic stem cells, embryos, brain, and germ cells from different spermatogenic stages (Table III and data not shown). However, the novel α subunit gustducin[28] was obtained by PCR amplification of rat tongue cDNA with MP19 and MP21 and two other degenerate oligonucleotide pairs (Table II). PCR amplification of cDNA from this novel tissue was presumably critical in obtaining gustducin, whose expression is restricted to taste

---

[24] M. P. Strathmann, T. M. Wilkie, and M. I. Simon, *Proc. Natl. Acad. Sci. U.S.A.* **87**, 6477 (1989).

[25] M. P. Strathmann and M. I. Simon, *Proc. Natl. Acad. Sci. U.S.A.* **88**, 5582 (1991).

[26] T. M. Wilkie, D. J. Gilbert, A. S. Olsen, X. N. Chen, T. T. Amatruda, J. R. Korenberg, B. J. Trask, P. de Jong, R. R. Reed, M. I. Simon, N. A. Jenkins, and N. G. Copeland, *Nature Genet.* **1**, 85 (1992).

[27] S. Yokoyama and W. T. Starmer, *J. Mol. Evol.* **35**, 230 (1992).

[28] S. K. McLaughlin, P. J. McKinnon, and R. F. Margolskee, *Nature (London)* **357**, 563 (1992).

TABLE III
Gα Subunits First Cloned by Polymerase Chain Reaction

Species[a]	Sequence[b]	PCR primers Sense	PCR primers Antisense	Clone sequence[c] Sense	Clone sequence[c] Antisense
M.m.	Gα$_{oB}$	KWIHCF	FLNKKD	*	*
M.m.	Gα$_q$	DVGGQR	FLNKKD	*	*
M.m.	Gα$_{11}$	DVGGQR	FLNKKD	*	*
		KWIHCF	FLNKKD	*	*
M.m.	Gα$_{12}$	DVGGQR	FLNKKD	*	*
M.m.	Gα$_{13}$	DVGGQR	FLNKKD	*	*
M.m.	Gα$_{14}$	KWIHCF	FLNKKD	*	*
M.m.	Gα$_{15}$	KWIHCF	FLNKKD	*	*
H.s.	Gα$_{16}$	KWIHCF	FLNKKD	*	*
R.n.	Gustducin	KWIHCF	FLNKKD	*	*
		HLFNSIC	VFDAVTD	*	*
		TIVKQM	FLNKQD	*	*
D.m.	Gα$_q$	KWIHCF	FLNKKD	*	*
D.m.	Gα$_f$	TDVGGQ	FLNKKD	YDVGGQ	FLNKYD
C.e.	gpa1	KWIHCF	FLNKKD	*	FLNKID
C.e.	gpa2	KWIHCF	TCATDT	*	*
C.e.	gpa3	KWIHCF	TCATDT	*	*
A.t.	gpa1	KWIHCF	FLNKID	KWIHLF	FLNKFD
N.c.	Gna1	KWIHCF	FLNKKD	*	*
N.c.	Gna2	KWIHCF	FLNKKD	*	*
D.d.	GPA4	KWIHCF	TCATDT	*	TCAVDT
D.d.	GPA5	KWIHCF	TCATDT	nd	nd
D.d.	GPA6	G(AT)GESGK	D(AV)(GA)GQR	nd	nd
D.d.	GPA7	G(AT)GESGK	D(AV)(GA)GQR	nd	nd

[a] M.m., Mouse; H.s., human; R.n., rat; D.m., Drosophila; C.e., Caenorhabditis elegans; A.t., Arabidopsis; N.c., Neurospora crassa; D.d., Dictyostelium discoideum.

[b] Asterisk indicates that the gene sequence corresponded to the primer sequences, exceptions are shown; nd, gene sequence not known.

[c] Amino acid and nucleotide sequences are referenced as follows: Gα$_{oB}$ [M. P. Strathmann, T. M. Wilkie, and M. I. Simon, *Proc. Natl. Acad. Sci. U.S.A.* **87**, 6477 (1990)], Gα$_{11}$, *Drosophila* and mouse Gα$_q$ [M. Strathman and M. I. Simon, *Proc. Natl. Acad. Sci. U.S.A.* **87**, 9113 (1990)], Gα$_{12}$ and Gα$_{13}$ [M. P. Strathman and M. I. Simon, *Proc. Natl. Acad. Sci. U.S.A.* **88**, 5582 (1991)], Gα$_{14}$ and Gα$_{15}$ [T. M. Wilkie, P. A. Scherle, M. P. Strathman, V. Z. Slepak, and M. I. Simon, *Proc. Natl. Acad. Sci. U.S.A.* **88**, 10049 (1991)], Gα$_{16}$ [T. T. Amatruda, D. A. Steele, V. Z. Slepak, and M. I. Simon, *Proc. Natl. Acad. Sci. U.S.A.* **88**, 5587 (1991)], gustducin [S. K. McLaughlin, P. J. McKinnon, and R. F. Margolskee, *Nature (London)* **357**, 563 (1992)], *Drosophila* Gα$_f$ [F. Quan, W. J. Wolfgang, and M. Forte, *Proc. Natl. Acad. Sci. U.S.A.* **90**, 4236 (1993)], *C. elegans* gpa1, gpa2, and gpa3 [M. A. Lochrie, J. E. Mendel, P. W. Sternberg, and M. I. Simon, *Cell Regul.* **2**, 135 (1991)], *Arabidopsis* gpa1 [H. Ma, M. F. Yanofsky, and E. M. Meyerowitz, *Proc. Natl. Acad. Sci. U.S.A.* **87**, 3821 (1990)], *Neurospora* Gna1 and Gna2 [G. E. Turner and K. A. Borkovich, *J. Biol. Chem.* **268**, 14805 (1993)], *Dictyostelium* GPA4 and GPA5 [J. A. Hadwiger, T. M. Wilkie, M. Strathmann, and R. A. Firtel, *Proc. Natl. Acad. Sci. U.S.A.* **88**, 8213 (1991)], and GPA6, GPA7, and GPA8 [L. Wu and P. N. Devreotes, *Biochem. Biophys. Res. Commun.* **179**, 1141 (1991)].

sensory cells.[28] Table III also lists novel G-protein α subunits that were cloned by PCR from invertebrates, fungi, plants, and *Dictyostelium*. Genomic DNA, rather than cDNA, was usually used as the PCR template to clone novel Gα genes from organisms with less complex genomes.

Materials and Reagents

Enzymes

    M-MLV (Maloney murine leukemia virus) reverse transcriptase: BRL (Gaithersburg, MD) or Pharmacia (Piscataway, NJ)
    AmpliTaq: Perkin-Elmer Cetus (Norwalk, CT)
    Restriction enzymes: New England Biolabs (Beverly, MA)
    T4 DNA ligase: New England Biolabs
    T4 DNA kinase: New England Biolabs
    Multiprime radiolabeling kit: Amersham (Arlington Heights, IL) or Strategene (La Jolla, CA)
    pGEM-T vector system I: Promega (Madison, WI)

Chemicals and Buffers

    Reverse transcriptase buffer (5×): 250 mM Tris-HCl (pH 8.3), 375 mM KCl, 50 mM dithiothreitol (DTT), 15 mM MgCl$_2$
    Random hexanucleotides [pd(N)6]: Pharmacia
    Deoxynucleotides (dATP, dCTP, dTTP, dGTP): Pharmacia
    Taq polymerase buffer (10×): 100 mM Tris-HCl (pH 8.3), 500 mM KCl, 15 mM MgCl$_2$, 0.01% (w/v) gelatin
    Oligonucleotide primers for PCR: prepared on an ABI (Foster City, CA) DNA synthesizer
    NuSieve agarose: FMC (Rockland, ME)
    E buffer (10×): 400 mM Tris-HCl, 330 mM sodium acetate, 10 mM EDTA, 12.5% (v/v) glacial acetic acid (pH 7.9)
    Restriction enzyme mix: 2 μl of 5× buffer (manufacturer's specifications), 0.4 μl enzyme 1, 0.4 μl enzyme 2, 5.2 μl distilled water
    Ligation mix: 1.5 ml of 10× ligase buffer, 1.5 μl distilled water, 0.75 μl 10 mM ATP, 0.75 μl T4 DNA ligase
    Ligase buffer (10×): 500 mM Tris-HCl (pH 7.8), 100 mM MgCl$_2$, 20 mM DTT, 10 mM ATP
    Cloning vector: Bluescript KS II (Strategene); pGEM-T vector (Promega)
    SOC medium: LB, 0.4% (w/v) glucose, 10 mM MgCl$_2$, 10 mM MgSO$_4$
    Denaturing solution: 0.5 M NaOH, 1.5 M NaCl
    Neutralizing solution: 0.5 M Tris-HCl, 1.5 M NaCl (pH 7.5)

SSC (20×): 3 M NaCl, 0.3 M Na Citrate (pH 7.0)
STET: 8% (w/v) Sucrose, 5% (v/v) Triton X-100, 50 m M Tris-HCl (pH 8.0), 50 m M EDTA
TE: 10 m M Tris-HCl, 0.25 m M EDTA, (pH 7.5)
Sequencing reagents (Sequenase sequencing kit or order separate reagents): U.S. Biochemical (Cleveland, OH)
Sequencing primers (M13 universal, M13 reverse, KS and SK if using the Stratagene Bluescript KS II cloning vectors): Strategene
Sequencing S-mix: 2 μl sequencing buffer, 1 μl of 0.1 M DTT, 0.5 μl labeling mix (dGTP or dITP), 0.5 μl [^{35}S]dATP, 0.25 μl Sequenase version 2.0, 0.5 μl dimethyl sulfoxide (DMSO)
Sequencing buffer (10×): 200 m M Tris-HCl (pH 7.5), 100 m M MgCl$_2$, 250 m M NaCl [^{35}S]dATP, >1000 Ci/mmol: Amersham
Sequencing stop solution: 95% (v/v) formamide, 20 m M EDTA, 0.05% (w/v) bromphenol blue, 0.05% (w/v) xylene cyanol
Nylon hybridization filters: Genescreen, Du Pont (Wilmington, DE); Hybond, Amersham

Equipment

Positive displacement pipettes: Baxter Scientific Products (McGaw Park, IL; SMI Digitron), Gilson (Middleton, WI; Microman M-25)
Microcentrifuge: Eppendorf
Speed-Vac (Savant, Farmingdale, NY)
Agarose gel electrophoresis equipment: laboratory shop
Sequencing gel electrophoresis equipment: BRL or Bio-rad (Richmond, CA)
PCR machine: Perkin-Elmer/Cetus or Ericomp
Flat-bottomed and round-bottomed microtiter dishes: Falcon

Optional

GeneAmp reaction tubes: Perkin-Elmer Cetus
Electronic digital repeat pipette: Rainin (Woburn, MA; EDPTM Motorized Microliter Pipettes EDP-25, EDP-100)
Electroporation equipment: Bio-Rad

Methods

cDNA Synthesis (First Strand Only)

The PCR requires single- or double-stranded DNA as a starting template to amplify a target sequence. To clone members of a multigene family that are expressed in a specific tissue, it is most convenient to use single-stranded cDNA prepared from that tissue as the input template.

cDNA may be synthesized from poly(A) or total RNA. Synthesis of the first strand of cDNA may be primed from an oligonucleotide specific to the G-protein α subunits, from oligo(dT), or from random hexadeoxynucleotide primers. The latter is by far the most versatile because cDNA from a single reaction may be used for PCR amplification of any target gene(s), irrespective of whether the PCR primers are located toward the 3' or 5' end of the cDNA sequence. All of the known mammalian G proteins have been amplified from cDNA synthesized using random primers according to the following protocol. In a reaction volume of 50 μl are mixed 2–5 μg total RNA from cells or tissues of interest, 1 μg random hexanucleotides [pd(N)6], 1 mM each deoxynucleotide, 1× reverse transcriptase buffer, and 400 U M-MLV reverse transcriptase. Incubate 90 min at 37°. Store at $-20°$.

Polymerase Chain Reaction Amplification of G-Protein α Subunits

Optimal conditions for the use of degenerate primers in the PCR must be empirically determined for each primer pair. In general, the greatest diversity of clones can be obtained by using the lowest possible annealing temperature that yields only a discrete band(s) of the expected size. The optimal annealing temperature will vary depending on the length of the target sequence and the ratio of A + T to G + C content among the degenerate oligonucleotides. In general, annealing temperatures between 37° and 45° are recommended for target sequences of 14–17 nucleotides; for target sequences of 24–27 nucleotides try 50°–65°. If the annealing temperature during the PCR is too low, nonspecific DNA fragments may appear as a band(s) of unexpected size or as an intense background smear following electrophoresis through an agarose gel. It is worth spending some time testing various PCR conditions before cloning and sequencing the PCR products. Suboptimal conditions will restrict the diversity and decrease the proportion of clones which are members of the G-protein α subunit multigene family. In this case, most or all of the clones may be unrecognizable, nonspecific sequences.

If the background is unacceptably high, first check that during the denaturing step the sample temperature is at least 92° but not more than 94°; temperatures in excess of 94° may denature *Taq* polymerase after several cycles, thus inhibiting DNA amplification. To suppress nonspecific DNA amplification, try increasing the annealing temperature and perhaps shortening the time that the samples are held there. If the background persists, then try reducing the concentration of the PCR primer. If all else fails, redesign the PCR primers to reduce the degeneracy, particularly at the 3' ends.

The PCR is extremely sensitive; therefore, it is imperative to avoid even minute contamination of the PCR reagents with DNA that contains sequence complementary to the primers. To avoid contamination, dispense the PCR template(s) only after all other aqueous components have been added to the reaction vessels. In addition, cotton-plugged micropipette tips or positive displacement pipette can be used in all procedures for the PCR. To verify that reagents are uncontaminated, routinely set up a control PCR in which no template DNA is added. When loading the PCR products onto an agarose gel for fractionation and subsequent reamplification, always leave an empty well between two different samples. If the isolated PCR products are to be reamplified, form the agarose gels on clean gel plates, and, if contamination is a problem, soak the plates in 0.1 M HCl and rinse with distilled water before pouring the gel. After the agarose gel has been stained, always place it on plastic wrap rather than directly onto an UV transilluminator that is in general use.

The conditions given below were used with the degenerate PCR primers indicated in Fig. 1 and listed in Table II. These conditions gave a discrete band of expected size with little or no background that included the PCR amplification products of all known G-protein α subunits.

To PCR amplify cDNA use a reaction volume of 10 μl containing 1 μl cDNA as prepared above, 9 μl PCR mix with 100 ng each primer (see detail of primer design), 200 μM each deoxynucleotide, 1× *Taq* DNA polymerase buffer, and 1 U AmpliTaq. Conduct 35 cycles: 1 min at 94° to denature, 1.5 min at 37°–65° to anneal, and 2 min at 72° to extend. Conclude the PCR with a 10-min extension at 72° at the end of the 35 cycles. All reaction volumes can be scaled up if desired, and a second round of PCR is generally recommended to increase the amount of DNA available for cloning and subsequent screening protocols.

1. Load 5 μl of the first round of PCR reactions in every second well of a 3% NuSieve agarose minigel submerged in 0.5× E buffer. DNA fragments between 100 and 1000 bp can be resolved in a 3% NuSieve gel.

2. Visualize the amplified DNA by submerging the agarose gel in an ethidium bromide solution (10 ng/ml) for 3 min at room temperature. Rinse the gel with water a few times to remove excess ethidium bromide, and illuminate the DNA–ethidium bromide complex with a long-wavelength UV source. Short-wavelength UV illuminators will nick the DNA. Use a single-edge razor blade to excise the band of interest, taking care to trim away excess agarose that does not contain DNA. Use extreme caution that the DNA fragments are not cross-contaminated while manipulating the small pieces of agarose. Place the isolated cubes of agarose into separate 0.75-ml microcentrifuge tubes.

3. Melt the agarose plug at 68° for 3 min.

4. Add 1 μl of the DNA that is contained in the molten agarose to 19 μl of PCR mix.

5. Reamplify the DNA according to the conditions that were used in the first round of amplification. For a reaction volume of 20 μl, combine 1 μl DNA (in NuSieve agarose, heated to 68° for 3 min) and 19 μl PCR mix containing 200 ng each primer, 200 μM each deoxynucleotide, 1× *Taq* buffer, and 2 U AmpliTaq. Perform the PCR for 35 cycles as above.

Cloning Amplified Polymerase Chain Reaction Products

The PCR-amplified DNA can be easily cloned into any one of a variety of bacterial plasmid vectors suitable for preparing double-stranded templates for DNA sequencing. High copy number plasmids are recommended because they usually increase the yield of DNA template. However, the multiple cloning sites (MCS) in these vectors are usually engineered into the α fragment of the *lacZ* gene. If expression of a *lacZ*/PCR fusion product were detrimental to the host, it could affect the cloning and sequencing of that PCR product. For this reason, it is recommended to transform a bacterial strain which expresses the *laciQ* repressor (e.g., *Escherichia coli* JM101) because expression of the *lacZ* transcription unit is suppressed in these strains. The protocol for cloning PCR products is as follows:

1. The PCR products can be isolated by electrophoresis through a NuSieve agarose gel in 0.5× E buffer. The second round of amplification should yield at least 100 ng of DNA.

2. Stain the gel with ethidium bromide, visualize the bands of DNA on a long-wave length UV illuminator, and isolate each band into individual 0.75-ml microcentrifuge tubes.

3. Melt the agarose cube(s) containing DNA by placing the microcentrifuge tube in a 68° water bath for 2 min.

4. Add 2 μl of the molten agarose to 8 μl of the restriction enzyme mix (see Chemicals and Buffers). Mix the contents and place the tube in a 37° water bath for 1–2 hr.

5. Remelt the agarose plug at 68° for 5 min. Add 0.5 μl of the appropriately cut vector DNA (1–20 ng). Mix and heat at 68° for an additional minute. Remove the microcentrifuge tube from the water bath and immediately add 4.5 μl of ligation mix. Mix the contents and incubate at room temperature for 1–2 hr.

6. Remelt the agarose plug at 68° for 2 min. Add 1.7 μl of 1 *M* NaCl and 30 μl of 95% (v/v) ethanol (room temperature). Mix the contents.

7. Spin tubes in the microcentrifuge for 5 min. Aspirate the ethanol

and wash the pellet twice with 70% ethanol (room temperature). Dry the pellet in a Speed-Vac or desiccation chamber.

8. To dissolve the DNA, add 7.5 µl water to the dried pellet and heat at 68° for 3 min.

9. Combine 5 µl of DNA and 40 µl of high-efficiency electrocompetent *E. coli* (5×10^8 colonies per mg supercoiled Bluescript KS+) and place on ice for 2 min. The cloning efficiency should be great enough to yield at least 200 colonies per ligation; electroporation generally yields thousands of transformants.

10. Electroporate and immediately add 800 µl SOC medium to the *E. coli* cells.

11. Incubate the cells for 15 to 60 min at 37° without shaking.

12. Plate 200 µl of the SOC culture. If desired, the remaining transformed *E. coli* may be grown up in 5 ml of LB-amp broth (100 µg/ml). Dispense 1 ml for storage at −70° in 7.5% DMSO to retain a library of the transformants.

We have also used the TA cloning method from Promega (pGEM-T Vector System I). In this case, 1 µl of the PCR reaction is mixed with 2 µl of pGEM-T vector following the manufacturer's instructions. If the PCR yields more than one band, the desired PCR fragments are purified from the agarose gel prior to ligation. All subsequent steps are followed, as described.

Screening Polymerase Chain Reaction Clones

Electroporation of a ligation mix can yield several thousand clones. A few of the clones may be picked at random and sequenced, but it is often more efficient to screen the population and sort clones based on hybridization patterns prior to DNA sequencing (e.g., see Table II). PCR amplification with *Taq* polymerase introduces sequence errors by nucleotide misincorporation. We have observed an error rate of about 1/500 bp in cloned PCR products. In addition to generating potential missense mutations in the open reading frame between PCR primers, this error rate allows a few clones to pass through the hybridization screen. For example, in a screen for novel members of the G_s class, 253 clones were identified as either $G\alpha_s$ of $G\alpha_{olf}$ among 262 total clones containing inserts. The remaining 9 clones were sequenced and identified as $G\alpha_s$, each with a single base pair substitution in the sequence complementary to the oligonucleotide hybridization probe (the mutagenic rate was calculated as 1/553 bp in these cloned PCR products). Thus, it may be advantageous to use a polymerase that has a lower intrinsic error rate during PCR amplification, such as New England Biolabs Vent polymerase.

Screening the PCR clones for known G-protein α subunits requires several duplicate hybridization filters. Therefore, transfer the original transformants to individual wells of four 96-well microtiter dishes and grow to stationary phase. Transfer aliquots of the microcultures to four replica filters each (Step 4a below), or dot all 384 colonies onto a single 3 by 6 inch filter using a 96-prong applicator in four staggered stamps transferred from the four 96-well microcultures (Step 4b below). Adherent bacteria can be lysed and the DNA cross-linked to the filter by UV irradiation. Sort clones prior to sequencing according to the hybridization patterns to radiolabeled, gene-specific oligonucleotide probes. Use one set of filters to hybridize with the degenerate oligonucleotides that were used in the original PCR (e.g., MP19 and MP21). Oligonucleotide probes are radiolabeled by T4 DNA kinase.[21] Clones which hybridize to the PCR primers are assumed to contain insert and then categorized according to the patterns of hybridization to the gene-specific oligonucleotide probes.

1. Dispense 150 µl LB-amp culture medium into each well of a 96-well flat-bottomed microtiter dish.
2. Pick isolated ampicillin-resistant colonies to inoculate individual wells in the microtiter dish.
3. Grow cultures until turbid (4–12 hr) on a shaker platform at 37°.
4a. Use a multichannel pipettor (8 to 12 channels) to transfer 10 µl of each culture in a 8 by 12 grid onto Genescreen nylon filters that have been placed on top of a damp pad of Whatman (Clifton, NJ) 3MM paper soaked in 2× SSC.
4b. Alternatively, clones can be picked directly to Hybond nylon filters placed onto LB-amp plates. Transfer bacterial colonies from four 96-well microtiter plates to the nylon filter using a 96-prong device in four staggered applications. Allow the colonies grow overnight at 30° on the filter.
5. Adherent bacteria can be lysed by floating the nylon filter on the surface of denaturing solution for 5 min.
6. Float the filters on the surface of neutralizing solution for 5 min.
7. Rinse the filters submerged in 2× SSC.
8. Air-dry the filters and either bake them at 80° for 1 hr or UV irradiate to covalently attach the DNA to the filter.
9. Hybridize the filters with the appropriate radiolabeled probes and wash according to conditions described by Sambrook et al.[21]

Rapid Boiling Minipreparations to Isolate Plasmid DNA for Sequencing

Plasmid DNA can be easily prepared from transformants which contain candidate clones. The following protocol is recommended when using the bacterial strains DH5, JM101, and MC1061; it is not recommended for strains RR1, HB101, or JM109. We also found that it was difficult to obtain good DNA sequence from minipreparation DNA if the cultures were grown too long.

1. Grow cultures in 2 ml of LB-amp for 8–12 hr at 37° (aeration is important; do not grow the cultures in sealed microcentrifuge tubes).
2. Transfer 1.5 ml to a microcentrifuge tube and spin in a microcentrifuge for at maximum speed for 25 sec at room temperature.
3. Pour off the supernatant, removing all residual LB.
4. Resuspend the cell pellet by vortexing in 300 µl STET.
5. Add 20–25 µl lysozyme made as a 10 mg/ml stock in water and kept frozen at −20°. Immediately mix by inverting the tubes no longer than 30 sec.
6. Place tubes in boiling water for 2 min.
7. Spin 5 min in the microcentrifuge.
8. Remove the mucoid pellet with a toothpick.
9. Add an equal volume (usually 250 µl) of 75% (v/v) 2-propanol/25% (v/v) 10 M ammonium acetate to the supernatant, mix vigorously, and spin 5 min in the microcentrifuge.
10. Rinse the pellet well with 70% (v/v) ethanol and dry in a Speed-Vac for 5 min.
11. Resuspend in 50 µl TE.

Sequencing Minipreparation DNA

DNA sequencing is done according to the manufacturer's protocol, except that DMSO is added to the sequencing reaction mix (10% by volume). Prepare the reaction mix (S-mix, see Chemicals and Buffers) when ready to sequence the plasmid DNA. To anneal the sequencing primer to the template, boil 9 µl of the minipreparation DNA with 1.1 µl DMSO and 1 µl sequencing primer at 20–40 ng/µl for 5 min. Place on ice or store frozen at −20° before sequencing. After the sequencing gel has been read, translate the DNA sequences of the PCR inserts into all reading frames. The proper open reading frame is obvious if the conserved amino acids are present in the appropriate positions within the PCR sequence.

Amplification by PCR under the conditions that we used introduced apparently random base pair substitutions about once per 400–500 nucleotides. At this rate, a significant fraction of individual isolates potentially contain missense mutations. However, nucleotide substitutions usually occur in the later PCR cycles,[29] and it is correspondingly rare to find two

[29] M. Krawczak, J. Reiss, J. Schmidtke, and U. Rosler, *Nucleic Acids Res.* **17**, 2197 (1989).

isolates of a cognate sequence which have the same nucleotide substitution. To minimize DNA sequence errors arising from PCR mutagenesis, we recommend sequencing pools of DNA templates from 3–5 individual isolates, when available. In this approach, a consensus DNA sequence is read on the gel. If the PCR fragment is short enough, both strands of each clone can be sequenced using the M13 universal and reverse primers. This method is fairly accurate; for example, the consensus DNA sequence of the PCR fragments of G-protein α subunits[23] agreed with the mouse cDNA clones that were later isolated.[7,24,25,30] This chapter has focused on cloning the α-subunit members of the heterotrimeric G-protein family. A more generic description of the approach applied the tyrosine kinase subfamily has been published.[31]

Conclusion

We have described a detailed method for differential hybridization screening of PCR products amplified from the G-protein α-subunit multigene family. These procedures are applicable to any multigene family exhibiting at least two short motifs of highly conserved amino acids that may support PCR amplification of a specific sequence domain. This approach provides a very efficient method for cloning and sequencing novel members of multigene families.

Acknowledgments

We thank Michael Strathmann and Narasimhan Gautum for numerous contributions, Kathy Borkovich and Michael Forte for providing amino acid sequences prior to publication, Carol Lee for screening and sequencing PCR clones, and members of the Microchemical Facility at Caltech (Pasadena, CA) for oligonucleotide synthesis and for coupling inosine to columns on special order. Some protocols presented in this chapter were taken with permission from Ref. 31. This work was supported in part by a National Institutes of Health postdoctoral fellowship to T.M.W. (GM11576), a CSIC Spansish postdoctoral fellowship to A.M.A., and a NIH grant to M.I.S. (GM34236).

[30] T. M. Wilkie, P. A. Scherle, M. P. Strathmann, V. Z. Slepak, and M. I. Simon, *Proc. Natl. Acad. Sci. U.S.A.* **88,** 10049 (1991).
[31] T. M. Wilkie and M. I. Simon, *Methods (San Diego)* **2,** 32 (1991).

[27] Microinjection of Antisense Oligonucleotides to Assess G-Protein Subunit Function

By CHRISTIANE KLEUSS, GÜNTER SCHULTZ, and BURGHARDT WITTIG

Introduction

Heterotrimeric G proteins represent a family of closely related regulatory proteins, composed of one α, β, and γ subunit each. About 20 different cDNA sequences (including splice variants) of α subunits are known, and at least 4 different cDNAs coding for G-protein β subunits and 6 cDNAs of γ subunits have been identified.[1,2] The G proteins are grouped into subfamilies classified according their α subunits. The functions of the various G proteins are highly diverse. For some subfamilies (G_s, G_i, G_o, G_q, and transducins), the physiological roles, namely, receptors stimulating the G protein and the G-protein-regulated effectors, are at least partially known; for others (e.g., G_{12}, G_{13}, and other pertussis toxin-insensitive G proteins) the functions remain largely unknown.

Assignment of a function to individual members of the G-protein family, to specific subunits, and to $\alpha\beta\gamma$-subunit heterotrimers is an important task in future studies on G-protein-mediated signal transduction. We have established a new approach to determine the functional identity of G proteins involved in hormonal regulations that can be studied on a single cell level. Antisense oligonucleotides selectively hybridizing with the mRNA of a G-protein subunit were microinjected into the nuclei of single cells. Following incubation to wait for reduction in the concentration of the target subunit, the resulting biological effects were measured electrophysiologically by the whole-cell modification of the patch-clamp method.[3] In a first experimental series, we studied antisense oligonucleotide effects on the inhibition of L-type voltage-dependent calcium channels by the receptor agonists somatostatin and carbachol in the rat pituitary cell line GH_3.

---

[1] M. I. Simon, M. P. Strathmann, and N. Gautam, *Science* **252**, 802 (1991).
[2] A. J. R. Hepler and G. Gilman, *Trends Biochem. Sci.* **17**, 383 (1992).
[3] J. Hescheler, this series, Vol. 238 [31].

Antisense nucleic acids (RNA or DNA) inhibit the expression of the respective targeted gene.[4,5] The actual mechanism is covered by two experimental hypotheses, one being the inhibition of translation at the ribosomal level, the other a destruction of the transcript by RNase H.[6-9] We interpret our results as a consequence of the latter mechanism since antisense oligonucleotides hybridizing with any part (including 3'-noncoding regions) of the mRNA were effective and since we used DNA oligonucleotides throughout which produce DNA–RNA hybrids susceptible to hydrolysis by RNase H.

General Considerations

Application of Antisense Nucleic Acids

Three methods of generating antisense oligonucleotides within the cell are generally used: (1) uptake of short DNA oligonucleotides added to the cell culture medium, (2) synthesis of antisense RNA from expression vector systems, and (3) microinjection. The former two were inappropriate for our purposes.

Cellular Uptake from Medium. Short nucleic acids (i.e., oligonucleotides) can be delivered to the cytoplasm of mammalian cells by simply adding the respective oligonucleotide to the culture medium.[10] In general, oligonucleotide uptake by cultured cells requires very high concentrations (5–200 μM) and is highly inefficient, with only a small fraction of the added oligonucleotide actually entering cells. Entry depends on the composition of the culture medium, the cell type, stage of the cell cycle,

---

[4] The simplest definition of the term antisense is an operational one. A nucleic acid, be it RNA or DNA, capable of hybridizing to an RNA transcript and/or the corresponding identical DNA strand (often named "coding strand") may be referred to as antisense. This definition is sufficient for the scope of this chapter. However, it is worth mentioning that in a situation where gene A is transcribed from one DNA strand and gene B from the corresponding one in a way that transcripts overlap B is antisense to A as well as A is to B.

[5] C. Hélène and J.-J. Toulmé, *Biochim. Biophys. Acta* **1049**, 99 (1990).

[6] C. A. Stein and J. S. Cohen, in "Oligodeoxynucleotides: Antisense Inhibitors of Gene Expression" (J. S. Cohen, ed.), p. 97. Macmillan, London, 1989.

[7] J. Minshull and T. Hunt, *Nucleic Acids Res.* **14**, 6433 (1986).

[8] P. Dash, I. Lotan, M. Krapp, E. R. Kandel, and P. Gollet, *Proc. Natl. Acad. Sci. U.S.A.* **84**, 7896 (1987).

[9] C. Boiziau, N. T. Thuong, and J.-J. Toulmé, *Proc. Natl. Acad. Sci. U.S.A.* **89**, 768 (1992).

[10] K. E. Brown, M. S. Kindy, and G. E. Sonenshein, *J. Biol. Chem.* **267**, 4625 (1992).

length of the oligonucleotide, and probably even the oligonucleotide sequence.[6,11]

Several different oligonucleotides are needed to evaluate the effect observed with a single antisense sequence. Aside from control experiments using "sense" and "nonsense" oligonucleotides, we validated the measured effects by using antisense oligonucleotides against 5'-noncoding, coding, and 3'-noncoding regions. If the different oligonucleotides do not reach their targets at comparable concentrations and within comparable times, such experiments are rendered meaningless. Moreover, incubating cells at micromolar concentrations of a highly negatively charged polymer, namely, the oligonucleotide, could unpredictably interfere with cellular signal recognition and transduction.

Nucleic Acid Transfer. As explained later, we needed short antisense oligonucleotides to guarantee selectivity in hybrid formation. Unfortunately, eukaryotic transcription systems mostly based on promoters for RNA polymerase II would add several hundred nucleotides of unwanted sequence to the 20–30 nucleotides of useful antisense RNA. Clipping of the extra nucleotides by the mRNA processing machinery of the nucleus would in turn trigger the addition of an equally unwanted poly(A) tail.[12] Despite such considerations in view of established knowledge, Watkins *et al.*[13] transcribed a short antisense oligonucleotide from a vector carrying a polymerase II promoter system and successfully suppressed expression of the $G\alpha_{i2}$ gene.

Microinjection. Nucleic acids can be delivered to the cytoplasm or nucleus of cells (oocytes, eukaryotic cells) by microinjection.[14-16] This technique is by far the most efficient among methods developed for the transfer of macromolecules into tissue culture cells. Microinjection is virtually independent of oligonucleotide length and sequence; the amount of intracellularly delivered oligonucleotide is well controlled, the oligonucleotide can be applied to the assumed site of action (nucleus or cytoplasm), and, finally, microinjection does not change the optimum culture environment.

---

[11] S. L. Loke, C. A. Stein, X. H. Zhang, K. Mori, M. Nakanishi, C. Subasinghe, J. S. Cohen, and L. M. Neckers, *Proc. Natl. Acad. Sci. U.S.A.* **86,** 3474 (1989).

[12] J. Sambrook, E. F. Fritsch, and T. Maniatis, in "Molecular Cloning: A Laboratory Manual," p. 16.5. Cold Spring Harbor Laboratory, Cold Spring Harbor, New York, 1989.

[13] D. C. Watkins, G. L. Johnson, and C. C. Malbon, *Science* **258,** 1373 (1992).

[14] E. G. Diacumakos, S. Holland, and P. Pecora, *Proc. Natl. Acad. Sci. U.S.A.* **65,** 911 (1970).

[15] J. E. Celis, A. Graessmann, and A. Loyter (eds.), "Transfer of Cell Constituents into Eukaryotic Cells." Plenum, New York, 1980.

[16] M. Graessmann and A. Graessmann, this series, Vol. 101, p. 482.

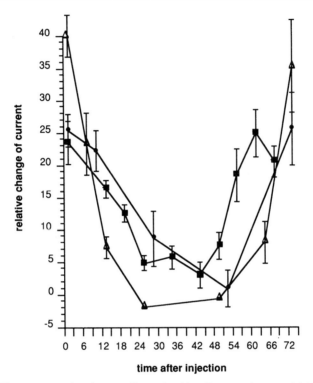

FIG. 1. Time course of antisense oligonucleotide effects on hormonal inhibition and stimulation of L-type calcium channels in GH_3 cells. Circles indicate the somatostatin-induced inhibition of currents in cells injected with an antisense oligonucleotide selective for α_o mRNA. Squares indicate carbachol-induced current inhibition in cells injected with an antisense oligonucleotide against mRNAs of all known β subunits. Triangles indicate thyrotropin-releasing hormone-induced current stimulation in cells injected with an antisense oligonucleotide against the mRNAs of all α_i subunits. (Adapted from Refs. 17–19).

Cell Division and Half-Life of Target Protein. Experiments with GH_3 cells showed an effective antisense oligonucleotide block between 1 and 2 days after injection[17–19] (Fig. 1). The time course of the block did not significantly differ for different α, β, and γ subunits. It appears that proteins with a slow turnover rate such as cell surface and adhesion molecules

[17] C. Kleuss, J. Hescheler, C. Ewel, W. Rosenthal, G. Schultz, and B. Wittig, *Nature (London)* **353**, 48 (1991).
[18] C. Kleuss, H. Scherübl, J. Hescheler, G. Schultz, and B. Wittig, *Nature (London)* **358**, 424 (1992).
[19] M. Gollasch, C. Kleuss, J. Hescheler, B. Wittig, and G. Schultz, *Proc. Natl. Acad. Sci. U.S.A.* **90**, 6265 (1993).

need at least 2 days to arrive at a sufficiently low concentration to make a biological effect measurable. This correspondingly requires a high intracellular stability of the injected oligonucleotide.

The effective oligonucleotide concentration is diminished by cellular nucleases[20] and cell divisions. Each cell cycle reduces the amount of injected oligonucleotide by half. The antisense oligonucleotide-mediated expression block will only be observed if the protein half-life is short and the antisense oligonucleotide concentration is maintained high. Whereas the turnover of a protein is biologically controlled and cannot be changed experimentally, the stability of oligonucleotides can be enhanced by a number of chemical modifications. In our hands, stabilization at least of both the starting and the terminating nucleotides by phosphothioate moieties was sufficient to prolong the stability in cases where unmodified oligonucleotides did not produce measurable effects. Cellular tolerance of phosphothioate-modified oligonucleotides was comparable to that of unmodified oligonucleotides (see below).

Experimental Procedures

Design of Suitable Antisense Oligonucleotides

The formation of hybrids depends on temperature, ion concentration, nucleic acid concentration, and length of hybridizable sequences; a characteristic and measurable parameter of the hybrid is its melting temperature T_m:

$$T_m = 79.8° + 18.5 \log M + 0.58(\%GC) + 11.8(\%GC)2 - 820/L$$

where M is the molarity of monovalent cations, %GC the percentage of G and C nucleotides in the DNA and L the length of the duplex in base pairs. This equation was modified from Schildkraut's equation to fit DNA–RNA heteroduplexes.[21] As the hybridization takes place in living cells, the incubation temperature, concentration of RNA, and ionic concentration are constant. Only changes in oligonucleotide concentration and length of the oligonucleotide may influence hybrid formation.[22] The optimum hybridization temperature T_h is about 20° lower than T_m. This brings the T_h of heteroduplexes close to the "natural" incubation tempera-

---

[20] E. Wickstrom, *J. Biochem. Biophys. Methods* **13**, 97 (1986).
[21] S. M. Freier, R. Kierzek, J. A. Jaeger, N. Sugimoto, M. H. Caruthers, T. Nelson, and D. H. Turner, *Proc. Natl. Acad. Sci. U.S.A.* **83**, 9373 (1986).
[22] B. L. Daugherty, K. Hotta, C. Kumar, Y. H. Ahn, J. Zhu, and S. Pestka, *Gene Anal. Tech.* **6**, 1 (1989).

ture of 37°, where the DNA strand is the short DNA oligonucleotide of 15–25 nucleotides and has a GC content between 40 and 50%.

Length. It can be assumed that the intracellular target of heteroduplex formation is the already spliced RNA population (mRNA), that an average mRNA contains 1500 nucleotides, and that a total of 20,000 different mRNA species are actually present per cell. Hence, total mRNA complexity is about 3×10^7 nucleotides, far lower than that of 4^{20} ($\sim 10^{12}$) combinations possible with four different nucleotides within a DNA oligonucleotide having a length of 20. Therefore, under ideal hybridization conditions, namely, no mismatches tolerated within a heteroduplex, DNA oligonucleotides of 15–25 nucleotides are highly selective for the respective targeted mRNA. Nevertheless, the products of partial hybridization shown in Fig. 2 can cause a great deal of nonselective targetting, since 8-base pair (bp) heteroduplexes have been proved stable under conditions approaching

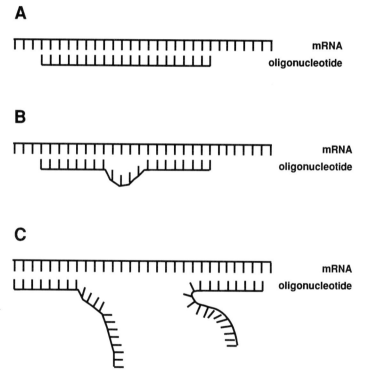

FIG. 2. Possible mRNA–antisense oligonucleotide heteroduplexes. (A) The desired product in which the antisense DNA oligonucleotide hybridizes in full length with the mRNA target. (B) Partial hybridization arising from internal mismatching. (C) Partial hybridization at different sites of the targeted mRNA or even nonrelated mRNAs.

the intracellular environment.[23] mRNA segments within such hybrids would be hydrolyzed by RNase H.

We therefore align the multiple sequences of the respective G-protein subunit families at the level of the amino acid sequences to detect a unique region within the sequence to be targeted. Then back-translation using the appropriate codon usage for the target cell line species is used to generate the nucleotide sequence (if the nucleic acid sequence has been experimentally determined for that species, back-translation can be omitted). A search algorithm for optimum PCR primers is subsequently used to find suitable antisense DNA oligonucleotides. This algorithm avoids sequence regions capable of forming dimers or foldback structures owing to self-complementarity and/or inverted repeats. Suggested oligonucleotide sequences are finally compared with all other G-protein sequences in the given context to detect possible matching stretches of minimum lengths with the respective oligonucleotide.[24] This takes care of the unwanted partial hybrids mentioned above. In our hands, antisense DNA oligonucleotides between 17 and 36 bp are efficient and selective for G-protein subunit suppression, using reasonable concentrations.

Preparation of Oligonucleotides. DNA oligonucleotides are synthesized chemically using the phosphotriester method. In cases where stabilization of the oligonucleotide is intended, phosphothioate modifications are introduced.[6,25] After deprotection, oligonucleotides are extracted with phenol/chloroform and precipitated with ethanol. No further purification is performed since oligonucleotides of the size range used are of minor variance of length. The oligonucleotide concentration is determined by UV absorption spectroscopy. Length variance can be checked by 5'-end labeling of the oligonucleotide with [γ-^{32}P]ATP followed by electrophoresis on sequencing gels. Oligonucleotides are dissolved in water at a concentration of 5 μM. Some cells may require buffered saline as the solvent; GH$_3$ cells tolerate water, and phosphate-buffered saline or "nuclear buffer" did not cause any difference in cell viability or hormone response. Before microinjection, the oligonucleotide solution is centrifuged for 10 min at 13,000 g to remove materials in suspension.

Concentration. The amount of injected oligonucleotide must be sufficiently high to block G-protein subunit expression. The G-protein subunits are parts of housekeeping proteins translated from 10–100 copies of mRNA

---

[23] K. R. Blake, A. Murakami, and P. S. Miller, *Biochemistry* **24**, 6132 (1985).

[24] V. Schöneberg, W. Vahrson, V. Priedemuth, and B. Wittig, "Analysis and Interpretation of DNA and Protein Sequences Using MacMolly® Tetra." Karoi-Verlag, Bielefeld, 1993.

[25] J. C. Marcus-Sekura, A. M. Woerner, K. Shinozuka, G. Zon, and G. V. Quinnan, *Nucleic Acids. Res.* **15**, 5749 (1987).

per subunit subtype and cell.[26] Even if 10 times more copies of nonspliced transcripts are present in the nucleus, 1000–5000 molecules of antisense DNA oligonucleotides should be sufficient to drive hybridization with all target mRNAs present in the cell. Supposing a nuclear injection volume of about 20 femtoliters (fl),[16] an antisense oligonucleotide solution of 0.8 μM will produce about 10,000 molecules per injected nucleus. High concentrations (70 μM) are deleterious for most cells, most likely because the injected polyphosphate competes with nucleic acids for binding to positively charged proteins. In our hands, GH_3 cells survive after injections with 30 μM antisense oligonucleotides. A solution containing 0.5 μM antisense DNA oligonucleotides is sufficient to suppress G-protein α subunits, whereas a 50 nM solution is no longer effective.

Nuclear Microinjection of Antisense Oligonucleotides

System for Microinjection. Microinjection has been reviewed in this series.[16] We describe the special considerations applying to the nuclear microinjection of antisense oligonucleotides and associated problems. We use a computer-operated automated injection system (AIS) commercially available from Zeiss (Oberkochen, Germany). Manually operated systems which are much less expensive have not been manageable in our hands. They require extensive practice times and/or a dedicated experimentalist before representative amounts of cells can be successfully injected.

The experimental conditions for injection depend mostly on the kind of cells to be injected. Pressure and injection time determine the injected volume. The optimal injection volume depends on the size of the nucleus and its rigidity. Because the nuclei inflate owing to the injected volume, successful injection can be judged more easily if the injection time is raised while injection pressure is reduced. We inject GH_3 cells for 0.1 to 0.3 sec at 20–30 hPa and at an angle of 60°. The angle by which the capillary penetrates the cell membrane may be reduced if flat cells are to be injected. In this way, minor variations in the slide thickness cause less jamming of the capillary.

Location of Injected Cells. Principally the AIS allows one to locate injected cells by the *XY* coordinate system. Since during the time course of our experiments cell divisions take place, thereby detaching cells from and reattaching cells to the surface, microinjected cells have to be localized by other means. Usually, all cells grown inside a marked area (frames) of 1 mm² are injected. Thin glass slides are marked using a rotating diamond; thick slides are marked by fluoric acid.[15,16] Alternatively, marked

---

[26] B. Lewin, *in* "Genes IV," p. 477. Cell Press, Cambridge, Massachusetts, 1990.

thin, circular slides of various grid width and a diameter of about 10 mm are commercially available (Eppendorf, Hamburg, Germany). Most cells will then detach and settle within the marked area. Neighboring areas are left empty to detect invading cells.

Capillaries Suitable for Injection. Capillaries suitable for nuclear injection are commercially available (Eppendorf). An outlet diameter of 0.2 μm fits the demands of oligonucleotide injections. If solutions with higher viscosity are to be injected or if cells often clog the capillary tip, wider openings should be used. We prefer capillaries supplied by Eppendorf. The preparation of glassware and conditions for pulling capillaries (practicing required) are described elsewhere.[15,16]

Cell Morphology. Cells to be microinjected require a certain shape and suitable growth behavior. Nuclear microinjection is ideally performed with cells harboring a giant nucleus, where the nucleus/cytoplasm ratio is high. Small nuclei of large cells are quite sensitive with respect to injection volume and cannot be targeted as easily as huge nuclei. Small nuclei may require regular phase-contrast microscopy instead of the inverted microscope preferentially used with the microinjection apparatus.

Cell Density. The appropriate density of cells depends on the detection assay. One frame should contain enough cells for one experiment. For electrophysiological detection, cells should be able to grow at the low density of 200–500 cells/mm^2 (about 10 cells are measured per experiment). Even at this low density, cells growing on top of one another or in dense clusters have to be removed with a needle to make every cell within the frame accessible to injection.

Cell Immobilization. Only adherent cells can be microinjected. If they do not adhere or if they become detached by the injection procedure, cells have to be immobilized by bipolar ionic compounds. For this purpose, slides or dishes on which cells are to be seeded for injection are pretreated with poly(L-lysine), gelatin, collagen, or Alcian blue.[27] It is important to assure that no interference with subsequent measurements occurs (e.g., fluorescence detection of Fura-2 cannot be performed on slides treated with Alcian blue).

Cell Stability. Some cells collapse after penetration with the capillary before any volume is injected. Failure to inject successfully can be due to missing the cell with the capillary, clogging of the capillary, low injection pressure, or cell disruption by excess infusion volumes. Successful injection is indicated by "popping up" of cells; they often exhibit a change in contrast with the additional volume received.

---

[27] R. G. Ham and W. L. McKeehan, this series, Vol. 58, p. 44.

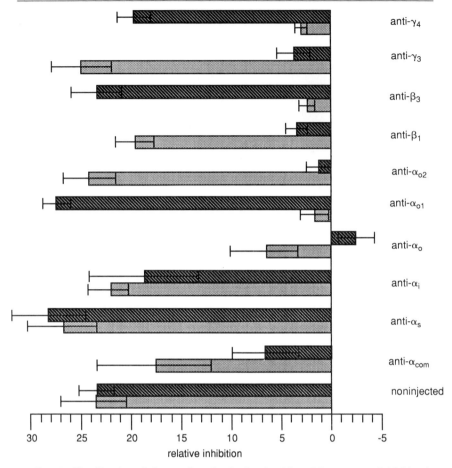

FIG. 3. Identification of G_o-protein subunits involved in calcium current inhibition in GH_3 cells. GH_3 cells were injected with antisense oligonucleotides directed against mRNAs encoding G-protein subunit subtypes. Relative current inhibition by hormones (1 μM somatostatin, shaded bars; 10 μM carbachol, hatched bars) is indicated in comparison to noninjected cells (mean values ± SEM; $n \geq 10$). The oligonucleotides injected specifically hybridize with the mRNA of all α_s subtypes (anti-α_s), all α_i subtypes (anti-α_i), all α_o subtypes (anti-α_o), α_{o1} (anti-α_{o1}), α_{o2} (anti-α_{o2}), β_1 (anti-β_1), β_3 (anti-β_3), γ_3 (anti-γ_3), and γ_4 (anti-γ_4). (Adapted from Refs. 17, 18, and 27).

Concluding Remarks

Using the above considerations on effective antisense oligonucleotides and the experimental conditions described, we succeeded in determining the functional subtype composition of G proteins involved in the hormonal regulation of the voltage-dependent L-type calcium channel in GH_3 cells; G_o forms composed of $\alpha_{o1}-\beta_3-\gamma_4$ and $\alpha_{o2}-\beta_1-\gamma_3$ were found to be involved in functional coupling of muscarinic m_4 and somatostatin receptors to calcium channels, respectively[17,18,28] (Fig. 3). Ongoing experiments in our laboratories indicate that our published approach can be generally applied to determine the functional differences of very similar members within a protein family. Whether injection into the nucleus is a necessary requirement for the biological effect observed has not been investigated thoroughly. However, preliminary results from the injection of rat pheochromocytoma PC-12 cells indicate that cytoplasmic injection of protected antisense oligonucleotides may also cause suppression of G-protein synthesis. Owing to the small size of the nucleus, the cytoplasm is often hit in these cells.

Besides patch-clamp recording of ion currents, other suitable detection methods to monitor suppression effects are single-cell fluorimetric measurements (e.g., of cytoplasmic calcium by the Fura-2 method) and immunocytochemical techniques. One obvious limitation of our approach is its restriction to single cells, thereby excluding assays depending on huge cell numbers, as, for example, required for enzymatic assays. Therefore, we are currently working on the development of the simultaneous "microinjection" of all cells on a petri dish, using ballistic methods.[29] Preliminary results indicate that DNA oligonucleotides can be delivered by use of oligonucleotide-coated gold particles into more than 10^6 cells at suitable concentrations.

Acknowledgments

The authors' studies reported herein were supported by grants of the Deutsche Forschungsgemeinschaft and the Fonds der Chemischen Industrie.

---

[28] C. Kleuss, H. Scherübl, J. Hescheler, G. Schultz, and B. Wittig, *Science* **259**, 832 (1993).
[29] R. S. Williams, S. A. Johnson, M. Riedy, M. J. DeVit, S. G. McElligott, and J. C. Sanford, *Proc. Natl. Acad. Sci. U.S.A.* **88**, 2726 (1991).

[28] Inactivation of G-Protein Genes: Double Knockout in Cell Lines

By RICHARD M. MORTENSEN and J. G. SEIDMAN

Introduction

Biochemical analysis of genetically altered cell lines has been important in determining the function of specific proteins. For example, the identification of mutant mouse lymphoma S49 cells which were resistant to stimulation of cAMP (cyc^- cells) was crucial in characterization of G_s activity. Until recently, mutant cell lines have been produced by random mutagenesis and then selection for a particular phenotypic change. This approach has limited applicability because selective conditions which allow the survival of only the desired mutationally altered cell are not easily identified. Technological advances in gene targeting by homologous recombination now enable the production of mutants in any desired gene.[1-5] Diploid cells contain two copies or alleles of each gene encoded on an autosome or non-sex chromosome. In most cases, both alleles must be inactivated to produce a phenotypic change in a mutant cell line. The production of mutationally altered cell lines has been described in which both alleles are inactivated by a technically difficult series of steps involving the production of two targeting vectors, two separate homologous recombination events, and selection.[6-8] Here we describe a simpler procedure, involving considerably less effort and time, that has been used to inactivate several α subunits of G proteins, as well as other genes.[9]

---

[1] T. Doetschman, N. Maeda, and O. Smithies, *Proc. Natl. Acad. Sci. U.S.A.* **85**, 8583 (1988).

[2] K. R. Thomas, T. S. Musci, P. E. Neumann, and M. R. Capecchi, *Cell (Cambridge, Mass.)* **67**, 969 (1991).

[3] S. L. Mansour, K. R. Thomas, and M. R. Capecchi, *Nature (London)* **336**, 348 (1988).

[4] P. L. Schwartzberg, S. P. Goff, and E. J. Robertson, *Science* **246**, 799 (1989).

[5] P. L. Schwartzberg, E. J. Robertson, and S. P. Goff, *Proc. Natl. Acad. Sci. U.S.A.* **87**, 3210 (1990).

[6] R. H. Te, E. R. Maandag, A. Clarke, M. Hooper, and A. Berns, *Nature (London)* **348**, 649 (1990).

[7] R. M. Mortensen, M. Zubiaur, E. J. Neer, and J. G. Seidman, *Proc. Natl. Acad. Sci. U.S.A.* **88**, 7036 (1991).

[8] A. Cruz, C. M. Coburn, and S. M. Beverley, *Proc. Natl. Acad. Sci. U.S.A.* **88**, 7170 (1991).

[9] R. M. Mortensen, D. A. Conner, S. Chao, L. A. Geisterfer, and J. G. Seidman, *Mol. Cell. Biol.* **12**, 2391 (1992).

Materials

Genomic libraries (BALB/c and 129Sv library) are from Stratagene (La Jolla, CA). G418, Dulbecco's modified Eagle's medium (DMEM, high glucose, with pyruvate), and agarose are from GIBCO (Gaithersburg, MD). Gancyclovir (GANC) can be obtained from Syntex (Palo Alto, CA). Fetal bovine serum (FBS) is obtained from Hyclone (Logan, UT). A Chinese hamster ovary (CHO) cell line overproducing leukemia inhibitory factor (LIF) is obtained from Genetics Institute (Cambridge, MA). Gelatin (porcine skin) is obtained from Sigma (St. Louis, MO). Genescreen is purchased from NEN Research Products (Boston, MA). Proteinase K is from American Bioanalytical (Natick, MA). Gene Pulser is a product of Bio-Rad (Melville, NY).

Experimental Procedures

The general procedure for the production of null mutants involves three basic steps: the creation of a targeting construct, production of a heterozygous clone by homologous recombination, and the selection of homozygous clones with both copies of the targeted gene inactivated (Fig. 1). This chapter gives general guidelines for the production of targeting constructs and inactivation of any desired gene. The steps are illustrated using results obtained during the production of cell lines lacking α_{i2} and α_{i3} genes.

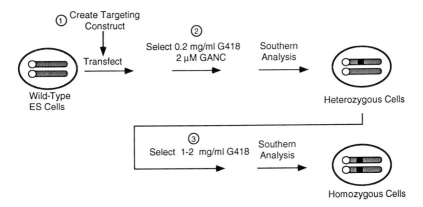

FIG. 1. Basic flow diagram for production of homozygous mutant cell lines with both copies of a gene inactivated. The three major steps are shown: (1) creation of a targeting construct, (2) selection at low G418 concentration for heterozygous colonies, and (3) selection of homozygous colonies at high G418 concentration.

Creation of Replacement Targeting Construct

A replacement targeting construct requires the assembly of several different DNA sequences: (1) A segment of the gene of interest is generally encoded on a bacteriophage or cosmid clone. The cloned segment should contain (a) homologous sequences which will be included in the construct and (b) sequences not included in the construct which will be used as a hybridization probe to screen for homologous recombinants. (2) A positive selectable marker such as the gene encoding neomycin phosphoribosyltransferase (*neo*) or hygromycin B phosphoribosyltransferase (*hyg*) is used to disrupt the gene. Both markers have been successfully used with this procedure. (3) A negative selectable marker such as herpes simplex virus thymidine kinase (HSV-TK) is used to enrich for homologous recombinants over random integrations of the construct.

A portion of the gene of interest is cloned from a genomic library. The cloned segment should be at least 5 kilobases (kb) and preferably greater than 10 kb long. The segment should be derived from DNA that is isogenic with the target DNA (i.e., from the same strain of animal or made from the cell line directly). The cloned segment must include regions which when interrupted will inactivate the gene. Typically an exon encoding an important region of the protein or upstream of such a region is selected as a site for inactivation. For G proteins, many options are available since the presumed GTP-binding site is distributed throughout the molecule. Interruptions of α-subunit genes as far downstream as exon 6 have been used successfully.

The targeting DNA is constructed so that the *neo* gene is embedded in the gene of interest, leaving regions of germ line sequence on either side of the *neo* gene. These regions of germ line sequence provide the substrates for homologous recombination. Generally, the homology regions should be greater than 1 kb on each side of the *neo* gene, with a total homology of 6 kb or greater. The degree of homology between the construct and the target genome can have a dramatic effect on the rate of homologous recombination in two ways. First, the DNA used to construct the targeting vector must be from the same species as the cell in which the mutation is to be introduced and should be isogenic with the target cell (this is not absolutely required, but increases the probability of success). Homologous recombination requires stretches of exact homology. Because different animal strains may differ just as individual outbred animals differ, there may be a mismatch on average every 500 base pairs (bp). This mismatch is sufficient to decrease the rate of homologous recombination dramatically.[10,11] Second, the longer the homologous regions, the

[10] C. Deng and M. R. Capecchi, *Mol. Cell. Biol.* **12**, 3365 (1992).
[11] R. H. Te, E. R. Maandag, and A. Berns, *Proc. Natl. Acad. Sci. U.S.A.* **89**, 5128 (1992).

higher the rates of recombination (within limits). The exact length at which creating longer constructs will not increase recombination rates is controversial.[10,12] Further, fidelity of recombination can be lower if the length of homology is shorter than 1 kb on a side.[13]

The *neo* coding sequence is expressed using a promoter active in embryonic stem (ES) cells. The construct pMC1-*neo*[3] is expressed well in ES cells, but the phosphoglycerate kinase (PGK) promoter characterized by McBurney and co-workers[14,15] driving *neo* yields 20- to 50-fold more colonies than pMC1-*neo*.[7] Both constructs originally contained a mutation in the *neo* gene which decreased the phosphoribosyltransferase activity. Constructs correcting the mutation have since been made. However, a single copy of PGK-*neo* without the mutation may give resistance to very high concentrations of G418, precluding the use of this single construct method. Use of the mutant form is therefore recommended (germline transmission occurs with the mutant form).

When constructing a targeting vector for the α_{i2} gene, *neo* was inserted into the *Bam*HI site in exon 6 (Fig. 2), and the α_{i3} gene was interrupted at the *Nco*I site at the initiating ATG in exon 1. Neither of the constructs were isogenic with the target DNA since they were derived from a BALB/c genomic library and the ES cells from the 129Sv murine strain. Genomic libraries from the 129Sv strain of mice are now commercially available.

Transfection of the targeting construct into ES cells results in random integration of the construct as well as targeting to the specific gene by homologous recombination. To enrich for homologous recombinant colonies, the HSV–TK gene is included outside the regions of homology. Cells lacking the TK gene can be selected by treatment of a cell culture with gancyclovir.[3] Random integration preferentially involves integration of the entire construct at the ends, thus retaining the TK gene, whereas homologous recombination involves a double crossover (on either side of the *neo* gene) leading to loss of the TK. The degree of enrichment varies but is generally 5- to 10-fold (α_{i2} 3- to 5-fold, α_{i3} 10-fold).

The complete targeting construct (100 μg at a time in 400 μl) is linearized by digestion with a restriction enzyme, preferably leaving the plasmid vector sequences attached to the TK gene. This will help preserve the activity of the TK gene if the construct is randomly inserted into the genome. The α_{i2} construct was linearized by digestion with *Xho*I.

[12] P. Hasty, P. J. Rivera, and A. Bradley, *Mol. Cell. Biol.* **11**, 5586 (1991).
[13] K. R. Thomas, C. Deng, and M. R. Capecchi, *Mol. Cell. Biol.* **12**, 2919 (1992).
[14] C. N. Adra, P. H. Boer, and M. W. McBurney, *Gene* **60**, 65 (1987).
[15] P. H. Boer, H. Potten, C. N. Adra, K. Jardine, G. Mullhofer, and M. W. McBurney, *Biochem. Genet.* **28**, 299 (1990).

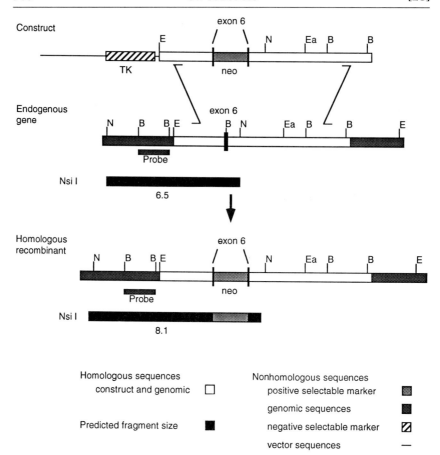

FIG. 2. Structure of the targeting construct used to inactivate the α_{i2} gene. The construct has approximately 7 kb of homology and the *neo* gene interrupting exon 6. Homologous recombination is detected by Southern analysis by digestion of genomic DNA with *Nsi*I and hybridization with a probe (*Bam*HI–*Bam*HI) outside the region of homology with the construct. Sizes of predicted hybridizing fragments (kb) are shown as solid bars. E, *Eco*RI; B, *Bam*HI; N, *Nsi*I; Ea, *Eag*I.

The targeting construct DNA is purified and sterilized by extraction with 1 volume of phenol–chloroform–isoamyl alcohol (25 : 25 : 1, v/v), and subsequent extraction with 1 volume of chloroform. The DNA is precipitated with 2 volumes of 95% ethanol and centrifuged in a microcentrifuge for 30 sec. The supernatant is removed, and the pellet is allowed to air-dry until only slightly moist, under sterile conditions. Add 100 μl of sterile water to dissolve the pellet. Check for complete digestion by agarose gel electrophoresis.

Before transfection of the construct, it is best to have verified the identity of an acceptable probe for detection of homologous recombination by Southern analysis. The probe can usually be obtained from sequences in the original genomic clone which were not used for the construct. By using a probe outside the sequences in the construct, only the endogenous gene and homologous recombinants will be detected, not the construct randomly integrated into the genome. The probe should detect no more than a few bands on a Southern blot of genomic DNA.

Transfection and Selection of Heterozygous Embryonic Stem Cells

The ES cells are cultured in DMEM containing 0.1 mM 2-mercaptoethanol, 15% FBS on feeder layers of irradiated mouse embryo fibroblasts (MEF) or in the presence of conditioned media (1 : 1000) containing LIF. The feeder layers or LIF prevent the ES cells from differentiating. All culture dishes are coated with 0.1% gelatin. Details of culture techniques for ES cells have been described by Robertson.[16]

Approximately 5×10^6 to 1×10^7 cells are harvested with 0.25% trypsin in 1 mM EDTA for approximately 5 min until cells are freed from the plate surface. Cells are pelleted by centrifugation and resuspended in electroporation medium (1 ml of 20 mM HEPES, 145 mM NaCl, and 0.1 mM 2-mercaptoethanol). Linearized sterile construct (1 pmol) is added, and the mixture is electroporated with a Bio-Rad Gene Pulser at 450 V and 250 μF. After electroporation the cells are allowed to stand for 10 min.

Approximately two-thirds of the cells are plated per 150-mm plate and one-third in a 100-mm plate. The culture medium is changed daily. After 24 hr, 0.2 mg/ml G418 is added to both plates and 2 μM GANC to the 150-mm plate only. Culturing is continued until single isolated colonies are visible (typically 1 week after electroporation). The number of colonies in the 150-mm and 100-mm plates are counted and compared. This ratio, corrected for the number of cells plated, gives the enrichment by GANC. Individual colonies in the 150-mm plate selected in G418 and GANC are picked with a pipette tip, trypsinized in a 35-μl drop for 5 min, and transferred to a 24-well gelatinized plate with 1 ml culture medium. When clones are grown, but not yet differentiating, half of the cells are passaged into a second 24-well plate and the remaining half is frozen in medium containing 20% dimethyl sulfoxide (DMSO), stored at $-70°$ overnight, and then in liquid nitrogen.

[16] E. J. Robertson, *in* "Teratocarcinomas and Embryonic Stem Cells: A Practical Approach (E. J. Robertson, ed.), p. 71. IRL Press, Oxford, 1987.

Screening for Homologous Recombinants

The ES cells in the 24-well plate are grown to near confluence (differentiation is unimportant at this stage). When grown, 300 μl of digestion buffer [1 mg/ml proteinase K, 20 mM Tris, pH 8.0, 10 mM NaCl, 10 mM EDTA, and 0.5% sodium dodecyl sulfate (SDS)] is added and the cells incubated at 55° overnight. Next, 150 μl saturated NaCl is added, and the mixture is vortexed vigorously. The solution turns milky white. DNA is precipitated by addition of 2 volumes of 95% ethanol. The solution should be clear except for DNA. *Note:* Some protocols directly precipitate with alcohol without addition of salt. We have usually found that the DNA pellet dissolves better if salt is first added. The DNA is dissolved in 30 μl water. The DNA concentration can be determined at this stage by measuring the absorbance at 260 nm.

Cells in which homologous recombination has occurred are identified by Southern blot analysis using a DNA probe from outside the targeting vector (see above). Approximately 10 μg DNA (or 5 μl if the DNA concentration was not determined) is digested with the appropriate restriction enzyme. The digested DNA is fractionated in a 1% agarose gel, transferred to a nylon membrane (Genescreen), and hybridized to the appropriate probe.

To screen for homologous recombination of the α_{i2} gene, genomic DNA was digested with *Nsi*I and hybridized with the probe indicated in Fig. 2. Homologous recombination yields a larger band if homologous recombination has occurred. Figure 3 shows the results with Southern analysis of a wild-type ES cell clone (lane A) and a heterozygous α_{i2}

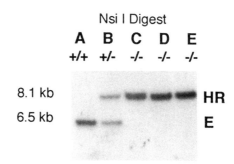

FIG. 3. Southern analysis of DNA isolated from ES cells: wild type (+/+), heterozygous for gene disruption with *neo* (+/−), and homozygous for the inactivated allele (−/−). E, Hybridization band from endogenous gene; HR, hybridization band from homologous recombinant gene.

homologous recombinant (lane B). Some results on the frequency of homologous recombination are shown in Table I.

Selection for Homozygous Clones

Once a heterozygous targeted clone has been obtained, a homozygous cell line using the mutant form of PGK-*neo* can be isolated from the heterozygous cells by selecting cells that are resistant to higher concentrations of G418 than the homozygous cells. Use of the PGK-*neo* (with wildtype coding sequence) may give heterozygous cells that can not be easily killed by even very high concentrations of G418 so that this selection method will be ineffective. Cells that have lost heterozygosity contain two interrupted alleles and therefore contain two *neo* genes, which make the cells more resistant to G418. The frozen stored sample of a heterozygous cell line is thawed and the clone expanded. The heterozygous cells are plated at 10^6 cells per 100-mm plate in each of three plates containing different concentrations of G418 (1.0, 1.5, and 2.0 mg/ml). The G418 must be neutralized to near pH 7.4 or the acidic conditions will be lethal to cells in the culture. The cells are cultured for 7–10 days with daily medium changes until single surviving colonies are detected. For some genes, higher G418 concentrations may be required, presumably depending on the relative expression of the neomycin resistance gene at different genomic sites. The relative resistance of the heterozygous cells to G418 will also depend on the promoter used to drive the expression of *neo* so that other promoters may require more or less G418. If cells overgrow plates and no single colonies are obtained, the cells should be replated using higher G418 levels (some clones have required greater than 10 mg/ml). The surviving clones are screened by Southern blot analysis exactly as performed for the isolation of the heterozygous cell line except that the

TABLE I
FREQUENCY OF HOMOLOGOUS RECOMBINATION OF α_{i2} AND α_{i3} IN EMBRYONIC STEM CELLS[a]

Gene	No. of cells transfected	No. of colonies G418^R	No. of colonies screened	Homologous recombinants
α_{i2}	1×10^7	250	65	29 (45%)
	5×10^6	135	52	35 (67%)
α_{i3}	5×10^6	68	64	24 (37%)
	1×10^7	102	35	10 (28%)

[a] Data are from four separate transfections in the D3 (ES) cell line.

Southern blot of the DNA derived from a double knockout (homozygous) clone will now completely lack the band corresponding to the gene segment found in the normal parent cell. Examples of Southern analysis of two clones homozygous for α_{i2} gene inactivation are shown in Fig. 3 (lanes C and D).

The frequency of homozygous clones differs among heterozygous clones since loss of heterozygosity appears to be a random event. Several separately derived heterozygous clones should be expanded in culture and subjected to selection with higher levels of G418 because there appears to be some clone-to-clone variation in resistance to the drug. The frequency of homozygous clones may also be influenced by the time in culture before selection since the homozygous clones are continuously produced and tend to accumulate with time. Typical results for the α_{i2} and the α_{i3} genes are shown in Table II.

Uses of Mutant Cell Lines and General Application of Method

Once mutant cell lines have been produced, the cells can be analyzed to confirm inactivation of the gene and to resolve phenotypic differences. Expression can be evaluated by Northern blot analyses of total RNA or Western blot analyses of proteins isolated from the cell line. The interruption of exon 6 in the α_{i2} gene lead to complete absence of detectable α_{i2} mRNA by Northern analysis.[9]

For ES cells lacking functional α_{i2} or α_{i3} genes, coupling of various receptors to different effectors can be analyzed provided that the re-

TABLE II
LOSS OF HETEROZYGOSITY IN EMBRYONIC STEM D3 CELL LINES

		[G418] (mg/ml)			
		1.0–1.5		2.0	
Gene	No. of cells plated[a]	No. of colonies	% Homozygotes (no. analyzed)	No. of colonies	% Homozygotes (no. analyzed)
α_{i2}	1×10^5	34	95% (20)	8	100% (8)
	1×10^5	54	47% (19)	34	53% (34)
α_{i3}	1×10^5	38	70% (17)	26	80% (15)
	1×10^5	12	8% (12)	0	

[a] Similar data have been published for the CCE and CC1.2 ES cell lines [R. M. Mortensen, D. A. Conner, S. Chao, L. A. Geisterfer, and J. G. Seidman, *Mol. Cell. Biol.* **12**, 2391 (1992)].

sponses are present in the wild-type cells. As so far no receptors known to couple through α_i have been identified in ES cells,[7] there are two approaches that can be taken. The first is that receptors can be heterologously expressed in the undifferentiated ES cells. Cell lines expressing the receptor would be developed in the wild-type and the null mutant cell background and tested for coupling to adenylyl cyclase (adenylate cyclase), phospholipase C, phospholipase A_2, mitogenesis, or other G-protein-mediated signal. This method requires the identification of the desired response in the wild-type cell line. Alternatively, ES cells offer a unique opportunity since they are capable of differentiating into any cell type. Many cell types can be produced in *in vitro* cultures, including beating cardiocytes, skeletal muscle cells, neurons, glial cells, and vascular endothelial cells. Provided that the phenotype can be analyzed with a single or a few cells, these differentiated cells can then be used to analyze the phenotype of the disrupted gene.

The ES cells can also differentiate within the entire organism by injection into normal blastocysts. Homozygous mutant cells offer no technical advantage over heterozygous cells if the goal is to obtain a mutant mouse line through germ line transmission. However, homozygous cells can be tagged by introducing a gene which has a histochemically detectable product (such as β-galactosidase). Studies of embryos derived from blastocysts injected with these tagged mutationally altered ES cells may reveal the role of the target gene during development.

The suitability of this technique for the production of types of cells, other than ES cells, that lack a particular gene is less certain. Because homologous recombination and spontaneous loss of heterozygosity are known to occur in other cell lines, these same methods have been used to produce cell lines lacking other functional genes. However, other types of immortalized cells (e.g., lymphocytes or fibroblasts) may not allow the targeting vector to undergo homologous recombination at the same frequency as found in ES cells. Furthermore, many other types of immortalized cells are polyploid or aneuploid; this method for the production double knockout cells may not be suitable for these cell types.

The method described here is a general one that can be used to produce a large variety of ES cell lines, each one of which will lack a particular gene. We recognize that this approach cannot be used to produce double knockout cell lines that bear genes that are required for cell viability. We have successfully used this method to produce cells that lack α_{i2}, α_{i3}, α-cardiac myosin heavy chain, β-cardiac myosin heavy chain, and T-cell α-receptor genes. We expect that future studies will lead to the production of many other such cell lines and that the characterization of these cell lines will lead to the definition of the role of different gene products in cell structure and function.

Acknowledgments

Research was supported by the Howard Hughes Medical Institute and the American Heart Association (Clinician Scientist Award to R.M.M.).

[29] Targeted Inactivation of the $G_{i2}\alpha$ Gene with Replacement and Insertion Vectors: Analysis in a 96-Well Plate Format

By UWE RUDOLPH, ALLAN BRADLEY, and LUTZ BIRNBAUMER

Introduction

It has become possible to inactivate genes in mammalian cells by introducing a mutation at a specific genomic locus. Murine embryonic stem cells (ES cells) have been isolated that can be genetically manipulated in cell culture[1] and reintroduced into blastocysts, giving rise to chimeric animals.[2,3] Chimeras may transmit the mutation to their offspring, which will then be heterozygous for the (in this example autosomal) mutation. The mutation can then be bred to homozygosity, enabling the analysis of the mutational phenotype both at the level of the whole animal (e.g., function of targeted genes in development) and at the cellular and molecular level by analyzing cell lines or tissues derived from animals homozygous for the desired mutation.

An important task in gene targeting experiments is to identify the rare targeted events over the background of random integrations. Most protocols include the use of a Neo cassette in the targeting vector disrupting an exon of the gene of interest and enabling a positive selection for cells that have stably integrated the transfected DNA into the genome. Various selection and screening procedures have been devised to enrich for or identify targeted clones. They include the positive–negative selection making use of the fact that most random integration events occur end to end, thus retaining a thymidine kinase (TK) marker [that can be

[1] K. R. Thomas and M. R. Capecchi, *Cell* (*Cambridge, Mass.*) **51,** 503 (1987).
[2] A. Bradley, M. Evans, M. H. Kaufman, and E. Robertson, *Nature* (*London*) **309,** 255 (1984).
[3] A. Bradley, in "Teratocarcinomas and Embryonic Stem Cells: A Practical Approach" (E. J. Robertson, ed.), p. 113. IRL Press, Oxford, 1987.

selected against with 1-(deoxy-2-fluoro-β-D-arabinofuranosyl)-5-iodouracil (FIAU) or gancyclovir] placed adjacent to the homologous sequences,[4] the use of enhancerless[5] or promoterless[6,7] Neo cassettes, which are expressed only when an enhancer or promoter is provided by the chromosomal locus at which the targeting vector integrates, the use of weak position-dependent Neo cassettes which, if expressed at the targeted locus, will provide an enrichment of targeted clones,[1] and the use of Neo cassettes devoid of a polyadenylation signal, which are thus dependent on the polyadenylation provided by the chromosomal locus.[1]

Targeting vectors can be designed as replacement or insertion vectors (Fig. 1). A replacement vector is linearized outside the region of homology. Two crossovers are thought to occur, thus replacing a segment of genomic DNA with the corresponding sequences of the targeting vector (usually including the Neo marker); the heterologous ends of the targeting vector are lost. An insertion vector, however, is linearized within the region of homology, either by a simple double-stranded cut or, if so desired, by a simultaneous introduction of a gap, and inserted in its entirety by single reciprocal recombination so that the genomic sequences contained in the targeting vector will be duplicated. Depending on the branch migrations and the type of resolution of the Holiday junctions that form as an intermediate at the moment of insertion, the mutation carried by the insertion vector, be it a single base change or a sizable DNA fragment, can be present in either, both, or none of the duplicates that result as a consequence of the insertion. Targeting with an insertion vector can also lead to target conversion,[8,9] giving rise to the same product that is expected to arise using a corresponding replacement vector.

General Strategy Considerations

The gene of interest should be cloned from a genomic library that is isogenic with the genome of the embryonic stem cell, in the present context to the AB-1 cell, in which the gene will be inactivated. The AB-1 cell line was derived from a male 129Sv mouse blastocyst.[10] Mixtures of four to

[4] S. L. Mansour, K. R. Thomas, and M. R. Capecchi, *Nature (London)* **336**, 348 (1988).
[5] M. Jasin and P. Berg, *Genes Dev.* **2**, 1353 (1988).
[6] J. M. Sedivy and P. A. Sharp, *Proc. Natl. Acad. Sci. U.S.A.* **86**, 227 (1989).
[7] P. L. Schwartzberg, E. J. Robertson, and S. P. Goff, *Proc. Natl. Acad. Sci. U.S.A.* **87**, 3210 (1990).
[8] G. M. Adair, R. S. Nairn, J. H. Wilson, M. M. Seidman, K. A. Brotherman, C. MacKinnon, and J. Scheerer, *Proc. Natl. Acad. Sci. U.S.A.* **86**, 4574 (1989).
[9] S. L. Pennington and J. H. Wilson, *Proc. Natl. Acad. Sci. U.S.A.* **88**, 9498 (1991).
[10] A. P. McMahon and A. Bradley, *Cell (Cambridge, Mass.)* **62**, 1073 (1990).

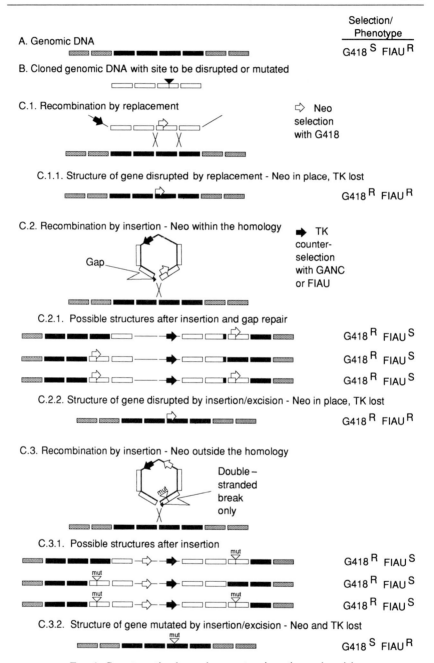

FIG. 1. Gene targeting by replacement or insertion and excision.

five 50-nucleotide long oligonucleotides are excellent screening probes for this purpose. The natural gene needs then to be characterized, and, based on the restriction map obained, a site is chosen to be disrupted by a Neo cassette, if a replacement type of targeting strategy is to be followed, or by either the disruption with a Neo cassette or the creation of a point mutation, if targeting is to be obtained by an insertion approach followed by excision of the duplicated homology by intrachromosomal recombination (insertion/excision, hit-and-run, or in-and-out strategy).

A replacement-type targeting vector (Fig. 1, C.1) is constructed so that the Neo cassette is located roughly in the middle of a fragment of the gene 5–10 kilobases (kb) long. The disrupted gene fragment is then flanked by one or two TK counterselection cassettes. The construct is linearized outside the region of homology (i.e., within the plasmid) and transfected by electroporation into the ES cells. The transfected ES cells are then plated onto a layer of feeder cells formed of mitomycin C-treated STO cells in medium containing leukemia inhibitory factor (LIF) that prevents ES cells from differentiating. Mitomycin C-treated SNL76/6 cells, which contain a LIF expression cassette in the genome, can also be used.[10] Incorporation of the disrupted homology by double reciprocal homologous recombination or gene conversion converts the cells from $G418^S$ to $G418^R$. If, as expected, the plasmid and flanking regions are lost, the cell will not carry a TK gene and will remain $FIAU^R$. If, however, the construct is inserted nonspecifically anywhere into the genome, or even into the homologous locus without losing the TK gene, the cell will be $FIAU^S$. Candidate cell clones targeted by homologous recombination are thus isolated in the presence of G418 and FIAU. Survivors are picked, expanded, and split. One part is frozen. The other will be analyzed by the mini-Southern technique that allows a single operator to analyze approximately 400 (or more) such cloned cell lines in 1 week.[11] Up to 1000 such clones should be analyzed with the aim of identifying several independent cell clones with a locus targeted by homologous recombination.

It may be that no positive cells are found. In this case the strategy can be redirected to using an insertion-type vector, for which there is evidence that the frequency of homologous recombination may be higher. Alternatively, another exon of the gene may be chosen as a target for disruption. An insertion-type vector could be the same vector as described above but which is linearized within the region of homology rather than outside (Fig. 1, C.2). A site about 1000 to 2000 bases from the Neo (or other) disruption, plus around 2500–4000 bases on both sides, should be

[11] R. Ramírez-Solis, J. Rivera-Pérez, J. D. Wallace, M. Wims, H. Zheng, and A. Bradley, *Anal. Biochem.* **201**, 331 (1992).

adequate as will be shown below for the targeting of the G protein α_{i2} locus. In this strategy the targeted ES cell is obtained in two steps. In the first, hit step, double-stranded break introduced into the region of homology promotes a single recombination event, leading to insertion of the construct into the genome with resulting duplication of the homology. The ES cells that have been hit in this way change from G418^S FIAU^R to G418^R FIAU^S. They are picked from plates grown in the presence of G418, expanded, and split into two portions. One is frozen for further selection; the other is analyzed by mini-Southern blotting to identify clones targeted by insertion. Targeted cell clones are then thawed, and each is analyzed for preservation of the potential to generate chimeric animals by injection into blastocysts followed by implantation into foster mothers and analysis of chimerism of the offspring through coat color inspection 10 days later (see below). Cells able to contribute to a high degree of chimerism are then subjected to a second round of selection in which clones are isolated that, by intrachromosal recombination, have lost the duplicated homology, together with the plasmid and the TK marker, and retained a disrupted gene with the Neo selection marker. This makes them G418^R and FIAU^R (growth in the presence of both G418 and FIAU: run step). Several surviving clones derived from each of the clones that had been targeted by insertion are then expanded and split into two portions, of which one is frozen for further use and the other analyzed by mini-Southern blotting to determine which has the exact predicted genomic structure: a targeted locus and the absence of plasmid, Neo, and TK sequences elsewhere in the genome.

It may be that the recombination frequency at the homologous locus is so low that it would be necessary to screen over 1000 Neo-resistant clones. In this case the introduction of a gap in addition to the double-stranded break will lead to a template-directed repair[12,13] and allow for a screen based on polymerase chain reaction (PCR) between the repaired gap, contributed by the homologous locus, and the disrupting Neo sequences, contributed by the targeting vector (Fig. 2). Such a screen was used to screen for G418^R clones arising from targeting $G_{i2}\alpha$ in AB-1 cells with pIV-1.

It may also be that rather than simply inactivating a gene by insertion of a disrupting selection marker one wishes to introduce a subtle mutation into the murine genome. This can also be accomplished with the use of an insertion vector by mutating the site of interest and placing both the Neo and the TK cassettes outside the region of homology as flanking

[12] P. Hasty, J. Rivera-Pérez, and A. Bradley, *Mol. Cell. Biol.* **11**, 5586 (1992).
[13] V. Valancius and O. Smithies, *Mol. Cell. Biol.* **11**, 4389 (1991).

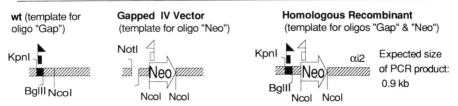

FIG. 2. Strategy for identification of clones based on PCR dependent on gap repair.

sequences that are part of the plasmid carrying the mutated homology.[14] In this case, after the hit step, cells change to G418^R FIAU^S. Loss of the duplicate during the run step will now entail loss of not only the TK but also the Neo resistance gene so that after the run step cells are again G418^S and FIAU^R (Fig. 1, C.3). In this way, one can get mice with no other foreign elements (like enhancers which are usually present in selectable markers) other than the subtle mutation in the genome.

Properly targeted ES cells, AB-1 cells in our case, obtained either by the one-step replacement procedure or the two-step hit-and-run procedure, are then thawed, grown, and injected into the cavity of 3.5-day-old blastocysts of a C57BL/6J mouse and implanted into 2.5-day pseudopregnant foster mothers that are F_1 offsprings of a C57BL/6J × CBA cross following standard techniques that are outside the scope of this chapter. Chimeric 129Sv–C57B/6J F_0 mice should be born. If the chimerism extends to the germ cells, it will be possible to obtain F_1 mice heterozygous for the mutation. By crossing F_0 chimeric males with C57BL/6J females, a crossbred background will be created. Agouti coat color in the F_1 animals indicates that the injected stem cells have contributed to the germ line of the chimeric founder animal. By crossing F_0 chimeric males with 129Sv females, an inbred strain will be created.

Here we describe the experimental details used to inactivate the gene encoding the α subunit of the G_{i2} G protein (α_{i2} or $G_{i2}\alpha$ gene) in the AB-1 embryonic stem cell line utilizing adaptations of techniques and approaches developed originally by Robertson, Bradley, and collaborators.[2,3,11,15]

Materials and Methods

Cell Lines. AB-1 embryonic stem cells, passage 13, and STO feeder cells can be obtained from Dr. Allan Bradley (Institute for Molecular

---

[14] P. Hasty, R. Ramírez-Solis, R. Krumlauf, and A. Bradley, *Nature (London)* **350**, 243 (1991).

[15] E. J. Robertson, in "Teratocarcinomas and Embryonic Stem Cells: A Practical Approach" (E. J. Robertson, ed.), p. 71. IRL Press, Oxford, 1987.

Genetics, Baylor College of Medicine, Houston, TX). Leukemia inhibitory factor (LIF)-producing Chinese hamster ovary (CHO) cells, cell line 8/24 720 LIF-D(.1), are obtained from Genetics Institute (Cambridge, MA).

Mouse Strains. Strain 129Sv and female F_1 offspring of a C57BL/6J × CBA cross have been provided by Dr. Allan Bradley. Strain C57BL/6J is purchased from the Jackson Laboratory (Bar Harbor, ME).

Plasmids. pNeo, a plasmid containing a 1.8-kb *Sal*I/*Xho*I fragment with the neomycin resistance gene flanked by a position-independent *Pol*II (long) promoter and a bovine growth hormone polyadenylation signal (*Pol*II–Neo–bpA cassette), and pKS(*Xho*)MC1TK, a Bluescript KS-derived plasmid carrying a 2.1-kb *Xba*I/*Sac*II fragment with the MC1TK transcription unit formed of an allele of the transcribed sequence of the herpes simplex virus (HSV) thymidine kinase (TK) gene, flanked upstream by a tandem repeat of a polyoma enhancer followed by the TK promoter and downstream by the HSV–TK polyadenylation signal (MC1TK cassette, see Ref. 1), were gifts from Drs. Philippe Soriano and Paul Hasty (Institute for Molecular Genetics, Baylor College of Medicine). pKS* and pKS# are derivatives of Bluescript KS in which the composition of the cloning cassettes are altered as shown in Fig. 3B. pSL1180 is purchased from Pharmacia (Piscataway, NJ). Bluescript KS(+) (pKS) was from Stratagene (La Jolla, CA).

Tissue culture media [Dulbecco's modified Eagle's medium (DMEM) high glucose, DMEM α medium], Hanks' balanced salt solution (HBSS), Dulbecco's phosphate-buffered saline (D-PBS), trypsin solution (0.25% trypsin, 1 m*M* EDTA) and pen/strep (penicillin/streptomycin) solution are from GIBCO (Grand Island, NY). Defined fetal bovine serum (FBS) is from Hyclone (Logan, UT). Penicillin and streptomycin are used at 100 U/ml and 100 μg/ml, respectively.

Tissue Culture

All cell lines are grown at 37° in a humidified atmosphere of air and 5% (v/v) CO_2.

Feeder Layers. Stocks of STO cells are grown in DMEM, 7.5% (v/v) FBS with high (4.5 g/liter) glucose and pen/strep (STO medium) by passaging every 5 days (1 : 20 splits). Prior to use they are grown to subconfluence (~6–7 × 10^6 cells per 100-mm dish), then treated with 10 μg/ml mitomycin C for 2–3 hr. Mitomycin C-containing medium is then aspirated. The cells of a 100-mm plate are washed three times with HBSS (GIBCO) and trypsinized with 2 ml of trypsin solution at room temperature for 2 min. The trypsinizing medium is diluted 3-fold with STO medium, and the cells are suspended by gentle up–down pipetting and collected by centrifugation in a clinical centrifuge. The STO cells are then resuspended in STO medium to a concentration of $0.4-0.5 \times 10^6$ cells/ml and added to gelatinized plates to form the feeder layer proper, at 8.5 ml per 100-mm dish or 150 μl per well of a 96-well titration plate (the latter distributed with a multichannel pipettor). Gelatinized plates (or wells) are prepared by addition of 0.1% (w/v) porcine gelatin (e.g., Sigma, St. Louis, MO, G1890), allowing the plates to stand at room temperature for at least 1 hr and discarding the gelatin solution by aspiration.

CHO-8/24 720 LIF-D(.1) Cells. The CHO-8/24 720 LIF-D(.1) cells are grown according to instructions provided by Genetics Institute in DMEM α medium (minus nucleosides), 10% FCS, 1 mM L-glutamine, pen/strep, and 0.1 μM methotrexate. After growing to confluence, cells are split twice weekly at 1:20 following trypsinization. Medium aspirated prior to trypsinization serves as a source of LIF.

AB-1 Cells. The AB-1 cells are subcultured onto fresh feeder cells (preferentially not older than 1 week) every 3–4 days 1:8 or 1:12 using "strong" trypsinization conditions (e.g., 15 min at 37°) and changing the medium every 1–2 days. The cells need to be refed about 2–4 hr before splitting. After each trypsinization the cells sould be disaggregated either by thoroughly pipetting up and down or, in the case of 100-mm plates, preferentially by using a plastic transfer pipette with a narrow tip (e.g., Saint-Amand Mfg. Co., 232-1S); this is conveniently done with a total volume of 6 ml in a 15-ml plastic tube. Disaggregation can be checked with a microscope. Embryonic stem cell medium (ES medium) is composed of DMEM with high glucose (4.5 g/liter), 2 mM L-glutamine, 10^{-4} M 2-mercaptoethanol, 15% defined FBS, pen/strep, and 200 μl/liter LIF-containing conditioned medium. The doubling time of AB-1 cells is approximately 20–24 hr. Approximately $50-60 \times 10^6$ AB-1 cells per 100-mm dish can be harvested from a "confluent" plate. Overconfluent plates tend to promote the differentiation of stem cells, as does insufficient disaggregation after trypsinization.

Plasmid Manipulations. Standard techniques described by Maniatis *et al.*[16] and Sambrook *et al.*[17] are used.

Electroporation and Plating for Selection

For the experiments reported below AB-1 cells are grown to around 60–80% confluence (as judged by the surface area covered with ES cells)

[16] T. Maniatis, E. F. Fritsch, and J. Sambrook, "Molecular Cloning: A Laboratory Manual." Cold Spring Harbor Laboratory, Cold Spring Harbor, New York, 1982.

[17] J. Sambrook, E. F. Fritsch, and T. Maniatis, "Molecular Cloning: A Laboratory Manual," 2nd Ed. Cold Spring Harbor Laboratory, Cold Spring Harbor, New York, 1989.

in ES medium, split 1:2, trypsinized on the following day, washed once with D-PBS, and resuspended in the same medium at 10^7 cells per 900 μl for electroporation in 0.4-cm cuvettes of the Bio-Rad (Richmond, CA) GenePulser. The DNA for electroporations is prepared by alkaline lysis and banded once in CsCl. Linearization is followed by phenol/chloroform extraction and analytical gel electrophoresis. The replacement vector pRV-6 is linearized at a *Xho*I site outside the region of homology; the insertion vector pIV-1 is cut at the engineered *Not*I site. For electroporation, pRV-6 or pIV-1 DNA (25 μg in 25 μl) is added to the AB-1 suspension and subjected to a pulse of 230 V at a capacitance of 500 μF. After electroporation the cells are kept at room temperature for 5–10 min, diluted 10-fold with ES medium, and then plated onto fresh STO feeder plates at a density of 10^7 electroporated cells per 100-mm dish. The plating efficiency of electroporated cells is approximately 70% of that of nonelectroporated cells. Selection with G418 (180 μg/ml active ingredient, GIBCO) and, when applicable, FIAU (0.2 μM; Bristol-Myers) is begun 1 or 2 days postelectroporation.

Mini-Southern Blotting

The procedure described is an adaptation of the method originally reported by Ramírez-Solis *et al.*[11]

Samples from Cells Grown in 96-Well Plates. Cells to be analyzed are grown to (over-)confluence in ES medium. The medium is aspirated and cells are washed once with HBSS. HBSS is aspirated and replaced with 50 μl of cell lysis buffer [100 mM NaCl, 50 mM Tris-HCl, pH 7.5, 10 mM EDTA pH 8.0, 0.5% sodium dodecyl sulfate (SDS), 0.8 mg/ml proteinase K (Boehringer-Mannheim, Indianapolis, IN; diluted from a fresh stock solution of 40 mg/ml)]. The microtitration plate is incubated in a closed, wet atmosphere overnight at 65°. DNA is then precipitated for 30 min at room temperature by addition of 50 μl 2-propanol followed by centrifugation at 2400 rpm (1200 g) for 5 min at room temperature in a swinging-bucket rotor (Beckman TH-4) fit with adaptors to hold 96-well microtitration plates. The liquid in the wells is discarded by slow and careful inversion of the plate. The precipitated DNA is washed three times with 70% ethanol by filling the drained wells using a wash bottle and emptying by inversion of plates. The contents of the wells are allowed to "dry" for 20 min at room temperature (see comments below). DNA is digested by addition in 30 μl of 10 units of restriction enzyme in appropriate buffer with 0.2 mg/ml bovine serum albumin (BSA), 1 mM spermidine, and 20 μg/ml RNase A (Boehringer Mannheim) and incubation overnight in a humid atmosphere at 37°. The DNA is further digested by a second addition of 10 units of restriction enzyme in 5 μl of the same buffer for 5 or more hr also at 37° in a wet atmosphere to prevent evaporation. Lower amounts of enzyme have been used successfully in experiments not reported here. The sample is then readied for electrophoresis by addition of 7 μl of 6× sample buffer (1× sample buffer is 10% (v/v) glycerol, 10 mM EDTA, and bromphenol blue). The complete sample is used for electrophoresis.

Mouse Tail Biopsies. Samples (0.5 to 1 cm long) are incubated overnight at 65° on a rocking bottom platform in 1.5-ml Eppendorf tubes with 300 μl cell lysis buffer (see above) plus 1 mg/ml proteinase K. The mixture is extracted with 300 μl phenol (optional), 300 μl phenol/chloroform (1:1), and 300 μl chloroform/isoamyl alcohol (24:1), and precipitated with 250 μl 2-propanol. The precipitated DNA is then treated as above, namely, washed twice with 70% ethanol, dried, digested, and prepared for electrophoresis with 6× sample buffer. Approximately 10–25% of the sample is used for electrophoresis.

Electrophoresis. Tandem electrophoresis is performed in a single 400 or 500 ml gel slab of 0.7% Seakem ME agarose (FMC, Rockland, ME; 20 cm wide × 25 cm long × 0.8 cm thick) made in 500 ml of 1× TAE (1× TAE is 40 mM Tris–acetate and 2 mM EDTA, pH 8.0) with 0.1 mg/ml ethidium bromide, fitted with two sets of sample wells separated by 12 cm. Wells are cast with custom-made 42-tooth combs that give wells with a capacity of about 40 μl. The buffer in the chamber (1× TAE) contains the same concentration of ethidium bromide as the gel.

Transfer. After electrophoresis the gel is photographed, and the areas of interest are cut out and soaked sequentially, twice for 5 min in 0.25 M HCl and once for 20 min in 0.4 M NaOH. A GeneScreenPlus nylon membrane (NEN, Boston, MA) is then placed on top of the gel slabs, and the DNA is transferred to the membrane by forced capillary flow at room temperature overnight in the presence of 0.4 M NaOH using standard procedures. The membrane is then submerged into 2× SSC (1× SSC is 150 mM NaCl, 15 mM sodium citrate)/0.2 M Tris-HCl, pH 7.5, and allowed to dry at room temperature.

Preparation of Hybridization Probes. Labeled copies of the probes, with specific activities of $0.5-1.5 \times 10^9$ counts/min (cpm)/μg, are prepared in 1.5-ml screw-capped microcentrifuge tubes by random priming with random hexamers[18] as follows. Template DNA, 100 ng in 12 μl, is heated to 100° for 3 min and chilled rapidly on ice. The following are then added in sequence after all the sample in the tube is collected at bottom with a short spin in the microcentrifuge: 6 μl of 5× labeling buffer [5× labeling buffer is made up of 2 parts 100 mM Tris-HCl, pH 8.0, 50 mM MgCl$_2$, 10

---

[18] A. P. Feinberg and B. Vogelstein, *Anal. Biochem.* **132**, 6 (1983).

mM dithiothreitol (DTT), 1 M sodium HEPES, pH 6.6, and 100 μM dATP, dTTP, and dGTP, and one part 50 A_{260} units (260 nm, 1 cm path) random hexamers (e.g., Pharmacia, 29-2166-01) in 654 μl of 0.1× TE (1× TE is 100 mM Tris-HCl, pH 8.0, 1 mM EDTA)], 11 μl [α-^{32}P]dCTP (>3000 Ci/mmol, 10 mCi/ml), and 1 μl Klenow fragment of DNA polymerase I (Boehringer Mannheim, labeling grade). The contents are mixed by up-and-down pipetting, and tubes are placed on ice for 5 min and incubated for 2 to 15 hr at 16°. Reactions are stopped by the addition of 5 μl of 400 mM EDTA, 5 μl of 1 mg/ml autoclaved herring or salmon sperm DNA, and 60 μl water. After mixing, the DNA is then precipitated with 50 μl of 7.5 M ammonium acetate and 375 μl absolute ethanol at $-70°$ for 15 to 30 min. The DNA is collected by centrifugation in a microcentrifuge for 15 min at 4°. The supernatant is collected in a separate tube, and the labeled DNA is resuspended in an equal volume of water. Radioactivity present in 2-μl aliquots of the supernatant and the resuspended DNA are used to calculate the labeling efficiency. The remainder of the labeled DNA is boiled, chilled, and added to the hybridization pouch.

Prehybridization/Hybridization/Wash. The membrane is placed in a sealable pouch (Kapak, Fisher) and wetted well with a prehybridization buffer composed of 1.5× SSPE (20× SSPE is 175.3 g/liter NaCl, 27.6 g/liter NaH$_2$PO$_4$ · H$_2$O, and 7.4 g/liter EDTA, pH 7.5), 1% SDS, 0.5% (w/v) fat-free milk powder (Carnation), and 200 μg/ml boiled single-stranded herring or salmon sperm DNA (prepared by autoclaving for 30 min). After 3 hr at 65°, excess liquid is removed, and 50–100 ng of labeled probe is added to a bag holding typically a membrane of 22 × 22 cm. Hybridization is allowed to proceed overnight at 65°. After hybridization the membrane is removed from the sealed bag and washed at increasing stringencies ending with 0.2× SSC with 0.5% SDS at 65° for 30 min. Kodak (Rochester, NY) XAR 5 films are exposed at $-70°$ with two intensifying screens.

Inactivation of $G_{i2}\alpha$ Gene by Construction of Insertion Vector IV-1 and Replacement Vector RV-6 to Inactivate Murine α_{i2} Gene by Disruption of Open Reading Frame with Neomycin Resistance Gene

Figure 3 depicts the detailed strategy followed for the construction of the insertion vector pIV-1 (also pIV-Ai2.x3). This vector contains a portion of the murine α_{i2} gene disrupted at a *Nco*I site in exon 3 by a 1.8-kb Neo cassette. The Neo cassette is flanked upstream by 6.7 kb of α_{i2} sequence and downstream by 3.2 kb of α_{i2} sequence. In the case of the insertion vector there is a 0.1-kb deletion 0.8 kb upstream from the *Nco*I site in exon 3 used to insert the Neo cassette, reducing length of the 5' homology to 6.6 kb. The plasmid backbone of the vector is pKS and contains in addition the MC1TK cassette that is used to counterselect for the loss of foreign sequences other than the Neo disruption. The replacement vector pRV-6 was constructed in similar manner but omitting the creation of the gap and addition of a NotI site.

Figure 4 shows diagrams of restriction fragment lengths expected from the pIV-1 vector, and from the homologous locus, before targeting, after insertion of the vector creating a duplicate with the Neo cassette in the 5' duplicate or in the 3' duplicate, and after excision of the duplicate together with plasmid and TK cassette sequences.

Selection of Targeted AB-1 Cell Clones

An overview of the general procedure used to select, propagate, and identify targeted AB-1 cell clones is presented in Fig. 5. If the strategy calls for using a replacement vector, the procedure entails electroporation, selection of G418-resistant colonies, freezing of candidate cell clones while the DNA is recovered from replica plates and analyzed, confirmation of the genomic structure of presumed targeted clones, thawing of the clones, expansion, and injection into embryos. The selection and expansion process prior to freezing takes approximately 3 weeks after electroporation. The expansion of the replicated cells and analysis by mini-Southern blotting takes another 2–3 weeks. The thawing and final expansion of the presumed targeted clones and second genomic analysis take another 2–3 weeks. A reasonable expectation is that it will take at least 2–3 months from electroporation to embryo injection if the targeting frequency is sufficiently high so that several targeted clones are obtained and there are no problems with the genomic sequences chosen for targeting. If the hit-and-run strategy is used, the procedure is extended by another 2–3 months.

Electroporation and First Plating. Electroporated cells are placed onto fresh feeder layer plates. The plates may be of the 100-mm size or 6-well plates. The former offer the advantage of easier medium changes, whereas the latter allow for an early partitioning of the cells and ensure proper identification of "independent" clones. Selection with G418 (180 μg/ml active ingredient, GIBCO) and, when applicable, FIAU (0.2 μM; Bristol-Myers) is begun 1 or 2 days postelectroporation. The medium is changed daily for the first week and every second day thereafter or when it turns yellow. Plates are inspected for the appearance of colonies, which are picked between days 9 and 14 after electroporation.

First Passage. We recommend that on the day prior to each passage fresh feeder layer plates be prepared, in this case 96-well feeder layer plates. Feeder layers that have been prepared more than 1 week in advance should not be used if blastocyst injection is planned. On the day of subcul-

A. Structure of αi2 gene and location of Exon 3 NcoI site

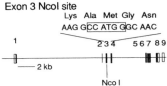

C. Partial restriction map and key restriction fragments from λEMBL3-αi2

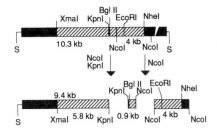

B. Cloning plasmids; Neo and TK plasmids

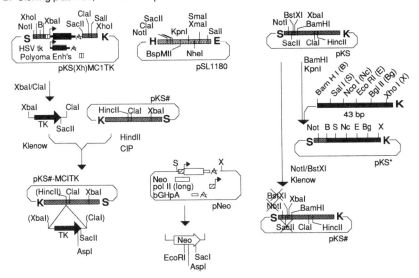

FIG. 3. Construction of the insertion vector pIV-1 (pIV-Ai2.x3). (A) Intron/exon structure of $G_{i2}\alpha$ gene. (B) Cloning plamids, and plasmids with TK and Neo cassettes, showing relevant restriction sites. (C) Relevant restriction map of the $G_{i2}\alpha$ insert isolated from a λEMBL3 murine (BALB/c) genomic DNA library (same scale as A). The *XmaI/NcoI* (exon 3) fragment includes exon 2 but not exon 1. The *NcoI* (exon 3)/*NcoI* fragment includes exon 4 but not exon 5. Hatched sequences are to be found in the final construct, black sequences are not. (D)–(F) Assembly of pIV-1. The composition of the *KpnI/BglII* linker with an internal *NotI* site was as follows: sense strand 5'-CAGCGGCCGCA-3', antisense strand 5'-GATCTGCGG-CCGCTGGTAC-3'. In (A)–(F) α_{i2} genomic sequences are drawn according to the scale shown (A); the linker, plasmid, TK cassette, and Neo cassette sequences are not.

D. Assembly of 5' Homology with 0.1 kb Deletion (Gap) and New Unique NotI Site

E. Assembly of 3' Homology and Disruption with Neo Selection Marker

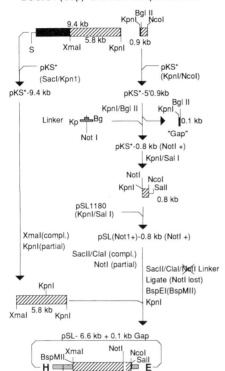
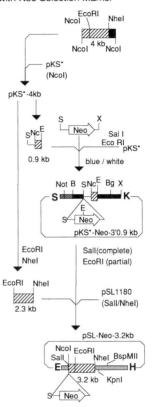

F. Final Assembly of Insertion Vector pIV-Ai2.x3 (pIV 1.0)

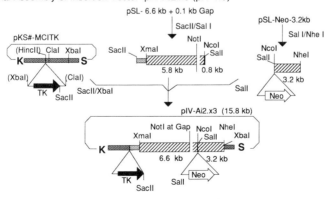

FIG. 3. (*continued*)

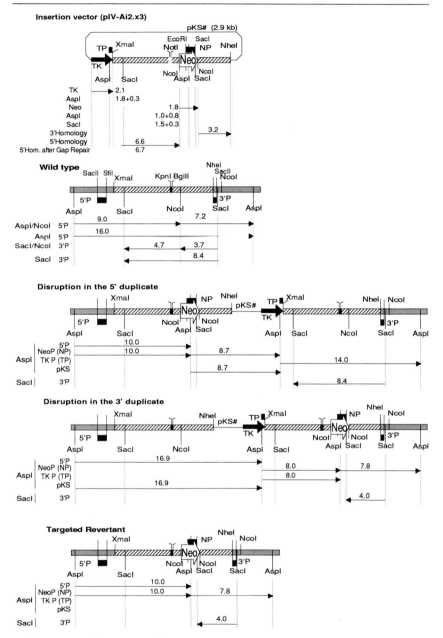

FIG. 4. Restriction maps of insertion vector pIV-1 and of the $G_{i2}\alpha$ locus before vector insertion, after vector insertion, and after excision of the duplicate. Hatched boxes represent $G_{i2}\alpha$ sequences present in both genome and insertion vector. Gray boxes represent $G_{i2}\alpha$

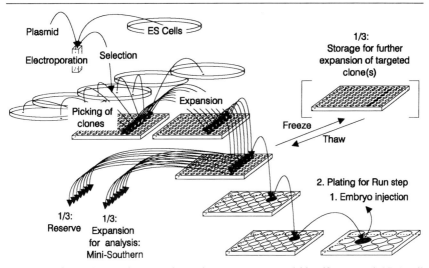

FIG. 5. General procedures used to select, propagate, and identify targeted AB-1 cell clones.

ture the STO medium is aspirated from the wells and ES medium is added at 150 μl/well.

The plate or well with the colony(ies) to be picked is washed with HBSS and flooded with HBSS (5 ml per 100-mm plate). Colonies are then rapidly aspirated from the plate held at an angle into the tip of a P-20 Pipetman (Gilson, Middleton, WI) or equivalent, set at 2 μl, and transferred into a 20-μl microdrop of trypsinizing solution that had been placed on the bottom of a well of an intermediary collecting 96-well microtitration plate. Each colony is placed in a separate well so that individual colonies from the primary selection process acquire an identifying coordinate in a 96-well plate. One operator can pick and process about 100 colonies in 2 hr. Cells in the trypsinizing solution are then diluted with 50 μl of ES medium removed from a positionally equivalent well of the recipient feeder

---

sequences not present in the insertion vector. Black boxes above or below represent restriction fragments used as probes. TP and NP represent TK and Neo probes, respectively; 5'P and 3'P, external 5' and 3' probes, respectively. The maps predict that on digestion with SacI and probing with 3'P, wild-type DNA as well as with DNA from a cell targeted in the hit step in such manner as to have the Neo cassette in the 5' duplicate only will yield a fragment of 8.4 kb. On the other hand, cells targeted in the hit step in such manner as to have the Neo cassette in the 3' duplicate, as well as cells that have undergone target conversion or have been carried through the run step, should yield a 4.0-kb fragment instead of the 8.4-kb fragment.

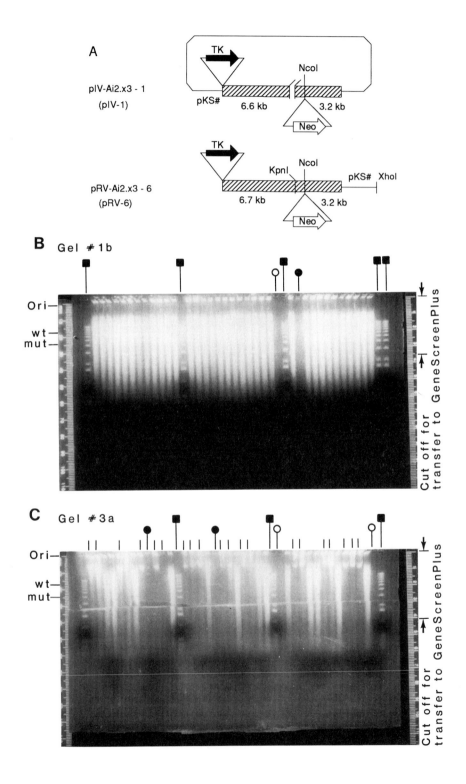

layer plate, disaggregated thoroughly by repeated up-and-down pipetting using a multichannel pipettor, and transferred into the well of the recipient feeder layer plate from which the 50 μl had been withdrawn.

Second Passage. In preparation, STO medium is removed from fresh (second) recipient 96-well feeder layer plates that had been prepared the day before, and replaced with 150 μl/well of ES medium. The cells to be passaged and now grown in G418- and FIAU-free ES medium (no drug selection) for 4 (or 5) days are inspected for cell density under the microscope. If close to confluence they will be split 1:2 into fresh wells (and thus duplicated); if not (as usually occurs) they will be passaged in their entirety to a new well. To this end the cells are refed 2–4 hr before splitting, media are aspirated, and the wells are rinsed once with HBSS. The cells are then trypsinized for 15 min at 37° with 50 μl trypsinizing solution. After dilution with 50 μl of ES medium taken from the positionally equivalent wells of the recipient feeder layer plate, the cells are thoroughly resuspended, and all or an aliquot are transferred back to the wells of the recipient feeder layer plate from which the medium had been taken.

Cells in this plate are expanded for 3–4 days to confluence, rinsed with HBSS, trypsinized with 50 μl trypsin solution, and diluted with 100 μl ES medium. Two 50-μl aliquots are transferred each to a separate gelatinized new recipient 96-well plate without feeder cells but with 150 μl ES medium, thus creating two replicas. The remainder of the cells (50 μl) are frozen after receiving 50 μl of 2× freezing medium [60% DMEM with high glucose, 20% FBS, and 20% dimethyl sulfoxide (DMSO)], followed by overlaying with 50 μl autoclaved light paraffin oil [EM Science (Merck), Gibbstown, NJ, PX0047-1]. To freeze, the plate is covered, sealed into a Kapak sealable pouch, and placed into a small Styrofoam box; the Styrofoam box is then placed overnight into a −70° freezer, where the cells can stay or from where they can be shifted to liquid N_2 if long-term storage is foreseen.

Analysis of Recombination Events in Isolated Cell Clones. Cells in the replica plates are grown to confluence for 3–4 days. The STO cells

---

FIG. 6. Structure of targeting vectors used and photographs of two separate *Sac*I digests. (A) Comparison of structures of replacement vector pRV-6 and insertion vector pIV-1. (B) DNA samples were from G418/FIAU-resistant cells obtained after transfection with replacement vector pRV-6 (gel 1b). (C) DNA samples are from G418-resistant cells obtained after transfection with insertion vector pIV-1 (gel 3a). Closed squares, 1 kb DNA ladder (BRL, Gaithersburg, MD); open circles, no DNA loaded; closed circles, no or insufficient DNA recovered from the clone; |, no or incomplete digestion of DNA.

are omitted because of the low split associated with the last passage which leaves sufficient cells to allow for continued growth and minimizes contribution of STO cell DNA to the analysis. One of the plates serves as a reserve in case of experimental failure. This plate may also be used for another restriction digest in order to confirm the structure of the targeted locus before the clones have to be defrosted or while they are being expanded. The cells on the reserve plate are lysed in the presence of proteinase K as described previously and then stored at 4°. The cells in the wells of the other plate are analyzed by mini-Southern blotting as described above.

Figure 6 shows the photographs of two *Sac*I digestions from two experiments. In one experiment (Fig. 6B), in which targeting of the $G_{i2}\alpha$ gene with the replacement vector pRV-6 was tested after selection with G418 (Variant C.1 of Fig. 1), the DNAs were from all 36 samples of a 96-well plate that were loaded and properly digested as seen by the rather homogeneous smear (or heterogeneous mixture) of the restriction fragments. In the other experiment (Fig. 6C), in which targeting of cells transfected with insertion vector pIV-1 and selected with G418 was tested, the digestions had failed in many of the wells. Failure could have been due to the quality of the enzyme batch used that time, overdrying of the DNA prior to digestion after rinsing with 70% ethanol, or too much residual ethanol in the wells at the time of the addition of restriction enzyme. In other cases, DNA was simply lost in the precipitation or one of the decantation steps.

Figure 7 shows an autoradiogram of a GeneScreenPlus membrane onto which we had transferred the *Sac*I DNA digests of 240 clones from two experiments. In one experiment we had transfected AB-1 cells with the replacement vector pRV-6 and then selected for G418 resistance (gels 1, 2, and 3; gel 1b of Fig. 6 is part of this experiment). From 144 samples loaded, 138 could be analyzed. Of those, 8 showed a restriction map consistent with recombination between the replacement vector and the homologous locus of the AB-1 genome. Any one of these clones would be a candidate for injection into blastocysts if targeting is confirmed with a 5'-flanking probe and with one or two more restriction enzymes, and if it is shown that it is devoid of cryptic Neo, TK, and plasmid sequences elsewhere in the genome. One additional random integration of the targeting vector may not cause a problem since it can be bred out later if there is no linkage to the target locus. In the other experiment we had transfected AB-1 cells with the insertion vector pIV-1 (Fig. 3 and Fig. 6A) and selected for insertional events by selecting for G418-resistant clones (gel 3a, Fig. 6C, is part of this experiment). Of 96 samples loaded, 56 could be successfully analyzed. Of those, two clones showed the restriction fragment ex-

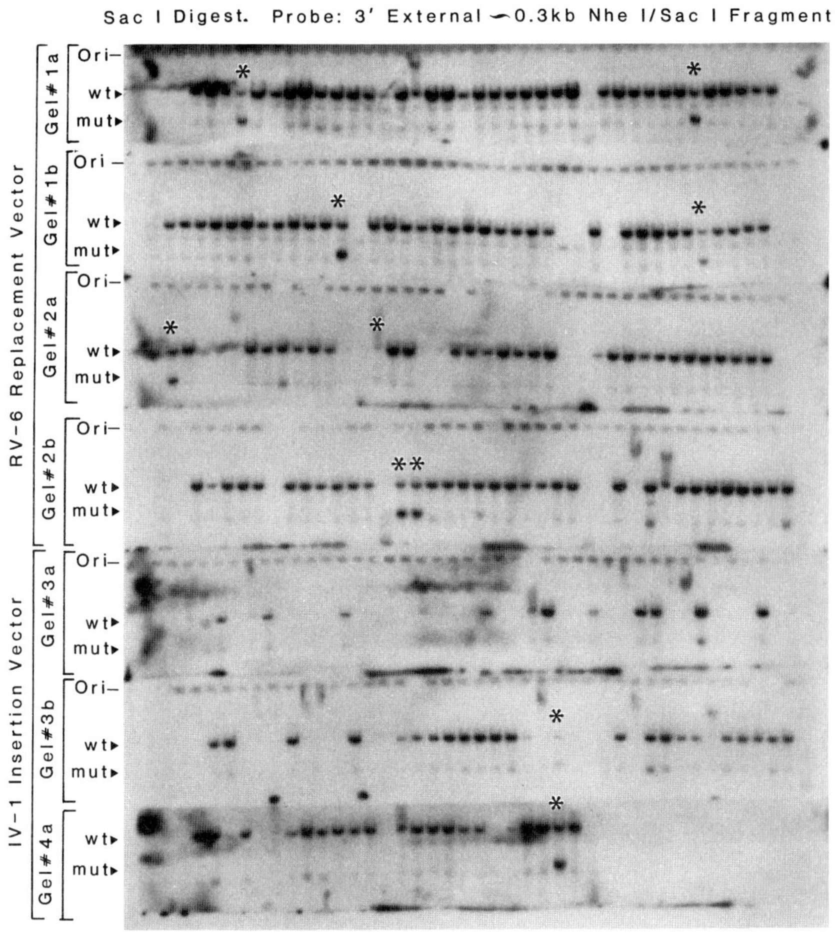

FIG. 7. Southern analysis of SacI digests using an external 3' $G_{i2}\alpha$ probe. Gels 1a through 2b are from G418/FIAU-resistant cells obtained after transfection with replacement vector pRV-6. Gels 3a through 4a are from G418-resistant cells obtained after transfection with insertion vector pIV-1. Asterisks (*) denote cell clones in which one of the $G_{i2}\alpha$ alleles has been targeted by homologous recombination.

pected from insertion at the homologous locus. There may have been more targeting events because the probe used does not test for recombinants in which the Neo marker is in the 5' duplicate, which frequently occurs with this construct (see Fig. 4).

Third and Fourth Passages: Cell Rescue (Thawing) and Expansion. The 96-well plate(s) with cell clones identified by mini-Southern blotting as possibly having been correctly targeted are taken from the $-70°$ freezer or from storage in liquid N_2, removed from the plastic bag in which they had been sealed, and placed into an air–CO_2 incubator at 37°, where they are allowed to thaw completely for 15–30 min. The entire contents of the desired wells are then transferred into the wells of a fresh 24-well feeder cell plate prepared as above and to which 1.5 ml of ES medium had been added just prior to addition of the thawed cells. Cells are then passaged every 3–4 days. The first passage (third passage since electroporation) is without splitting, simply a trypsinization at 37° for 15 min with 0.5 ml of trypsinizing solution, followed by dilution to 1/2 with 0.5 ml of ES medium from a new 24-well plate with fresh feeder cells and 1 ml of fresh ES medium, followed by transfer of the total into the well from which the ES medium had been taken. The next passage (fourth since electroporation) is into a 6-well plate with fresh feeder cells and 2.5 ml ES medium.

Genomic DNA for further analysis of targeted clones can be prepared from 6-well plates, lysing the cells on the plate with lysis buffer (see above) for 20 min at room temperature and then adding proteinase K to a concentration of 0.8 mg/ml. Subsequently the DNA is extracted as described above for the tail biopsies.

Preparation for Injection into Blastocysts

Cells from a 6-well (or 24-well) plate are trypsinized 1 or 2 days after plating to ensure that the plates are not overgrown, in which case the viability of the cells may be decreased. Further experimental details of embryonic stem cell handling as well as the embryo injection aspect of the work can be found in Robertson[15] and Bradley.[3] Use of the PCR based on gap repair to detect homologous recombination at the $G_{i2}\alpha$ gene locus can be found in Rudolph et al.[20] Examples of Southern blots obtained on analyzing several cells isolated with a $G_{i2}\alpha$ gene targeted with the pIV-1 vector can be found in Rudolph et al.[19,20]

Acknowledgments

We are especially indebted to Dr. Paul Hasty, and members of the Bradley laboratory for sharing expertise. We thank Dr. Philippe Soriano for pNeo with the PolII–Neo–bpA cassette, Dr. Paul Hasty for the pKS(Xho)MC1TK cassette, and Dr. David Nelson for the BALB/c-derived λEMBL3 library. U.R. was supported by a fellowship from the Deutsche Forschungsgemeinschaft. L.B. is supported by grants from the National Institutes of Health. A.B. is an Associate Investigator of the Howard Hughes Medical Institute.

[19] U. Rudolph, P. Brabet, J. Kaplan, P. Hasty, A. Bradley, and L. Birnbaumer, *J. Recept. Res.* **13,** 619 (1993).
[20] U. Rudolph, P. Brabet, P. Hasty, A. Bradley, and L. Birnbaumer, *Transgenic Res.* **2,** 345 (1993).

[30] G-Protein Assays in *Dictyostelium*

By B. EWA SNAAR-JAGALSKA and PETER J. M. VAN HAASTERT

Introduction

The eukaryotic microorganism *Dictyostelium* provides a convenient model to investigate signal transduction. *Dictyostelium* is a soil amoeba which feeds on bacteria. Exhaustion of the food supply induces a developmental program: cells aggregate to form a multicellular structure in which they differentiate to either dead stalk cells or viable spores. In this system extracellular cAMP acts as a hormonelike signal inducing chemotaxis, morphogenesis, and cell differentiation.[1]

The sensory transduction mechanism involving extracellular cAMP resembles hormone signal transduction in mammalian cells.[2] cAMP is detected by surface receptors, which activate several G proteins and second messenger enzymes including adenylyl cyclase (adenylate cyclase), guanylyl cyclase (guanylate cyclase), and phospholipase C. In contrast to higher organisms, *Dictyostelium* has a short and simple life cycle. Cells are easy to grow in large quantities, and the small haploid genome makes it easier to perform both classic as well as molecular genetics.[3] Many mutants have been isolated that are defective in cell aggregation, chemotaxis, morphogenesis, or differentiation. Several mutants appear to be defective in signal transduction; in some cases the mutated genes have been identified as G proteins,[4–7] the regulatory subunit of protein kinase A (PKA),[8] or a cGMP phosphodiesterase.[9–11] Multiple

---

[1] W. F. Loomis, "The Development of *Dictyostelium discoideum*." Academic Press, New York, 1982.
[2] P. J. M. Van Haastert, P. M. W. Janssens, and C. Erneux, *Eur. J. Biochem.* **195,** 289 (1991).
[3] P. Devreotes, *Science* **5,** 242 (1989).
[4] F. Kesbeke, B. E. Snaar-Jagalska, and P. J. M. Van Haastert, *J. Cell Biol.* **107,** 521 (1988).
[5] B. E. Snaar-Jagalska, F. Kesbeke, M. Pupillo, and P. J. M. Van Haastert, *Biochem. Biophys. Res. Commun.* **156,** 757 (1988).
[6] M. Pupillo, A. Kumagai, G. S. Pitt, R. A. Firtel, and P. N. Devreotes, *Proc. Natl. Acad. Sci. U.S.A.* **86,** 4892 (1989).
[7] A. Kumagai, M. Pupillo, R. Gundersen, R. Miake-Lye, P. N. Devreotes, and R. A. Firtel, *Cell (Cambridge, Mass.)* **57,** 265 (1989).
[8] M. N. Simon, O. Pelegrini, M. Veron, and R. R. Kay, *Nature (London)* **356,** 171 (1992).
[9] F. M. Ross and P. C. Newell, *J. Gen. Microbiol.* **127,** 339 (1981).
[10] P. J. M. Van Haastert, M. M. Van Lookeren Campagne, and F. M. Ross, *FEBS Lett.* **147,** 49 (1992).
[11] P. C. Newell and G. Lin, *BioEssays* **14,** 373 (1982).

genes that code for components of the signal transduction cascade have been cloned, including genes for four surface cAMP receptors,[12–14] eight G-protein α subunits,[6,7,15–17] one G-protein β subunit, two adenylyl cyclases,[18,19] one phospholipase C,[20] and over twenty protein kinases.[21–28] In *Dictyostelium* inactivation of a gene is easily obtained by homologous recombination. This has been done for a number of the genes listed above, and more cell lines are currently under construction. Cells lacking a specific protein may allow the identification of the function of that protein in signal transduction.

Biochemical assays are indispensable tools for the analysis of both wild-type and mutant cells. In this chapter we describe G-protein assays used in *Dictyostelium*. Because G proteins act as a transducer between surface receptor and effector enzymes, we describe first *in vivo* assays for cAMP binding to surface receptors and cAMP-induced second messenger responses. Subsequently, we describe assays to investigate the interactions between the cAMP receptor, G protein, and effector enzymes *in vitro*. These interactions are unstable in cells and especially in lysates or membranes. It is important to complete the experiments as soon as possible after preparation of the cells or lysates. Therefore, we have divided each procedure into steps that can be completed before lysing the cells, the assay itself, and procedures to complete the experiment.

---

[12] P. S. Klein, T. J. Sun, C. L. Saxe III, A. R. Kimmel, R. L. Johnson, and P. N. Devreotes, *Science* **241,** 1467 (1988).
[13] C. L. Saxe III, R. L. Johnson, P. N. Devreotes, and A. R. Kimmel, *Dev. Genet.* **12,** 6 (1991).
[14] A. R. Kimmel and P. N. Devreotes, personal communication.
[15] L. Wu and P. N. Devreotes, *Biochem. Biophys. Res. Commun.* **179,** 1141 (1991).
[16] J. A. Hadwiger, T. M. Wilkie, M. Strathmann, and R. A. Firtel, *Proc. Natl. Acad. Sci. U.S.A.* **88,** 8231 (1991).
[17] M. Pupillo and P. N. Devreotes, manuscript in preparation.
[18] P. Lilly, L. Wu, D. L. Welker, and P. N. Devreotes, *Genes & Dev.* **7,** 986 (1993).
[19] G. S. Pitt, N. Milona, J. Borleis, K. C. Lin, R. R. Reed, and P. N. Devreotes, *Cell* (*Cambridge, Mass.*) **69,** 305 (1992).
[20] A. L. Drayer and P. J. M. Van Haastert, *J. Biol. Chem.* **267,** 18387 (1992).
[21] R. Mutzel, M. L. Lacombe, M. N. Simon, J. De Gunzburg, and M. Veron, *Proc. Natl. Acad. Sci. U.S.A.* **84,** 6 (1987).
[22] J. L. Tan and J. A. Spudich, *Mol. Cell. Biol.* **10,** 3578 (1990).
[23] E. Bürki, C. Anjard, J. C. Scholder, and C. D. Reymond, *Gene* **102,** 57 (1991).
[24] B. Haribabu and R. P. Dottin, *Proc. Natl. Acad. Sci. U.S.A.* **88,** 1115 (1991).
[25] S. Ravid and J. A. Spudich, *Proc. Natl. Acad. Sci. U.S.A.* **89,** 5877 (1992).
[26] C. Anjard, S. Pinaud, R. R. Kay, and C. D. Reymond, *Development* (*Cambridge, UK*) **115,** 785 (1992).
[27] S. K. O. Mann, W. M. Yonemoto, S. S. Taylor, and R. A. Firtel, *Proc. Natl. Acad. Sci. U.S.A.* **89,** 10701 (1992).
[28] P. N. Devreotes, M. Veron, J. G. Williams, personal communication.

General Methods and Materials

General Materials

[2,8-^{3}H]cAMP (34 Ci/mmol; 1.26 TBq/mmol), Amersham (Arlington Heights, IL)
[2,8-^{3}H]cGMP (25 Ci/mmol; 0.91 TBq/mmol), Amersham
[x-^{3}H]Ins(1,4,5)P$_3$ (40 Ci/mmol; 1.48 TBq/mmol), Amersham
[^{35}S]GTPγS (1320 Ci/mmol; 48.8 TBq/mmol), New England Nuclear (Boston, MA)
[γ-^{32}P]GTP (43 Ci/mmol; 1.59 TBq/mmol), New England Nuclear
cAMP, 2'-deoxy-cAMP, (Sp)-cAMPS, cGMP, ATP, GTP, AppNHp, ATPγS, GTPγS, GTPβS, Boehringer (Mannheim, Germany)
Dithiothreitol (DTT)
Silicone oils AR 20 and AR 200 Wacker-Chemie (Munich, Germany)
Scintillator 299, Packard
Nuclepore polycarbonate filters, 3 μm pore size

Dictyostelium Culture Conditions

The strains that have been used in G-protein studies are wild-type NC4, axenic strains AX3 or AX4, several mutant strains derived from NC4, and transformants derived mainly from AX3. Axenic strains are grown at 22° in liquid medium containing, per liter, 14.3 g peptone, 7.15 g yeast extract, 10 g glucose, 0.49 g KH$_2$PO$_4$, and 1.36 g Na$_2$HPO$_4 \cdot$ 12H$_2$O (HG5 medium; an economic version of HL5 medium[29]). Nonaxenic strains are grown in coculture with *Klebsiella aerogenes* on solid medium containing 3.3 g/liter glucose, 3.3 g/liter peptone, and 10 mM Na$_2$HPO$_4$/KH$_2$PO$_4$ buffer, pH 6.5 (PB). Axenically grown strains are harvested in the late logarithmic phase and bacterial grown strains just before clearing of the bacterial lawn. Cells are washed three times in PB by repeated centrifugation at 300 g for 3 min and resuspended in PB.

The vegetative cells acquire aggregation competence by starvation. Cells are starved either by shaking in suspension for 4–6 hr (10^7 cells/ml in PB at 150 rpm and 22°), or by incubation on nonnutrient agar (1.5% agar in PB; 2 × 10^8 cells per plate). These plates are incubated at 22° for 4–6 hr or at 6° for 16 hr.

Preparation of Membranes

The following is a general procedure for the rapid preparation of lysates or membranes that preserve the interaction between G proteins and sur-

---

[29] S. M. Coccuci and M. Sussman, *J. Cell Biol.* **45**, 399 (1970).

face receptors or effector enzymes. For some assays the crude lysate and/or other buffers are used.

1. *Dictyostelium* cells (NC4, AX3, or other) are grown and harvested as described above.
2. After starvation cells are washed twice with PB, once with AC^- buffer (40 mM HEPES–NaOH, 0.5 mM EDTA, pH 7.7), and resuspended in AC^+ buffer (AC^- supplemented with 250 mM sucrose) to a density of 1–2 × 10^8 cell/ml. Cells and solutions are kept at 0° during the procedure.
3. Homogenization is performed by pressing the cell suspension through a Nuclepore filter (pore size 3 μm).
4. The lysate is centrifuged at 14,000 g for 5 min at 4°; the pellet is washed once with PB, and the final pellet is resuspended in PB to a density equivalent to 2 × 10^8 cells/ml. The membranes are kept on ice during the experiment, which is completed within 1 hr after membrane preparation.

Isotope Dilution Assays for cAMP, cGMP, and Inositol 1,4,5-Trisphosphate

Isotope dilution assays are used for the determination of cAMP, cGMP, or inositol 1,4,5-trisphosphate [Ins(1,4,5)P_3] levels. The assays are based on the competition between a fixed quantity of radioactive tracer and the unlabeled second messenger from the sample for binding to a specific antibody or receptor. The amount of labeled tracer bound to the binding protein is inversely related to the amount of second messenger in the sample. Commercial assay kits are available for each ligand. We have used the cGMP radioimmunoassay, the cAMP-binding protein, and the Ins(1,4,5)P_3-binding protein kits from Amersham. For optimal use, we have modified the protocols of these commercial kits as follows: the reagents are dissolved as described by the manufacturer, but in the assay all volumes are divided by five. When many assays are to be performed, it is suggested to prepare assay kits in the laboratory.

Outline for Preparation of Isotope Dilution Assay Kits. The main ingredients in all kits are radioactive tracer of high specific activity (>25 Ci/mmol; >1 TBq/mmol) and a protein that binds the radioactive tracer with high specificity and affinity. The binding proteins are diluted in the appropriate buffers such that about 30% of the radioactive tracer is bound.

The anti-cGMP antiserum is prepared in rabbits by immunization of cGMP coupled to bovine serum albumin (BSA) as described by Steiner et al.[30] The concentration of cGMP that induces half-maximal displacement of tracer cGMP should be below 10 nM. Specificity is analyzed using different concentrations of cAMP, GTP, GDP, GMP, and ATP. A good 20-ml bleed provides sufficient materials for about 40,000 assays.

The regulatory site of cAMP-dependent protein kinase type I is used as the cAMP-binding protein, and it is isolated from beef muscle as described.[31] In 3 days the binding protein is isolated from 500 g fresh beef shoulder sufficient for 200,000 assays.

The Ins(1,4,5)P$_3$-binding protein is isolated from beef liver or adrenal glands as described elsewhere.[32] In 4 hr sufficient material is isolated from 500 g liver for 20,000 assays.

Materials

cGMP assay buffer: 200 mM K$_2$PO$_4$, 10 mM EDTA, 10 mM EGTA, pH 7.0

cAMP assay buffer: 100 mM K$_2$PO$_4$, 10 mM EDTA, 2 mg/ml BSA, 3 mM NaN$_3$, pH 7.0

Ins(1,4,5)P$_3$ assay buffer: 100 mM Tris-HCl, 4 mM BSA, 4 mM EDTA, pH 9.0

Tracers: 100,000 counts/min (cpm) [cGMP and Ins(1,4,5)P$_3$] or 400,000 cpm (cAMP) per milliliter assay buffer

60% saturated ammonium sulfate precipitate of calf serum, dissolved in cGMP assay buffer to the original volume

cGMP antiserum, diluted in the calf serum solution (see above)

cAMP-binding protein, diluted in cAMP assay buffer

Ins(1,4,5)P$_3$-binding protein, undiluted

60% saturated ammonium sulfate, on ice

Charcoal: 1.25 g activated charcoal and 0.5 g BSA in 25 ml cAMP assay buffer, on ice

Procedure for cGMP

1. Incubate at 0° in 1.5-ml tubes 20 μl tracer, 20 μl water, standard or sample, and 20 μl serum.
2. Terminate the incubation after 2 hr by addition of 0.5 ml ammonium sulfate. Incubate for 5 min.
3. Centrifuge the tubes for 2 min at 4° at 14,000 g; aspirate the supernatant.
4. Finish by adding 100 μl water. Dissolve the pellet and add 1.3 ml scintillator.

[30] A. L. Steiner, C. W. Parker, and D. M. Kipins, *J. Biol. Chem.* **247**, 1106 (1972).
[31] A. G. Gilman and F. Murad, this series, Vol. 38, p. 49.
[32] A. A. Bominaar and P. J. M. Van Haastert, this series, Vol. 238 [16].

Procedure for cAMP

1. Incubate at 0° in 1.5-ml tubes 20 µl tracer, 20 µl water, standard or sample, and 20 µl cAMP-binding protein.
2. Terminate the incubation after 2 hr by addition of 60 µl charcoal suspension. Incubate for 1 min.
3. Centrifuge the tubes for 2 min at 4° at 14,000 g; transfer 90 µl of the supernatant to a scintillator vial.
4. Finish by adding 2-ml scintillator.

Procedure for Inositol 1,4,5-Trisphosphate

1. Incubate at 0° in 1.5-ml tubes 20 µl tracer, 20 µl water, standard or sample, and 20 µl Ins(1,4,5)P$_3$-binding protein.
2. Terminate the incubation after 10 min by centrifuging the tubes for 2 min at 4° at 14,000 g. Aspirate the supernatant.
3. Finish by adding 100 µl water. Dissolve the pellet and add 1.3-ml scintillator.

Calculation. The binding of radioactive tracer is determined in the absence of cold ligand (C_0), in the presence of excess cold ligand (Bl), and at different concentrations of ligand or in samples (C_x). The relationship between picomoles ligand and measured C_x is given by the following equation:

$$\text{pmole} = [(C_0 - Bl)/(C_x - Bl) - 1] \times X$$

The value of X is determined from a standard curve with known amounts of ligand.

Comments. Samples derived from experiments are generally lysed by perchloric acid (PCA) and neutralized with potassium bicarbonate (see later). The pH of the samples should not be above pH 7.5 for cAMP and cGMP assays, and not below pH 6.5 for the Ins(1,4,5)P$_3$ assay.

cAMP and cGMP are more stable before than after neutralization. Samples are neutralized on the day of the isotope dilution assay; reassay of the same samples on subsequent days is possible, but may lead to lower values and an increase of the standard deviation, especially for the cAMP determinations.

Stock solutions are made in water; cAMP is stable in phosphate and HEPES buffers, but it is deaminated in Tris buffers after storage for a few weeks at $-20°$. cGMP does not show this problem.

Results. A summary of the primary data from the three isotope dilution assays is presented in Table I. Each assay has a characteristic ratio of maximal/minimal binding; these values determine the range of ligand concentrations that can be determined accurately. The cAMP and cGMP

TABLE I
PRIMARY DATA FOR ISOTOPE DILUTION ASSAYS, RECEPTOR-STIMULATED RESPONSES *in Vivo*, AND RECEPTOR- AND G-PROTEIN-STIMULATED ENZYME ACTIVITIES *in Vitro*

Incubation	cGMP cpm	cGMP pmol	Ins(1,4,5)P$_3$ cpm	Ins(1,4,5)P$_3$ pmol	cAMP cpm	cAMP pmol
Standard curve						
Input radioactivity	2000		2000		8000 ± 500	
Standards						
0 pmol	620 ± 31		767 ± 8		2417 ± 224	
0.1 pmol	511 ± 23		ND^a		2012 ± 186	
0.25 pmol	403 ± 32		635 ± 12		1726 ± 163	
0.5 pmol	310 ± 20		545 ± 16		1346 ± 129	
1 pmol	227 ± 19		463 ± 15		1007 ± 98	
2 pmol	185 ± 11		369 ± 20		675 ± 63	
4 pmol	148 ± 8		315 ± 14		461 ± 53	
10 pmol	ND		248 ± 7		303 ± 26	
Blank	95 ± 7		196 ± 10		178 ± 16	
X^b	0.376		0.895		0.572	
Cell stimulation						
Unstimulated	424 ± 21	0.22 ± 0.04	572 ± 53	0.54 ± 0.22	2056 ± 192	1.1
Stimulated	194 ± 11	1.56 ± 0.21	528 ± 38	0.72 ± 0.15	591 ± 56	22.4
Enzyme activities						
t$_0$ effector enzyme	425 ± 11	0.21 ± 0.02	568 ± 43	0.53 ± 0.20	2525 ± 227	0
Basal effector enzyme	320 ± 18	0.49 ± 0.07	272 ± 9	6.50 ± 0.90	1530 ± 148	9.7
GTPγS-stimulated activity	233 ± 15	1.02 ± 0.15	251 ± 5	9.36 ± 1.75	515 ± 49	82.6
cAMP-stimulated activity	ND		246 ± 3	10.25 ± 0.65	1145 ± 103	19.1
cAMP,GTPγS-stimulated activity	ND		247 ± 6	10.10 ± 1.81	362 ± 34	162

^{a} ND, not determined.
^{b} X, correction factor obtained from a standard curve.

assays allow one to determine the concentration of unlabeled cAMP or cGMP at a wide range of concentrations, whereas the range for the Ins(1,4,5)P$_3$ assay is more narrow. Each assay has also a characteristic affinity for the respective ligand, which determines the absolute concentrations that can be determined. The cGMP and Ins(1,4,5)P$_3$ assays are more sensitive than the cAMP assay. The accuracy of all assays is sufficient to result in a standard deviation of determined concentrations of less than 10%.

cAMP Binding and cAMP-Induced Responses *in Vivo*

cAMP Binding Assays

Principle. *Dictyostelium* cells possess surface receptors that bind cAMP with high specificity and affinity. These receptors may have different kinetic forms, which are probably related to the interaction with G proteins. The receptor–cAMP complex dissociates very fast with halftimes as short as 1 sec.[33,34] Therefore, the binding assay requires the separation of bound and free cAMP without washing of the cells. This can be achieved either by pelleting of the cells and aspiration of the supernatant, or by centrifuging cells through silicone oil. The latter method is based on the density of the oil, which must be denser than buffer, but less dense than the cells. After centrifugation the oil separates the cells in the pellet and the unbound cAMP in the supernatant.

cAMP receptors are a heterogeneous mixture of different forms, not only with respect to the interaction with G proteins and the state of phosphorylation, but also because a substantial portion of the receptors are not assessable for cAMP binding. A portion of the receptors are cryptic and can be exposed by polyvalent ions. Another portion of the receptors are sequestered as an intermediate during cAMP-induced down-regulation of the receptors.[35] We have observed that in nearly saturated ammonium sulfate, binding of cAMP to cells is increased substantially and shows a more homogeneous population. Interactions with G proteins are lost, and cryptic as well as sequestered receptors bind cAMP. In addition, ammonium sulfate largely retards the dissociation of the receptor–cAMP complex. This not only increases the affinity of the receptor, but also allows the extensive washing of cells. Three cAMP-binding assays are described and compared.

---

[33] P. J. M. Van Haastert and R. J. W. De Wit, *J. Biol. Chem.* **259,** 13321 (1984).
[34] P. J. M. Van Haastert, R. J. W. De Wit, P. M. W. Janssens, F. Kesbeke, and J. De Goede, *J. Biol. Chem.* **261,** 6904 (1986).
[35] P. J. M. Van Haastert, *Biochim. Biophys. Acta* **845,** 254 (1985).

Solutions and Materials

Aggregation competent cells at 10^8 cells/ml in PB

Radioactive binding mixture containing, per milliliter, 4 μl [^{3}H]cAMP stock, 50 μl 1 M DTT, and 946 μl PB

Mixture of silicone oils: the mixture for 20° is AR 20 : AR 200 at 2 : 1; the mixture for 0° is 11 : 4

90% saturated ammonium sulfate in PB: Prepare saturated ammonium sulfate at room temperature, place on ice until equilibrated, and dilute with water

BSA in water at 1 mg/ml

Sucrose, 20% (w/v) in PB

Procedure A: Phosphate Buffer Pellet Assay. The cells are pelleted at the end of the binding reaction. This method is a fast and simple way to detect equilibrium binding of exposed cAMP receptors.

1. Preparation: Label 1.5-ml plastic tubes and add 10 μl radioactive binding mixture and 10 μl water (or unlabeled cAMP, see modifications).
2. Assay: Add 80 μl of the cell suspension. Incubate for 30–45 sec (room temperature) or for 2 min at 0°. Centrifuge the tubes for 2 min at 4° at 14,000 g. Aspirate the supernatant.
3. Finish: Resuspend the pellets in 100 μl of 0.1 M acetic acid. Add 1.3-ml scintillator and determine the radioactivity.

Procedure B: Ammonium Sulfate Pellet Assay. The binding reaction is performed in nearly saturated ammonium sulfate. The cells are pelleted at the end of the binding reaction. This method is a fast and simple way to detect the total number of cAMP receptors.

1. Preparation: Label tubes and add 10 μl radioactive binding mixture, 10 μl water, and 880 μl of 90% saturated ammonium sulfate. Place tubes on ice.
2. Assay: Add 80 μl of cells and 20 μl of BSA. Incubate for 5 min at 0°. Centrifuge the tubes for 2 min at 4° at 14,000 g. Aspirate the supernatant.
3. Finish: Resuspend the pellets in 100 μl of 0.1 M acetic acid. Add 1.3-ml scintillator and determine the radioactivity.

Procedure C: Silicone Oil Assay. The binding reaction occurs in phosphate buffer, and cells are centrifuged through silicone oil. This method is more laborious, but it has lower nonspecific binding and allows the determination of nonequilibrium binding.

1. Preparation: Label 1.5-ml plastic tubes and add 10 μl radioactive binding mixture and 10 μl water. Label a second series of tubes; add 10 μl sucrose and 200 μl silicone oil mixture.

2. Assay: Add 80 μl of the cell suspension. Transfer the incubation mixture to a tube containing the silicone oil. Centrifuge the tube with cells and silicone oil for 20 sec at 4° at 14,000 g.
3. Finish: Place tubes at −20° until frozen or longer. Cut the tip of the tube containing the sucrose and cell pellets; this is done most easily with a scalpel. Transfer the tip to a scintillation vial. Add 100 μl water and 2-ml scintillator.

Modifications. The incubation mixtures contain 10 μl water, which can be replaced by different compounds. For nonspecific binding 10 μl of 1 mM cAMP is used. For the analysis of the affinity and number of binding sites by Scatchard analysis different concentrations of radioactive and cold cAMP are used. The standard conditions described above contain 10 μl of 100 nM [^{3}H]cAMP. Generally we use 10 μl of radioactive cAMP at 20, 50, 100, and 300 nM, and 300 nM [^{3}H]cAMP with 700, 1700, 4700, 9700, and 19,700 nM unlabeled cAMP. Compared to these concentrations, the final concentrations in the binding reaction are 10-fold lower in procedures A and C, and 100-fold lower in B.

Comments. cAMP induces the activation of adenylyl cyclase and subsequent secretion of synthesized cAMP. This should be prevented, because it dilutes the radioactive cAMP. One method is to complete the binding reaction before secretion starts, which is 45 sec at 20° and about 2 min at 0°; binding equilibrium is reached within 30 sec. Another method is to inhibit adenylyl cyclase activation with 5 mM caffeine, which is included as 50 mM in the radioactive reaction mixture. cAMP secretion does not occur in the ammonium sulfate assay. Moreover, the timing of the experiment does not allow one to incubate more than three tubes simultaneously in assays A and C.

During centrifugation the temperature of the silicone oil may increase, thereby decreasing its density. This may result in the floating of the silicone oil on top of the supernatant. This problem is especially important for incubations at 0° with a centrifuge operating at room temperature. Careful preparation of the silicone oil mixture and short centrifugation times should eliminate the necessity to perform the assay in a cold room.

The BSA in the ammonium sulfate assay serves to attach the cells to the wall of the tube, which facilitates the aspiration of the supernatant. It slightly increases nonspecific binding.

Results. The binding data from a typical experiment are presented in Table II. The ammonium sulfate assay provides high binding with relatively low nonspecific binding, resulting in the highest ratio of specific to nonspecific binding. This assay also has the lowest standard deviation. The phosphate buffer pellet assay and the silicone oil assay have approxi-

TABLE II
PRIMARY DATA FOR cAMP AND GTPγS BINDING TO CELLS AND MEMBRANES

Assay/condition	Binding (cpm)		Ratio
	Nonspecific	Specific	
cAMP binding to cells			
Input 40,000 cpm			
PB pellet assay	640 ± 91	2163 ± 311	3.38
PB silicone oil assay	247 ± 25	1846 ± 209	7.47
Ammonium sulfate assay	605 ± 55	5195 ± 127	8.59
cAMP binding to membranes			
Input 20,000 cpm			
Control	224 ± 12	1014 ± 47	4.53
100 μM GTPγS	231 ± 20	314 ± 38	1.36
GTPγS binding to membranes			
Input 80,000 cpm			
Control	841 ± 56	8551 ± 232	10.17
1 μM cAMP	847 ± 72	12715 ± 319	15.01

mately the same level of specific binding. However, the pellet assay has a significantly higher level of nonspecific binding.

Each assay has different applications. The ammonium sulfate assay measures the total number of receptors, irrespective of their functional status (exposed versus cryptic or sequestered). The method is very accurate and sensitive. We have noticed that receptor levels determined by ammonium sulfate binding or by Western blots are similar. The phosphate buffer pellet assay is a fast and convenient assay to determine the level of exposed and functional receptors. The assay is reasonably accurate and sensitive. The phosphate buffer silicone oil assay is more cumbersome, but it has the advantage of immediate separation of bound and unbound cAMP, thereby allowing one to assay for nonequilibrium binding. The method has the additional advantage of the lowest nonspecific binding. Applications of each of these assays have been published.[4,36,37]

The ammonium sulfate assay has the additional advantage that bound cAMP dissociates very slowly, thereby allowing washing of the cells to further reduce nonspecific binding. This property has been exploited for photoaffinity labeling of the receptor.[38]

[36] P. J. M. Van Haastert, M. Wang, A. A. Bominaar, P. N. Devreotes, and P. Schaap, *Mol. Biol.* **3,** 603 (1992).
[37] B. E. Snaar-Jagalska, P. N. Devreotes, and P. J. M. Van Haastert, *J. Biol. Chem.* **263,** 897 (1988).
[38] P. N. Devreotes and J. A. Sherrin, *J. Biol. Chem.* **260,** 6378 (1985).

cAMP-Induced Second Messenger Responses

Principle. Stimulation of aggregation competent *Dictyostelium* cells with cAMP leads to the activation of several effector enzymes and the formation of the second messengers cAMP, cGMP, and $Ins(1,4,5)P_3$. Three nearly identical protocols allow the determination of these responses. Stimulated cells are lysed at the desired time by perchloric acid. After neutralization, the levels of the second messengers are determined by specific isotope dilution assays. For cAMP-induced cAMP accumulation we use the analog 2'-deoxy-cAMP as stimulus, because this compound has high affinity for the surface receptor and low affinity for cAMP-dependent protein kinase which is used as the cAMP-binding protein in the isotope dilution assay.

Solutions and Materials

Aggregation competent cells at 5×10^7 cells/ml in PB
Stimulus solution: 5 μM cAMP in PB for the cGMP and $Ins(1,4,5)P_3$ response, and 50 μM 2'-deoxy-cAMP and 50 mM DTT for the cAMP response
3.5% (v/v) perchloric acid (PCA)
$KHCO_3$ (50% saturated at 20°)

cGMP and Inositol 1,4,5-Trisphosphate Response

1. Preparation: Label 1.5-ml tubes and add 20 μl stimulus solution. Add 100 μl PCA to the t_0 samples.
2. Assay: Add 80 μl of the cell suspension to the tubes. Stop the reaction after 0, 3, 6, 9, 12, 15, 20, and 30 sec by the addition of 100 μl PCA. Shake and place samples on ice for at least 10 min or store at $-20°$.
3. Finish: Neutralize by adding 50 μl $KHCO_3$ solution. Let stand to allow CO_2 to escape (shake carefully). Centrifuge for 2 min at 4° at 14,000 g. Use 20 μl of the supernatant in the isotope dilution assays (see above).

cAMP Response. The assay is performed as described above, except that the stimulus solution differs and the times of the reaction are 0, 0.5, 1, 1.5, 2, 3, and 5 min.

Modifications. The assay described above uses independent stimulations. This procedure is optimal for the determination of the magnitude of the response. In experiments where the time course is more important than the magnitude of the response, we stimulate 900 μl of cells with 100 μl of stimulus (twice as concentrated as described above), and at the desired times 100-μl samples are transferred to tubes containing 100 μl PCA. In these experiments the t_0 sample is taken just before stimulation.

To catch the early time points, cells are stimulated while vortexed, and the same pipette is used to add the stimulus and to withdraw the samples.

The protocols determine intra- and extracellular levels of second messenger. For the cAMP response it is often relevant to distinguish between intra- and extracellular levels. Just before termination of the reactions, cells are centrifuged for 5 sec at 14,000 g, the supernatant is transferred to a tube containing 100 μl PCA, and 100 μl PCA is added to the pellet.

Comments. Prior to stimulation air is bubbled through the cell suspension for 10 min at a rate of about 15 ml air/ml suspension/min.

Results. The primary data from a typical experiment are presented in Table I. The magnitude of the responses are very different, with small responses and high basal levels for Ins(1,4,5)P_3 and large responses and low basal levels for cAMP and cGMP. All assays have a small standard deviation. These assays have been used to determine the cGMP response, the cAMP response, and the Ins(1,4,5)P_3 response under a variety of conditions and in mutant cell lines.[4,39]

Receptor–G Protein–Effector Enzyme Interactions *in Vitro*

All *in vitro* assays are performed with lysates or membranes, the preparation of which was described above.

GTP Inhibition of cAMP Binding

Principle. The effect of guanine nucleotides on agonist binding to the surface receptor is a useful indicator of receptor–G protein interaction. Addition of guanine nucleotides reduces the apparent affinity, but not the number of receptors.[40] *Dictyostelium* membranes are incubated with subsaturating concentrations of [^{3}H]cAMP in the presence of GTPγS. Bound [^{3}H]cAMP is separated from free [^{3}H]cAMP by centrifugation, and radioactivity associated with membrane pellets is measured. The method provides a very convenient and accurate assay for the interaction from G protein to receptor.

Solutions and Materials

Membranes resuspended in PB to a density equivalent to 10^8 cells/ml
Radioactive binding mix containing 50 nM [^{3}H]cAMP, 50 mM DTT, and 20 mM PB

---

[39] A. A. Bominaar, F. Kesbeke, B. E. Snaar-Jagalska, D. J. M. Peters, P. Schaap, and P. J. M. Van Haastert, *J. Cell Sci.* **100,** 825 (1991).

[40] P. J. M. Van Haastert, *Biochem. Biophys. Res. Commun.* **124,** 597 (1984).

1% sodium dodecyl sulfate (SDS)
1 mM cAMP, and 0.3 mM GTPγS

Procedure

1. Preparation: Label 1.5-ml tubes and add 10 μl radioactive binding mixture, 10 μl water or GTPγS, and 10 μl water or excess cAMP. Place the tubes on ice.
2. Assay: Make membranes (see above). Add 70 μl membranes to the tubes and incubate for 5 min at 0°. Centrifuge for 2 min at 4° at 14,000 g. Aspirate the supernatant.
3. Finish: Add 100 μl of 1% SDS, and mix until the pellet is dissolved. Add 1.3-ml scintillator.

Modifications. The concentration of [^{3}H]cAMP in the assay is 5 nM. For Scatchard analysis different concentrations of radioactive and cold cAMP are used as described in the binding assays for cells (see above).

For nonequilibrium binding, the reaction is terminated by rapid centrifugation through silicone oil as described for nonequilibrium binding to cells. The same mixture of silicone oils can be used as for cells. Because nonspecific binding is very low in the silicone oil assay, this method is also preferred for membranes with very low levels of cAMP receptors.

Comments. Membranes pass through silicone oil more slowly than do cells. This requires longer centrifugation times (about 30 sec). Accurate tuning of the silicone oil mixture and centrifugation time allows tubes to be centrifuged at room temperature. For rapid kinetics the workshop has made a swing-out rotor that fits in an Eppendorf microcentrifuge.

Results. Primary data are presented in Table I. Binding of 5 nM [^{3}H]cAMP to membranes is higher than binding to the equivalent number of cells. The increased binding is due to an enhanced affinity and not to an increase of the number of binding sites. GTP, GDP, GTPγS, and GDPβS reduce cAMP binding, owing to a decrease of the affinity of the receptor for cAMP.[40] This assay has been used extensively to investigate the effect of guanine nucleotides on the transition of different kinetic forms of the cAMP receptor. In these experiments the rate of association and dissociation of the receptor was determined in the absence or presence of guanine nucleotides using the silicone oil assay with a microcentrifuge swing-out rotor.[33] The equilibrium assay as well as the nonequilibrium assay were used to identify the defect of mutant *fgdA* at the interaction between cAMP receptor and a G protein.[4] Later studies have shown that the *fgdA* gene is the α subunit of G_2.[5-7]

cAMP Stimulation of GTPγS Binding

Principle. The G proteins are activated by the exchange of bound GDP for GTP. In *Dictyostelium* cAMP binding to cell surface receptors promotes release of bound GDP and permits binding of GTP. Receptor stimulation of G proteins can be measured as the cAMP-stimulated binding of GTP or GTPγS to G proteins. Membranes are incubated with [^{35}S]GTPγS in the presence of cAMP. Bound [^{35}S]GTPγS is separated from free [^{35}S]GTPγS by centrifugation, and radioactivity associated with membrane pellets is measured. This procedure provides a convenient and accurate assay for the interaction from receptor to G protein.

Solutions and Materials

Membranes resuspended in PB to a density equivalent to 10^8 cells/ml
Binding mix containing 2 nM [^{35}S]GTPγS, 1 mM ATP, 30 mM MgCl$_2$, and 20 mM PB
1% SDS
1 mM cAMP, and 1 mM GTP

Procedure

1. Preparation: Label 1.5-ml Eppendorf tubes and add 10 μl radioactive binding mixture, 10 μl water or cAMP, and 10 μl water or GTP. Place tubes on ice.
2. Assay: Prepare membranes (see above). Add 70 μl of membranes. Incubate for 30 min at 0°. Centrifuge for 3 min at 4° at 14,000 g. Aspirate the supernatant.
3. Finish: Add 100 μl of 1% SDS; mix until the pellet is dissolved. Add 1.3-ml scintillator.

Comments. The timing of the experiment allows one to incubate 12 tubes simultaneously.

Results. The primary data for [^{35}S]GTPγS binding to membranes is shown in Table II. Equilibrium binding is enhanced 40–80% by cAMP. Detailed analysis[41] has revealed that [^{35}S]GTPγS binding is relatively slow, with half-maximal association at 2 nM [^{35}S]GTPγS after 10 min at 0°; equilibrium is reached after 30 min of incubation. Scatchard analysis of [^{35}S]GTPγS binding showed two forms of binding sites with, respectively, high (K_d 0.2 μM) and low (K_d 6.3 μM) affinity. cAMP does not affect the rate of binding but enhances the affinity and number of the high-affinity sites, whereas the low-affinity sites are not affected by cAMP.

[41] B. E. Snaar-Jagalska, R. J. W. De Wit, and P. J. M. Van Haastert, *FEBS Lett.* **232**, 148 (1988).

cAMP Stimulation of GTPase Activity

Principle. G proteins have low intrinsic GTPase activity. Surface receptors promote the exchange of GDP for GTP in G proteins. The enhanced occupancy of the G protein with GTP consequently results in stimulated GTPase activity. This can be demonstrated in crude membranes from *Dictyostelium* when appropriate conditions are used to measure high-affinity GTPase activity. [γ-^{32}P]GTP is used at submicromolar concentrations. Nonspecific nucleotide triphosphatases are inhibited by the ATP analog AppNHp. Redistribution of radioactivity among guanine and adenine dinucleotides by nucleoside diphosphate kinase is prevented by a nucleoside triphosphate regeneration system and by ATPγS. Under these conditions the release of [^{32}P]P$_i$ from [γ-^{32}P]GTP is suppressed to 8–12% of added [γ-^{32}P]GTP, and stimulation by cAMP becomes detectable. Release of [^{32}P]P$_i$ from [γ-^{32}P]GTP in the absence of membranes is 0.5–2% of added [γ-^{32}P]GTP.

Solutions and Materials

Membranes prepared as described above except that the membrane pellet is washed once with 10 mM triethanolamine hydrochloride, pH 7.4 (TAA), containing 0.5 mM EDTA and the final pellet resuspended in TAA to the equivalent of 1×10^8 cells/ml

[γ-^{32}P]GTP, 0.1 μCi/assay (~60000 cpm/assay)

GTP at 0.1, 1, and 500 μM, and cAMP at 100 μM

Activated charcoal: 5% (w/v) in 50 mM sodium phosphate buffer, pH 2.0

2.5× Reaction mixture; the composition is given in Table III

Procedure

1. Preparation: Label tubes and place on ice. Add 40 μl reaction mixture, 10 μl [γ-^{32}P]GTP, and 10 μl GTP and 10 μl cAMP or water.
2. Assay: Prepare membranes as described earlier (see note above). Preincubate the tubes for 5 min at 25°. Start the reaction by addition of 30 μl membranes from ice; vortex and conduct the assay for 3 min at 25°. Terminate the reaction by the addition of 0.6 ml ice-cold activated charcoal in sodium phosphate buffer; place the tubes on ice.
3. Finish: Centrifuge the tubes at 4° for 5 min at 14,000 g and take 0.4 ml of the supernatant for determination of radioactivity.

Modifications. Total GTPase is detected in the absence of added nonradioactive GTP, whereas low-affinity GTPase is determined in the presence of 50 μM GTP. High-affinity GTPase is defined as the difference between total GTPase and low-affinity GTPase activity.

TABLE III
REACTION MIXTURE FOR GTPase ASSAY

Component[a]	Stock concentration	Volume (μl)	Final concentration in assay
TAA	1 M	125	50 mM
AppNHp	10 mM	50	0.2 mM
DTT	100 mM	250	10 mM
EGTA	10 mM	25	0.1 mM
MgCl$_2$	100 mM	50	2 mM
ATPγS	10 mM	25	0.1 mM
Creatine kinase[b]	1 mg		0.4 mg/ml
Creatine phosphate[b]	4 mg		5 mM
BSA	20 mg/ml	250	2 mg/ml
Water		225	

[a] All reagent stock solution are stored at $-20°$ unless otherwise noted.
[b] Weight reagents are added.

Comments. The reaction is started and terminated at 10-sec intervals to obtain 36 incubations with 3-min incubation times in all tubes. Aspiration of the supernatant should be done carefully.

Results. A typical experiment measuring the release of [^{32}P]P$_i$ from [γ-^{32}P]GTP is presented in Table IV. Total GTPase in this experiment produces 4980 cpm [^{32}P]P$_i$. In the presence of 50 μM GTP, low-affinity GTPase is detected which amounts to 1896 cpm. The difference, 3084 cpm, represents high-affinity GTPase. The release of [^{32}P]P$_i$ is routinely measured at 3 min of incubation. The relationship between membrane protein and GTP hydrolysis is linear in the range of 10–40 μg membrane protein per assay for incubation at 25° for 3 min.

TABLE IV
RECEPTOR-STIMULATED GTPase ACTIVITY IN MEMBRANES

	GTP hydrolyzed (cpm)		
	Concentration of GTP		
Condition	0.01 μM	50 μM	Difference
Input 60,000 cpm			
No enzyme	900 ± 300		
Enzyme	4980 ± 285	1896 ± 43	3084 ± 319
Enzyme + 1 μM cAMP	6724 ± 427	2040 ± 174	4684 ± 461

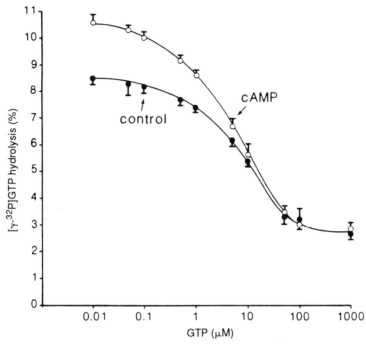

FIG. 1. Total GTPase activity in membranes at different GTP concentrations in the absence (●) and presence (○) of 3 μM cAMP.

The hydrolysis of [γ-^{32}P]GTP at different concentrations of GTP in the absence and presence of 3 μM cAMP is shown in Fig. 1. Hydrolysis of [γ-^{32}P]GTP is potently reduced by increasing concentrations of unlabeled GTP. At all GTP concentrations below 50 μM, cAMP increases [γ-^{32}P]GTP hydrolysis. Both curves reach a plateau at about 50 μM GTP. These observations indicate that *Dictyostelium* membranes contain a high-affinity, cAMP-sensitive GTPase with an apparent K_m value of about 6.5 μM and a low-affinity cAMP-insensitive GTPase with a K_m above 1 mM. cAMP stimulates high-affinity GTPase by increasing the affinity for GTP from 6.5 to 4.5 μM without a change of V_{max}.[42]

GTPγS Stimulation of Adenylyl Cyclase in Vitro

Principle. Adenylyl cyclase in *Dictyostelium* lysates is activated by the surface cAMP receptor and G protein. This assay combines two previously described methods. Lysis is performed according to Theibert and

[42] B. E. Snaar-Jagalska, K. H. Jakobs, and P. J. M. Van Haastert, *FEBS Lett.* **236**, 139 (1988).

Devreotes.[43] Enzyme activity is measured using the method of Van Haastert et al.[44] The receptor agonists 2'-deoxy-cAMP and GTPγS are present during lysis. The lysate is subsequently incubated for 5 min at 0°. Adenylyl cyclase is assayed using nonradioactive ATP. The reaction is terminated by adding excess EDTA and boiling the samples. The cAMP produced is determined by the isotope dilution assay (see above).

Solutions and Materials

Aggregation competent *Dictyostelium* cells: Starve cells on nonnutrient agar for 16 hr at 6° and shake for an additional hr at 22° at a density of 10^7 cells/ml in PB; collect cells by centrifugation and adjust to a density of 1.5×10^8 cells/ml in ice-cold PB

3× Lysis buffer: 30 mM Tris, pH 8.0, 6 mM MgCl$_2$

2× Lysis buffer: 20 mM Tris, pH 8.0, 4 mM MgCl$_2$

DTT at 0.5 M, ATP at 0.1 M

EDTA at 0.1 M, pH 8.0

GTPγS at 1 mM, 2'-deoxy-cAMP at 1 mM

1 ml assay mixture: 40 μl of 0.5 M DTT, 10 μl of 0.1 M ATP, 340 μl of 3× lysis buffer, 610 μl water

Procedure

1. Preparation: Label 1.5-ml tubes. Add 20 μl of assay mixture to all tubes; add 10 μl EDTA to control tubes (t_0 incubations). Prepare 1-ml syringes with a Nuclepore membrane inserted between the gauge and the needle.
2. Assay: Mix 100 μl of cells with 100 μl of 2× lysis buffer at 0°. Immediately lyse the cells by pressing them through a Nuclepore filter with 3-μm pores. Collect the lysates in tubes at 0° and keep on ice for 5 min. To measure adenylyl cyclase, add 20 μl of lysate to the tubes containing 20 μl of assay mixture. Incubate for 5 min at 20°. Terminate the reaction by the addition of 10 μl of 0.1 M EDTA, pH 8.0.
3. Finish: Boil the samples for 2 min. Store at $-20°$ or conduct the cAMP assay directly. Assay cAMP levels using the isotope dilution assay (see above; note that samples do not have to be neutralized).

Modifications. Stimulated adenylyl cyclase is measured by adding 30 μM GTPγS or 50 μM 2'-deoxy-cAMP (final concentrations) to the cells just prior to cell lysis.

---

[43] A. Theibert and P. N. Devreotes, *J. Biol. Chem.* **261**, 15121 (1986).
[44] P. J. M. Van Haastert, B. E. Snaar-Jagalska, and P. M. W. Janssens, *Eur. J. Biochem.* **162**, 251 (1987).

Comments. At the end of the adenylyl cyclase reaction the tubes contain 20 μl assay mixture, 20 μl lysate, and 10 μl EDTA and have been boiled. It is convenient to continue with the cAMP isotope dilution assay in these tubes by adding 40 μl radioactive cAMP and 40 μl cAMP-binding protein.

Results. Primary data for the adenylyl cyclase assay are presented in Table I. Basal adenylyl cyclase activity is easily detected. GTP and GTP analogs produce significant (up to 17-fold) activation of adenylyl cyclase in lysates of *Dictyostelium*. Activation is enhanced 2- to 4-fold by cAMP. Detailed experiments[43,44] have shown that maximal activation occurs when GTPγS is present in the lysate directly after cell lysis and the lysate is preincubated with GTPγS for 5 min prior to assay. Stimulation by cAMP is optimal when added 1 min before lysis. This and related assays have been used to show that activation of adenylyl cyclase requires a soluble protein defective in mutant *synag7*. Furthermore, the role of specific cAMP receptor gene products and G proteins in the activation and desensitization of adenylyl cyclase has been investigated.[45]

GTPγS Regulation of Phospholipase C

The regulation of phospholipase C by stimulatory and inhibitory G proteins is discussed in detail elsewhere.[32] The primary results of a phospholipase C assay and the effects of cAMP and GTPγS are presented in Table I for comparison with the other assays.

GTPγS Stimulation of Guanylyl Cyclase

Principle. Guanylyl cyclase in *Dictyostelium* lysates can be activated by GTPγS. Because guanylyl cyclase has a catalytic site for GTP, it is a priori difficult to prove that GTPγS stimulation of guanylyl cyclase is mediated by a G protein. Evidence has been obtained by the antagonizing effect of GTPβS. The assay follows the same principle as the adenylyl cyclase assay. Cells are lysed in the presence of GTPγS, and enzyme activity is measured using a radioimmunoassay to quantify the cGMP formed from unlabeled GTP. The procedure has been optimized for detecting magnesium-dependent guanylyl cyclase activity.[46] There are important factors to be considered: (1) Lysis buffer should not contain sucrose or calcium ions because these strongly inhibit enzyme activity, and (2) the

---

[45] M. Pupillo, R. Insall, G. S. Pitt, and P. N. Devreotes, *Mol. Biol. Cell* **3**, 1229 (1992).
[46] P. M. W. Janssens, C. C. C. De Jong, A. A. Vink, and P. J. M. Van Haastert, *J. Biol. Chem.* **264**, 4329 (1989).

lysate should be incubated immediately with substrate GTP because enzyme activity is very unstable.

Solutions and Materials

Aggregation competent *Dictyostelium* cells: Starve cells on nonnutrient agar for 16 hr at 6° and shake for an additional hour at 22° at a density of 10^7 cells/ml in PB; collect cells by centrifugation, then wash and resuspend the cells in 40 mM HEPES–NaOH, pH 7.0, to a density of 1.5×10^8 cells/ml

$2\times$ Lysis buffer: 40 mM HEPES–NaOH, 6 mM MgSO$_4$, 2 mM EGTA, 0.2 mM AppNHp, pH 7.0

DTT at 0.5 M, GTP at 10 mM

GTPγS at 1 mM

1 ml assay mixture: 20 μl of 0.5 M DTT, 60 μl of 10 mM GTP, 920 μl water

3.5% (v/v) PCA

Procedure

1. Preparation: Label 1.5-ml tubes. Add 20 μl of assay mixture to all tubes; add 20 μl PCA to control tubes (t_0 incubations). Prepare 1-ml syringes with Nuclepore membrane inserted between the gauge and the needle.
2. Assay: Mix 100 μl of cells with 100 μl of $2\times$ lysis buffer at 0°. Immediately lyse the cells by pressing them through a Nuclepore filter with 3-μm pores. Collect the lysates in tubes at 0°. Start the guanylyl cyclase assay at 30 sec after lysis by addition of 20 μl of lysate to the tubes containing 20 μl of assay mixture. Incubate for 1 min at 22°. Terminate the reaction by the addition of 20 μl PCA.
3. Finish: Boil the samples for 2 min. Store at $-20°$ or complete by assaying cGMP levels using the radioimmunoassay (see above).

Modifications. Stimulated guanylyl cyclase is measured by adding 100 μM GTPγS (final concentration) to the cells just prior to cell lysis.

Comments. Guanylyl cyclase activity can be measured with Mg^{2+} (as described here) and with Mn^{2+} as cofactor. At physiological metal ion concentrations, Mg^{2+}-dependent activity is relevant. This activity is unstable, strongly inhibited by sucrose and Ca^{2+}, and membrane associated. The Mn^{2+}-dependent enzyme activity is not regulated by GTPγS or surface receptors, is not inhibited by sucrose or Ca^{2+}, and is found in membrane-associated and cytosolic fractions.

Results. Primary data from the guanylyl cyclase assay are presented in Table I. Basal enzyme activity is easily detected, and GTPγS produces significant activation of guanylyl cyclase in lysates of *Dictyostelium*.

Discussion

Biochemical, genetic, and molecular evidence indicates that cAMP receptor-mediated events in *Dictyostelium* are regulated through $\alpha\beta\gamma$ heterotrimeric G proteins, similar to those found in mammals.[47] Cyclic AMP is the extracellular signal in *Dictyostelium*, to be compared with the hormone in mammalian cells. Cyclic AMP is detected by surface receptors that have the classic seven putative transmembrane spanning domains of G-protein-linked receptors that include rhodopsin, the α,β-adrenergic receptors, the serotonin receptor, and the muscarinic acetylcholine receptor. The effector enzymes in *Dictyostelium* are adenylyl cyclase, guanylyl cyclase, and phospholipase C; the second messengers interact with target enzymes, such as protein kinases, calcium channels, and cytoskeletal components. The two main cellular effects of extracellular cAMP in *Dictyostelium* are chemotaxis, bringing the amoeboid cells into a multicellular structure, and cell type-specific gene expression to induce cell differentiation in this system.

Several standard assays of G-protein function have been described in this chapter. At the receptor level it was shown that GTP and analogs induce a decrease of cAMP binding to the surface receptors. Conversely, cAMP binding to the receptor stimulates binding of GTPγS and high-affinity GTPase activity in membranes. In addition GTP or nonhydrolyzable analogs can stimulate adenylyl cyclase, guanylyl cyclase, or phospholipase C *in vitro*. These methods provide powerful tools to analyze the phenotype of the increasing number of mutants and transformants. Recent advances in molecular genetics such as the polymerase chain reaction, antisense technology, targeted gene replacement, and restriction enzyme-mediated integration (REMI) have opened the possibility of isolating many mutants with defects in signal transduction as well as the mutated genes. The ability to combine molecular approaches with biochemistry and cell biology makes *Dictyostelium* useful in studying signal transduction and the regulation of development at the molecular level.

---

[47] M. I. Simon, M. P. Strathmann, and N. Gautam, *Science* **252**, 802 (1991).

[31] Fluorescence Assays for G-Protein Interactions

By RICHARD A. CERIONE

Introduction

Understanding the molecular mechanisms by which GTP-binding proteins (G proteins) shuttle between cell surface receptors and biological effectors requires approaches for reconstituting the interactions between the different protein components of the system and readouts for directly monitoring these interactions. In many cases, the protein–protein interactions responsible for different steps of a receptor signaling cascade have been monitored by assaying activities that occur downstream from the protein interactions of interest. These assays have provided useful mechanistic information [e.g., regarding the role of G-protein subunit dissociation in activation (reviewed in Ref. 1)] and have highlighted regions on the G-protein α subunit that are felt to be involved in G-protein–effector coupling.[2,3] Nonetheless, to understand the detailed nature of the binding domains that participate in the receptor–G-protein and G-protein–effector interactions that make up a signaling pathway, it will be necessary to develop direct readouts for these interactions. It is expected that these readouts will provide mechanistic information that will complement structural information that becomes available through nuclear magnetic resonance (NMR) and X-ray crystallographic techniques.

Over the past several years, our laboratory and others[4-12] have set out to establish fluorescence spectroscopic approaches for monitoring the

---

[1] A. G. Gilman, *Annu. Rev. Biochem.* **56**, 615 (1987).
[2] S. B. Masters, K. A. Sullivan, R. T. Miller, B. Beiderman, N. G. Lopez, J. Ramanchandran, and H. R. Bourne, *Science* **241**, 448 (1988).
[3] C. H. Berlot and H. R. Bourne, *Cell (Cambridge, Mass.)* **68**, 911 (1992).
[4] T. Higashijima, K. M. Ferguson, P. C. Sternweis, E. M. Ross, M. D. Sinigel, and A. G. Gilman, *J. Biol. Chem.* **262**, 752 (1987).
[5] T. Higashijima, K. M. Ferguson, M. D. Smigel, and A. G. Gilman, *J. Biol. Chem.* **262**, 757 (1987).
[6] W. J. Phillips and R. A. Cerione, *J. Biol. Chem.* **263**, 15498 (1988).
[7] P. M. Guy, J. G. Koland, and R. A. Cerione, *Biochemistry* **29**, 6954 (1990).
[8] W. J. Phillips and R. A. Cerione, *J. Biol. Chem.* **266**, 11017 (1991).
[9] J. W. Erickson and R. A. Cerione, *Biochemistry* **30**, 7112 (1991).
[10] W. J. Phillips and R. A. Cerione, *J. Biol. Chem.* **267**, 17032 (1992).
[11] N. O. Artemyev, H. M. Rarick, J. S. Mills, N. P. Skiba, and H. E. Hamm, *J. Biol. Chem.* **267**, 25067 (1992).
[12] T. Wensel and L. Stryer, *Biochemistry* **29**, 2155 (1990).

GTP-binding/GTPase cycle of G proteins and for monitoring the interactions of G-protein α and $\beta\gamma$ subunits with one another and with the receptor and effector proteins. For the most part, we have used the vertebrate phototransduction system as a model for these studies, since it is possible to isolate each of the components of this system in milligram quantities and in highly purified form. This system is comprised of the receptor protein rhodopsin, the G protein transducin, and the effector enzyme, the cyclic GMP phosphodiesterase; light activation of rhodopsin stimulates the exchange of GDP for GTP on transducin, and it leads to the stimulation of cyclic GMP hydrolysis by the phosphodiesterase. It is expected that the development of fluorescence approaches for studying the phototransduction system will be generally applicable to a number of G-protein signaling systems, especially as expression systems (e.g., baculovirus) are developed for the large-scale generation of various signaling proteins. In this chapter, we describe techniques that have been used to monitor the GTP-binding and the GTPase activities of G-protein α subunits through changes in intrinsic tryptophan fluorescence and the use of extrinsic reporter groups to read out G-protein interactions with other signal transduction components.

General Considerations

Intrinsic Tryptophan Fluorescence

Previous work has shown that the intrinsic tryptophan fluorescence of the α subunits of G proteins is highly sensitive to the state of guanine nucleotide occupancy.[4-7] Specifically, the α subunits of the G_o protein and retinal transducin each contain two tryptophan residues and undergo a 2-fold enhancement in intrinsic tryptophan fluorescence as a result of the GDP-GTP exchange event. The methods that have been used to monitor the changes in the tryptophan fluorescence of the α_o subunit that accompany the binding of the GTP analog GTPγS (together with the divalent metal ion activator, Mg^{2+}), or the interactions of $AlCl_3$ with NaF (which form the activating agent AlF_4^-), have been described in detail.[13] It also has been possible to use the tryptophan fluorescence of the α subunit of transducin (α_T) to examine the effects of a receptor (rhodopsin) on the rates of GTP binding and GTP hydrolysis.[6,7]

The experimental approaches are described in some detail, below. However, there are some important factors that are briefly considered here. One has to do with the difficulty of detecting changes in the fluores-

[13] T. Higashijima and K. M. Ferguson, this series, Vol. 195, p. 321.

cence of one protein (α_T) during interactions with other proteins that contain a number of tryptophan residues (i.e., rhodopsin contains five tryptophan residues, and the $\beta\gamma_T$ subunit complex contains six tryptophan residues). It is the fact that both the receptor protein (rhodopsin) and the $\beta\gamma$ subunit complex can act catalytically to promote GTP binding to multiple α_T subunits that enables the problem of "background tryptophan fluorescence" to be circumvented. In the fluorescence experiments, it is possible to use ratios of 1 rhodopsin molecule and 1 $\beta\gamma_T$ subunit complex per 50 α_T molecules and still achieve full, immediate activation of the total α_T pool. Thus, it is possible to minimize the contributions of the rhodopsin and $\beta\gamma_T$ tryptophan residues to the total tryptophan signal.

A related factor has to do with the need to establish detergent-free systems to monitor receptor–G-protein interactions. This typically has required the use of reconstituted phospholipid vesicles that contain the receptor and the G protein in sufficient amounts so that all of the vesicles contain both protein components (otherwise the receptor–G-protein coupling would be ineffective because some lipid vesicles might contain only the receptor while others would contain only the G protein). The lipid vesicles can cause a significant degree of light scattering and detract from the tryptophan fluorescence signal. However, the vertebrate vision system offers a significant advantage because it is possible to establish systems where the free α_T and $\beta\gamma_T$ subunits are added directly to cuvettes containing reconstituted rhodopsin vesicles, under conditions where the α_T concentration greatly exceeds that of rhodopsin and where light scattering is minimized.

Extrinsic Fluorescence Reporter Groups

An alternative approach for monitoring key events during the GTP-binding/GTPase cycle of a G protein involves the use of fluorescent reporter groups. There are two types of strategies that can be employed. One involves the use of fluorophores that are sensitive to the local environment. These reagents serve as valuable environmentally sensitive probes that can undergo a change in fluorescence (either in emission wavelength or quantum yield) as an outcome of changes in tertiary structure (conformation). A second strategy is specifically directed toward monitoring protein–protein interactions and involves resonance energy transfer.[14,15] In this case, a donor–acceptor pair is chosen so that the fluorescence emission spectrum of the donor fluorophore overlaps the absorption spec-

[14] L. Stryer, *Annu. Rev. Biochem.* **47**, 819 (1978).
[15] G. G. Hammes, "Enzyme Catalysis and Regulation," p. 20. Academic Press, New York, 1982.

trum of the acceptor chromophore. Because the efficiency of energy transfer is dependent on the distance separating the donor and acceptor moieties, measurements of the extent of energy transfer between a fluorescent donor on one protein molecule and an acceptor chromophore on a second protein molecule provide a direct means for following protein–protein interactions. Given that the interactions are being assessed by changes in the effective distance between the interacting protein molecules, the environmental sensitivity of the fluorescent labels is not of primary importance.

Two important considerations when using fluorescent reporter groups for monitoring protein–protein interactions are resolution of the free unreacted probe from the protein and determination of the stoichiometry and specificity of the protein labeling. Because most proteins are labeled under conditions where there are millimolar concentrations of the fluorescent reagents and micromolar (or even lower) concentrations of the protein, it is extremely important that the unreacted fluorescent probes are effectively removed. Under these labeling conditions, the presence of even a small amount of the unreacted probe can contribute significantly to the background fluorescence and make it difficult to detect changes in the fluorescence of the labeled protein. It also is important to identify the site(s) of labeling on the protein molecule. If a number of sites are labeled by a fluorescent reagent (and the microenvironment of only one of the sites changes in response to conformational changes), then a small signal-to-noise (S/N) ratio may make it difficult to detect fluorescence changes that accompany protein conformational changes (as an outcome of ligand interactions or protein–protein interactions). Perhaps an even greater potential problem is that the labeling of a protein may affect (inhibit) biological activity. In the case of signal transduction, this may be reflected by an inability of the labeled protein to interact with other proteins in a signaling cascade. In this case, the labeled protein would not be useful for monitoring protein–protein interactions. In the section on the use of extrinsic reporter groups, presented below, we describe a case where a particular labeling procedure did not inhibit the ability of the α_T subunit to interact with its biological effector protein (the cyclic GMP phosphodiesterase) but did inhibit the ability of α_T to stimulate the effector enzyme activity.

Intrinsic Fluorescence: Studies of Receptor-Stimulated GTP Binding and GTPase Activity

Incorporation of Rhodopsin into Phospholipid Vesicles

A number of methods can be used to incorporate rhodopsin into phospholipid vesicles. For the purposes of assaying rhodopsin-stimulated GTP binding to transducin, as monitored by changes in the tryptophan fluorescence of the α_T subunit, we have found the following method to be most convenient. The first step of the protocol involves incubating 50 μl of a 17 mg/ml solution of phosphatidylcholine (Sigma, St. Louis, MO, type II-S; sonicated to clarity using any standard bath-type sonicator) with 25 μl of a 17% (w/v) solution of octylglucoside (Calbiochem, La Jolla, CA) and 225 to 325 μl of 10 mM sodium HEPES (pH 7.5), 100 mM NaCl, 1 mM dithiothreitol (DTT), and 5 mM MgCl$_2$ for 30–45 min on ice. Rhodopsin, purified by the procedure outlined by Litman,[16] except that 10 mM CHAPS is substituted for octylglucoside, is then added to the lipid–detergent mixture in the dark. Specifically, 100- to 200-μl aliquots from a stock solution containing 30 μM rhodopsin (as determined from the absorbance of the retinal at 498 nm, using an extinction coefficient of 42,700 M^{-1} cm^{-1})[16] is added to the reconstitution incubation to make a final volume of 500 μl. The entire protein–lipid–detergent mixture is then added (under red light) to a 1 ml column of Extracti-gel resin (Pierce Chemical Co., Rockford, IL). The Extracti-gel column is prepared by pouring the resin (~0.45 mL) into a 1-ml syringe stoppered with glass wool. The resin is preequilibrated by washing with 2 ml of 10 mM sodium HEPES (pH 7.5), 100 mM NaCl, 1 mg/ml bovine serum albumin (this is to minimize the nonspecific adsorption of rhodopsin to the resin) and then with 10 mM sodium HEPES (pH 7.5), 100 mM NaCl, 1 mM DTT, and 5 mM MgCl$_2$ (elution buffer). The rhodopsin-containing lipid vesicles are then eluted from the resin by washing the column with 2 ml of elution buffer.

The above protocol can be easily modified to prepare lipid vesicles that are more highly concentrated in rhodopsin. For example, 16 μl of the soybean phosphatidylcholine can be incubated with 9 μl of octylglucoside and 75 μl of 10 mM sodium HEPES (pH 7.5) and 100 mM NaCl for 30 min on ice. Then, as much as 50 μl of rhodopsin (from a 30 μM stock solution) can be added to the lipid–octylglucoside incubation. This mixture (final volume 150 μl) is then added to 0.5 ml of dried Extracti-gel resin in a 1-ml syringe. The Extracti-gel resin is pretreated as described above, and the rhodopsin-containing lipid vesicles are then eluted from the resin using 250 μl of the elution buffer.

Using these reconstitution approaches, we have determined efficiencies of incorporation of rhodopsin into the lipid vesicles of approximately 50% (54 ± 2% SE, $n = 12$). These values were determined using iodinated rhodopsin (prepared using chloramine-T as outlined in Greenwood et al.[17]). Based on our studies of a number of hormone receptor and polypeptide growth factor receptors, we have determined that this reconstitution proto-

---

[16] B. J. Litman, this series, Vol. 81, p. 150.
[17] F. C. Greenwood, W. M. Hunter, and J. S. Glover, Biochem. J. **89**, 114 (1963).

col yields a random (50:50) orientation of the receptor molecules. Thus (typically) approximately 50% of the receptors are incorporated into the liposomes so that the cytoplasmic domains face the inside of the vesicle and 50% of the receptors are oriented so that the cytoplasmic domains face the outside. In the case of reconstituted rhodopsin, it is the latter that are able to couple to the α_T and $\beta\gamma_T$ subunits that are added to the lipid vesicles (see below).

Reconstitution of Rhodopsin–α_T–$\beta\gamma_T$ Systems: Measurements of Intrinsic Tryptophan Fluorescence

The α_T and $\beta\gamma_T$ subunit complexes used in these studies are obtained in the following manner. The holotransducin molecules are extracted from rod outer segments using GTP, and then the α_TGDP complex, which results from GTP hydrolysis, is resolved from the $\beta\gamma_T$ subunits by Blue Sepharose chromatography as outlined in detail by Phillips et al.[18] The α_TGTPγS species are prepared by first extracting the holotransducin with GTPγS.

The effects of light-activated rhodopsin on the GTP-binding or GTPase activity of transducin can be monitored via changes in the intrinsic tryptophan fluorescence of the α_T subunit as follows. Typically, 5 μl of the purified α_TGDP species (from an ~500 μg/ml stock solution), in a buffer that contains 10 mM Tris-HCl (pH 7.5), 6 mM MgCl$_2$, 1 mM DTT, 500 mM KCl, 25% glycerol, and 5 μl of the $\beta\gamma_T$ complex (~500 μg/ml stock), in the Tris/MgCl$_2$/DTT/glycerol buffer lacking 25% glycerol, are added to 5 μl of the phosphatidylcholine vesicles containing rhodopsin (prepared as outlined in the previous section, in 10 mM sodium HEPES, 100 mM NaCl, 20 mM MgCl$_2$, and 1 mM DTT) plus 125 μl of 10 mM sodium HEPES (pH 7.5), 5 mM MgCl$_2$, 1 mM DTT, and 10 μl of guanine nucleotide (typically from 1–10 μM stock solutions). These additions are made in the dark. A corrected fluorescence spectrum for the α_T subunit is obtained by first scanning the excitation spectrum [we typically have used an SLM 8000c spectrofluorometer (SLM, Urbana, IL) in the ratio mode] from 220 to 320 nm (emission 335 nm) or the emission spectrum from 300 to 400 nm (excitation 280 nm) for a mixture of rhodopsin-containing vesicles and the $\beta\gamma_T$ subunit complex in the presence of 10 mM sodium HEPES (pH 7.5), 5 mM MgCl$_2$, 1 mM DTT, and in some cases 0.5 μM to 1 μM guanine nucleotide (final concentration). This (control) spectrum is stored in an IBM personal computer that is interfaced with the fluorometer, and then the α_TGDP subunit is added to cuvettes (either in the light or the dark) and the fluorescence (excitation or emission) spectrum rescanned. The

[18] W. J. Phillips, S. Trukawinski, and R. A. Cerione, *J. Biol. Chem.* **264**, 16679 (1989).

net fluorescence contributed by the α_T subunit is then determined by subtracting the control fluorescence (contributed by rhodopsin and $\beta\gamma_T$) from the total fluorescence measured on the addition of α_T to the rhodopsin-containing vesicles.

The fluorescence measurements can be made either in 0.3×0.3 cm cuvettes or in 1×1 cm cuvettes (in which case the total volume of the reaction mixture has been increased by $\sim$10 fold, i.e., from a volume of $\sim$0.15 ml to a final volume of $\sim$1.5 ml). The advantage of the latter is that the mixing of the different components is easier, whereas the advantage of the former is that less total (protein) sample is necessary, although the adequate, rapid mixing of these samples is more problematic.

The kinetics for the rhodopsin- and GTP- (or GTPγS-) induced enhancement of the tryptophan emission of the α_T subunit can be measured by first incubating the rhodopsin-containing lipid vesicles with the purified α_T subunit and the $\beta\gamma_T$ complex (with these protein components being mixed together as outlined above) for 5–10 min at room temperature and in room light (to activate the rhodopsin). The reactions then are initiated by the addition of 10 μl of a stock solution of guanine nucleotide. An example is shown in Fig. 1. It typically is easier to perform the kinetic experiments using a large volume reaction mixture ($\sim$1.5 ml) in order to facilitate mixing. The fluorescence emission can be monitored for 1-sec intervals when using an SLM 8000 spectrofluorometer (emission 335 nm, excitation 280 nm). As is the case when determining the excitation or emission spectra for the α_T subunit (as outlined above), the kinetic measurements can be made by first determining the fluorescence emission (at a single emission wavelength) for rhodopsin-containing vesicles and for the α_T and $\beta\gamma_T$ subunit complexes, prior to the addition of guanine nucleotide, and then subtracting this measurement from the time-dependent fluorescence emission measured after the addition of guanine nucleotide.

Studies Using Fluorescent Reporter Groups

Environmentally Sensitive Fluorophores as Probes for Protein–Protein Interactions

As indicated above, environmentally sensitive fluorophores, when covalently attached to proteins, can serve as useful reagents for monitoring conformational changes that occur as an outcome of protein–ligand or protein–protein interactions. In the receptor–G-protein-mediated signaling pathway that constitutes vertebrate vision, we have taken advantage of the fact that the β subunit of the retinal G protein transducin can be covalently modified with reporter groups without having any effect on

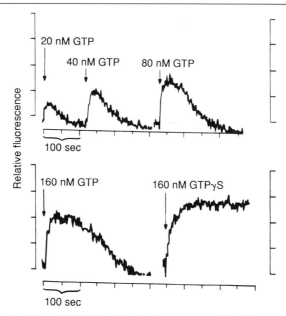

FIG. 1. Rhodopsin- and GTP-stimulated enhancement of the fluorescence of the α_T subunit. Rhodopsin-containing phosphatidylcholine vesicles (30 nM rhodopsin), $\beta\gamma_T$ (130 nM), and α_T (200 nM) were incubated in room light for 10 min at room temperature. GTP or GTPγS was added to the final concentrations indicated, and the change in the fluorescence emission (excitation 280 nm, emission 335 nm) was continuously monitored. (Reprinted with permission from Guy et al.[7] Copyright 1990, American Chemical Society.)

the normal function of the $\beta\gamma$ subunit complex. Thus, we have used fluorescently labeled $\beta\gamma_T$ subunit complexes to read out directly the interactions of $\beta\gamma_T$ with light-activated rhodopsin and with the GDP-bound α_T subunit.

Modification of Protein. In the case of the $\beta\gamma_T$ subunit complex, the purified protein (prepared by Blue Sepharose chromatography of a purified holotransducin complex as outlined in Ref. 18) in 10 mM sodium HEPES, pH 7.5, 6 mM MgCl$_2$, 1 mM DTT, 25% (v/v) glycerol is first dialyzed versus a "modification buffer" (20 mM sodium HEPES, pH 7.5, 5 mM MgCl$_2$, 0.15 M NaCl, 20% glycerol) to remove the DTT from the storage buffer (the DTT will interfere with the reaction of the cysteine residues with the fluorescent cysteine reagents). The dialyzed $\beta\gamma_T$ subunit complex (typically 0.5 ml of 0.1–0.5 mg/ml stock solution) is then mixed with the fluorescent reagent. When modifying the protein cysteine residues, there are a number of possible choices including 2-(4'-maleimidylanilino)naphthalene-6-sulfonic acid (MIANS), various iodoacetamidoaminonaphthalene derivatives (IANS or IAEDANS), the acryloyl derivative (acrylodan), and dansylaziridine. All these fluorescent cysteine reagents become more highly fluorescent in a hydrophobic environment. Although the choice of a specific fluorescent probe to some degree depends on the specific protein (and the types of reaction conditions that can be tolerated and the types of protein–protein interactions that are being monitored), we have found that MIANS serves as an excellent "all-purpose" environmentally sensitive, fluorescent cysteine reagent.

The modification of the $\beta\gamma_T$ subunit complex with MIANS is performed by incubating the dialyzed protein (typically 0.5 ml of a 0.25 mg/ml stock solution) with MIANS (final concentration of 0.5–1 mM; prepared by adding an aliquot from a 0.1 M MIANS stock solution in dimethylformamide). The reaction is allowed to proceed for 1 hr at room temperature. The MIANS-labeled $\beta\gamma_T$ is then resolved from the unreacted MIANS by hydroxyapatite chromatography. As indicated above, the effective removal of the unreacted fluorescent reagent is an extremely important step since the presence of even a small percentage of the total reagent can contribute significantly to the fluorescent background. Although dialysis is an option for some reagents, we have not found dialysis to be effective in removing many probes (particularly fluorescein reagents and pyrene derivatives). Hydroxyapatite chromatography, however, has worked well. Typically, a 0.5 ml (bed volume) column is poured and equilibrated with the storage buffer for the particular protein (e.g., $\beta\gamma_T$). The reaction mixture (0.5 ml) is then applied to the resin, and the column is washed with several bed volumes (4 ml) of the protein storage buffer containing 20 mM sodium phosphate. The MIANS-labeled $\beta\gamma_T$ is then eluted from the hydroxyapatite resin using the storage buffer plus 0.1 M phosphate. When using this chromatographic procedure, we do not detect any free MIANS (as monitored by the fluorescence emission of the probe at 430 nm). The stoichiometry of incorporation of the fluorescent reagent into the protein can be determined by absorbance measurements using the extinction coefficient of the probe (in the case of MIANS this is 17,000 at 325 nm) and from either an extinction coefficient of the protein (absorbance at 280–290 nm) or a protein determination (e.g., using the Cu^+/bicinchoninic acid complex; Pierce).

Fluorescence Studies with Labeled Protein: Examples from Experiments with Fluorescently Labeled $\beta\gamma_T$

Given the fluorescence properties of the MIANS probe, there are two types of experimental approaches that can be taken to obtain information regarding protein–protein interactions. One of these involves the fact that the MIANS fluorescence is strongly influenced by the hydrophobicity of the microenvironment. If as an outcome of a protein–protein interaction the accessibility of the MIANS moiety to solvent is reduced, either via a change in the tertiary structure of the protein or because one of the proteins (within the protein complex) actually buries the probe, then this will be reflected by an enhancement in the MIANS fluorescence and a shift toward shorter wavelengths in the emission maximum.

We have found that the MIANS–$\beta\gamma_T$ species is an excellent reporter group for the interactions of the $\beta\gamma_T$ subunit complex with the α_TGDP species and with the photoreceptor rhodopsin. In both cases, the fluorescence emission of the MIANS moiety is enhanced by 1.5- to 2-fold and the emission maximum shifted by approximately 5 nm. These changes make it possible to titrate the interactions of the $\beta\gamma_T$ complex both with the α_T subunit[8] and rhodopsin,[10] and to perform kinetic experiments along the lines described for the α_T subunit (when using changes in the intrinsic tryptophan fluorescence of α_T as the readout; see above). In this case, because the interaction of the α_TGDP subunit with the MIANS–$\beta\gamma_T$ species results in an enhancement of the MIANS fluorescence, it is possible (in the presence of reconstituted rhodopsin) to add guanine nucleotides (e.g., GTPγS) and monitor a time-dependent reversal of the MIANS fluorescence, which occurs as an outcome of the dissociation of the α_TGTPγS species from the MIANS-labeled $\beta\gamma_T$ (see Fig. 2).

Another approach that can be taken with the MIANS-labeled $\beta\gamma_T$ is to monitor its interaction with rhodopsin or the α_T subunit via resonance energy transfer. The absorbance maximum of the MIANS moiety is at approximately 325 nm which overlaps well with the emission maxima of the protein tryptophan residues (emission ~320–350 nm). In fact, the MIANS moiety effectively quenches the tryptophan emission of the $\beta\gamma_T$ complex to the extent that very little intrinsic fluorescence is measured for the MIANS-labeled $\beta\gamma_T$ (<10% of the tryptophan fluorescence of the unlabeled protein). The addition of either the α_TGDP species or light-activated rhodopsin to the MIANS-labeled $\beta\gamma_T$ (but not the α_TGTPγS complex) then results in an additional quenching of the protein emission (~30%), that is, this represents a quenching of the fluorescence of the α_T subunit and the photoreceptor. However, this readout has its limitations when performing studies in multicomponent reconstituted systems, since, under these conditions, the background tryptophan fluorescence makes it difficult to detect changes resulting from resonance energy transfer between specific tryptophan residues and appropriate acceptor chromophores. As outlined below, there are more direct approaches for using resonance energy transfer to read out protein–protein interactions.

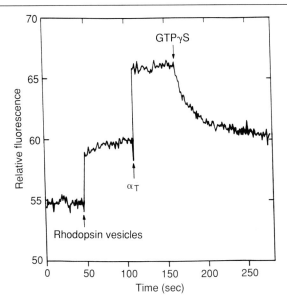

FIG. 2. Effects of activation of the α_T subunit on α_T- and rhodopsin-induced enhancement of MIANS–$\beta\gamma_T$ fluorescence. The MIANS–$\beta\gamma_T$ complex (85 pmol) was diluted into 125 μl of 20 mM HEPES, pH 7.5, 5 mM MgCl$_2$, and 1 mM DTT, and the MIANS emission (excitation 322 nm, emission 420 nm) was monitored at 1 determination per second. At the times indicated, phosphatidylcholine vesicles containing rhodopsin (90 pmol), the α_T subunit (140 pmol), and GTPγS (final concentration 1.5 mM) were added and mixed while continuing to monitor the MIANS fluorescence. (Data from Phillips and Cerione.[10])

Resonance Energy Transfer between Donor–Acceptor Pairs as Readout for Protein–Protein Interactions

Resonance energy transfer offers an alternative approach for monitoring protein–protein interactions and in fact often represents a more direct readout (relative to approaches involving the use of environmentally sensitive reporter groups). The idea is to label one protein with an environmentally insensitive donor fluorophore and a second protein with an acceptor chromophore that has an absorption spectrum that overlaps the emission spectrum of the fluorescent donor. The increase in the proximity of the donor fluorophore and the acceptor chromophore, which occurs as an outcome of the protein–protein interaction, then increases the probability of a transfer of excitation energy from the donor to the acceptor. We have used resonance energy transfer to monitor the interactions of an activated α_T subunit with the cyclic GMP phosphodiesterase (PDE)[9] and to monitor the interactions of the γ_{PDE} subunits with the larger α_{PDE} and β_{PDE} subunits of the effector enzyme. An example of the experimental methods underlying these studies is presented below.

Modification of Proteins. To monitor the binding of an activated α_T subunit to the cyclic GMP PDE, we have labeled the PDE with donor fluorophores that react with cysteine residues [i.e., iodoacetamidofluorescein (IAF)], using procedures similar to those described above. The α_T subunit is labeled at a reactive lysine residue with a fluorescent isothiocyanate reagent [eosin isothiocyanate (EITC)]. Typically, 500 μl of a 0.5 mg/ml solution of a purified preparation of the α_T subunit (isolated by the elution of holotransducin from rod outer segments using GTPγS and then purified by Blue Sepharose chromatography as described in Ref. 18) is mixed with EITC (10 μl of a 25 mM stock solution in dimethyl sulfoxide). The reaction is allowed to proceed for 4 hr at 22° in 10 mM sodium HEPES (pH 7.5). We have shown previously that these reaction conditions are appropriate for labeling the α_T subunit at a single reactive amino group with a stoichiometry of 1 label per protein molecule. However, it is likely that optimal reaction conditions will be quite different for different proteins, and, in general, the reaction of amino (lysine) groups occurs best at higher pH values (e.g., when using a borate buffer at pH 8.3–8.5). In the case of the α_T subunit, once the modification has been performed, the reaction mixture is applied to a 600-μl hydroxyapatite column preequilibrated with 10 mM K$_2$HPO$_4$. This column is washed extensively with 10 mM phosphate buffer until the eluate is free of any detectable (unreacted) EITC, and then the EITC–α_TGTPγS is eluted with 100 mM phosphate and subsequently dialyzed versus 150 mM NaCl, 5 mM MgCl$_2$, 20 mM sodium HEPES, 1 mM DTT, 0.05% NaN$_3$ (w/w), and 50% glycerol (usually overnight). The stoichiometry of the labeled α_TGTPγS can be determined by measuring the absorbance of the EITC (ε_{max} = 86,000 M^{-1} cm^{-1} at 523 nm).

Fluorescence Resonance Energy Transfer Measurements

As discussed above, the basic rationale underlying the resonance energy transfer approach is to monitor the interaction between one protein containing a donor fluorophore (in this case IAF-labeled PDE) and a second protein containing an acceptor chromophore (EITC–α_TGTPγS) by following a quenching of the donor emission. These measurements (like those outlined in the preceding sections) can be made using a SLM 8000c spectrofluorometer operated in the ratio mode. When monitoring EITC–α_TGTPγS/IAF–PDE interactions, we typically fix the excitation wavelength at 460 nm and scan the emission from 480 to 600 nm. Typically, 200 μl of a buffer containing 120 mM NaCl, 30 mM KCl, 2 mM MgCl$_2$, and 20 mM sodium HEPES, pH 7.4, is made 10–100 nM in IAF–PDE by the addition of small aliquots of a stock IAF–PDE solution (1–5 μl). Increasing aliquots (typically 1–5 μl) of the EITC–α_TGTPγS from a stock solution are then added with microcapillary pipettes. The solutions are mixed [using a small mixing apparatus (e.g., SLM Aminco magnetic stirrer control) that rests under the cuvette compartment of the fluorimeter], and emission spectra are recorded after each addition of EITC–α_TGTPγS. The peak (IAF–PDE) fluorescence intensity, for each addition of the acceptor (EITC–α_TGTPγS), is corrected for dilution and recorded. These experiments are performed under conditions where inner filter effects are negligible (i.e., using solutions that have over 0.01 absorbance units at the excitation wavelength). It should be noted that it is relatively straightforward to assess the contributions of inner filter effects to the total quenching of the donor fluorescence. For example, in the case of the IAF–PDE/EITC–α_TGTPγS donor–acceptor pair, the inner filter effects due to the absorbance of the excitation energy by the acceptor can be determined by adding free EITC to samples containing IAF–PDE.

The extents of energy transfer, as a function of added acceptor molecules, can be determined by titrating increasing concentrations of the accceptor into the cuvette solution that contains a fixed concentration of the donor. An example for the IAF–PDE/EITC–α_TGTPγS donor–acceptor pair is presented in Fig. 3. In this experiment, a 30% quenching of the donor emission was recorded at saturating levels of the acceptor (EITC–α_TGTPγS). Once the extent of energy transfer between the donor and acceptor probes has been determined, various control experiments should be considered to verify that the readout being used reflects a specific interaction between the protein components of interest. In the case of an interaction between an activated α_T subunit and the effector enzyme (PDE), the fluorescence changes (i.e., quenching) that reflect the interaction should be specific for the activated form of the α_T subunit and should not be observed when using the GDP-bound form of α_T. Figure 3 shows that this is the case when monitoring the energy transfer between the IAF-labeled PDE and an EITC–α_TGDP species. A second control for the IAF–PDE/EITC–α_TGTPγS interactions involves trypsin treatment of the PDE. It has been well documented that trypsin selectively digests the γ_{PDE} subunits and that these subunits are responsible for binding the activated α_T species. Thus, it is expected that trypsin digestion of the γ_{PDE} subunits would eliminate the energy transfer between the IAF–PDE and the EITC–α_TGTPγS, and the results in Fig. 3 show that this in fact is the case. When the energy transfer data are plotted in this manner, it should be possible to fit the data to a binding model. The solid line shown in Fig. 3 represents the best fit to a simple bimolecular reaction between the PDE and the activated α_T subunit (see Ref. 9) and yields an apparent

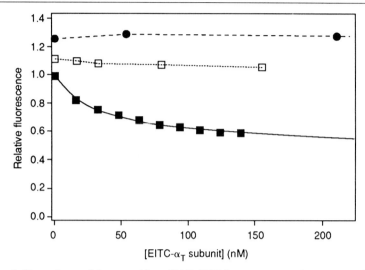

FIG. 3. Dependence of the quenching of IAF–PDE fluorescence on the concentration of EITC–α_T. Data from titrations of IAF–PDE with EITC–α_TGTPγS (■) and EITC–α_TGDP (□) are plotted as the relative emission intensities of IAF–PDE (at 518 nm) measured in the presence and absence of different α_T species and represent the raw fluorescence data from three different samples. The solid curve drawn through the data points represents the best fit for a single class of binding sites for the interaction of EITC–α_TGTPγS and IAF–PDE (see Ref. 9). (●) Quenching of IAF–PDE fluorescence by EITC–α_TGTPγS following trypsin treatment. IAF–PDE samples were trypsinized as previously described [S. Kroll, W. J. Phillips, and R. A. Cerione, J. Biol. Chem. **264**, 4490 (1989)], resulting in a level of activation of the cGMP hydrolysis activity that was identical with that of the unmodified enzyme. The IAF–PDE concentration in these titrations determined by a protein assay of the stock IAF–PDE solution was 22 nM. (Reprinted with permission from Erickson and Cerione.[9] Copyright 1991, American Chemical Society.)

dissociation constant of 20 nM (which is similar to other estimates that have been made for the binding interaction between α_T and the effector enzyme).

Conclusions

It seems likely that fluorescence spectroscopic approaches will continue to prove useful for directly monitoring the protein–protein interactions that underlie different signal transduction pathways. Whereas rhodopsin–transducin and transducin–PDE interactions have served as prototypes for receptor–G-protein and G-protein–effector coupling, other protein–protein interactions are now being identified as being important for the regulation of various receptor–G-protein-coupled signaling pathways. These include the interactions of serine/threonine kinases with receptors [e.g., the phosphorylation of the β-adrenergic receptor by the specific β-adrenergic receptor kinase (BARK) or the phosphorylation of rhodopsin by the rhodopsin kinase], the interactions of G-protein $\beta\gamma$-subunit complexes with receptors[8,19] and serine/threonine kinases (e.g., BARK[20,21]), and the identification of new G-protein–effector interactions (G_q with phospholipase C-β[22,23]). In addition, various other receptor-coupled signaling systems are becoming more amenable to fluorescence spectroscopic approaches, one in particular being the superfamily of growth factor receptor tyrosine kinases. In these systems receptor–receptor interactions (both homodimer and heterodimer formation[24]) are key events for receptor activation, and the interactions between activated receptors and phosphosubstrates that contain *src* homology regions (i.e., SH-2 domains[25]) are likely to represent key (early) events in mitogenic signaling. The development of high sensitivity readouts for these different binding events should represent a key step toward the development of rational pharmacological and therapeutic strategies aimed at modulating these interactions and thereby altering different biological responses linked to pathological states.

[19] W. J. Phillips, S. C. Wong, and R. A. Cerione, *J. Biol. Chem.* **267**, 17040 (1992).
[20] K. Haga and T. Haga, *J. Biol. Chem.* **267**, 2222 (1992).
[21] J. A. Pitcher, *et al., Science* **257**, 1264 (1992).
[22] A. V. Smrcka, J. R. Hepler, K. O. Brown, and P. C. Sternweis, *Science* **251**, 804 (1991).
[23] S. J. Taylor, H. Z. Chae, S. G. Rhee, and J. H. Exton, *Nature (London)* **350**, 516 (1991).
[24] A. Ullrich and J. Schessinger, *Cell (Cambridge, Mass.)* **61**, 203 (1990).
[25] L. C. Cantley, K. R. Avger, C. Carpenter, B. Duckworth, A. Graziani, R. Kapeller, and S. Soltoff, *Cell (Cambridge, Mass.)* **64**, 281 (1991).

[32] Specific Peptide Probes for G-Protein Interactions with Receptors

By HEIDI E. HAMM and HELEN M. RARICK

Introduction

The studies described in this chapter attempt to delineate regions on G-protein α subunits involved in interactions with receptors. The main model system to be employed is the light receptor rhodopsin and the rod G protein transducin (G_t). The visual signaling pathway is similar to that of other transduction cascades involving receptor–G-protein–effector coupling. Light activation of rhodopsin leads to the formation of metarhodopsin II (Meta II), which displays sites on the cytoplasmic surface for interaction with heterotrimeric G_t. Meta II catalyzes a conformational change in the G_t α subunit ($α_t$) which opens its nucleotide binding pocket and decreases the affinity of GDP. "Empty-pocket" G_t displays transient high affinity for rhodopsin which normally is terminated when GTP binds and causes a second conformational change, leading to decreased affinity of $α_t$ for both rhodopsin and $βγ$. In the absence of GTP, the molecular basis for the high-affinity interaction between receptor and G protein can be examined in great detail.

The use of synthetic peptides to study sites on $α_t$ of interaction with rhodopsin originated with the epitope mapping of monoclonal antibodies against $α_t$ which block its interaction with rhodopsin.[1,2] Monoclonal antibody 4A (MAb 4A) binding to proteolytic fragments of $α_t$[3,4] and inhibition of $α_t$ interaction with MAb 4A by synthetic peptides[5] provided evidence that the carboxy and amino termini of $α_t$ contribute to the epitope. The synthetic peptides that were evaluated in the epitope mapping studies also blocked rhodopsin–G_t interactions,[5] suggesting that MAb 4A had its effect by binding to amino- and carboxyl-terminal regions on $α_t$ involved in interactions with rhodopsin.

Synthetic Peptides: A Powerful Approach to Mapping Protein–Protein Interactions

The preliminary studies demonstrated that synthetic peptides are useful reagents for mapping sites of protein–protein interaction. Synthetic peptides corresponding to interfacial regions of proteins may be competitive inhibitors of the protein–protein interaction, or they may mimic the biological effect of one peptide on another (Fig. 1). Both effects are illustrated in this chapter in studies of receptor–G-protein interaction, and elsewhere in studies of G-protein–effector interaction.[6]

Once the interfacial regions have been mapped, analog peptides with single amino acid substitutions can provide information on residues critical for protein–protein interaction. Structural analysis of the interaction using

[1] H. E. Hamm and M. D. Bownds, *J. Gen. Physiol.* **84,** 265 (1984).
[2] H. E. Hamm, D. Deretic, K. P. Hofmann, A. Schleicher, and B. Kohl, *J. Biol. Chem.* **262,** 10831 (1987).
[3] D. Deretic and H. E. Hamm, *J. Biol. Chem.* **262,** 10831 (1987).
[4] M. R. Mazzoni, J. A. Malinski, and H. E. Hamm, *J. Biol. Chem.* **266,** 14072 (1991).
[5] H. E. Hamm, D. Deretic, A. Arendt, P. A. Hargrave, B. Koenig, and K. P. Hofmann, *Science* **241,** 832 (1988).
[6] H. M. Rarick, N. O. Artemyev, J. S. Mills, N. P. Skiba, and H. E. Hamm, this series, Vol. 238 [2].

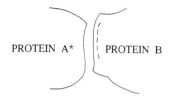

PEPTIDES FROM THE INTERFACE CAN

1) BLOCK THE INTERACTION

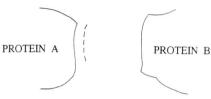

2) SIMULATE THE INTERACTION

FIG. 1. Synthetic peptides are useful reagents for mapping sites of protein–protein interaction. Synthetic peptides corresponding to interfacial regions of interaction with neighboring proteins can occupy their binding sites on neighbor protein surfaces and thus act as competitive inhibitors of the protein–protein interaction. Some proteins can cause conformational changes on neighboring proteins. Interfacial peptides can sometimes mimic the effect of the parent proteins binding to the neighbor protein binding site and cause the same conformational changes.

transferred nuclear Overhauser effect nuclear magnetic resonance (Tr NOE NMR) provides dynamic three-dimensional structural information on the biologically active conformation of the peptide when bound to its receptor site on a partner protein, conformational changes in the peptide during biological activity, and point-to-point interactions with the binding site on the partner protein. Insights into the active, bound conformation can be used in the design and synthesis of peptides containing nonnative amino acids, cyclic peptides, and nonpeptide mimetics which may have higher affinity and increased biostability. Such agents could specifically disrupt the interaction and might serve as starting points for drug design.

Practical Aspects of Synthetic Peptide Studies of Protein Interfacial Regions

When using synthetic peptides for examining functional effects in a biological system, there are two important aspects to consider, namely, the choice of peptides and proper purification and characterization of peptides. Peptide choice involves choosing an appropriate length of sequence, the appropriate area on the protein that is to be mapped, and the hydrophilicity of the peptide. Hydrophilicity analysis will provide peptides that are soluble under the assay conditions. Additionally, many sites of protein–protein interaction involve the exposed, hydrophilic regions on proteins. The second aspect is proper purification of each peptide and evaluation of the purified peptide. This will minimize side effects from incompletely cleaved peptides and other contaminants of the peptide synthesis process.

Choice of Peptides

A variety of strategies can be employed to select interfacial peptides. The most thorough would employ overlapping peptides from the entire primary sequence of the protein of interest. However, this approach, besides being prohibitively costly, would yield peptides from the protein core which not only would have a low probability of being part of an interaction site with another protein, but also would probably be insoluble. Hydrophilicity analysis provides a simple and quite accurate assessment of hydrophilic, and thus surface-exposed, residues. There is evidence that interaction sites with partner proteins are located on the surface of a protein when it is folded in its three-dimensional structure. Hopp and Woods[7] found that known antigenic residues in several proteins were correlated to hydrophilic residues of the proteins. A soluble protein folds so that mostly hydrophilic residues are located on the surface and hydrophobic residues are buried, resulting in minimized exposure of hydrophobic side chains to the aqueous medium.[8] More selective approaches for peptide selection might employ preexisting knowledge about likely sites of interaction of a protein with neighbor proteins, such as epitope analysis of antibodies that block the interaction, chemical modification, fluorescence, and cross-linking studies of interacting sites.

In the case of α subunits of G proteins, for identifying the receptor binding domain, epitope mapping studies had implicated the amino terminus and an internal region within the carboxyl terminus of α_t. Several

---

[7] T. P. Hopp and K. R. Woods, *Proc. Natl. Acad. Sci. U.S.A.* **78**, 3824 (1981).
[8] G. A. Grant, "Synthetic Peptides–A User's Guide." Freeman, New York, 1992.

lines of evidence show that the carboxyl-terminal region of α subunits directly interacts with receptors. Pertussis toxin (PT) ADP-ribosylation of Cys^{347} of α_t blocks interaction with activated rhodopsin (R*)[9] without changing other functional properties of G_t.[10] When G_t is tightly bound to R* and the GDP has dissociated from α_t, PT is unable to ADP-ribosylate α_t,[11] suggesting that access to the substrate site is blocked because it forms part of the interaction surface of α_t and R*. A monoclonal antibody specific for α_t, MAb 4A, shows similar behavior. It blocks G_t interaction with rhodopsin (and its light activation) by interaction with sites on the amino- and carboxyl-terminal regions,[2,4,5] and binding to α_t is blocked by G_t binding to R*.[11] The *unc* mutation in α_s, involving replacement of an arginine by a proline near the carboxy terminus, uncouples it from the β-adrenergic receptor.[12,13]

The next consideration for choice of peptides is hydrophilicity of the peptide. The entire α_t sequence has been subjected to relative hydrophilicity/hydrophobicity prediction analysis[14] and synthetic peptides chosen based on predicted hydrophilicity. Using these criteria, several α_t peptides from the amino and carboxyl termini have been chosen for examining the rhodopsin-binding domain. The amino acid sequences of these peptides are α_t-1-23, GAGASAEEKNSRELEKKLKEDAE; α_t-8-23, EKNSRELEKKLKEDAE; α_t-311-328, DVKEIYSHMTCATDTQNV; α_t-acetyl 311-329-amide, acetyl-DVKEIYSHMTCATDTQNVK-amide; and α_t-340-350, IKENLKDCGLF. The other important consideration in choosing hydrophilic peptides is that they are most likely to go into the aqueous buffer solutions used in the functional assays. A peptide that is insoluble in the assay buffer solutions should not be used because the results obtained may not be functionally relevant. The question of peptide solubility is discussed elsewhere.[6]

Peptide Synthesis, Purification, and Evaluation

Peptides are synthesized by the solid-state Merrifield method on an automated synthesizer in the Protein Sequencing/Synthesis Laboratory at the University of Illinois at Chicago (UIC). The peptides are purified

---

[9] C. VanDop, G. Yamanaka, F. Steinberg, R. D. Sekura, C. R. Manclark, L. Stryer, and H. R. Bourne, *J. Biol. Chem.* **259**, 23 (1984).
[10] L. Ramdas, R. M. Disher, and T. G. Wensel, *Biochemistry* **30**, 11637 (1991).
[11] H. E. Hamm, *Adv. Second Messenger Phosphoprotein Res.* **24**, 76 (1990).
[12] K. A. Sullivan, R. T. Miller, S. B. Masters, B. Beiderman, W. Heideman, and H. R. Bourne, *Nature (London)* **330**, 758 (1987).
[13] T. Rall and B. A. Harris, *FEBS Lett.* **224**, 365 (1987).
[14] J. Kyte and R. F. Doolittle, *J. Mol. Biol.* **157**, 105 (1982).

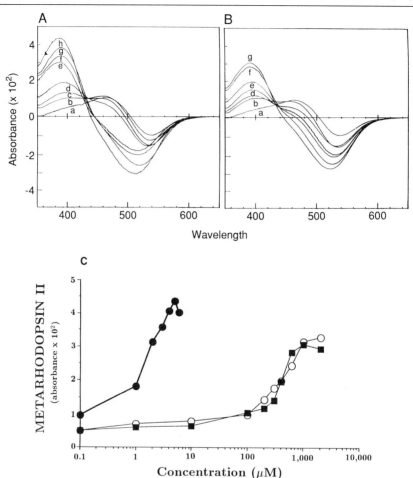

FIG. 2. (A) Spectroscopic assay of G_t stabilization of Meta II. Increasing concentrations of G_t shift the absorbance difference spectra from Meta I (peak at 480 nm) to Meta II (peak at 380 nm). Rod outer segment (ROS) membranes were washed four times in 10 mM Tris-HCl (pH 7.5), 1 mM dithiothreitol, and 1 mM EDTA to remove all G_t and other peripheral membrane proteins. The absorbance spectrum of ROS membranes (5 μM rhodopsin) in 200 mM NaCl, 10 mM MOPS (pH 8.0), 2 mM $MgCl_2$, 1 mM dithiothreitol, and 10 μM phenylmethylsulfonyl fluoride (PMSF) (a) or buffer with increasing concentration of G_t (b, 0.05 μM; c, 0.5 μM; d, 1 μM; e, 2 μM; f, 3 μM; g, 4 μM; h, 5 μM) was measured in the dark at 4° with an SLM Aminco DW2000 spectrophotometer. The ROS membranes were then exposed to light bleaching 5% of the rhodopsin with an actinic flash. After 1 min the absorbance spectrum was again measured, and the difference spectrum was computed. (B) Synthetic peptide α_t-340–350 causes the same shift in the absorbance difference spectra from Meta I to Meta II as G_t. Difference spectra were determined as above in the presence of 0 (a) or increasing concentrations (b, 0.1 mM; c, 0.2 mM; d, 0.3 mM; e, 0.4 mM; f, 0.6

by reversed-phase high-performance liquid chromatography (HPLC). It is to be stressed that synthetic peptides should not be tested for functional effects unless they are determined to be pure. Purity of all peptides should be checked by the following: (1) fast atom bombardment (FAB) mass spectrometry, (2) analytical HPLC, and (3) amino acid analysis. Only those peptides with a single peak corresponding to the predicted molecular weight on a FAB mass spectrogram should be used for testing functional effects. If more than one peak is present, then the peptide should be purified further. Otherwise, it is unclear whether a particular functional effect is due to the peptide, a different peptide sequence synthesized in error, or a contaminant. It is to be noted that the upper limit of mass determination for FAB mass spectrometry is about 3000 daltons. To enable detection of the purity of the synthetic peptide, most are approximately 20 and no more than 30 residues in length.

Sites on α_t of Interaction with Rhodopsin

The interaction of G_t with rhodopsin can be directly measured by binding to and stabilization of metarhodopsin II. Light-excited rhodopsin rapidly transits a series of spectral intermediates and arrives at an equilibrium between the tautomeric forms metarhodopsin I (Meta I, A_{max} 480 nm) and Meta II (A_{max} 380 nm).[15] Under conditions of the G-protein binding assay (pH 7.5–8 and 0°–8°), the Meta I ⇌ Meta II equilibrium is shifted strongly to Meta I in the absence of the G protein. The G protein binds to Meta II and pulls the Meta I ⇌ Meta II equilibrium to Meta II. The resulting formation of the Meta II–G-protein complex, called the "extra Meta II," is monitored spectrophotometrically.[16] Figure 2A shows visible and ultraviolet absorption difference spectra induced by light exposure of

[15] R. G. Matthews, R. Hubbard, P. K. Brown, and G. Wald, *J. Gen. Physiol.* **47**, 215 (1963).
[16] D. Emeis, H. Kühn, J. Reichert, and K. P. Hofmann, *FEBS Lett.* **143**, 29 (1982).

---

mM; g, 1 mM) of α_t-340–350. Lack of complete stabilization of Meta II by the peptide may reflect an ability of the peptide to bind weakly to Meta I as well as more strongly to Meta II [M. Watanabe, M. B. Lazarevic, H. Hamm, and M. M. Rasenick, *Neuroscience (Oxford)* **18**, 356 (1991)], thus resulting in less complete stabilization of Meta II. (C) G_t and peptide dose response. The dose-dependent increase in Meta II caused by G_t (●), α_t-340–350 (■), or an Arg-substituted analog, α_t-340–350K341R (○), is plotted as a function of concentration. The analog peptide was as potent as the native peptide, and it was used in NMR experiments described elsewhere [E. A. Dratz, J. E. Furstenau, C. G. Lambert, D. L. Thierault, H. M. Rarick, T. Schepers, S. Paylian, and H. E. Hamm, *Nature (London)* **363**, 276 (1993)]. Meta II was measured as the difference between absorbance at 380 and 420 nm on the difference spectrum. SEM error bars are within the area of the symbols.

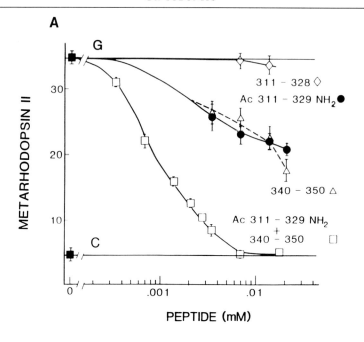

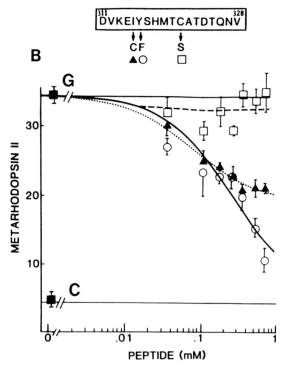

rhodopsin in washed rod outer segment (ROS) membranes in the presence of increasing concentrations of heterotrimeric G_t. The spectral changes are as expected for the loss of Meta I and the gain of Meta II in the presence of increasing amounts of G_t. Figure 2C shows the amount of Meta II stabilized as a function of G_t concentration. The intensity of the absorbance signal at 380 nm is proportional to the number of rhodopsin–G_t complexes and thus provides a direct assay of the protein–protein interaction.

Synthetic peptides from the regions of α_t that form the interface with rhodopsin can directly bind to Meta II and compete with G_t.[5] Stabilization of Meta II by G_t was used as a direct assay of interaction. Two peptides that had been found to block MAb 4A binding to G_t[5] also blocked interaction with rhodopsin (Fig. 3A). These sequences were at the carboxyl terminus (α_t-340–350) and just internal to the carboxyl terminus (α_t-311–328). The internal peptide was a more effective competitor after acetylation and amidation (α_t-acetyl-311–329-amide), whereas truncated versions were only partially effective.[5] An analog peptide to α_t-311–328 which substituted Cys^{321} with Ser was no longer a competitor, suggesting that this amino acid is important in G_t–rhodopsin interaction (Fig. 3B). When α_t-acetyl-311–329-amide and α_t-340–350 are added together to the assay, they are 20-fold more potent than either peptide alone (Fig. 3A). Thus, there is a marked potentiation of effects when the peptides are presented in combination.

Amino-terminal peptides, α_t-1–23 and α_t-8–23, but not α_t-1–15, were also able to block interaction with rhodopsin (Fig. 4), suggesting either that the amino terminus is directly involved in rhodopsin binding (as has been previously suggested[17,18]) or that the peptide blocked α–$\beta\gamma$ interaction and consequentially blocked G_t interaction with rhodopsin. None of the peptides is able to block directly α_t–$\beta\gamma$ interaction measured either in sucrose density gradient experiments or in studies of $\beta\gamma$-mediated stimu-

---

[17] V. N. Hingorani and Y.-K. Ho, *J. Biol. Chem.* **263**, 19804 (1988).
[18] T. Higashijima and E. M. Ross, *J. Biol. Chem.* **266**, 12655 (1991).

FIG. 3. Effect of synthetic peptides that are part of the MAb 4A epitope on rhodopsin–G_t interaction measured using the extra Meta II assay as shown in Fig. 2. The filled squares show the amount of Meta II in the presence (G) and absence (C) of G protein. The dose-dependent effects of peptides on G-protein-stabilized Meta II are shown. (A) Cooperative effects of two peptides, α_t-acetyl-311–329-NH_2 and α_t-340–350, on rhodopsin–G_t interaction. At the peptide concentration range shown, α_t-311–328 has nearly no effect. (B) Effects of single amino acid substitutions in α_t-311–328 on Meta II stabilization.

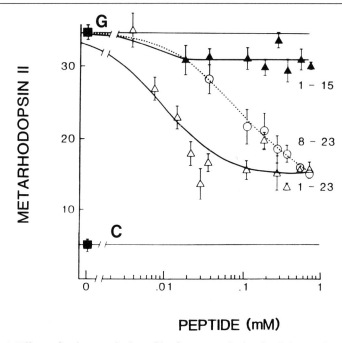

FIG. 4. Effects of amino-terminal peptides from α_t on rhodopsin–G_t interaction. Methods are similar to those in Fig. 2. α_t-1–23 is the most potent at inhibiting the interaction, whereas α_t-8–23 is less potent. α_t-1–15 has no significant effect on the interaction.

lation of pertussis toxin ADP-ribosylation of α_t.[19] Thus, our data suggest that the amino terminus of α_t is also involved in interactions with rhodopsin.

In the absence of G_t, α_t-340–350 dose-dependently stabilized metarhodopsin II (Fig. 2B, C), mimicking the heterotrimeric G_t. It is apparent that increasing amounts of the α_t-340–350 peptide cause identical spectral changes as the full heterotrimeric G_t protein and therefore stabilize indistinguishable Meta II products. The stabilization of Meta II saturates at about 1 mM peptide, and half-saturation occurs at about 300 μM (Fig. 2C).

Conformation of Receptor-Bound Peptide α_t-340–350

The conformation of α_t-340–350 when bound to rhodopsin was determined by two-dimensional transferred NOE NMR.[20] The free peptide has

[19] H. E. Hamm, unpublished observations, 1989.
[20] E. A. Dratz, J. E. Furstenau, C. G. Lambert, D. L. Thierault, H. M. Rarick, T. Schepers, S. Paylian, and H. E. Hamm, *Nature (London)* **363,** 276 (1993).

very little structure, but the rhodopsin-bound peptide has a well-defined structure. One particularly intriguing aspect of the structure is an apparent β turn at the carboxyl terminus of $α_t$, involving Cys^{347} (the target for pertussis toxin ADP-ribosylation), Gly^{348}, Leu^{349}, and Phe^{350}. Because β turns are commonly found in regions of molecular recognition, the importance of this turn for the biological activity of the peptide was tested. It is known that glycine residues are frequently found in the second or third positions of β turns. Replacement of Gly with any L-amino acid should disrupt the turn formed in the bound state. However, replacement of Gly^{348} with the enantiomorphic D-alanine is consistent with the turn structure, but would be predicted to break other common types of secondary structure.

Peptide analogs replacing Gly^{348} with D-Ala or L-Leu were synthesized and tested for the ability to bind to and stabilize Meta II. Figure 5 shows the effect of these peptides on stabilization of Meta II. The D-Ala analog is nearly as potent as the native peptide in stabilizing Meta II, whereas the L-Leu analog has greatly reduced activity. The combined structural and functional data provide evidence for a type II' β turn between Cys^{347} and Phe^{350}, which is required for peptide binding to and stabilization of Meta II.[20]

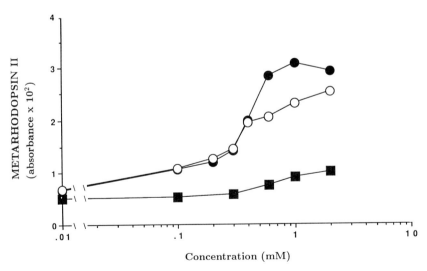

FIG. 5. Effect of residues 340–350 and single amino acid substitutions. Meta II stabilization by $α_t$-340–350 (●) and analogs at the Gly^{348} position (D-Ala, ○; L-Leu, ■) was measured as described in Fig. 2.

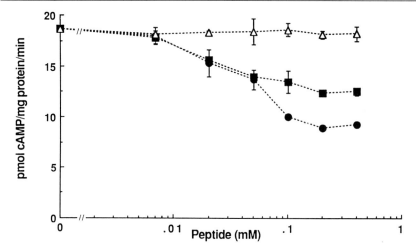

FIG. 6. Effects of peptides from homologous regions of α_s on isoproterenol-stimulated adenylyl cyclase in permeable C6 glioma cells. Effects of peptides α_s-354–372 (■), α_s-384–394 (●), and peptide α_i-345–355 (△) on the adenylyl cyclase activity of C6 cells stimulated by 10 μM each of isoproterenol and GTP are shown. Adenylyl cyclase was measured as described [S. Hatta, M. M. Marcus, and M. M. Rasenick, *Proc. Natl. Acad. Sci. U.S.A.* **83**, 5439 (1986)]. Data are means ± SEM of triplicates from a representative experiment which was repeated three times.

Sites on G_s of Interaction with β-Adrenergic Receptor

It would be important to determine whether short synthetic peptides from interfacial regions of G-protein α subunits contain within their sequence the information for encoding specificity of interaction with cognate receptors, or rather whether they would interact with any of the G-protein-coupled receptors. To address this question, a comparative approach was taken. Synthetic peptides from the homologous regions of the α subunit of G_s were synthesized. The sequence of the amino-terminal peptide from α_s related to α_t-8–23 is α_s-15–34, EEKAQREANKKIEKQLQKDK; the carboxyl-terminal 11 amino acids in α_s, α_s-384–394, are QRMHLRQY-ELL; and the internal region near the carboxyl terminus corresponding to α_t-311–329 is α_s-354–372, DGRHYCYPHSTCAVDTENIR. These α_s peptides were tested for the effects on the β-adrenergic receptor adenylyl cyclase (adenylate cyclase) transmembrane signaling complex in C6 glioma cells. Synthetic peptides from the carboxyl-terminal regions of α_s block β-adrenergic receptor-mediated stimulation of adenylyl cyclase[20] (Fig. 6). They have no effect on fluoride- or Mn^{2+}-stimulated adenylyl

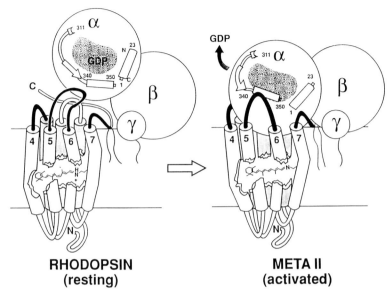

RHODOPSIN
(resting)

META II
(activated)

FIG. 7. Model of the structural basis of rhodopsin–G_t interaction. A relatively low-affinity interaction is proposed in the dark (resting) state. Light activation of rhodopsin opens up additional sites of interaction, and there is a large increase in affinity between the two proteins (activated state). The amino- and two carboxyl-terminal regions of α_t that block the protein–protein interaction are shown as potential rhodopsin interaction sites. The three filled regions on the cytoplasmic surface of rhodopsin that block the interaction[23] are shown as complimentary interacting regions. This six-point high-affinity interaction catalyzes removal of GDP from the nucleotide binding site on α_t.

cyclase activity.[21] In addition, the carboxyl-terminal α_s peptides stabilize the active receptor conformation that has high affinity for agonists,[22] similar to the stabilization of Meta II by α_t peptides. The effects are specific, since α_i peptides have no effect on β-adrenergic receptor–G_s interaction (Fig. 6), whereas an α_s peptide has no effect on rhodopsin–G_t interaction (data not shown).

In contrast to the G_t–rhodopsin interaction, a synthetic peptide corresponding to the amino-terminal region of α_s (α_s-15–34) had no effect on β-adrenergic receptor–G_s coupling (data not shown). This finding suggests either that the amino-terminal region of α_s is not involved in interaction

[21] M. B. Lazarevic, M. Watanabe, M. M. Rasenick, and H. E. Hamm, *Neuroscience* (*Oxford*) **18**, 356 (1991).
[22] M. Watanabe, M. B. Lazarevic, H. Hamm, and M. M. Rasenick, *Neuroscience* (*Oxford*) **18**, 356 (1991).
[23] B. Koenig, A. Arendt, J. H. McDowell, M. Kahlert, P. A. Hargrave, and K. P. Hofmann, *Proc. Natl. Acad. Sci. U.S.A.* **86**, 6878 (1989).

with the β-adrenergic receptor or that this peptide was unable to form the appropriate conformation for binding.

These studies have allowed the determination of certain sites on G proteins of interaction with their cognate receptors. At least three regions on α_t bind rhodopsin, and they have synergistic effects on binding. This suggests a complex interaction with rhodopsin, including at least two sequences juxtaposed in the tertiary structure of the protein to form a binding site with rhodopsin. This information, and other information on receptor–G-protein interaction mechanisms, based on a variety of studies, is summarized in Fig. 7.[23]

The above findings suggest that short peptide sequences may carry a surprising degree of information about specific interaction with cognate receptors. These data suggest that peptides or peptide mimetics might be useful tools to modulate directly specific receptor–G-protein interactions. The structural insights provided by NMR studies of the active, rhodopsin-bound conformation of α_t-340–350 may provide a useful approach to drug design of receptor–G-protein modulators.

[33] Vaccinia Virus Systems for Expression of Gα Genes in S49 Cells

By FRANKLIN QUAN and MICHAEL FORTE

Introduction

The G proteins are a family of heterotrimeric guanine nucleotide-binding proteins that couple the receptors for a large array of extracellular signals to intracellular second-messenger-generating effector proteins. Studies of the mechanisms underlying G-protein-coupled signaling have been directed, in large part, toward defining the functional domains of the α subunit, the primary mediator of accurate information flow from receptors to effectors. To this end, models of Gα structure based on the three-dimensional X-ray crystal structures of members of related families of GTP-binding proteins have been proposed.[1,2] Subsequently, domains responsible for high-affinity guanine nucleotide binding and hydrolysis

---

[1] S. Masters, R. Stround, and H. Bourne, *Protein Eng.* **1,** 47 (1986).
[2] S. Holbrook and S.-H. Kim, *Proc. Natl. Acad. Sci. U.S.A.* **86,** 1751 (1989).

and those responsible for interacting with receptors, effectors, and the $\beta\gamma$ subunit complex have all been experimentally localized within the primary amino acid sequence of a variety of α subunits with varying degrees of precision.[3]

The stimulatory G protein (G_s) activates adenylyl cyclase (adenylate cyclase) in response to a variety of hormonal stimuli and is well characterized in terms of its activity and mechanism of action. As a result, this subunit has been used extensively in studies of α-subunit structure and function. These studies have been greatly facilitated by the identification and characterization of somatic cell mutants of the murine lymphoma cell line S49 that are defective in a variety of $G_s\alpha$ functions. Significantly, in one such mutant S49 line, the cyc^- variant, the activity of modified or chimeric $G_s\alpha$ subunits can be assayed in the absence of background activity from endogenous $G_s\alpha$ subunits since this line lacks endogenous $G_s\alpha$ mRNA and protein.[4] In most studies, the expression of heterologous $G_s\alpha$ cDNAs in these cells has required the generation of stable transformants using recombinant retrovirus vectors since S49 cells are relatively refractile to standard cell transfection techniques.[5] This procedure requires multiple transfection and selection steps and is time consuming and labor intensive.

In this chapter we describe the use of vaccinia viruses (VV) for the transient expression of mammalian and *Drosophila* $G_s\alpha$ subunits in S49 cyc^- cells.[6] Vaccinia viruses are lytic DNA viruses with a broad host cell range and have been used for the heterologous expression of a variety of cDNAs. The biology of these viruses has been described in detail elsewhere.[7,8] The transcription of VV genes takes place in the cytoplasm of infected cells and is carried out by viral enzymes. As a result of this cytoplasmic replication mode, the only requirement for the expression of foreign cDNAs is that they be cloned downstream of a high-efficiency vaccinia promoter. Additional regulatory sequences are not required in the cDNAs to be expressed.

Materials

Wild-type vaccinia virus (WR strain) is obtained from Dr. Gary Thomas (Vollum Institute, Portland, OR). S49 cyc^- cells are obtained from Dr.

---

[3] C. Berlot and H. Bourne, *Cell (Cambridge, Mass.)* **68**, 911 (1992).
[4] B. Harris, J. Robishaw, S. Mumby, and A. Gilman, *Science* **229**, 1274 (1985).
[5] S. Masters, K. Sullivan, R. Miller, D. Beiderman, N. Lopez, J. Ramachandran, and H. Bourne, *Science* **241**, 448 (1988).
[6] F. Quan, L. Thomas, and M. Forte, *Proc. Natl. Acad. Sci. U.S.A.* **88**, 1902 (1991).
[7] D. Hruby, G. Thomas, E. Herbert, and C. A. Franke, this series, Vol. 124, p. 295.
[8] A. Karschin, B. A Thorne, G. Thomas, and H. A. Lester, this series, Vol. 207, p. 408.

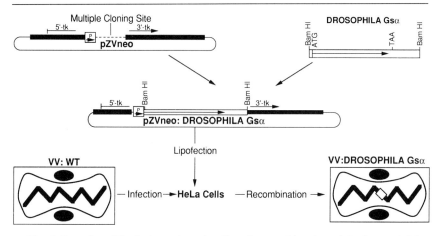

FIG. 1. Overview of the design and construction of a recombinant vaccinia virus containing the *Drosophila* G$_s\alpha$ gene. For details of the construction of the recombinant insertion plasmid, infection and transfection procedures, and selection of recombinant VV, see text.

Henry Bourne (University of California, San Francisco). Media and sera are from GIBCO (Grand Island, NY). Lipofectin is obtained from Bethesda Research Laboratories (Gaithersburg, MD, Cat. No. 8292A). Nitrocellulose (HA85, 0.45 mm) is from Schleicher and Schuell (Keene, NH). Colony/plaque screen is from Du Pont/New England Nuclear (Boston, MA). ^{125}I-conjugated protein A is from Amersham (Arlington Heights, IL).

Construction of Recombinant Vaccinia Viruses

Overview

Owing to the large size of the VV genome [187 kilobases (kb)], conventional molecular biological techniques cannot be used to generate recombinant viruses. Instead, recombinant VV are constructed by homologous recombination using vectors containing nonessential VV genomic sequences (Fig. 1). The insertion plasmid pZV*neo* has previously been described.[9] This plasmid contains the nonessential VV thymidine kinase (*tk*) gene interrupted by a high-efficiency VV promoter (constitutive p7.5K promoter) followed by a multiple cloning site (MCS). The viral *tk* sequences target the site of homologous recombination to a nonessential region of the VV genome and also provide the information necessary for the proper processing of the chimeric transcript. pZV*neo* also carries the

[9] J. S. Hayflick, W. J. Wolfgang, M. A. Forte, and G. Thomas, *J. Neurosci.* **12,** 705 (1992).

Tn5 aminoglycoside transferase gene under the control of the VV p11 promoter and thus confers neomycin resistance. This plasmid also contains the appropriate sequences for replication and antibiotic (ampicillin) selection in *Escherichia coli*.

To generate a VV recombinant, cells are first infected with wild-type virus at a low multiplicity of infection (m.o.i.). The recombinant insertion vector is then introduced by lipofection. Homologous recombination results in the insertion of the cDNA into the VV genome at a specific location. A crude virus stock is prepared, and recombinant viruses are detected and purified from the wild-type virus background by successive rounds of plaque hybridization. Once a recombinant VV has been isolated and purified, a high-titer viral stock is generated for expression studies.

Standard Vaccinia Virus Infection Protocol

The appropriate dilution of the partially purified (see below) VV stock is made up in 1.5 ml of phosphate-buffered saline containing 1 mM $MgCl_2$ (PBS/M) and 0.01% (w/v) bovine serum albumin (PBS/MB). Confluent monolayers in 10-cm plates are washed once with warm PBS/M. The virus inoculum is added and the cells incubated for 30 min at room temperature with gentle rocking every 10 min. The virus is removed by aspiration and the cells refed with culture medium. Plaques appear after 48–72 hr at 37°.

Homologous Recombination Procedure Using Lipofectin

Recombinant insertion vector is purified by cesium chloride gradient centrifugation. The homologous recombination procedure is performed using HeLa cells. HeLa cells are maintained in Dulbecco's modified minimal essential medium (DMEM) containing 10% (v/v) fetal calf serum (FCS) and 25 μg/ml gentamycin. For the recombination procedure, cells are grown in 35-mm culture dishes until 50–80% confluent. Cells are infected for 30 min with wild-type VV at a m.o.i. of 1. During this time period, the lipofectin–DNA complex is generated. Recombinant insertion vector (5.0 μg) is resuspended in 50 μl of sterile water. The lipofectin reagent (20 μl) is diluted with 30 μl of sterile water in a polystyrene tube. The DNA is added to the lipofectin and mixed gently to avoid the formation of clumps. Complex formation is allowed to proceed at room temperature for 15 min. After the removal of the virus inoculum the cells are washed 3 times with serum-free DMEM, and 1 ml of serum-free DMEM is added to the plate. The lipofectin–DNA complex is added to the cells dropwise. The plate is then gently swirled and incubated at 37° for 3 hr. One milliliter of DMEM plus 10% (v/v) FCS is added and the cells incubated at 37° for an additional 21 hr. A crude virus stock is then prepared. Cells are scraped into 15-ml conical centrifuge tubes and pelleted at low speed in a clinical centrifuge (Sorvall, Boston, MA, Rt6000B, 1000 rpm, 5 min at room temperature). The cells are washed with 5 ml of PBS/M and resuspended in 1.0 ml of PBS/M. Virus is released by three cycles of freeze/thawing and the crude stock stored at $-70°$.

Detection of Recombinant Viruses and Plaque Purification

Confluent monolayers of African green monkey BSC40 cells in 10-cm plates are used for recombinant VV detection and plaque purifications. The BSC40 cells are grown in DMEM containing 10% (v/v) heat-inactivated FCS and 25 μg/ml gentamicin. Prior to infecting cells, crude virus stocks are treated with an equal volume of 0.25% (w/v) trypsin in Versene for 12 min at 37°. The reaction is terminated by the addition of cold PBS/MB. The infectious titer of the crude virus stock is assumed to be approximately 10^7 plaque-forming units (pfu)/ml. Standard VV infections are carried out so as to yield 10, 100, and 1000 plaques/plate. After 2 days, plaques are visualized by staining with 0.01% (w/v) neutral red in PBS at 37°. The medium is then aspirated and an 82-mm colony/plaque screen filter gently pressed against the monolayer using a Kimwipe moistened with PBS/MB. Gloved hands are used to hold the filter in place in order to prevent smearing of plaques. The filter is then carefully removed using forceps and placed plaque side up on Whatman (Clifton, NJ) 3MM paper wetted with PBS/M in a 10-cm petri dish. The pattern of plaques is then replicated onto an 82-mm nitrocellulose filter by pressing it against the colony/plaque screen filter as described above. A single-hole paper punch (4 mm diameter), sterilized in ethanol, is then used to punch an asymmetric pattern of holes around the periphery of the filters. The filters are separated using forceps, and the nitrocellulose filter, to be used for subsequent virus recovery, is stored face up in a 10-cm petri dish on PBS/M-wetted Whatman 3MM paper at $-70°$.

The colony/plaque screen filter is processed for plaque hybridization. The filter is successively floated plaque side up on denaturing solution (0.5 N NaOH/1.5 M NaCl, 10 min), neutralizer (3.0 M NaCl/0.5 M Tris-HCl, pH 7.5, two times for 2 min each time) and then washed in 2× SSC (1× SSC is 0.15 M NaCl, 15 mM trisodium citrate). After being air dried or baked *in vacuo* at 80° for 30 min, the filter is treated with 0.1 M Tris-HCl, pH 8.0/0.15 M NaCl/10 mM EDTA/0.2% (w/v) sodium dodecyl sulfate (SDS) containing 50 μg/ml proteinase K at 55° for 30 min. Prehybridization of the filter is carried out in 50% (v/v) formamide/1 M NaCl/ 10% (w/v) dextran sulfate/1% (w/v) SDS/100 μg/ml denatured salmon sperm DNA for 2–4 hr at 42°. A radiolabeled probe specific for the inserted cDNA is then added to the prehybridization mixture. The probe is allowed to hybridize overnight. The filter is washed ($2\times$ SSC/0.1%, w/v, SDS, 65°, 30 min) and the recombinant plaques identified by autoradiography.

Recombinant plaques are recovered from the nitrocellulose filter. The filter is removed from the $-70°$ freezer, thawed, and the paper punch holes used to align the filter with the film. A razor blade is used to cut the positive plaques out of the filter, and the filter slice is placed in 200 μl of PBS/MB. A crude virus stock is generated by three cycles of freeze/thawing and six 10-sec cycles of indirect sonication. This stock is used for the next round of plaque purification. Aliquots of 5 and 20 μl are added to 1 ml of PBS/MB and plated out on confluent monolayers of BSC40 cells in 10-cm dishes. Plaque lifts and hybridizations are repeated until all plaques are positive for hybridization and a single plaque can be cleanly isolated.

Final Purification of Recombinant Viruses

An agar overlay is used to complete the plaque purification of recombinant VV. A monolayer of BSC40 cells in a 10-cm dish is infected with a dilution of the crude recombinant VV stock in order to yield 25–200 plaques. The agar overlay is prepared by combining 5 ml of a 1.5% (w/v) agarose (SeaPlaque, FMC, Rockland, ME) solution in water (at 48°) with 5 ml of $2\times$ DMEM containing 10% (v/v) FCS. After virus adsorption, the overlay is added and allowed to harden for 15 min, and the cells are incubated at 37° for 48 hr. The cells are stained by adding a second overlay consisting of 5 ml of 1% (w/v) agarose in water (at 48°) containing 200 μl of 1% (w/v) neutral red. Staining is allowed to proceed for 2–3 hr or until the plaques (which appear clear) become visible. A long-stem Pasteur pipette, containing 200 μl of PBS/M, is used to transfer agarose plugs containing well-isolated plaques to 4-ml polypropylene tubes. The virus is released by three freeze/thaw cycles. Monolayers of BSC40 cells in 24-well plates are then infected with 100 μl of the crude virus stock and refed. After 2 days, a crude stock is made from the infected cells and a portion used for slot-blot hybridization in order to confirm that a recombinant VV has been isolated.

Preparation of Partially Purified Vaccinia Virus Stock

After the isolation of a recombinant VV, a high-titer stock is generated for expression experiments. The 24-well crude virus stock is estimated to have a titer of approximately 1×10^7 pfu/ml. Four 15-cm plates of BSC40 cells are infected with the virus stock at a m.o.i. of 0.005 pfu/cell. Cells should be harvested after 48–72 hr, before becoming detached from the plate. The cells, still in culture medium, are scraped into 50-ml conical tubes, pelleted in a clinical centrifuge at low speed (1000 rpm, 5 min), and washed once with PBS/M. The cells are resuspended in 10 ml of cold 10 mM Tris-HCl, pH 9.0. All steps subsequent to this are carried out at 4°. The cells are broken with 25 strokes of a Dounce homogenizer (Kontes Pestle A). The homogenate is transferred to a 15-ml conical centrifuge tube and nuclei and unbroken cells removed by low-speed centrifugation (1000 rpm, 5 min at room temperature); the supernatant is saved. The pellet is then rehomogenized and reextracted in 10 ml of 10 mM Tris-HCl, pH 9.0. The supernatants are combined and overlayed onto a 16-ml pad of 36% (w/v) sucrose/10 mM Tris-HCl, pH 9.0, in a 38-ml ultracentrifuge tube. The virus particles are sedimented through the sucrose pad by centrifugation at 18,000 rpm for 80 min at 4° (Beckman SW28 rotor, Richmond, CA). After centrifugation the solution is aspirated away from the visible virus pellet and the pellet resuspended in 1.0–1.5 ml of 10 mM Tris-HCl, pH 9.0. To obtain a uniform suspension of virus particles, the pellet is homogenized with 7 strokes of a Dounce homogenizer. The virus stock is stored at −70° in small aliquots. The stock is stable at this temperature for several years. Once thawed an aliquot is not refrozen but is stored at 4°. The virus stock is stable at that temperature for up to 1 month.

The titer of the partially purified virus stock is of the order of 10^{10} pfu/ml. The actual titer of the stock must be obtained by plaque counting using 10-cm plates of confluent BSC40 cells. Cells are infected with dilutions of the virus stock calculated to yield approximately 10, 100, and 1000 plaques per plate. After 2 days the medium is removed and 1 ml of 0.5% (w/v) methylene blue in 50% (v/v) methanol is added. After at least 10 min at room temperature the plate is washed with 10% (v/v) methanol and the plaques counted.

Infection of S49 cyc⁻ Cells with Recombinant Vaccinia Viruses

The S49 cyc^- cells are grown in DMEM containing 10% (v/v) heat-inactivated horse serum and 25 μg/ml gentamycin. To infect these cells efficiently with vaccinia viruses it is necessary to use more stringent infection conditions. Infections are carried out using a high m.o.i. and are performed at higher temperatures for longer periods. S49 cyc^- cells are first counted and washed once with PBS/M. The cells are then resuspended in PBS/MB at a density of approximately 3×10^7 cells/ml and added to 10-cm petri dishes. Infections are carried out at a m.o.i. of 30–40. A suitable dilution of virus is made up in PBS/MB and added directly to the cells. The infection is then carried out at 31° for 2 hr. Sufficient medium [DMEM plus 10% (v/v) heat-inactivated horse serum plus 25 μg/ml gentamicin] is added directly to the infection mixture such that the final cell density is approximately 6×10^6 cells/ml. Cells are incubated at 37° for 14–18 hr.

Preparation of Membranes

All steps in preparing membranes are carried out at 4°. Infected cells are washed twice with PBS/M and resuspended in 5 ml of 20 mM Tris-HCl, pH 7.5/2.5 mM MgCl$_2$/1 mM EDTA (TME). Cells are disrupted by Dounce homogenization and nuclei and unbroken cells pelleted in a clinical centrifuge at low speed (1000 rpm, 10 min). The supernatant is saved and the pellet reextracted with 5 ml of TME. Membranes are pelleted from the combined supernatants by centrifugation at 34,000 rpm for 1 hr (Beckman 50Ti rotor). The membrane pellet is resuspended in a minimal volume of TME and protein concentrations determined using the BCA method (Pierce Chemical Co., Rockford, Il).

Expression of Gα Subunits in S49 cyc^- Cells

As a test of the utility of the vaccinia system for the expression of G$_s\alpha$ subunits in S49 cyc^- cells, recombinant VV were generated that coded for G$_s\alpha$ subunits from both mammalian and invertebrate sources. A cDNA encoding the 52-kDa form of rat G$_s\alpha$ in pGEM2 was obtained from Reed.[10] cDNAs encoding the 51- and 48-kDa forms of *Drosophila* G$_s\alpha$, as well as a constitutively active 48-kDa subunit, have been described.[6,11,12] The constitutively active 48-kDa *Drosophila* subunit was generated by site-directed mutagenesis and is an α subunit in which Gln-215 has been replaced by Leu. After infecting S49 cyc^- cells with the recombinant VV, the expression of Gα subunits in S49 cyc^- cells was assessed by Western blot analysis of membrane proteins. Membrane proteins were prepared as described above. Western blots were probed with an affinity-purified rabbit antibody (designated RM), directed against a synthetic peptide (Arg-Met-His-Leu-Arg-Gln-Tyr-Glu-Leu-Leu) specific for the carboxy terminus of vertebrate and *Drosophila* G$_s\alpha$ subunits.[13] Cross-reacting proteins were visualized using ^{125}I-conjugated protein A.

[10] D. Jones and R. Reed, *J. Biol. Chem.* **265**, 14241 (1987).
[11] F. Quan, W. Wolfgang, and M. Forte, *Proc. Natl. Acad. Sci. U.S.A.* **86**, 4321 (1989).
[12] F. Quan and M. Forte, *Mol. Cell. Biol.* **10**, 910 (1990).
[13] W. Simonds, P. Goldsmith, J. Codina, C. Unson, and A. Spiegel, *Proc. Natl. Acad. Sci. U.S.A.* **86**, 7809 (1989).

Western Analysis of $G_s\alpha$ Expression in S49 cyc^- Cells

The results of a typical infection are shown in Fig. 2. Wild-type S49 cells express multiple forms of $G_s\alpha$ (lane 1, Fig. 2), and, as expected, no immunoreactive protein is seen in S49 cyc^- cells (lane 6). In addition, no immunoreactive protein is seen in S49 cyc^- cells infected at a high m.o.i. with wild-type VV (lane 7, Fig. 2). However, expression of $G_s\alpha$ subunits is seen in S49 cyc^- cells infected with recombinant VV (lanes 2–5, Fig. 2). Immunoreactive proteins of the expected molecular weight are seen in each case. The *Drosophila* proteins are expressed at similar (lanes 3–5, Fig. 2) but reduced levels relative to the endogenous subunits of the wild-type S49 cells. In contrast, the rat $G_s\alpha$ subunit is expressed at higher levels (lane 2, Fig. 2). The difference in expression levels of the vertebrate and invertebrate subunits may be a reflection of different patterns of codon usage in mammalian and invertebrate genes or differences in mRNA stabilities owing to differences in untranslated regions of the mRNAs.

Summary

In this chapter, we describe the use of recombinant VV to express $G\alpha$ subunits efficiently in S49 cyc^- cells, a murine lymphoma cell line deficient in endogenous $G_s\alpha$ activity. Additional studies have shown that VV infection does not interfere with G-protein-coupled signal transduction events.[6]

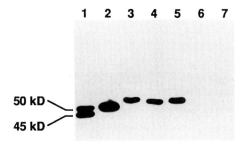

FIG. 2. Expression of $G_s\alpha$ subunits in S49 cyc^- cells using recombinant VV. Membrane proteins (50 μg/lane) prepared from infected or mock-infected cells were separated on a SDS/11% polyacrylamide gel, transferred to nitrocellulose, and probed with the RM antibody. The S49 cyc^- cells were infected with wild-type VV (lane 7) or recombinant VV encoding rat $G_s\alpha$ (lane 2) or *Drosophila* $G_s\alpha$ (51-kDa form, lane 3; 48-kDa form, lane 4; constitutively active Q227L form, lane 5). Membranes were also prepared from uninfected S49 wild-type (lane 1) and cyc^- (lane 6) cells. (Reprinted from Quan *et al.*,[6] with permission.)

In particular, the VV-expressed rat $G_s\alpha$ subunit is able to mediate efficient receptor-dependent and receptor-independent activation of adenylyl cyclase in cyc^- cells. This system should therefore be useful for rapidly assessing the activity of modified or chimeric $G_s\alpha$ subunits.

Section II

$G\beta\gamma$ Subunits

[34] Purification of T$\beta\gamma$ Subunit of Transducin

By JOËLLE BIGAY and MARC CHABRE

Introduction

Like other heterotrimeric G proteins, transducin, the G protein of retinal rods, is made up of a nucleotide-binding subunit Tα and a dimeric T$\beta\gamma$ subunit formed of two undissociable β- and γ-peptides. The transducin T$\beta\gamma$ subunit contains only the β_{36}-peptide, and not the β_{35}-peptide found in other heterotrimeric G proteins.[1,2] The γ-peptide of transducin is farnesylated (C_{15} polyisoprenoid),[3] whereas the γ-peptides of other G-proteins are geranylgeranylated (C_{20} polyisoprenoid).[2] This accounts for the extractability of transducin from bovine retinal rod outer segment (ROS) membranes in the absence of detergent. T$\beta\gamma$ is associated with Tα in the inactive Tα_{GDP}–T$\beta\gamma$ holoenzyme. This association is required for the catalysis by photoactivated rhodopsin of GDP/GTP exchange in Tα. On this exchange, T$\beta\gamma$ dissociates from Tα_{GTP}.

Preferential Extraction of T$\beta\gamma$ Subunits from Rod Outer Segment Membranes

The nucleotide dependence of the subunit dissociation is the basis of a procedure of sequential extraction of Tα and T$\beta\gamma$ subunits from illuminated bovine ROS membranes, which has been described in [11] of this volume. In brief, Tα_{GTP} (or T$\alpha_{GTP\gamma S}$) is first extracted preferentially from an illuminated ROS membrane pellet by elution in isotonic buffer in the presence of GTP (or GTPγS). T$\beta\gamma$, which remains mostly membrane bound in isotonic buffer, is eluted subsequently in hypotonic buffer without adding further nucleotides (see [11]). This last extract contains typically 80% T$\beta\gamma$, 20% Tα, and little free nucleotide. A nucleotide removal step is thus not required before the purification of this T$\beta\gamma$-enriched extract on an ion-exchange column.

---

[1] E. J. Neer and D. E. Clapham, *in* "G Proteins" (R. Iyengar and L. Birnbaumer, eds), p. 41. Academic Press, San Diego, 1990.
[2] M. I. Simon, M. P. Strathann, and N. Gautam, *Science* **252,** 802 (1991).
[3] Y. Fukada, H. Ohguro, T. Saito, T. Yoshizawa, and T. Akino, *J. Biol. Chem.* **264,** 5937 (1989).

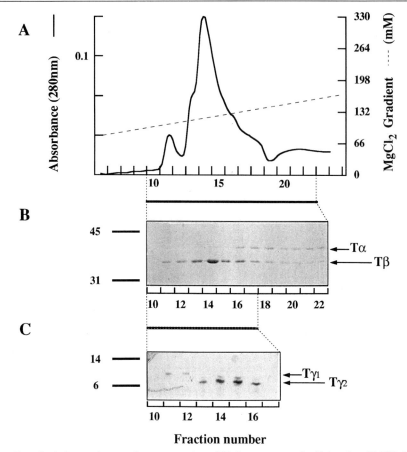

FIG. 1. Anion-exchange chromatography of Tγβ extract on the Polyanion SI HR 5/5 column (Pharmacia). (A) Elution profile; (B) Tα and Tβ subunits visualized on a 12% polyacrylamide gel; (C) Tγ subunits visualized on a urea–polyacrylamide gel (12.5% acrylamide, 1.25% bisacylamide, 8 M urea). (B) and (C) Molecular weight markers ($\times 10^{-3}$) are shown on the left-hand side.

Tβγ Subunit Purification

The weak anion-exchange column Polyanion SI HR 5/5 from Pharmacia (Piscataway, NJ) that we use for the purification of Tα (see [11]) is also best suited for the purification of Tβγ, but with a different salt gradient. The column is equilibrated with buffer A [20 mM Tris, pH 7.5, 0.1 mM phenylmethylsulfonyl fluoride) (PMSF), 5 mM 2-mercaptaethanol (2-ME)]. The protein extract, filtered through a Millex-GV 0.22-μm filter unit (Millipore, Bedford, MA), is deposited on the column and then eluted with a $MgCl_2$ linear gradient (0–0.33 M, 4.4 mM/min, flow rate 0.5 ml/min). With this salt gradient, T$\beta\gamma$ elutes first, in a sharp peak whose major fractions are not contaminated by Tα, which is already a minor component of the extract and elutes later as a lower and more diffuse peak (not shown). Pooled T$\beta\gamma$ fractions from previous T$\alpha_{GTP\gamma S}$ subunit purification batches on the Na_2SO_4 gradient can be purified after concentration/dilution (buffer A) on a Centricon 30 microconcentrator (Amicon, Danvers, MA) up to a final dilution factor of about 1/10 (less than 10 mM Na_2SO_4). About 30 μg of pure T$\beta\gamma$ can be obtained after treatment of 1 mg rhodopsin containing membranes.

Separation of Modified Tγ Subunits

With some transducin extracts, particularly when the preliminary washing step in Hypo buffer, prior to the GTP addition, had been deleted, the elution profile of the final T$\beta\gamma$ extract often appears split in two distinct peaks (Fig. 1A). Both peaks contain the same Tβ component (Fig. 1B), but the first minor peak contains a Tγ component which migrates more slowly on urea–polyacrylamide gels (Fig. 1C). Such a separation of two T$\beta\gamma$ components with distinctive γ subunits has been described by Fukada et al.,[3] who separated T$\beta\gamma_1$ plus T$\beta\gamma_2$ (first peak) and pure T$\beta\gamma_2$ (second peak) on a DEAE-Toyopearl 650S (Toyo Soda Mfg. Co., Japan) column, then purified the T$\beta\gamma_1$ complex from T$\beta\gamma_2$ onto a Mono Q HR 5/5 column (Pharmacia). They identified the slower migrating Tγ_1 as a minor defarnesylated component and Tγ_2 as a farnesylated and partly carboxymethylated γ subunit.[4,5]

[4] Y. Fukada, T. Takao, H. Ohguro, T. Yoshizawa, T. Akino, and Y. Shimonishi, *Nature (London)* **346**, 658 (1990).
[5] H. Ohguro, Y. Fukada, T. Takao, Y. Shimonishi, T. Yoshizawa, and T. Akino, *EMBO J.* **10**, 3669 (1991).

[35] Adenylyl Cyclase Assay for $\beta\gamma$ Subunits of G Proteins

By JIANQIANG CHEN, DONNA J. CARTY, and RAVI IYENGAR

Introduction

Transmission of signals through G proteins can be achieved by the use of either α or $\beta\gamma$ subunits.[1,2] It had been generally believed that α

[1] J. R. Hepler and A. G. Gilman, *Trends Biochem. Sci.* **17**, 383 (1992).
[2] L. Birnbaumer, *Cell (Cambridge, Mass.)* **71**, 1069 (1992).

subunits were the prime signal transmitters. Only a few systems such as certain types of K$^+$ channels[3] and the effector(s) in the yeast pheromone pathway[4] were thought to be regulated by $\beta\gamma$ subunits. However, it has now been shown that $\beta\gamma$ subunits can by themselves regulate the activity of a number of mammalian effectors such as certain types of adenylyl cyclase (adenylate cyclase)[5,6] and phospholipaseC.[7,8] Thus it has become increasingly obvious that $\beta\gamma$ subunits will play a central role in many signaling pathways.

In deciphering the role of $\beta\gamma$ subunits it is useful to have simple assay systems that can be used to assess the biological activity of the purified or expressed $\beta\gamma$ subunits. Both adenylyl cyclases as well as phospholipases C offer excellent readout systems for $\beta\gamma$ subunits. Gilman and co-workers have shown that type 2 and 4 adenylyl cyclases are stimulated by $\beta\gamma$ subunits in the presence of activated α_2 subunits.[5,6] Giershick and co-workers have shown that phospholipase C from HL-60 cells and neutrophils can be stimulated by $\beta\gamma$ subunits.[7] The use of the cytosolic HL-60 cell phospholipase C to measure $\beta\gamma$ activity is described elsewhere.[9] In this chapter we describe a very simple adenylyl cyclase assay that can be used to assess the activity of purified $\beta\gamma$ subunits. The assay is based on the original observations by Katada *et al.*[10] that $\beta\gamma$ subunits stimulate the mouse lymphoma S49 *cyc*$^-$ cell membrane adenylyl cyclase. This assay works on crude S49 *cyc*$^-$ cell membranes and does not require the presence of any other adenylyl cyclase stimulants for the expression of $\beta\gamma$ activity.

Preparation of S49 *cyc*$^-$ Cell Membranes

The S49 *cyc*$^-$ cells are grown in Dulbecco's modified Eagle's medium (DMEM) supplemented with 10% heat-inactivated horse serum. The S49 cells grow as suspension cultures and can be grown in spinner flasks or in T-75 or T-175 bottles. Growth of S49 cells is very sensitive to the pH of the medium, and it is best to maintain a slightly acidic pH (reddish orange). We routinely do this by adding sterile 1 N HCl to the culture medium. Cells should be harvested at a density of 2–3 × 10^6 cells/ml.

[3] D. E. Logothetis, *et al.*, *Nature (London)* **325**, 321 (1987).
[4] C. Dietzel and J. Kurgan, *Cell (Cambridge, Mass.)* **50**, 1001 (1987).
[5] W.-J. Tang and A. G. Gilman, *Science* **254**, 1500 (1991).
[6] B. Gao and A. G. Gilman, *Proc. Natl. Acad. Sci. U.S.A.* **88**, 10178 (1991).
[7] M. Camps, H.-H. Lee, D. Park, C.-W. Lee, and K.-H. Lee, *Eur. J. Biochem.* **206**, 821 (1992).
[8] D.-K. Jhon, D. J. Yoo, and S. G. Rhee, *J. Biol. Chem.* **268**, 6654 (1993).
[9] P. Giershick and M. Camps, this series, Vol. 238 [14].
[10] T. Katada, G. M. Bukoch, M. D. Smigel, M. Ui, and A. G. Gilman, *J. Biol. Chem.* **259**, 3586 (1984).

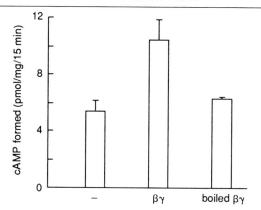

FIG. 1. Effect of bovine brain βγ subunits on the S49 cyc^- adenylyl cyclase. Adenylyl cyclase was assayed in the presence of 10 mM MgCl$_2$. When present the concentration of βγ subunits was 300 mM. The final concentration of Lubrol-PX in the assay was 0.02%. Values are means ± S.D. of triplicate determinations.

Cells are collected by low-speed centrifugation (10 min, 900 g) and washed with ice-cold Puck's saline G without divalent cation. After this wash, all further manipulations are in the cold (4°). It is essential that all centrifuge tubes, cylinders, etc., be precooled prior to use.

The cells are lysed in 25 mM sodium HEPES, 1 mM EDTA, and 120 mM NaCl, pH 8.0. Typically, cells from 1 liter of medium are suspended in 100 ml of lysis buffer. Cells are lysed by nitrogen cavitation in a Parr bomb. Cells are allowed to equilibrate for 20 min at 400 psi after which the pressure is rapidly released. The lysate is then collected and centrifuged at 900 g for 10 min. The supernatant from the first centrifugation is then centrifuged at 40,000 g for 30 min. The 40,000 g pellet is then resuspended in 10 ml of 25 mM sodium HEPES, 1 mM EDTA, pH 8.0, and washed by centrifugation at 40,000 g for 30 min. The wash procedure is repeated twice. The final pellet is resuspended in the wash buffer to a final concentration of 1–2 mg/ml protein, divided into aliquots, and snap-frozen in a dry ice–acetone bath. The membranes can be stored at −70° without loss of activity for at least 6 months. The membranes can be thawed to 4° only once and should be assayed soon (10–15 min) after thawing. Prolonged maintainence on ice or refreezing leads to inactivation of adenylyl cyclase catalytic activity.

Preparation of βγ Subunits

Protocols for the purification of βγ subunits are described elsewhere in this volume.[11,12] It should be noted that most tissue sources will yield

---

[11] T. Katada, K. Kontani, A. Inanobe, I. Kobayashi, Y. Ohoka, H. Nishina, and K. Takahashi, this volume [10].

[12] J. Bigay and M. Chabre, this volume [34].

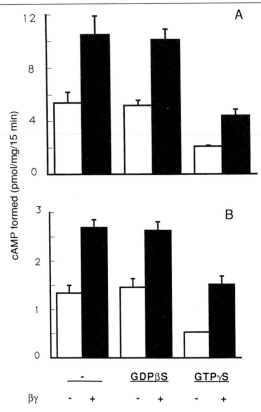

FIG. 2. Stimulation of S49 cyc^- adenylyl cyclase by bovine brain $\beta\gamma$ subunits in the presence of GDPβS or GTPγS. The S49 cyc^- adenylyl cyclase was assayed at either 10 mM (A) or 2 mM (B) MgCl$_2$ with no other additions (−) or in the presence of 100 μM GDPβS or GTPγS in the presence or absence of 300 mM added $\beta\gamma$ subunits. Values are means ± S.D. of triplicate determinations.

heterogeneous mixtures of $\beta\gamma$ subunits. In our laboratory $\beta\gamma$ subunits are prepared from bovine brain. During elution of the heptylamine column with cholate we invariably obtain free $\beta\gamma$ subunits at higher (~3%) concentrations of cholate. These $\beta\gamma$ subunits, which are about 80–90% pure, are pooled and concentrated on a DEAE-Sephacel column. Further purification is obtained on a Mono Q column (Pharmarcia, Piscataway, NJ) using conditions previously described.[13] On the salt gradient the $\beta\gamma$ subunits elute very early. At this stage very pure mixtures of $\beta\gamma$ subunits can be obtained. There do not appear to be present any measurable amounts of

[13] E. Padrell, D. J. Carty, T. M. Moriarity, J. D. Hildebrandt, E. M. Landow, and R. Iyengar, *J. Biol. Chem.* **266**, 9771 (1991).

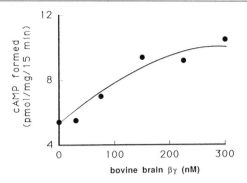

FIG. 3. Effect of varying concentrations of added $\beta\gamma$ subunits on the S49 cyc^- adenylyl cyclase. Varying amounts of $\beta\gamma$ subunits were added to the assay to achieve the indicated concentrations in the final assay mixture. The $\beta\gamma$ subunits were always added in a volume of 10 μl, and the final concentration of Lubrol-PX was always 0.02%. Adenylyl cyclase activity was measured in the presence of 10 mM MgCl$_2$. Values are means of triplicate determinations. The coefficient of variance was less than 10%.

α subunits as assessed by either silver staining or by ADP-ribosylation assays. The $\beta\gamma$ subunits are stored in 25 mM sodium HEPES, 1 mM EDTA, 0.1% Lubrol, 100 mM NaCl, and 20 mM 2-mercaptoethanol at a final concentraion of 60–100 μg/ml.

$\beta\gamma$-Stimulated Adenylyl Cyclase Assay

Standard adenylyl cyclase assay conditions are used. The assay mixture contains 0.1 mM [α-^{32}P]ATP [1000–2000 counts/min (cpm)/pmol], 25 mM sodium HEPES, 1 mM EDTA, pH 8.0, indicated amounts of Mg^{2+}, and an ATP-regenerating system consisting of 20 mM creatine phosphate, 0.2 mg/ml creatine phosphokinase, and 0.02 mg/ml myokinase. The $\beta\gamma$ subunits and cyc^- membranes are added to yield a firal volume of 50 μl. About 5–10 μg of cyc membranes are used. The β- subunits are diluted at least 5-fold so that the final concentration of Lubrol-PX in the assay is no more than 0.02%. The assay mixture is incubated for 15 min at 32°. At the end of the incubation the assay is terminated and the [^{32}P]cAMP formed is quantified by the method of Salomon et al.[14]

Results

Addition of 300 nM $\beta\gamma$ subunits results in a 2-fold stimulation of the basal adenylyl cyclase activity of cyc^- membranes. If the $\beta\gamma$ subunits are

[14] Y. Salomon, C. Londos, and M. Rodbell, *Anal. Biochem.* **58,** 541 (1974).

boiled prior to use in the assay, no stimulation is observed (Fig. 1). Although the extent of stimulation by $\beta\gamma$ subunits is relatively modest, the stimulation is reproducibly observed. Typically we observe a 2-fold stimulation. The stimulation can be observed both at low (2 mM) and higher (10 mM) Mg^{2+} concentrations (Fig. 2). Addition of saturating concentrations of GDPβS or GTPγS does not appear to affect significantly the extent of stimulation by exogenously added $\beta\gamma$ subunits at either low or high Mg^{2+} concentrations (Fig. 2). Addition of GTPγS, however, results in lower activities, presumably owing to the activation of G$_i\alpha$. Thus, it appears best to assay the $\beta\gamma$ subunits without any added guanine nucleotides. Stimulation by added $\beta\gamma$ subunits is dependent on the concentration of $\beta\gamma$ subunits used and is saturable (Fig. 3). The concentration required to obtain half-maximal stimulation is around 50 nM. We have not used forskolin in our assays, but essentially similar stimulations with $\beta\gamma$ subunits are also observable when forskolin is used.[10]

The mechanism by which $\beta\gamma$ subunits stimulate the S49 *cyc*⁻ cell membrane adenylyl cyclase is not known. However, the stimulation by $\beta\gamma$ subunits is observable only in the absence of G$_s\alpha$. Addition of $\beta\gamma$ subunits to wild-type S49 cell membranes results in inhibition rather than stimulation.[10] Thus, it does not appear likely that the adenylyl cyclase stimulated by $\beta\gamma$ subunits in *cyc*⁻ membranes is like type 2 or 4. Irrespective of the type of adenylyl cyclase stimulated, the S49 *cyc*⁻ cell membranes offer a very simple assay system to ascertain the biological activity of $\beta\gamma$ subunits.

Acknowledgment

Research was supported by National Institutes of Health Grants CA-44998 and DK-38761.

[36] Synthesis and Use of Biotinylated βγ Complexes Prepared from Bovine Brain G Proteins

By JANE DINGUS, MICHAEL D. WILCOX, RUSSELL KOHNKEN, and JOHN D. HILDEBRANDT

Introduction

Many extracellular signals exert their intracellular effects by means of signal-transducing, GTP-binding G proteins.[1-3] These proteins are heterotrimeric, each composed of an α, a β, and a γ subunit. The α subunit, which binds and hydrolyzes GTP, historically has been assumed to be the most important of the three subunits. Multiple types of the α subunit exist that associate with different receptors and effectors. Subunit dissociation into an isolated α subunit and a βγ complex is thought to be an integral part of the mechanism of G-protein activation. The β and γ subunits remain noncovalently associated in a tight complex that is now thought to have regulatory roles as varied as that of the α subunits.[4-9] The isolated α and βγ components are both capable of directly regulating downstream effector systems. In spite of its importance, subunit dissociation has been difficult to study directly, and most studies[10-13] have used indirect approaches such as analysis of sedimentation behavior or ADP-ribosylation of the α subunit (which requires a bound βγ). Even then, GTP effects on subunit dissociation have been difficult to demonstrate *in vitro*, even though this is assumed to occur *in vivo*,[1-3] and nonhydrolyzable

---

[1] L. Birnbaumer, J. Abramowitz, and A. M. Brown, *Biochim. Biophys. Acta* **1031**, 163 (1990).
[2] M. Freissmuth, P. J. Casey, and A. G. Gilman, *FASEB J.* **3**, 2125 (1989).
[3] L. Birnbaumer, *FASEB J.* **4**, 3178 (1990).
[4] T. Katada, G. M. Bokoch, M. D. Smigel, M. Ui, and A. G. Gilman, *J. Biol. Chem.* **259**, 3586 (1984).
[5] R. A. Cerione, C. Staniszewski, M. G. Caron, R. J. Lefkowitz, J. Codina, and L. Birnbaumer, *Nature (London)* **318**, 293 (1985).
[6] W. Tang and A. G. Gilman, *Science* **254**, 1500 (1991).
[7] J. D. Hildebrandt and R. E. Kohnken, *J. Biol. Chem.* **265**, 9825 (1990).
[8] C. L. Jelsema and J. Axelrod, *Proc. Natl. Acad. Sci. U.S.A.* **84**, 3623 (1987).
[9] A. D. Federman, B. R. Conklin, K. A. Schrader, R. R. Reed, and H. R. Bourne, *Nature (London)* **356**, 159 (1992).
[10] B. M. Denker, E. J. Neer, and C. J. Schmidt, *J. Biol. Chem.* **267**, 6272 (1992).
[11] E. J. Neer, L. Pulsifer, and L. G. Wolf, *J. Biol. Chem.* **263**, 8996 (1988).
[12] S. E. Navon and B. K. Fung, *J. Biol. Chem.* **262**, 15746 (1987).
[13] T. Katada, M. Oinuma, and M. Ui, *J. Biol. Chem.* **261**, 8182 (1986).

GTP analogs and persistent activators such as AlF_4^- do induce subunit dissociation *in vitro*.[14]

This chapter describes a method developed for directly studying the association and interaction of the α subunits with the $\beta\gamma$ complex.[15] Modified (biotinylated) $\beta\gamma$ is immobilized on agarose beads, which allows a straightforward binding assay to be performed with α subunits. Biotinylated $\beta\gamma$ (b$\beta\gamma$) is prepared by treating intact bovine brain G protein with NHS-biotin, activating with AlF_4^-, and separating the subunits on an ω-aminooctyl-agarose column. The b$\beta\gamma$ is immobilized on streptavidin-agarose. This allows the association of the various α subtypes with $\beta\gamma$ to be studied, and it permits the investigation of factors affecting the interaction of the subunits with one another. Because intact G protein is biotinylated, the binding site(s) on $\beta\gamma$ for α is protected from modification. Although the α subunit is heavily biotinylated with this procedure, $\beta\gamma$ is minimally modified and appears fully functional. The b$\beta\gamma$ is further purified by anion-exchange chromatography. This highly purified, biotinylated $\beta\gamma$ appears to maintain all the functional properties of unmodified $\beta\gamma$.[15]

$\beta\gamma$ Biotinylation and Purification

Preparation of Bovine Brain G Protein

Heterogeneous bovine brain G protein is used as the starting material for the synthesis of biotinylated $\beta\gamma$. The proteins are purified as previously described,[15] based on modification of the purification described by Sternweis and Robishaw.[16] Bovine brain membranes are extracted with cholate buffers (Fig. 1, lane Ext) and the extract chromatographed on a DEAE-Sephacel (Pharmacia, Piscataway, NJ) column. Fractions containing the peak of GTP-binding activity (Fig. 1, lane DEAE) are pooled, concentrated, and separated on Ultrogel AcA 34 (IBF Biotechnics, Columbia, MD). The peak of G protein from this column (Fig. 1, lane AcA) is further purified on an octyl-agarose (Sigma, St. Louis, MO) column. The major peak from this column is bovine brain G protein (Fig. 1, lane Oct); it is pooled, concentrated to approximately 4 mg/ml, exchanged into Lubrol buffer (see Table I), and stored at $-80°$ in aliquots. This protein is used as the starting material for the synthesis of biotinylated $\beta\gamma$.

---

[14] J. K. Northup, M. D. Smigel, P. C. Sternweis, and A. G. Gilman, *J. Biol. Chem.* **258**, 11369 (1983).
[15] R. E. Kohnken and J. D. Hildebrandt, *J. Biol. Chem.* **264**, 20688 (1989).
[16] P. C. Sternweis and J. D. Robishaw, *J. Biol. Chem.* **259**, 13806 (1984).

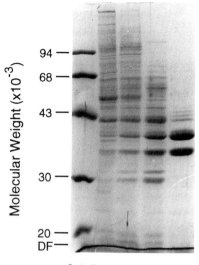

FIG. 1. Analysis of steps in the purification of bovine brain G protein used as the starting material for the synthesis of biotinylated $\beta\gamma$. Samples (5 mg) were run on 11% SDS–polyacrylamide gels and stained with Coomassie blue. Abbreviations are described in the text. The material lane Oct is the starting material for the preparation of biotinylated $\beta\gamma$.

Solutions and Buffer Preparation

The stock solutions used for making up the buffers necessary for these procedures are listed in Table I. All solutions, with the exceptions of dithiothreitol (DTT) and GDP, are stored at 4°; DTT and GDP are stored at −80°. The NHS-LC-biotin solution is made just prior to use. Lubrol is no longer commercially available, but Thesit is suggested to be a suitable substitute. Cholate is recrystallized 6 times prior to use,[17] and Lubrol is deionized on mixed-bed resin prior to use.[17] All chemicals and reagents are from Sigma with the exception of CHAPS and NHS-LC-biotin, which are from Pierce (Rockford, IL), and horseradish peroxidase (HRP)-conjugated streptavidin, which is from Vector (Burlingame, CA).

Buffers required for all included procedures are listed in Table I. They are routinely made up and used within 48 hr. They are also filtered prior to use, especially those for the FPLC (fast protein liquid chromatography) steps. DTT and GDP are always added just prior to use, DTT being added after filtration. Buffers containing $AlCl_3$, $MgCl_2$, and NaF (AMF) should

---

[17] J. Codina, W. Rosenthal, J. D. Hildebrandt, L. Birnbaumer, and R. D. Sekura, this series, Vol. 109, p. 446.

TABLE I
STOCK SOLUTIONS AND BUFFERS FOR PREPARATION OF BIOTINYLATED $\beta\gamma$

Solution	Composition
Stock solutions and reagents	
Sodium HEPES, pH 8.0	1 M
Tris-HCl, pH 8.0	1 M
Tris-HCl, pH 6.8	1 M
$MgCl_2$	1 M
NaF	0.5 M
$AlCl_3$	0.1 M
NaCl	5 M
Dithiothreitol (DTT)	1 M
Sodium cholate, pH 8.0	10%
Sodium CHAPS, pH 8.0	10%
Lubrol	10%
Nonidet P-40 (NP-40)	5%
GDP	10 mM
Methanol	—
Glycerol	—
Mercaptoethanol	—
NHS-LC-biotin	20 mM in DMSO
Bovine serum albumin	25 mg/ml
Ethanolamine, pH 8.0	200 mM
Streptavidin-agarose beads	—
HRP-conjugated streptavidin	—
4-Chloro-1-naphthol	30 mg/ml in methanol
H_2O_2	30%
Buffers for $\beta\gamma$ biotinylation, activation, and purification	
HED	20 mM Sodium HEPES, pH 8.0, 1 mM EDTA, 1 mM DTT
TED	20 mM Tris-HCl, pH 8.0, 1 mM EDTA, 1 mM DTT
AMF	10 mM $MgCl_2$, 10 mM NaF, 50 μM $AlCl_3$
Buffer 1	HED, 100 mM NaCl, 10 μM GDP
Lubrol buffer	HED, 0.1% Lubrol, 100 mM NaCl, 10 μM GDP
Buffer 2	TED, 100 mM NaCl, 10 μM GDP, 0.125% cholate
Activation buffer	TED, AMF, 100 mM NaCl, 10 μM GDP, 0.125% cholate
Wash buffer	TED, AMF, 300 mM NaCl, 10 μM GDP, 0.125% cholate
Elution buffer A	TED, AMF, 200 mM NaCl, 10 μM GDP, 0.125% cholate
Elution buffer B	TED, AMF, 50 mM NaCl, 10 μM GDP, 1.0% cholate
Mono Q buffer 1	TED, 100 mM NaCl, 0.7% CHAPS
Mono Q buffer 2	TED, 500 mM NaCl, 0.7% CHAPS
Mono Q wash buffer	20 mM Tris, pH 8.0, 1 M NaCl
Buffers for b$\beta\gamma$ analysis and binding assay	
Lubrol buffer	HED, 0.1% Lubrol, 100 mM NaCl, 10 μM GDP
TBS	50 mM Tris-HCl, pH 8.0, 150 mM NaCl
TBSB	50 mM Tris-HCl, pH 8.0, 150 mM NaCl, 2.5 mg/ml BSA
TBSBN	50 mM Tris-HCl, pH 8.0, 150 mM NaCl, 2.5 mg/ml BSA, 0.05% NP-40
Staining solution	20% Methanol, 80% TBS, 6 mg/ml 4-chloro-1-naphthol, 0.03% H_2O_2
Sample buffer	62.5 mM Tris-HCl, pH 6.8, 3% SDS, 5% mercaptoethanol, 5% glycerol

also be made immediately before use. To avoid precipitation, AlCl$_3$ should always be added last.

Columns

Buffer Exchange Column. A gel-filtration column is used to separate the biotinylated protein from the free biotin, and to exchange the buffer and detergent. A 25-ml Ultrogel AcA 202 column (2.5 × 10 cm) is preequilibrated with at least 3 column volumes of buffer 2 (Table I). This column, equilibrated with the appropriate buffer, is also used when other buffer exchange is required. Pharmacia PD-10 or other premade desalting columns might be used when the volume is small, but usually the volume is large enough to preclude the use of such columns.

Hydrophobic Column. A 5-ml ω-aminooctyl-agarose column (1 × 10 cm), packed with buffer 2, is used for the primary separation of the biotinylated βγ complex from the α subunits. Either agarose or Sepharose-substituted resins obtained from Sigma or Pharmacia have been used for this purpose. However, octyl-agarose does not seem to be as efficient as ω-aminooctyl-agarose for subunit separations. The flow rate for packing and washing is 15 ml/hr. Just prior to use, the column is equilibrated with 3 volumes of activation buffer, and it should always be washed with Buffer 2 after use.

Anion-Exchange Column. The biotinylated βγ is further purified on a Mono Q HR 5/5 column using a Pharmacia LKB FPLC system. Other high-performance anion-exchange media such as Waters (Milford, MA) Accell QMA should also work for this separation but have not been specifically investigated. The column is washed with Mono Q wash buffer prior to each separation run, then equilibrated with Mono Q buffer 1.

Procedure

Note: All containers used for any protein solutions should be polyethylene or polypropylene plastic, or siliconized glass.

Step 1: Biotinylation. The purified G protein (4-10 mg) should be in Lubrol buffer (Table I). The protein is diluted to 1 mg/ml with buffer 1, and the final Lubrol concentration is adjusted to 0.05%. If necessary, 10% Lubrol is added to adjust the final concentration to 0.05%.

The NHS-LC-biotin reagent (Pierce) is added to the G protein to a final concentration of 1 mM from a 20 mM stock in dimethyl sulfoxide (DMSO) (11.13 mg/ml), made just prior to use. The protein is mixed well and incubated at 23° (room temperature) for 30 min. Ethanolamine is then added to stop the reaction, to a final concentration of 10 mM, from a 200 mM stock, pH 8.0. The mixture is incubated on ice for 10 min. Under these conditions, the α subunit is multiply biotinylated, but the βγ complex is biotinylated only to a stoichiometry that assures its nearly complete binding to streptavidin-agarose.[15]

Step 2: Buffer Exchange. The crude biotinylated G protein is passed over the Ultrogel AcA 202 column previously described to remove the unconjugated biotin and to exchange the protein into buffer 2. The protein-containing fractions are pooled, and the protein concentration is determined.

Step 3: G-Protein Activation. The exchanged, biotinylated G-protein pool is diluted to a final protein concentration of 0.05 mg/ml, using activation buffer (for the ω-aminooctyl-agarose column). A 10- to 20-fold dilution of the pool is usually required. The final buffer composition must be the same as the activation buffer (20 mM Tris, pH 8.0, 1 mM EDTA, 1 mM DTT, 100 mM NaCl, 10 μM GDP, 0.125% cholate, 10 mM MgCl$_2$, 10 mM NaF, 50 μM AlCl$_3$), so it is necessary to adjust the concentration of AlCl$_3$, MgCl$_2$, and NaF. The AlCl$_3$ should be added last. The diluted protein is activated by incubating at 32° for 20 min, then put on ice.

Step 4: ω-Aminooctyl-Agarose Chromatography. The following protocol is designed for amounts of protein from 4 to 10 mg, and the described volumes of the wash and eluant buffers should be adequate. However, the actual amount of buffer used may vary since the absorbance at 280 nm is monitored as the column is eluted, and the next elution is not begun until the baseline is flat. Because changes in NaCl and cholate concentrations affect the background absorbance, the absorbance will not necessarily return to the original value. All fractions from the column are collected into EDTA to prevent aggregation and precipitation of the separated subunits, particularly the βγ complex. The final concentration of EDTA should be 1 mM greater than the concentration of MgCl$_2$. For convenience, EDTA is added to the tubes in the fraction collector before beginning the column. The sequence for eluting the column is given below.

a. The activated G protein is loaded, using a flow rate of 15 ml/hr, and large fractions (~5 ml) of the flow-through are collected. The small amount of protein in the flow-through fractions is heavily biotinylated, inactive α, which is discarded.

b. The column is now washed with 10 ml activation buffer. The fraction size is changed to 1 ml for this and all subsequent elutions, and fractions are continued to be collected into EDTA.

c. The column is eluted with 20 ml of wash buffer. This elutes most of the heavily biotinylated α and some βγ which appears to be irreversibly bound to the α.

d. The column is eluted with 20 ml elution buffer A, or until the baseline stabilizes.

e. The biotinylated $\beta\gamma$ is eluted with 40 ml of elution buffer B, collecting 1-ml fractions until the end.

f. When use of the column is finished, AMF is washed off the ω-aminooctyl-agarose column with 20 ml of buffer 2.

All protein-containing fractions are analyzed on sodium dodecyl sulfate (SDS)–11% polyacrylamide gels stained with Coomassie blue. Most of the biotinylated α is present in the wash eluate, and the biotinylated $\beta\gamma$ (b$\beta\gamma$) is in the eluate of elution buffer B. Figure 2B shows a graph of the absorbance at 280 nm, and the corresponding SDS–polyacrylamide gel electrophoresis (SDS–PAGE) analysis of the protein-containing fractions

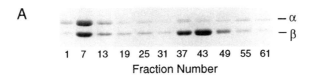

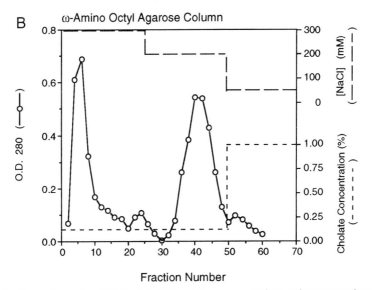

FIG. 2. Chromatography of biotinylated G protein on an ω-aminooctyl-agarose column. (A) Analysis by SDS–PAGE of every sixth fraction across the chromatogram in (B).

is shown in Fig. 2A. The b$\beta\gamma$ is always contaminated to some degree with α. This is presumably due to the presence of modified trimer which is not able to be activated and dissociated. (Unmodified, activated G-protein subunits chromatographed on an ω-aminooctyl-agarose column using this method separate well.) Furthermore, attempts at reactivation and repeated chromatography of these subunits have not been successful in improving the separation, and this protein is assumed to be aggregated, inactive trimer.

Step 5: Protein Concentration and Buffer Exchange. The b$\beta\gamma$ peak from the ω-aminooctyl-agarose column is pooled and concentrated to approximately 1 mg/ml using an Amicon (Danvers, MA) ultrafiltration cell with a YM10 membrane. This protein is then chromatographed on an Ultrogel AcA 202 column of appropriate size preequilibrated with Mono Q buffer A in order to remove cholate and exchange the protein into the starting buffer for the Mono Q column.

Step 6: Anion-Exchange Chromatography. The partially purified b$\beta\gamma$ peak from the ω-aminooctyl-agarose column is chromatographed on a Pharmacia Mono Q anion-exchange column using a Pharmacia LKB FPLC system. This separates the b$\beta\gamma$ from the nondissociated trimer. (The ω-aminooctyl-agarose column can also be run with the FPLC, and when done with a continuous gradient much less α is present in the $\beta\gamma$ eluate. Nevertheless, this modification of the method does not yield homogeneous biotinylated $\beta\gamma$ without a second chromatography step.) The exchanged, pooled peak from the ω-aminooctyl-agarose column is loaded onto the equilibrated Mono Q column using a flow rate of 0.5 ml/min; the back pressure should not exceed 1 mPa (150 psi). After loading, the column is washed with Mono Q buffer A until the baseline stabilizes. The protein is eluted at a flow rate of 0.5 ml/min with a continuous 20-ml gradient from 0 to 25% Mono Q buffer B (100 to 200 mM NaCl in TED with 0.7% CHAPS), then a 5-ml gradient from 25 to 100% Mono Q buffer B (200 to 500 mM NaCl in TED with 0.7% CHAPS). All protein-containing fractions are analyzed on 11% SDS–PAGE gels. Figure 3 shows the chromatogram and corresponding SDS–PAGE analysis of the fractions. The b$\beta\gamma$ is eluted in the shallow part of the gradient, and it contains no detectable α subunit (see Fig. 4). The steep part of the gradient elutes the α and nondissociable trimer.

Step 7: Final Preparation and Storage of Biotinylated $\beta\gamma$. The b$\beta\gamma$ peak is pooled, concentrated to approximately 1 mg/ml using an Amicon ultrafiltration cell with a YM10 membrane, exchanged into Lubrol buffer, and stored in small aliquots at $-80°$. A typical yield is 1–2 mg per 5 mg of G-protein starting material.

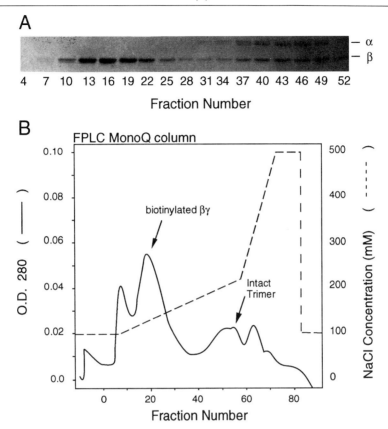

FIG. 3. Chromatography of the bβγ pool from the ω-aminooctyl-agarose column on an FPLC Mono Q column. (A) Analysis by SDS–PAGE of every third fraction across the chromatogram (and gradient) shown in (B).

Analysis of Biotinylated βγ

The bβγ, at various stages in the purification, is analyzed for protein composition and for the presence of biotin. Duplicate 11% SDS–PAGE gels are run with samples of the G-protein starting material, the biotinylated G protein, the ω-aminooctyl-agarose pool, and the final purified bβγ. One of the gels is stained with Coomassie blue (Fig. 4A); the other is transferred to nitrocellulose,[18] blocked with TBSBN (Table I), and washed three times with TBS. The nitrocellulose is incubated with a 1 : 1000 dilu-

[18] H. Towbin, T. Staehelin, and J. Gordon, *Proc. Natl. Acad. Sci. U.S.A.* **76,** 4350 (1979).

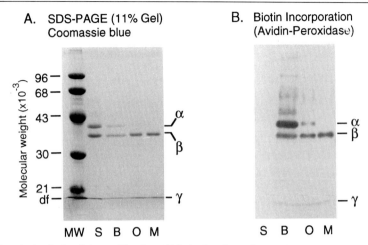

FIG. 4. Analysis of the purification of biotinylated $\beta\gamma$. Comparison of the stages in the purification of b$\beta\gamma$ by Coomassie blue staining (A) and HRP–streptavidin staining for biotin (B) is shown. S, Starting G protein; B, biotinylated G protein; O, ω-aminooctyl-agarose-purified $\beta\gamma$, M, Mono Q-purified $\beta\gamma$.

tion of HRP-conjugated streptavidin in TBSBN for 30 min, then washed with TBS three times. The biotinylated bands are visualized by incubation in staining solution. Before biotinylation, the intact G protein (lanes S, Fig. 4) does not contain any biotin. After biotinylation, both the α and β subunits of the intact G protein (lane B, Fig. 4A) are biotinylated (lane B, Fig. 4B), the α subunit more heavily. Any α remaining in the pool from the first column (Fig. 4B, lane O) should be removed after chromatography on the second column (Fig. 4B, lane M).

Subunit Binding Assays

Streptavidin-Agarose Bead Preparation

The binding assays routinely use 10 μl of streptavidin-agarose beads per sample. The number of samples can be fairly large (40 or more); however, each sample must be run on a gel, and, for comparison purposes, it is preferable that all samples of a given set be run on the same gel. Nevertheless, gels made at the same time and handled identically will give good reproducibility, and duplicate sets of samples are often run on different gels.

Streptavidin-agarose beads are blocked in order to prevent nonspecific binding by preincubating the beads with bovine serum albumin (BSA).

The required amount of beads (10 μl times number of samples, plus about 10% to allow for error) are placed in a 15- or 50-ml polypropylene tube with 20 volumes of TBS (Table I) and washed twice, centrifuging in a clinical centrifuge for 1 min at setting 3 to sediment the beads. The beads are blocked with 10 volumes of TBSB for 1 hr and then washed four times with Lubrol buffer.

Binding of Biotinylated βγ to Streptavidin-Agarose

The washed beads are resuspended in 5 volumes of Lubrol buffer with the desired amount of bβγ (usually 0.5 μg per 10 μl beads) in a polypropylene or polyethylene tube of appropriate size with a tight-fitting cap. The beads are incubated with the bβγ for 1 hr at room temperature with continuous gentle mixing on a Labquake (Labindustries, Berkeley, CA) rocker shaker, then washed twice with 10 volumes of Lubrol buffer. The beads are resuspended in 4 volumes of buffer to allow them to be pipetted for the assay.

α-Subunit Binding Assay

Binding assays are carried out in Lubrol buffer, usually in a final volume of 200 μl in a 500-μl Eppendorf-type microcentrifuge tube. The α protein is added to each tube first, along with buffer, and the required samples are set up (e.g., a series of α concentrations or a series for a time course), in duplicate or triplicate. Beads, with or without bound bβγ, are added last as a suspension of 10-μl beads in a total volume of 50 μl, resuspending the beads immediately prior to the removal of each aliquot. Controls routinely include samples without α and samples using blocked and washed streptavidin-agarose beads without bound bβγ. The samples are incubated at room temperature for 30 min, or as required, slowly mixing on a Labquake rocker shaker. The reaction is stopped by the rapid sedimentation of the beads in a Picofuge microcentrifuge (Stratagene, La Jolla, CA) for 10 sec. The supernatants are carefully removed using a curved Pasteur pipette to avoid disturbance of the bead pellet, then discarded. The pellet contains the bound α, and no washing steps are done so as not to disturb any binding established during the incubation. However, some α is nonspecifically retained in the void volume of the beads, so it is imperative that the amount of α associated with control beads (without bβγ) be determined. This nonspecific binding is subtracted from the total values during the quantitation of the experiment (see below). The amount of nonspecific binding can range from significant to undetectable, depending on the concentration of α subunit in the assay. Nonspecific binding remains negligible up to about 5 μg α/ml (or 1 μg in 200 μl).

The samples are prepared for analysis by adding 4 volumes (40 μl) of sample buffer to the 10 μl of beads and heating to 95° for 5 min. After cooling on ice, the samples are sedimented again, this time at 14,000 rpm for 4 min, and the supernatants with the solubilized proteins are analyzed on 11% SDS–PAGE gels. In addition, α and $\beta\gamma$ standards with known amounts of protein are run on each gel in order to quantitate the protein (see below). The gels are stained with Coomassie blue.

Properties and Characterization of Biotinylated $\beta\gamma$

To determine the proportion of b$\beta\gamma$ able to bind to streptavidin-agarose, as well as the binding capacity of streptavidin-agarose, known amounts of b$\beta\gamma$ are incubated with a constant amount (10 μl) of beads. The amount of b$\beta\gamma$ bound is compared to the total (amount originally added) by running both on a gel. A binding assay is set up as described above. Aliquots (10 μl) of washed, blocked streptavidin-agarose beads are resuspended in 5 volumes of Lubrol buffer and incubated with a range of b$\beta\gamma$ concentrations (usually 1 to 10 μg/ml or 0.2 to 2.0 μg per sample). The beads are incubated with the b$\beta\gamma$ for 30 min at room temperature with continuous gentle mixing on a Labquake shaker, then washed twice with Lubrol buffer. Forty microliters of sample buffer is added to the beads, and the samples are heated to 95° for 5 min, then cooled on ice, centrifuged, and the supernatants analyzed on an 11% gel. A range of b$\beta\gamma$ concentrations, equivalent to the starting amounts in the binding assay samples, are also run on the same gel for comparison and quantitation. The amount of b$\beta\gamma$ bound to the beads is quantitated by comparing each lane of bound to the comparable lane with the total (see below). Generally, the lower amount of b$\beta\gamma$ gives the best estimate of the fraction of the protein that can bind to streptavidin-agarose, whereas the larger amounts of added b$\beta\gamma$ give the best estimate of binding capacity of the streptavidin-agarose. This capacity varies with the batch of material and with the age. The binding of b$\beta\gamma$ to streptavidin-agarose is saturable, but, under limiting conditions, 80% or more of the b$\beta\gamma$ is capable of binding (Fig. 5A).

Experiments similar to those described above have also been used to determine the specificity of b$\beta\gamma$ binding to streptavidin-agarose, as well as the specificity of α binding to b$\beta\gamma$ beads. Binding of purified b$\beta\gamma$ to streptavidin-agarose is dependent on $\beta\gamma$ being biotinylated (Fig. 5B, lane 3); unbiotinylated $\beta\gamma$ does not bind to the beads (Fig. 5B, lane 1). This binding is blocked if, prior to the addition of b$\beta\gamma$, the beads are incubated with free biotin (Fig. 5B, lane 2). Additionally, b$\beta\gamma$ bound to streptavidin-agarose does not readily dissociate, as addition of excess exogenous biotin does not result in loss of the bound b$\beta\gamma$ from the beads. Finally, isolated

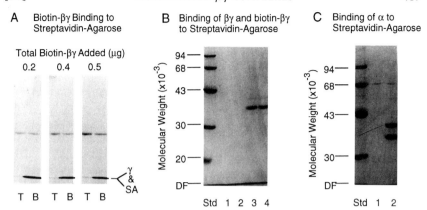

FIG. 5. Determination of α and bβγ binding to streptavidin-agarose. (A) Binding of different amounts of bβγ (0.2, 0.4, and 0.5 μg) to a constant amount of streptavidin beads (10 μl); T, total, B, bound. (B) Binding of βγ and bβγ to streptavidin-agarose; lane 1, unbiotinylated βγ; lane 2, bβγ with excess biotin added before binding; lane 3, bβγ only; lane 4, bβγ with excess biotin added after binding. (C) Binding of α to streptavidin-agarose in the absence (lane 1) and presence (lane 2) of bβγ.

α subunits bind to streptavidin-agarose only if bβγ is prebound (Fig. 5C, lane 2); α does not bind specifically to beads which do not have bβγ bound.

Analysis and Quantitation of Results

The amount of α subunit bound to bβγ in each sample is determined by densitometric scanning of the Coomassie blue-stained gels of the samples using an XRS Omni-Media 6cx scanner (X-ray Scanner Corp., Torrance, CA) connected to a Macintosh IIci computer. The gels are scanned using software provided by XRS Omni-Media, and the scanned images are then analyzed using the National Institutes of Health (NIH) Image 1.40 program. The gel should be well destained and, just prior to scanning, equilibrated in water for 30 min. The data generated from scans of duplicate gels with replicate sets of samples agree quite well provided the gels are stained and destained in parallel.

The relevant areas of the gel (with the bands of interest) are scanned in two dimensions by reflectance using a gray-scale (8-bit) scanning mode with neutral brightness at 75 dots per inch (dpi), the maximum allowable density which Image 1.40 will accommodate. (It may be necessary to adjust the brightness depending on the contrast of the gel.) Considerably better results, with greater sensitivity, are obtained when Coomassie blue-stained gels are scanned by reflectance than when scanned by transmission.

Scanned images are analyzed using Image 1.40. Each band is carefully outlined using the freehand tracing tool, being certain to include all of the band, and the area (A_s) and average density (D_s) are obtained. (These are two of many standard measurements offered by the Image software.) Background (or baseline) measurements are made for each spot individually. An area of the scanned gel within the same lane as the sample band is circled as described above, and the area (A_b) and average density (D_b) are recorded. This area does not need to be the same size as the sample band, but the average density must reflect that of the background surrounding the appropriate band. The calculated value that is most linear with protein concentration (Fig. 6) is the integrated density (ID) of the band calculated as follows:

$$ID = A_s(D_s - D_b) \qquad (1)$$

(Note: this calculated integrated density is different from the parameter of the same name in the Image software, which makes a different assumption about background density.) This procedure is repeated for all the relevant bands on the gel. Measurements and calculations are done for the standards, namely, known amounts of α or $\beta\gamma$ run on the same gel. These

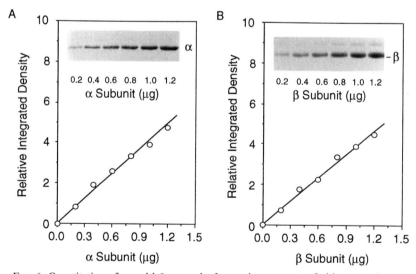

FIG. 6. Quantitation of α and b$\beta\gamma$ on gels. Increasing amounts of either α or $\beta\gamma$ were run on 11% SDS–polyacrylamide gels. After Coomassie blue staining, the gels were scanned on an XRS Omni-Media 6cx scanner and analyzed using the NIH Image 1.41 software. The linearity of integrated density (defined in text) with amounts of α (A) or $\beta\gamma$ (B) added is shown. In each case, the inset shows the bands which were scanned.

are used to generate a standard curve (*ID* versus microgram of protein; see Fig. 6). For both α and β there is a linear relationship between amount of protein and *ID* over the range of amounts used in the binding assays.

The above determined values for the bound α are graphed versus the known parameter (α concentration, time, salt concentration, or whatever) to generate a curve. If detectable α is present in the control samples (the beads without bβγ), those bands are measured and quantitated in the same way. A binding curve can be generated after subtracting the control curve. We are using this binding assay and quantitation method to examine the reassociation kinetics of α binding to βγ.

[37] Design of Oligonucleotide Probes for Molecular Cloning of β and γ Subunits

By CHRISTINE GALLAGHER and NARASIMHAN GAUTAM

Introduction

Molecular cloning of cDNAs encoding distinct G-protein β- and γ-subunit types has shown that these subunits comprise families of proteins in mammalian tissues. cDNAs encoding four different β subunits and seven different γ subunits have now been identified. All the cDNAs were isolated using appropriate oligonucleotide probes. Two broadly different approaches have been used to isolate the cDNAs. The first method involves screening a cDNA library with oligonucleotide mixtures specific to the amino acid sequence of a protein. The second method is based on the polymerase chain reaction (PCR).[1] In this method oligonucleotide mixtures specific to the amino acid sequence of a protein or specific to conserved amino acid residues among members of a family of proteins are used as PCR primers for amplifying a portion of the cDNA.

Cloning cDNAs for G-Protein β and γ Subunits for Which Amino Acid Sequence Information Exists

Analysis with specific antibodies has revealed that purified G-protein βγ protein complexes from different sources are mixtures of β and γ subtypes. In addition, subtypes exist that are distinct from those for which

---

[1] R. K. Saiki, S. Scharf, F. Faloona, K. B. Mullis, G. T. Horn, H. A. Erlich, and N. Arnheim, *Science* **230**, 1350 (1985).

cDNAs have been identified. If a partial amino acid sequence of one of these novel subtypes is determined, one of the methods below can be used to isolate the corresponding cDNA. These methods initially involve the synthesis of oligonucleotides specific to the amino acid sequence. Different methods can then be used to isolate the cDNA with these oligonucleotides. Two of these methods are described.

Method 1: Oligonucleotide Probes for Isolating cDNAs by Plaque Hybridization

Until the advent of the PCR, a common method employed to isolate cDNAs for purified proteins was to screen a cDNA library with oligonucleotide probes specific to the amino acid sequence of the protein. As several amino acids are encoded by more than one codon, probes are synthesized as mixtures containing all the possible base combinations. This results in poor specificity for the target cDNA. The following alterations can improve the likelihood of success with this technique. (a) Two batches of oligonucleotides specific to different portions of the amino acid sequence may be used. The library is initially screened with one of these probes. Clones that hybridize to the first probe are screened with the second probe to eliminate "false positives." (b) Less degenerate oligonucleotides can be synthesized in the following ways. (i) Substitution of deoxyinosine (I) at positions where there is 4-fold degeneracy may be undertaken. (ii) A particular codon for an amino acid might occur more frequently in a mammalian species compared to other codons for the same amino acid. In such cases, the oligonucleotide can be made with that codon alone. (iii) Because deoxyguanosine (G) and deoxythymidine (T) can base pair weakly, they can be incorporated at positions that are 4-fold degenerate.

Methods for labeling the oligonucleotides with radioactive or nonradioactive prercursors and screening a cDNA library are available in considerable detail from a variety of sources including previous volumes of these series.[2]

Method 2: Isolating cDNAs Using Methods Based on Polymerase Chain Reaction

Design of Primers Specific to Amino Acid Sequence of Protein. An alternative approach to the one described above is based on the PCR and has several advantages over the previous method. Oligonucleotide primers, sense and antisense, are synthesized specific to the amino acid sequence such that they will generate a PCR-amplified product that will

[2] C. G. Miyada and R. B. Wallace, this series, Vol. 154, p. 94.

encode a few amino acids in between the regions specific to the primers. The length of cDNA amplified between the primers depends on the amino acid sequence information available for synthesizing primers.

One of the advantages of the PCR approach is that the effect of several different primer combinations on amplification can be examined rapidly since many reactions can be performed at the same time. In the case of primers that are specific to sequences containing serine or leucine, the extent of degeneracy can be reduced by synthesizing the primers in two batches and using them separately. Because these amino acids are each encoded by two families of codons, one batch of oligonucleotides is made with two related codons and another batch with the remaining four codons.

Bases that specify restriction sites can be added at the 5' ends of the primers. This makes cloning the amplified fragment into a plasmid easier. We prefer to design primers without such sites since the same sites could potentially occur in the amplified DNA. The length of the oligonucleotide primers is determined by the amino acid sequence. The optimal length is 17 bases. In cases where there are several amino acids that are specified by a single codon, it might be possible to increase the length of the oligonucleotide primers without increasing degeneracy of the mixture. Very highly degenerate oligonucleotide mixtures have been used successfully for amplifying cDNAs.[3] In general, however, it is preferable to minimize the extent of degeneracy. We attempt to keep degeneracy at or below 256-fold.

Template. The template for the PCR can be any one of the following. (a) Single-stranded cDNA can be made by reverse transcribing poly(A)$^+$ RNA (whole RNA can also be used at higher concentrations) with one of the following primers: (i) oligo(dT) (12–16 bases), (ii) the antisense PCR primer, or (iii) hexamers with random sequences. cDNA synthesized using random hexamers is more representative of the full-length cDNA in comparison to that using oligo (dT), especially in the case of relatively long messages. While transcribing regions considerably upstream of the poly(A)$^+$ tail, reverse transcriptase can terminate prematurely, yielding a cDNA population that lacks sequences at the 5' end. This might exclude the region specific to the primers. cDNA made with random hexamers or with the antisense primer is more likely to contain the nucleic acid sequences complementary to the primers. (b) Genomic DNA may also be used. Potential problems with using genomic DNA as a template are the presence of pseudogenes or introns in the region that is being amplified. Because the PCR is performed with primers to a specific amino acid

---

[3] S. J. Gould, S. Subramani, and I. E. Scheffler, *Proc. Natl. Acad. Sci. U.S.A.* **86**, 1934 (1989).

sequence, the length of the amplified fragment can be predicted. This may help in identifying the presence of an intron from the altered length of the amplified fragment.

Procedure. The PCR primers (17 bases long) are designed specific to the amino acid sequence of a β or γ subtype such that cDNA amplified between the primers is long enough to specify at least three amino acids.[4] This approach requires that sequence information on at least 15 contiguous amino acids from the purified protein be available. Poly(A)$^+$ RNA is reverse transcribed after annealing the primer (random hexamers or the antisense PCR primer) by heating at 90° for four min and chilling in ice. In a 50-μl reaction containing 50 mM Tris-HCl, pH 8.3, 75 mM KCl, 10 mM dithiothreitol (DTT), and 3 mM MgCl$_2$, RNA [0.3 μg of poly(A)$^+$ RNA or 5- to 10-fold more total RNA] is reverse transcribed with 600 U of MMLV (Moloney murine leukemia virus) reverse transcriptase (BRL Gaithersburg, MD), 500 μM each deoxynucleotide, and 200 ng of primer. The reaction is incubated at 37° for 90 min. DNA from the reaction is precipitated with ethanol and resuspended in 20 μl of water. One microliter of this is used for each PCR amplification. Standard PCR mixtures are in 100 μl with 5 to 7 μM primers for 30 cycles. Reaction mixes are denatured at 94° for 5 to 10 min prior to beginning the PCR cycles. Denaturation during the cycles is for 30 sec to 1 min at 94°. Extension and annealing conditions will depend on the primers. In general, for degenerate mixtures of primers, annealing is at about 45° for 1 min. The extension time at 72° will depend on the length of the DNA to be amplified. Amplified DNA is purified and cloned into a plasmid as described in the procedure given below for cloning novel β and γ subunit types using the PCR.

Our analysis indicates that primers can amplify the correct cDNA even though there are several mismatches between the primer and the cDNA. Thus, the nucleic acid sequence of the fragments that have been isolated using the PCR may not be identical to the cDNA clone. When the amplified fragment is used to screen a cDNA library for the full-length clone, it is advisable to lower the stringency of hybridization conditions in order to increase the efficiency with which the probe hybridizes to the cDNA clone.

Cloning cDNAs for Novel Members of G-protein β- and γ-Subunit Families

The size and the extent of diversity in the β- and γ-subunit families are unclear at present. The conservation across species of differences in primary structure among different members of these families as well as

[4] N. Gautam, M. Baetscher, R. Aebersold, and M. I. Simon, *Science* **244,** 971 (1989).

the distinct expression patterns indicate that these families are large. For the molecular cloning of novel members of these families two different approaches can be used.

Cloning Related cDNAs Using Low-Stringency Hybridization Conditions

Previously characterized cDNAs for different β and γ subunits can be used as probes to screen a library of cDNAs under hybridization conditions that favor hybridization to related but distinct nucleic acid sequences. Although this method may be successful in the case of proteins that are highly homologous (e.g., β subunits), it is less effective in the case of proteins that are structurally diverse (e.g., γ subunits) since the cDNAs will be less homologous to one another. Protocols for altering the conditions for hybridization are available from several sources including previous volumes of this series.[5]

Polymerase Chain Reaction-Based Methods for Isolating Novel Members of Families

cDNAs for three novel γ subunits and one β subunit have been identified using an approach that is now widely used for isolating members of families of genes. Oligonucleotide primers for the PCR are synthesized specific to the amino acids that are conserved between members of a family of related proteins. Because putative new members of the family are likely to have the same amino acids conserved, the primers are capable of amplifying the cDNAs for novel members.

Design of Primers. Primers are synthesized as mixtures of oligonucleotides specific to two different portions of the β- and γ-subunit amino acid sequences that are highly conserved among family members. If possible, conserved regions that flank more divergent sequences are chosen. This allows us to identify an amplified molecule as a novel cDNA without any ambiguity. Because the γ-subunit family is more diverse structurally, performing separate PCR amplifications using primers specific to more than one conserved block of residues will improve the probability of amplifying a novel cDNA. Figures 1 and 2 show the amino acid sequences of the G-protein γ- and β-subunit types. Regions of homology that are potentially useful for synthesizing PCR primers are shown. Apart from being conserved among the different subtypes, these regions also contain few leucines or serines. This helps reduce the extent of degeneracy in the oligonucleotide mixture.

[5] G. A. Beltz, K. A. Jacobs, T. H. Eickbush, P. T. Cherbas, and F. C. Kafatos, this series, Vol. 100, p. 266.

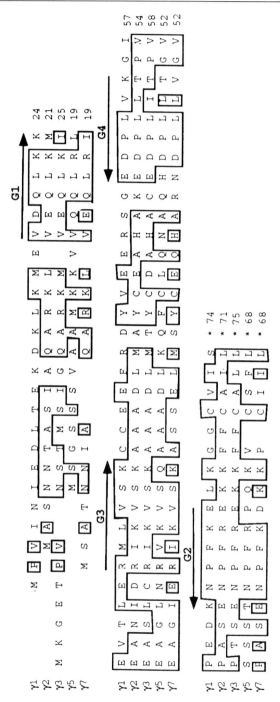

FIG. 1. Amino acid sequences of G-protein γ-subunit types. Identical residues are boxed. The sequences are from the following sources: γ_1, J. B. Hurley, H. K. W. Fong, D. B. Teplow, W. J. Dreyer, and M. I. Simon, *Proc. Natl. Acad. Sci. U.S.A.* **81**, 6948 (1984); γ_2, N. Gautam, M. Baetscher, R. Aebersold, and M. I. Simon, *Science* **244**, 971 (1989); γ_3, N. Gautam, J. K. Northup, H. Tamir, and M. I. Simon, *Proc. Natl. Acad. Sci. U.S.A.* **87**, 7973 (1990); γ_5, K. J. Fisher and N. N. Aronson, *Mol. Cell. Biol.* **12**, 1585 (1992); γ_7, J. J. Cali, E. A. Balcueva, I. Rybalkin, and J. D. Robishaw, *J. Biol. Chem.* **267**, 24023 (1992). Amino acid sequences shown are derived from cDNAs. Only the fully characterized sequences that have been published are shown. Arrows indicate regions that are useful for synthesizing PCR primers. Sense and antisense primers can be used in different combinations. These regions flank the more divergent amino acid sequences, and this facilitates identification of a novel cDNA. Primers corresponding to G_1 and G_2 amplify γ_2, γ_3, γ_4, and γ_6 [N. Gautam, J. K. Northup, H. Timir, and M. I. Simon, *Proc. Natl. Acad. Sci. U.S.A.* **87**, 7973 (1990); C. Gallagher and N. Gautam, unpublished (1991)].

Base substitutions can be used to reduce the extent of degeneracy as discussed earlier. Primers containing substitutions of different kinds can potentially amplify different members of the family with differing efficiencies. Substitutions can be useful for altering the proportion of different members of the family in an amplification reaction. This prevents cDNAs that have been previously isolated from swamping out potentially novel cDNAs. We design primers that are 17 bases long. If the amino acid sequence specific to the primers contains residues that are encoded by single codons, longer primers can be made without increasing degeneracy. We do not include restriction sites at the ends of the primers since there is a chance that the same sites may be present inside a novel cDNA.

Template. The probability of identifying a new cDNA can be increased by using single-stranded cDNAs specific to various mammalian tissues. Because the levels of expression of a novel gene in different tissues is unknown, this method enables the isolation of cDNAs unique to specific tissues. Table I shows the number of γ-subunit cDNA clones amplified by similar PCR experiments using the DG17 and DG18 primers as described below. The cDNAs for different subtypes were isolated with varying frequencies in different mouse tissues. This indicates that in various tissues a particular subtype is expressed at different levels relative to other subtypes.

Procedure. The following protocol has been used to amplify at least four members of the γ subunit family, namely, γ_2, γ_3, γ_4, and γ_6, from cDNA specific to various mouse tissues (Table I).[6] The nucleotide sequence of the PCR primers are as follows; DG17 (sense), GTIGAICA(A/G)CTIAA(A/G)AT; DG18 (antisense), GG(A/G)TT(T/C)TCI(G/C)(A/T)IG(C/T)IGG. The primers are specific to amino acids 16–21 and 55–60 of γ_2 (G_1 and G_2 in Fig. 1). The single-stranded cDNA is obtained by reverse transcription of total RNA using random hexamers. Two micrograms of the RNA from several mouse tissues is heated with 1 μg of primer to 68° for 5 min, then chilled on ice.[7] Reverse transcription is performed as described earlier. The components of the reaction are as described in the previous procedure section. One microliter of this reaction is used for the PCR with 1 μM of each PCR primer. The PCR is carried out in 50 μl for 35 cycles (30 sec at 94°, 30 sec at 72°, and 30 sec at 46°). Then 15 μl of the PCR mix is separated on a 1.4% (w/v) agarose gel. DNA fragments of approximately 140 base pairs (bp) are amplified from all tissues.

[6] N. Gautam, J. K. Northup, H. Tamir, and M. I. Simon, *Proc. Natl. Acad. Sci. U.S.A.* **87,** 7973 (1990).
[7] T. M. Wilkie and M. I. Simon, *Methods (San Diego)* **2,** 32 (1991).

```
                      B1
              1                                                          50
Beta1   MSELDQLRQE AEQLKNQIRD ARKACADATL SQITNNIDPV GRIQMRTRRT
Beta2   ....E..... ....R..... .....G.S.. T...AGL... ..........
Beta3   .G.M...... ....KK..A. .....A.V.. AELVS..EV. ..V.......
Beta4   .S.L...... ....RN..Q. .....N.A.. VQIT.NMDS. ..I.......

                      B2
              51                                                         100
Beta1   LRGHLAKIYA MHWGTDSRLL VSASQDGKLI IWDSYTTNKV HAIPLRSSWV
Beta2   .......... .......... .......... .......... ..........
Beta3   .......... ....A...K. .......... V......... ..........
Beta4   .......... ....Y..... .......... .........M ..........


              101                                                        150
Beta1   MTCAYAPSGN YVACGGLDNI CSIYNLKTRE GNVRVSRELA GHTGYLSCCR
Beta2   .......... F......... ....S..... .........P ..........
Beta3   .......... .........M ....N..S.. ...K.....S A.........
Beta4   .......... Y.........I C......T.. .D.R.....A G.........

                      B3
              151                                                        200
Beta1   FLDDNQIVTS SGDTTCALWD IETGQQTTTF TGHTGDVMSL SLAPDTRLFV
Beta2   .....I.... .......... .......VG. A..S...... .....G.T..
Beta3   .....N.V.. .......... ......KTV. V..T..C... AVS..FNL.I
Beta4   ....G..I.. .......... .......... ...S...... .....LKT.V


              201                                                        250
Beta1   SGACDASAKL WDVREGMCRQ TFTGHESDIN AICFFPNGNA FATGSDDATC
Beta2   .......I.. ....DS.... ..I....... .VA.....Y. .T........
Beta3   .......A.. ....EGT... ..T....... .IC.....E. IC......S.
Beta4   .......S.. ..I.D.M... S....I.... .VS...S.Y. FA........

                      B4
              251                                                        300
Beta1   RLFDLRADQE LMTYSHDNII CGITSVSFSK SGRLLLAGYD DFNCNVWDAL
Beta2   .......... .LM....... ......A..R .......... .....I...M
Beta3   .......... .ICF..ES.. .........L .......... .....V..S.
Beta4   .......... .LLY..DN.. .........K .......... ....S...AL

                      B5
              301                                       341
Beta1   KADRAGVLAG HDNRVSCLGV TDDGMAVATG SWDSFLKIWN *
Beta2   .G........ .......... .......... .......... *
Beta3   .SE.V.I.S. .......... .A........ .......... *
Beta4   .GG.S.V.A. .......... .D........ ......R... *
```

FIG. 2. Amino acid sequences of G-protein β-subunit types. Dots indicate identity with the amino acid above. The sequences are from the following sources: β_1, H. K. W. Fong, J. B. Hurley, R. S. Hopkins, R. Miake-Lye, M. S. Johnson, R. F. Doolittle, and M. I. Simon, *Proc. Natl. Acad. Sci, U.S.A.* **83,** 2162 (1986); β_2 H. K. W. Fong, T. T. Amatruda, B. W. Birren, and M. I. Simon, *Proc. Natl. Acad. Sci. U.S.A.* **84,** 3792 (1986); β_3, M. A. Levine, P.M. Smallwood, P. T. Moen, L. J. Helman, and T. G. Ahn, *Proc. Natl. Acad. Sci. U.S.A.* **87,** 2329 (1990); β_4, E. Von Weizacker, M. Strathmann, and M. I. Simon, *Biochem. Biophys. Res. Commun.* **183,** 350 (1992). Arrows indicate regions that are useful

TABLE I
AMPLIFIED FRAGMENTS OF cDNAs FROM DIFFERENT MOUSE TISSUES OBTAINED WITH PRIMERS DG17 AND DG18[a]

Tissue	γ_1	γ_2	γ_3	γ_4	γ_6	Total
Brain	—[b]	7 (0.21)	26 (0.79)[c]	—	—	33
Heart	—	8 (0.42)	7 (0.37)	4 (0.21)	—	19
Intestine	—	—	—	3	—	3
Kidney	—	7 (0.27)	6 (0.23)	11 (0.42)	2 (0.08)	26
Liver	—	—	2	3	1	6
Lung	—	5 (0.36)	7 (0.50)	2 (0.14)	—	14
Retina	—	1	2	5	—	8
Skeletal muscle	—	3 (0.23)	5 (0.38)	5 (0.38)	—	13
Testes	—	—	3	—	—	3

[a] cDNAs were identified as described in the text by single lane DNA sequencing. Only the γ_6 nucleic acid sequence was determined by electrophoresing all four reactions.
[b] None isolated.
[c] The proportion of clones relative to the total is shown in parentheses only in those cases where more than a total of 10 clones were examined.

Five microliters of the reaction mix is PCR amplified in 50 μl using the same conditions as before. [In cases where a specific DNA fragment from a mixture of fragments needs to be reamplified, the products of the first PCR are separated on a low melting point agarose gel (NuSieve from FMC, Rockland, ME), the appropriate band excised, and 1–5 μl of the melted DNA/agarose amplified in a PCR using the same conditions but in a larger volume[7].] The PCR-reamplified DNA is blunt ended as follows. Thirty microliters of the reaction mix is increased to a total volume of 60 μl with water, 10 μl of 10× Klenow DNA polymerase buffer, deoxynucleoside triphosphates, and Klenow DNA polymerase. The reaction is incubated at room temperature for 30 min. Three units of T4 DNA polymerase is then added and the reaction incubated for 3–5 min at 37°. DNA is extracted with phenol precipitated with ethanol. [We have used a product from Qiagen (Chatsworth, CA) and Bio 101 (La Jolla, CA) to purify PCR-amplified DNA.] The DNA pellet is resuspended in polynucleotide kinase buffer containing ATP and phosphorylated with polynucleotide kinase. The entire reaction mix is separated on a 3% low melt agarose gel (NuSieve

---

for synthesizing PCR primers. Sense and antisense primers can be used in different combinations. Primers corresponding to B$_1$ and B$_2$ were used to isolate a portion of the β_4 cDNA [E. Von Weizacker, M. Strathmann, and M. I. Simon, *Biochem. Biophys. Res. Commun.* **183,** 350 (1992)].

from FMC), the band of approximately 140 bp excised, and the melted agarose ligated to a plasmid that has been cut with a blunt end generating restriction enzyme. We use plasmids that allow blue/white colony selection.

Although we blunt the ends of the amplified DNA and clone it into a plasmid, other alternatives are (1) to synthesize the primers with restriction sites at the ends (for most common enzymes 2–3 bases in addition to the site will help the enzyme work) and (2) to use commercially available cloning systems that enable PCR-generated DNA to be ligated directly with high efficiency. Plasmid DNA is purified from bacterial transformants and the nucleotide sequence determined.

The PCR-amplified DNA is a mixture of cDNA fragments specific to different members of the γ-subunit family including those that have already been characterized. Several hundred clones may therefore need to be screened before a novel cDNA is identified. To screen large numbers of clones rapidly we set up sequencing reactions (U.S.B. Biochemical, Cleveland, OH) on microtiter plates for the deoxyguanosine (G) lanes only. Because each sample occupies only one lane, 96 clones can be examined on a single sequencing gel (14 × 17 inches). The band pattern on the autoradiograms is compared to the pattern from the "G lane" of known clones electrophoresed using the same conditions. Novel patterns can be rapidly identified from this comparison. Once a potentially novel clone is identified, all four (G, A, T, C) reactions are performed and electrophoresed to determine the nucleic acid sequence.

If the primers were designed with two different restriction sites at their 5' ends, all the PCR cDNA fragments when cloned into a plasmid will be oriented the same way. When examining the pattern from sequencing gels, it is an advantage to have all cDNAs oriented in the same direction. When the cDNAs are cloned into a plasmid using sites that generate blunt ends, each clone generates two distinct patterns.

Cloning cDNA Directly from Library Using Polymerase Chain Reaction

We have used the PCR to isolate cDNAs directly for two different γ-subunit types, namely, γ_2 and γ_3, from a mouse cDNA library. Figure 3 outlines this method. Once a novel member of the β- or γ-subunit family is identified using the method described earlier, this method can be used to isolate the full-length cDNA.

Method

Six aliquots of a mouse brain cDNA library in the vector λ Zap (Stratagene) are amplified separately on agarose plates (see Fig. 3). Each aliquot con-

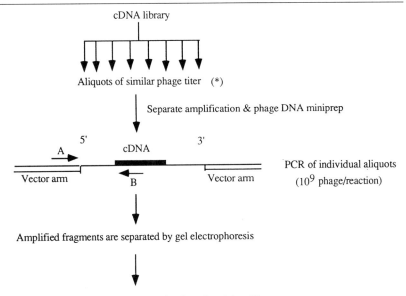

FIG. 3. A, Cloning γ_2 and γ_3 from a mouse cDNA library. Sense PCR primer specific to the vector. B, Antisense PCR primer specific to the cDNA. The number of aliquots and the phage titer (*) will vary depending on the abundance of the cDNA of interest. The filled box represents the protein coding portion of the cDNA. Primer B is specific to the protein coding portion of the cDNA. The strategy shown can also be used to isolate the 3' end of the cDNA using appropriate primers. Full-length cDNAs can be obtained by designing the primers such that the 5' and 3' amplified fragments have an overlap. Amplifying a mixture of these fragments by the PCR will yield a full-length product.

tains approximately 1.5×10^5 phage. The phage DNA is purified by mixing 0.9 ml of the phage lysate with 0.5 ml of DE-52 anion-exchange resin. The mixture is inverted 20–30 times and centrifuged at 14,000 rpm in a microcentrifuge for 5 min. The supernatant is phenol extracted and the DNA precipitated with a mixture of 10 M ammonium acetate/2-propanol (1:3 v/v). The precipitated DNA is washed in 70% ethanol and resuspended in 70 μl of water. Then 5–10 μl (the equivalent of $\sim 10^9$ phage) from each of the aliquots is used separately as a template for the PCR. To isolate the 5' portion of the γ_2 cDNA a primer specific to λ Zap (LZ1: CGCTCTAGACTAGTGGATC) and primers specific to the γ_2 cDNA (BG28: GTAGGCCATCAAGTCAGCAG specific to amino acids ADLMAY of γ_2) or the γ_3 cDNA (CG16: GTATGTCATCAGGTCTGCTG specific to amino acids ADLMTY of γ_3) (see Fig. 1) are used. Each PCR is 100 μl in volume containing 700 ng of each primer. The conditions are

1 min at 94°, 1 min at 58° and 2 min at 72°. The PCR is performed over 35 cycles.

DNA of approximately 200–500 bp is amplified from several aliquots. These fragments are subcloned into plasmids and the nucleic acid sequence determined. Using this method we have rapidly isolated mouse cDNAs for γ_2 and γ_3 that include 5' noncoding regions. The original clones for the cDNAs were from bovine tissues. Because the coding portions of different γ subunits are significantly homologous to one another, amplified DNA that includes noncoding regions is a more specific probe in experiments such as *in situ* hybridization.

This method can also be used to amplify rare cDNAs. For instance, a much larger number of aliquots of the same titer of a library can be screened in the same way. The PCR tends to amplify abundant messages. Fractionating the library increases the ratio of rare to abundant clones in individual aliquots and prevents the rare cDNA from being swamped out during amplification. Further advantages are that several million clones can potentially be screened more easily and rapidly than traditional plaque hybridization methods, and the size of the isolated clone is more rapidly estimated. Once overlapping novel cDNA fragments covering the 5' and 3' regions are identified, PCR primers specific to these fragment can be used to isolate a full-length cDNA.

Acknowledgments

N.G. is an Established Investigator of the American Heart Association. This work was supported by grants from the National Institutes of Health and Monsanto–Searle.

[38] Characterization of Antibodies for Various G-Protein β and γ Subunits

By ALEXEY N. PRONIN and NARASIMHAN GAUTAM

Introduction

Antibodies to the β and γ subunits of G proteins have been invaluable in the studies that have begun to dissect the structure and function of these subunits.[1–4] The cDNAs for four β subunit types[5–10] and for several

---

[1] W. F. Simonds, J. E. Butrynski, N. Gautam, G. Unson, and A. M. Spiegel, *J. Biol. Chem.* **266**, 5363 (1991).

[2] K. H. Muntz, P. C. Sternweis, A. G. Gilman, and S. M. Mumby, *Mol. Biol. Cell* **3**, 49 (1992).

γ subunits[11-17] have now been characterized. The most useful antisera are those raised against peptides specific to the amino acid sequence encoded by cDNAs for different β and γ subunit types. Depending on the region of the protein chosen for synthesizing a corresponding peptide, these antibodies are capable of reacting with a particular subunit type specifically or with several different subunits. Of an immunoblot of a mixture of different subunit types, specific peptide antisera can distinguish between different members of the β and γ subunit families of proteins. Although the antisera are raised against a peptide corresponding to a relatively small portion of the protein, several of the antisera against the G-protein β and γ subunits are capable of immunoprecipitating the βγ complex. They can also be used for immunocytochemical localization of the β or γ subunits present in transient transfectants.

Choosing Amino Acid Sequences for Peptide Synthesis

The source for the amino acid sequence peptide can be (1) a peptide derivative of the protein or (2) a partial or full-length nucleic acid sequence of the cDNA encoding the protein. Comparison of the amino acid sequence

---

[3] A. Pronin and N. Gautam, *Proc. Natl. Acad. Sci. U.S.A.* **89**, 6220 (1992).

[4] T. Murakami, W. F. Simonds, and A. M. Spiegel, *Biochemistry* **31**, 2905 (1992).

[5] K. Sugimoto, N. Toshihide, T. Tanabe, H. Takahashi, M. Noda, N. Minamino, K. Kangawa, H. Matsuo, T. Hirose, S. Imayama, and S. Numa, *FEBS Lett.* **191**, 235 (1985).

[6] H. K. W. Fong, J. B. Hurley, R. S. Hopkins, R. Miake-Lye, M. S. Johnson, R. F. Doolittle, and M. I. Simon, *Proc. Natl. Acad. Sci. U.S.A.* **83**, 2162 (1986).

[7] H. K. W. Fong, T. T. Amatruda, B. W. Birren, and M. I. Simon, *Proc. Natl. Acad. Sci. U.S.A.* **84**, 3792 (1986).

[8] B. Gao, A. G. Gilman, and J. D. Robishaw, *Proc. Natl. Acad. Sci. U.S.A.* **84**, 6122 (1987).

[9] M. A. Levine, P. M. Smallwood, P. T. Moen, L. J. Helman, and T. G. Ahn, *Proc. Natl. Acad. Sci. U.S.A.* **87**, 2329 (1990).

[10] E. Von Weizacker, M. P. Strathmann, and M. I. Simon, *Biochem. Biophys. Res. Commun.* **183**, 350 (1992).

[11] J. B. Hurley, H. K. W. Fong, D. B. Teplow, W. J. Dreyer, and M. I. Simon, *Proc. Natl. Acad. Sci. U.S.A.* **81**, 6948 (1984).

[12] K. Yatsunami, B. V. Pandya, D. D. Oprian, and H. G. Khorana, *Proc. Natl. Acad. Sci. U.S.A.* **82**, 1936 (1985).

[13] N. Gautam, M. Baetscher, R. Aebersold, and M. I. Simon, *Science* **244**, 971 (1989).

[14] J. D. Robishaw, V. K. Kalman, C. R. Moomaw, and C. A. Slaughter, *J. Biol. Chem.* **264**, 15758 (1989).

[15] N. Gautam, J. K. Northup, H. Tamir, and M. I. Simon, *Proc. Natl. Acad. Sci. U.S.A.* **87**, 7973 (1990).

[16] K. J. Fisher and N. N. Aronson, *Mol. Cell. Biol.* **12**, 1585 (1992).

[17] J. J. Cali, E. A. Balcueva, I. Rybalkin, and J. D. Robishaw, *J. Biol. Chem.* **267**, 24023 (1992).

of interest with the amino acid sequence of other members of the family shows the regions of identity and diversity. Figures 1 and 2 show the regions on the amino acid sequences of the β and γ subunits that have been chosen for peptide synthesis. Tables I and II provide a list of antisera

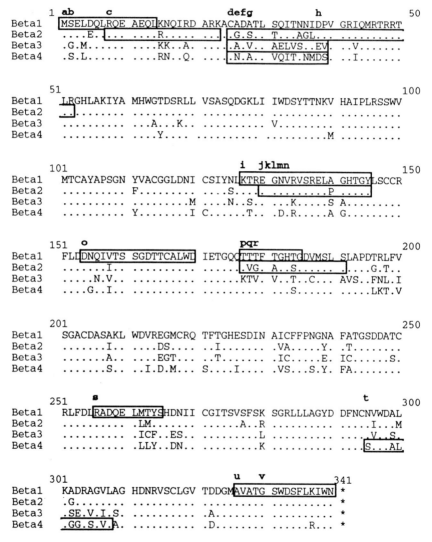

FIG. 1. Alignment of amino acid sequences of the β-subunit types. Dots indicate identity with the amino acid above. Peptides are indicated alphabetically above the sequences (see the first column of Table I). Some boxes include more than one peptide. Refer to Table I for more details on the peptides.

```
         a   bcd      e  f       g
Gamma1   M P V I N I E D L T E K D K L K M E V D Q L K K    24
Gamma2           M A S N N T A S I A Q A R K L . E . . . M  21
Gamma3   M K G E T P V . S . M . . G . . . . M . . . . . I  25

                                       h
Gamma1   E V T L E R M L V S K C C E E F R D Y V E E R S G E D P L V K G I  57
Gamma2   . . N I D . I K . . . A A A D L M A . C . A H A K . . . . L T P V  54
Gamma3   . . S L C . . . . . . . . . . . . . T . . D . . . C . . . . I . .  58

         i j         k
Gamma1   P E D K N P F K E L K G G C V I S  *  74
Gamma2   . A S E . . . R . K . F F . A . L  *  71
Gamma3   . T . . . . . . . . . . . . . L .  *  75
```

FIG. 2. Alignment of amino acid sequences of three γ-subunit types. Dots indicate identity with the amino acid above. Peptides are indicated alphabetically above the sequences (see the first column of Table II). Some boxes include more than one peptide. Refer to Table II for more details on the peptides.

raised against the peptides highlighted in Figs. 1 and 2. Regions such as the NH$_2$-terminal portions of both the β and γ subunits show higher diversity and are obvious choices for synthesizing peptides to generate antisera specific to a subunit type.

It is not clear that there are any general rules in the selection of an appropriate region of a protein excepting those that are dictated by the experiment where the peptide antiserum is to be used. For instance, a single amino acid difference can prevent a peptide antiserum from reacting with a related protein (e.g., the antiserum BN1 can selectively react with the $β_1$ protein although there is only a single amino acid difference between the peptide that it is directed against and the corresponding region of the $β_2$ protein). In other cases such as antiserum B4-1, the antiserum can react with both the $β_4$ and $β_1$ proteins although there are several differences in the amino acid sequence of the two proteins in the region where the peptide was made (Fig. 1 and Table I). The characteristics of an antiserum can also be changed by affinity purification with the peptide or protein antigen. The antisera KT and MS (Fig. 1 and Table I) recognize the native $β_1γ_1$ complex when purified over a column containing the immobilized native protein but not when purified over a column containing the peptide. Also, it is not clear whether the lengths of peptides have any effect on the reactivity of the antisera. Peptides that have been used successfully to raise antisera against the β and γ subunits vary in length from 10 to 18 amino acids. In our experience, therefore, the choice of a peptide and the reactivity of the antibodies directed against it are highly empirical.

An experiment portrayed in Fig. 3 shows the specificity of peptide-directed antisera in comparison to an antiserum raised against the whole βγ complex of a G protein. When the βγ complex of transducin (G_t) is partially digested with trypsin, it gives rise to an NH$_2$-terminal portion of

TABLE I
ANTISERA TO β SUBUNITS

Peptide	Symbol	Amino acid sequence	Position	Specificity[a]	Blotting	Precipitation	Cytochemistry	Refs.[g]
a	BN1	MSELDQLRQEAEQL	1–14 β_1	β_1 $(1:10^6)$[b]	+	+[c]	ND[d]	1–3
b	MS	MSELDQLRQE	1–10 β_1	β_1 [β_2]	+	+	ND	4, 5
c	K521	RQEAEQLRNQIRDARK	8–23 β_2	β_1, β_2	+	ND	ND	6
d	B2N1	CGDSTLTQITAGLDP	25–39 β_2	β_2 [β_1] (1:800)	+	ND	ND	2
e	J887	CGDSTLTQITAGLD	25–38 β_2	β_2	+	ND	ND	6
f	B3-1	CADVTLAELVSGLEV	25–39 β_3	β_3 [β_1] (0.5 μg/ml)	+	ND	ND	2
g	B4-1	CNDATLVQITSNMDS	25–39 β_4	β_4, β_1[e]	+	ND	ND	3
h	8135	DPVGRIQMRTRRTLR	38–52 β_2	β_1, β_2	+	ND	ND	7
i	KT	KTREGNVRVS	127–136 β_1	β_1, β_2	+	+	ND	5, 8
j	U49	EGNVRVSRELAGHTGY	131–145 β_1	β_1[f]	+	ND	+	6, 9, 11
k	AS11	EGNVRVSRELAGHTGY	131–145 β_1	β_1, β_2, β_4 (1:2000)	+	ND	ND	10, 3
l	K523	EGNVRVSRELPGHTGY	131–145 β_2	β_2	+	ND	ND	6
m	8132	EGNVRVSRELAGHTGY	132–145 β_1	β_1	+	+	ND	7
n	8129	EGNVRVSRELPGHTGY	132–145 β_2	β_2	+	+	ND	7
o	BR1	DNQIVTSSGDTTCALWD	154–170 β_1	β_1, β_2 (1:1000)	+	ND	ND	3
p	BP1-16	TVGFAGHSGDVMSLS	177–191 β_2	β_2	+	ND	ND	1
q	AS36	TVGFAGHSG	177–185 β_2	β_2	+	ND	ND	10
r	AS28	TTTFTGHTG	177–185 β_1	β_1	+	ND	ND	10
s	RA	RADQELMTYS	256–265 β_1	β_1	+	+[c]	+[c]	3, 5
t	B4-2	SVWDALKGGRSGVL	295–308 β_4	β_4 [β_1] (1:600–1000)	+	ND	ND	3
u	BC1	AVATGSWDSFLKIWN	326–341 β_1	β_1, β_2 (1:1000)	+	ND	ND	3
v	SW	GSWDSFLKIWN	330–340 β_1	β_1, β_2	+	+	ND	5

[a] The antiserum or antibodies can react specifically on an immunoblot react with the subunit type shown. Cross-reaction with the subunit type shown in brackets takes place usually under conditions such as high concentrations of the antiserum or antigen. Reactivity against the β_3 and β_4 proteins is estimated from immunoblotting proteins from cells transiently transfected with either the β_3 or β_4 cDNA.

[b] Dilutions are shown as ratios for antisera and μg/ml for affinity purified antibodies. Appropriate concentrations are for immunoblotting and are shown only for those antisera and antibodies used in our laboratory. For others, examine appropriate references.

[c] Preliminary experiments with cell lines expressing β1 alone or together with a γ subunit indicate that affinity purified antibodies react with the β subunit alone but not the $\beta\gamma$ complex.

[d] ND, Not determined.

[e] This antiserum is poorly characterized. Reactivity is to the β_1 protein and β_4 peptide.

[f] This is the affinity-purified antibody reactivity. Antiserum to the same peptide (k) reacts nonspecifically with three different β subunits. More details are in the text.

[g] Key to references: (1) T. Amatruda, N. Gautam, H. K. W. Fong, J. K. Northup, and M. I. Simon, J. Biol. Chem. **263**, 5008 (1988); (2) A. Pronin and N. Gautam, Proc. Natl. Acad. Sci. U.S.A. **89**, 6220 (1992); (3) A. N. Pronin and N. Gautam, unpublished results; (4) P. Goldsmith, K. Rossiter, A. Carter, W. F. Simonds, C. G. Unson, R. Vinitsky, and A. M. Spiegel, J. Biol. Chem. **263**, 6476 (1988); (5) T. Murakami, W. F. Simonds, and A. M. Spiegel, Biochemistry **31**, 2905 (1992); (6) B. Gao, S. Mumby, and A. G. Gilman, J. Biol. Chem. **262**, 17254 (1987); (7) S. F. Law, D. Manning, and T. Reisine, J. Biol. Chem. **266**, 17885 (1991); (8) W. F. Simonds, J. E. Butrynski, N. Gautam, G. Unson, and A. M. Spiegel, J. Biol. Chem. **266**, 5363 (1991); (9) S. Mumby, R. A. Kahn, D. R. Manning, and A. G. Gilman, Proc. Natl. Acad. Sci. U.S.A. **83**, 265 (1986); (10) K. D. Hinsch, I. Tychowiecka, H. Gausepohl, R. Frank, W. Rosenthal, and G. Schultz, Biochim. Biophys. Acta **1013**, 60 (1989); (11) K. H. Muntz, P. C. Sternweis, A. G. Gilman, and S. M. Mumby, Mol. Biol. Cell **3**, 49 (1992).

TABLE II
ANTISERA TO γ SUBUNITS

Peptide	Symbol	Amino acid sequence	Position	Specificity	Blotting	Precipitation	Cytochemistry	Refs.[d]
a	CG	MKGETPVNSTMSIG	1–14 γ_3	γ_3 (0.2–0.4 μg/ml)	+	+	+	1–3
b	GN	MPVINIEDLTEKDKL	1–15 γ_1	γ_1 (0.2–0.4 μg/ml)	+	+	+	1, 3, 4
c	KG	KGETPVNSTMSIG	2–14 γ_3	γ_3	+	ND[c]	ND	5
d	PV	PVINIEDLTEK	2–12 γ_1	γ_1	+	+	ND	6
e	BG	MASNNTASIAQARKL	1–15 γ_2	γ_2 (0.1–0.2 μg/ml)	+	ND	+	1, 3, 4
f	SN	ASNNTASIAQAR	2–13 γ_2	γ_2	+	ND	ND	5
g	X-263	NNTASIAQARKLVEQLKM	4–21 γ_2	γ_2	+	ND	+	7
h	—[a]	DLMAYCEAHAK	36–46 γ_2	γ_2	+	ND	ND	8
i	PE	PEDKNPFKELK	58–68 γ_1	γ_1	+	+	ND	6, 9
j	SE	PASENPFREK	55–64 γ_2	γ_2, γ_3	+	ND	ND	9
k	GC	KNPFKELKGGGVIS[b]	61–74 γ_1	γ_1 (1 : 400–600)	+	ND	ND	10

[a] Not designated.
[b] Cys-71 has been substituted with Gly to prevent KLH from coupling to the cysteine (see text).
[c] Reacts with unisoprenylated but not isoprenylated γ1.
[d] Key to references: (1) N. Gautam, J. K. Northup, H. Tamir, and M. I. Simon, *Proc. Natl. Acad. Sci. U.S.A.* **87**, 7973 (1990). (2) A. N. Pronin, and N. Gautam, *FEBS Lett.* **328**, 89 (1993); (3) A. Pronin, C. Gallagher, and N. Gautam, unpublished results; (4) A. Pronin and N. Gautam, *Proc. Natl. Acad. Sci. U.S.A.* **89**, 6220 (1992); (5) W. F. Simonds and A. M. Spiegel, unpublished results (1992); (6) T. Murakami, W. F. Simonds, and A. M. Spiegel, *Biochemistry* **31**, 2905 (1992); (7) K. H. Muntz, P. C. Sternweis, A. G. Gilman, and S. M. Mumby, *Mol. Biol. Cell* **3**, 49 (1992); (8) J. D. Robishaw, V. K. Kalman, C. R. Moomaw, and C. A. Slaughter, *J. Biol. Chem.* **264**, 15758 (1989); (9) W. F. Simonds, J. E. Butrynski, N. Gautam, G. Unson, and A. M. Spiegel. *J. Biol. Chem.* **266**, 5363 (1991); (10) A. N. Pronin and N. Gautam, unpublished results.

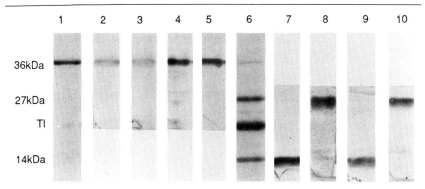

FIG. 3. Immunoblot of the β_1 protein (from the purified $\beta\gamma$ complex of transducin) with different peptide-specific antisera both before and after treatment with trypsin (37° for 30 min). Lane 1 shows a Coomassie blue-stained gel with the β_1 protein. Lanes 2–5 show immunoblots of β_1 protein probed with the following antisera: 1, antiserum to the native $\beta\gamma$ complex from transducin; 2, BC1; 3, BN1; 4, BR1. (See Fig. 1 and Table I for details regarding antisera.) Lane 6 shows a Coomassie blue-stained gel with tryptic products of β_1 protein, of approximately 27 and 14 kDa. TI, Trypsin inhibitor. In lanes 7–10, showing immunoblots of tryptic products, the sequence of antisera used were the same as above. Coomassie blue-stained proteins (~0.75 μg) were resolved on a 12–22% continuous gradient SDS–polyacrylamide gel. The immunoblot is from a 15% gel (~0.3 μg of $\beta\gamma$ complex without trypsin and ~0.75 μg with trypsin).

14 kDa and a COOH portion of 27 kDa. An antiserum raised against the whole $\beta\gamma$ complex reacts only with the NH$_2$-terminal portion (Fig. 3, lanes 2 and 7). The peptide antisera react specifically with the portion of the protein to which the appropriate peptides correspond. BN1 reacts with the NH$_2$-terminal portion of the β_1 protein (lane 9, Fig. 3) whereas BC1 and BR1 react with the COOH-terminal 27-kDa portion within which the amino acid sequences of both the peptides specific to these antisera lie (see Fig. 1 and Table I).

To facilitate the coupling of the peptide to the carrier protein KLH (keyhole limpet hemocyanin), a cysteine is added to the NH$_2$ or COOH terminus of the peptide. All the peptides that we have used have a Cys at the COOH terminus in comparison to others that contain Cys at the NH$_2$ terminus. Because peptides with a Cys at both positions have been successfully used to raise antisera, the position of the Cys does not seem to be important. What is the effect of cysteines inside the peptide sequence? Although we have changed the Cys to an alternate amino acid so that the Cys to which KLH is coupled is the only Cys in the peptide (GC), this is probably unnecessary since two other peptides that possess a cys residue within the peptide sequence have been used successfully to generate antisera [BR1 and an antiserum against γ_2 (peptide h, Table II)]. The peptide is coupled to KLH basically using the protocol of Green et al.[18] This method has also been outlined in detail in an earlier chapter in this series on peptide antisera to the α and β subunits of G proteins.[19]

Raising Antisera in Rabbits

We have used the protocol given below to raise antisera to peptides in rabbits.

1. Mix the following: 250 μl of peptide solution containing 1.5–2.0 mg peptide in guanidine–PBS (137 mM NaCl, 6 M guanidine hydrochloride, 10 mM phosphate buffer, pH 7.5), 3.8 ml PBS, and 4.0 ml Freund's adjuvant (complete for the first injection and incomplete for the remaining injections).
2. Sonicate in ice to obtain a thick white emulsion.
3. Inject two rabbits as follows: once with 0.5 ml, left leg intramuscular; once with 0.5 ml, right leg intramuscular; and six times with 0.2 ml subcutaneous. Boost with a similar solution as above with some modifications [125 μl of peptide solution, 3.8 ml PBS, and 4.0 ml Freund's adjuvant (incomplete)]: five times with 0.4 ml, subcutaneously into the back.

It is preferable to inject more than one rabbit with the antigen since the responses of rabbits vary. Apart from the variation in the titer of the antisera the specificity of antisera to the same peptide from two different rabbits can also vary. It is therefore recommended that the antisera from different rabbits be stored separately. We follow the following schedule:

1. Prebleed before the first injection (~5 ml).
2. Boost with an injection each week for the first 2 weeks.
3. From the third week on test bleed (5 ml) every alternate week.
4. Continue boosting every week until the seventh week and then boost every alternate week.
5. If antisera are present in a test bleed, bleed approximately 25–30 ml every alternate week.

Serum is isolated and stored using standard procedures. We have used the services of a company (Cocalico Biologicals, Reamstown, PA) to raise some antisera. Injections and bleeds are performed by the company according to the schedule provided.

---

[18] N. Green, S. M. Mumby, and A. G. Gilman, this series, Vol. 195, p. 215.
[19] S. M. Mumby, J. C. Voyta, B. Edwards, and I. Bronstein, Clin. Chem. **34**, 1157 (1988).

Affinity Purification of Antibodies

Although antisera are usually sufficiently active and specific on immunoblots, purified antibodies are required for immunoprecipitation and, especially, immunocytochemical experiments. We have used Affi-Gel 10 (for basic or neutral peptides; use Affi-Gel 15 for acidic peptides) from Bio-Rad (Richmond, CA) to purify antibodies. The peptide is first coupled to the Affi-Gel column and the antiserum passed through the column. Antibodies bound to the peptide are then eluted using appropriate conditions. We have used the following protocol to purify several antibodies (T. Amatruda, University of Minnesota, personal communication).

Reagents

Affi-Gel 10 or 15
100 mM NaHCO$_3$, pH 8.3
Synthetic peptide
1 M Ethanolamine, pH 8.0
TBST: 10 mM Tris-HCl, pH 8.0, 150 mM NaCl, 0.05% Tween 20
Buffer A: 100 mM Tris HCl, pH 8.0, 500 mM NaCl, 0.5% bovine serum albumin (BSA), 0.1% Triton X-100, 0.05% sodium azide
Buffer B: 10 mM sodium phosphate, pH 7.0, 50 mM NaCl
500 mM sodium phosphate, pH 8.0.
500 mM sodium phosphate, pH 7.4
500 mM glycine, pH 2.5

Procedure

1. Desalt the peptide using a desalting column. Approximately 70% of the peptide preparations is assumed to be the peptide.

2. Rinse (2.5 ml of gel slurry per 2.5 mg of peptide; 2.5 ml of slurry makes ~2 ml of gel bed) the Affi-Gel in 10–20 volumes of cold deionized water. Rinse rapidly within about 10 min since Affi-Gel is unstable in water.

3. Incubate the gel in NaHCO$_3$ buffer at 3–4 ml per milliliter of gel bed with the peptide. Mix by shaking for 4 hr at 4° or 1 hr at room temperature.

4. Block the remaining reactive groups in the gel using ethanolamine at 100 μl/ml of gel bed for 30 min at room temperature.

5. Spin the gel down on a tabletop centrifuge at low speed (e.g., 1000 rpm for less than 10 min).

6. Wash the gel as follows: (i) wash twice with NaHCO$_3$ buffer, 10–15 ml/2 ml gel bed; (ii) wash in approximately 30 column volumes of TBST buffer; (iii) resuspend gel in buffer A, shake for 5 min, and then spin the gel down as before.

7. Add 5–10 ml of antiserum containing 0.05% sodium azide and shake gently overnight at 4°.

8. Spin the gel down. Save an aliquot of the supernatant to examine whether the antibody has bound to the column.

9. Wash the gel as follows: (i) wash twice with buffer A, 10–15 ml; (ii) wash in approximately 30 column volumes of TBST buffer; (iii) wash the gel in buffer B (~5 column volumes).

10. Elute the antibody with 1 ml of 100 mM glycine, pH 2.5, containing 0.1% BSA into 1.5-ml microcentrifuge tubes containing 300 μl of 500 mM sodium phosphate, pH 8.0. Elute with 6 ml of glycine and pool into three fractions. Dialyze the fractions against 1–2 liters of 20 mM sodium phosphate, pH 7.4.

11. Store antibody at $-20°$, preferably with 0.05% sodium azide.

Dot Blots to Test Antisera/Antibodies

Dot blots are useful for rapid screening of antisera from test bleeds or for examining the reactivity of purified antibodies. For these purposes, the peptide is normally the antigen dotted onto the membrane. Dot blots can also be used for a quick examination of reactivity against the whole protein. When performed in the absence of sodium dodecyl sulfate (SDS), a dot blot is not a good indicator of whether the antibodies will react with the protein in its native conformation since it is unclear whether a protein retains its native conformation after air drying on a membrane.

The protocol is as follows: equilibrate a nitrocellulose-based membrane in immunoblot transfer buffer (see section on immunoblotting below). Air dry the membrane and then allow 1 μl of the appropriate dilution of the peptide or protein to blot onto the membrane. Do this by ejecting the 1 μl in more than one fraction, allowing each fraction to air dry before adding the next fraction. This keeps the antigen from diffusing over a larger area. Several samples to be probed with the same antiserum/antibody can be dotted on a piece of membrane approximately 3–4 cm^2 after appropriate labeling with a pencil. Several such membranes can be handled simultaneously using plastic boxes with dividers. The standard immunoblotting protocol (given below) is used to process the membranes. For testing a bleed for the presence of the antiserum the following protocol usually is sufficient (see section on immunoblotting for details): blocking for 20–30 min; treating with antiserum (1 : 100 dilution) for 30 min; and treating with secondary antibodies conjugated to alkaline phosphatase for 20 min with three 5-min washes with TBST after exposure to the primary or secondary antibodies.

Sodium Dodecyl Sulfate–Polyacrylamide Gel Electrophoresis

Proteins are separated on SDS–polyacrylamide gels according to Laemmli with some modifications to improve the resolution of the β and γ subunits. To resolve the γ subunits better, the resolving gel is made in a buffer with a pH of 8.9 and the running buffer with a pH of 8.6. To resolve the different β and γ subunits on the same gel we use a resolving gel that is 12% (w/v) in the top half and 17% (w/v) polyacrylamide in the bottom half (Fig. 4A).

The resolution of the γ subunits can be further improved by separation on a 16% SDS–polyacrylamide gel containing 6–8 M urea (Fig. 4B). Prestained molecular weight markers covering 2350–46,000 (Amersham, Arlington Heights, IL) are used to monitor electrophoresis. Gels of 125 × 150 × 1 mm are run at constant current of 15 mA at room temperature. After the proteins enter the resolving gel, the gels are run at a constant power of 4.5 W. The tracking dye reaches the bottom in 4–5 hr.

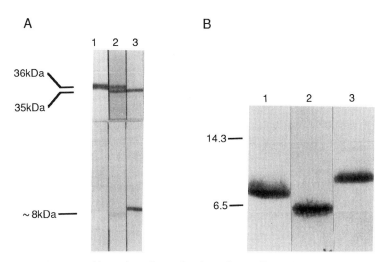

FIG. 4. (A) Immunoblots of total proteins from GH3 cell membranes (~5 μg protein/lane) probed with different antisera to the β and γ subunits of G proteins. BN1 reacts only with the 36-kDa $β_1$ protein (lane 1); AS11 reacts with both the $β_1$ and the 35-kDa $β_2$ proteins (lane 2); B2N1 reacts only with the $β_2$ protein. The γ subunits (bottom of blots) have been probed with antisera specific to different γ subunits. GN1, which is specific to the γ subunit of transducin, $γ_1$, does not react with any protein in GH3 cell membranes. BG reacts with the $γ_2$ protein, and CG is specific to $γ_3$. The proteins were separated on a 12%/17% SDS–polyacrylamide gel system described in the text. (B) Immunoblots of total homogenate from QT6 cell transfectants containing different γ-subunit cDNAs under the control of a viral promoter (~15 μg protein/lane). Lane 1, Proteins from cells containing the $γ_1$ cDNA probed with the GN1 antiserum. Lane 2, Proteins from cells containing the $γ_2$ cDNA probed with the BG antiserum. Lane 3, Proteins from cells containing the $γ_3$ cDNA probed with the CG antiserum. Proteins were separated on a 8 M urea/SDS gel system. 6 M urea/SDS gels provide better resolution.

Immunoblotting

Proteins are transferred from the gels to Immobilon P membranes (Millipore, Bedford, MA) using a semidry electrophoretic transfer cell (Bio-Rad). Gels are initially equilibrated for 15 min in transfer buffer [48 mM Tris, 39 mM glycine, pH 9.2, 0.0375% (w/v) SDS, 20% (v/v) methanol]. The Immobilon P membrane is soaked in methanol for 10 sec and then immersed in water for 1–2 min. It is then equilibrated in transfer buffer for 5–10 min. Gels are transferred for 50 min at 20 V. After the transfer the membrane is rinsed in water and stained with 0.01% amido black in 10% methanol/2% acetic acid. Excess dye is removed by washing with 50% methanol/7% acetic acid. Once the lanes containing proteins are visible, the membrane is cut into appropriate pieces. Incubating in blocking solution [5% (w/v) BSA in TBS (20 mM Tris-HCl, pH 7.5, 250 mM NaCl, 0.05%, v/v, Tween 20)] for 1–1.5 hr at 45° removes the Amido black stain from the membrane. The membrane is gently agitated in the presence of appropriate antibodies or antiserum diluted in TBS containing 0.2% BSA, usually overnight at room temperature. It is then washed with several changes of TBS for approximately 1 hr and incubated with the anti-rabbit mouse antibodies conjugated to alkaline phosphatase diluted 1:5000 in TBS with 0.2% BSA. After washing with TBS twice for 15 min each time, the membrane is stained with nitro blue tetrazolium (NBT) and 5 brom-4-chlor-3 indoylphosphate (BCIP) using a standard protocol.

Visualizing Immunoblots

All our experiments have been performed using alkaline phosphatase and its color-generating substrates (NBT and BCIP) to visualize antibody presence (Promega, Madison, WI).[20] We find this highly sensitive (~1–10 ng of protein can be detected on an immunoblot), rapid, and safe. Recently, we have used a detection system based in a chemiluminescant substrate (ECL, Amersham) and found this to be even more sensitive.

Specificity of Peptide-Directed Antisera

Several approaches can be used to examine the specificity of a newly generated antiserum. Dot blots of the peptide antigen in comparison to other peptides (e.g., peptides corresponding to other nonidentical portions of the same protein or peptides specific to other members of the family with a distinct amino acid sequence) may be performed. Although this can indicate the specificity of the antiserum, it is more important to examine the

---

[20] A. N. Pronin and N. Gautam, *FEBS Lett.* **328,** 89 (1993).

reactivity of the antiserum on an immunoblot containing the whole protein. The ability to react with a protein of the appropriate molecular weight and not with related but distinct proteins indicates specificity. The ability to block the reaction with the peptide (~50–100 µg/ml in the solution containing the antibody) is usually conclusive demonstration of the specificity of the reaction. Another method that estabishes the specificity of reaction with a protein belonging to a family such as the G-protein β and γ subunits is specific reactivity against a family member that has been expressed in a heterologous system such as cell lines or bacteria (e.g., Fig. 4B). This is possible only in cases where the cDNA encoding the protein is available. It is important to note that the ability of an antiserum to react specifically with a member of a family may be dependent on the concentration at which it is used. For instance the BN1 antiserum cross-reacts with the β_2 protein at dilutions of less than 1:500, but at very low dilutions (see Table I, peptide a) it reacts specifically with the β_1 protein. Specificity of antisera raised to the same peptide can also differ as in the case of AS11 (Table I, peptide k) and U49 (Table I, peptide j). AS11 reacts nonspecifically with three different β-subunit types, whereas the affinity-purified U49 reacts specifically with the β_1 protein. It is not clear whether this difference is due to the characteristics of the antisera from different animals or whether it is due to affinity purification.

Transfectants for Immunochemical Experiments

Quail fibroblasts (QT6 cells) transiently transfected with different β and γ cDNAs are used in many of our experiments. Cells are transfected with vector DNA containing appropriate cDNAs. Usually 2 days after transfection cells are used for immunoblotting, immunoprecipitation, or for immunocytochemistry. For immunocytochemistry cells are removed from the dishes with PBS and 1 mM EDTA and plated on collagen-coated coverslips. Cells were allowed to settle down for 3 hrs to overnight, then fixed and stained.

Immunoprecipitation

Several of the antibodies to both the β and γ subunits can be used for immunoprecipitating the $\beta\gamma$ complex (see Tables I and II) from purified G proteins. We have used these antibodies to precipitate β or γ subunits expressed transiently in QT6 cells after labeling with [^{35}S]methionine.[20] All steps are carried out at 0°–4°.

1. Cells are harvested and washed with buffer, 10 mM Tris-HCl, pH 8.0, 135 mM NaCl, 5 mM KCl, 1 mM EDTA.

2. Cells are broken in a buffer containing 10 mM Tris-HCl, pH 8.0, 0.5 mM EDTA by freezing and thawing twice.

3. DNase I is added and samples incubated for 10 min on ice. The cell homogenate is centrifuged for 15 min at 15,000 g.

4. The supernatant is carefully removed, and the membrane pellet is suspended in extraction buffer (20 mM Tris-HCl, pH 8.0, 200 mM NaCl, 0.5% Triton X-100, 1 mM EDTA) and incubated for 20 min.

5. The supernatant fraction is supplemented with NaCl and Triton X-100 to the same concentration as the extraction buffer and used for immunoprecipitation. Affinity-purified antibodies are used in amounts sufficient for complete binding of expressed protein as estimated according to molecular weight (~2–5 μg of antibodies in our experiments).

6. Samples are incubated with antibodies for 1 hr.

7. Then 20 μl of a 50% suspension of protein A-Sepharose (Pharmacia, Piscataway, NJ) in 20 mM Tris-HCl, pH 7.5, 200 mM NaCl, 1 mM EDTA, 0.1% Tween 20 is added and incubated for 1 hr with periodic shaking.

8. The samples are centrifuged, the supernatant removed, and the resin washed twice with 0.75 ml buffer containing 20 mM Tris-HCl, pH 7.5, 200. mM NaCl, 1 mM EDTA, 0.1% Tween 20.

9. Proteins are eluted with 50 μl of 2× SDS–polyacrylamide gel electrophoresis sample buffer [125 mM Tris-HCl, pH 6.8, 200 mM dithiothreitol (DTT), 2% SDS, 20% glycerol] and incubated for 10 min at 65°. Usually 15 μg of total cell protein is added to the samples prior to electrophoresis as a carrier.

Immunoprecipitation after Denaturation

For some experiments it is necessary to be able to precipitate protein in the presence of SDS. For instance, proteins that are resistant to extraction from membranes with nonionic detergents can be extracted with SDS and then immunoprecipitated. This approach is also useful for precipitating proteins that are recognized by an antibody in the denatured state but not when in the native state. We have immunoprecipitated the β_1 protein extracted with SDS from membranes basically using the protocol above with the following modifications. Membranes that have been extracted with a buffer containing Triton X-100 (see Step 4, above) are extracted again with the same buffer containing 1% SDS and the supernatant used for immunoprecipitation. Samples are heated at 65° for 10 min, then chilled and diluted by the addition of 4 volumes of cold buffer (20 mM Tris-HCl, pH 7.5, 200 mM NaCl, 1 mM EDTA, 1% Triton X-100, 0.1% Tween 20). Antibodies are added to the samples at appropriate concentrations and treated as in the protocol above (after Step 5).

Immunocytochemistry

Peptide antibodies to G-protein β and γ subunits have been used to localize the β and γ subunits expressed transiently in cell lines.[2] We have used the following method from Muntz et al.[2] to localize the G-protein β and γ subunits expressed in QT6 cells using peptide-specific antibodies.

1. Two days after transfection the transfected cells on coverslips are fixed on ice for 20 min using 10% (v/v) formalin/PBS with 1 mM MgCl$_2$ (pH 7.4).

2. The cells are washed four times in PBS/1 mM MgCl$_2$/0.1% (w/v) saponin (Sigma, St. Louis, MO) and then incubated in 5% BSA (Amersham) for 15 min at 37°. All subsequent washes and incubations are in the same buffer excepting the final wash.

3. Cells are incubated overnight on a rotary shaker at room temperature in an appropriate dilution of the antibodies and then washed four times in buffer.

4. This is followed by incubation for 15 min at 37° in a 1:20 dilution of normal goat serum (GIBCO, Grand Island, NY) followed by a 1-hr incubation in fluorescein-conjugated goat anti-rabbit IgG (Sigma).

5. The cells are then washed five times in the buffer containing 0.1% saponin followed by two washes in PBS plus 1 mM MgCl$_2$.

6. Treated cells are mounted in Fluoromount G (Fisher, Pittsburgh, PA) with 1% 1,4-diazabicyclo[2.2.2]octane (Sigma) to prevent photobleaching.

Coverslip Preparation for Immunocytochemistry

Because the cells for immunocytochemistry are grown on coverslips, it is important that the coverslips are treated appropriately so that cells adhere to them. We have used the following method to treat coverslips. Round glass coverslips 0.07 mm in thickness (Biophysica Technologies, Baltimore, MD) are soaked overnight in concentrated nitric acid to generate OH groups on the surface. They are then rinsed twice in two changes of deionized water in an ultrasonic cleaner for 15 min. After this, they are dipped in a 10% (v/v) dilution of γ-aminopropyltriethoxysilane (Sigma) which has been warmed to 70° at pH 3.0. They are again rinsed in the ultrasonic cleaner and dried at 100° for 30 min. The coverslips are then treated for 1 hr in 2.5% glutaraldehyde/PBS, pH 7.0, and rinsed again for 15 min in the ultrasonic cleaner. Under a sterile hood a few slivers of autoclaved broken coverslips are placed on the bottom of a 35-mm culture dish (this makes it easier to remove the coverslips later on). The cleaned glass coverslips are placed in the culture dishes and allowed to dry for 30 min after which they are coated with 2 ml of a 1:10 dilution of Vitrogen (Celtrix Labs, Palo Alto, CA) in 0.01 N HCl and allowed to dry overnight.

It is important not to expose the coverslips to UV light at this point. Rinse several times in sterile water and allow to dry. Sterilize under UV light for 10 min. The treated coverslips may be stored for 1 month at 4°. Warm to room temperature prior to plating cells.

Acknowledgments

We thank Dr. William F. Simonds for providing unpublished information. This work supported by grants from the National Institutes of Health and Monsanto–Searle. N.G. is an Established Investigator of the American Heart Association.

[39] Preparation, Characterization, and Use of Antibodies with Specificity for G-Protein γ Subunits

By JANET D. ROBISHAW and ERIC A. BALCUEVA

Introduction

The G proteins are composed of α, β, and γ subunits, with each of these subunits having several different subtypes.[1-3] Thus, the number of possible associations of these subunits to form functionally distinct G proteins is enormous. The production of subtype-specific antibodies is an important tool in determining the associations and functions of particular α, β, and γ subunits. The production of antibodies for the various α and β subtypes of the G proteins was the focus of a chapter by Mumby and Gilman in a previous volume in this series.[4] Accordingly, in this chapter, we focus on the production of antipeptide antibodies for the different γ subtypes and on special requirements for the use of these antibodies in detecting the various γ subtypes. These antibodies have been shown to be specific and fairly versatile for use in detecting the various γ subtypes by immunoblotting and immunostaining procedures.

---

[1] A. G. Gilman, *Annu. Rev. Biochem.* **56**, 615 (1987).
[2] M. I. Simon, M. P. Strathman, and N. Gautam, *Science* **252**, 802 (1991).
[3] J. J. Cali, E. A. Balcueva, I. Rybalkin, and J. D. Robishaw, *J. Biol. Chem.* **267**, 24023 (1992).
[4] S. M. Mumby and A. G. Gilman, this series, Vol. 195, p. 215.

Preparation of Antibodies

Selection of Peptide Sequences

When selecting a peptide that will produce an antibody of the desired specificity, the length and degree of conservation of the amino acid sequence are prime considerations. The minimum length of a peptide for antibody production is considered to be 10 amino acids. In this regard, a peptide based on a stretch of 8 amino acids that was conserved among the various γ subunits failed to produce antibodies that recognized these proteins. Because conserved regions of the γ subunits are limited to stretches of 8 amino acids or less, it has not been possible to produce antibodies that react indiscriminately with the majority of the γ subunits, as has been possible for the α and β subunits of the G proteins. On the other hand, it has been possible to generate a series of antibodies that react specifically with each of the known γ subunits by choosing peptide sequences from relatively nonconserved regions of these proteins (Fig. 1 and Table I). As hydrophilic amino acid residues are more likely to be exposed on the surface of proteins for recognition by the antipeptide antibodies, we have targeted regions containing an abundance of hydrophilic residues. In addition, we include regions containing proline residues, whenever possible, since they were found to be particularly useful in directing the specificity of the antibodies (e.g., B-17 and A-67 in Table I).

When indicated, the peptides are synthesized with a cysteine residue at the amino terminus to facilitate conjugation to the carrier proteins (Table I). The peptides are synthesized using a solid-phase peptide synthesizer (Applied Biosystems, Foster City, CA; Model 430-1a), as described by the manufacturer. The purities of the peptides are assessed using C_8 or C_{18} reversed-phase high-performance liquid chromatography, and the sequences of the peptides are confirmed using amino acid analysis. Alternatively, when indicated, the peptides are synthesized on a heptalysine backbone, as described by Tam and co-workers[5] (Table I).

Peptide Conjugation

The peptides themselves are not particularly immunogenic, necessitating their conjugation to carrier proteins prior to injection. In general, peptides are conjugated to keyhole limpet hemocyanin (KLH; Sigma, St. Louis, MO), as outlined by Green et al.[6] (Table I). For conjugation,

---

[5] K. J. Chang, W. Pugh, S. G. Blanchard, J. McDermed, and J. P. Tam, Proc. Natl. Acad. Sci. U.S.A. **85**, 4929 (1988).

[6] N. H. Green, A. Alexander, S. Olson, T. Alexander, T. Shinnick, J. Sutcliff, and R. Lerner, Cell (Cambridge, Mass.) **28**, 477 (1982).

γ7------------MSATNNIAQARKLVEQLRIEAGIE
γ5------------MSGSSSVAAMKKVVQQLRLEAGLN
γ3-MKGETPVNSTMSIGQARKMVEQLKIEASLC
γ2--------MASNNTASIAQARKLVEQLKMEANID
γ1---MPVINIEDLTEKDKLKMEVDQLKKEVTLE

γ7-RIKVSKASSELMSYCEQHARNDPLLVGVPA
γ5-RVKVSQAAADLKQFCLQNAQHDPLLTGVSS
γ3-RIKVSKAAADLMTYCDAHACEDPLITPVPT
γ2-RIKVSKAAADLMAYCEAHAKEDPLLTPVPA
γ1-RMLVSKCCEEFRDYVEERSGEDPLVKGIPE

γ7-SENPFKDKKP-CIIL
γ5-STNPFRPQKV-CSFL
γ3-SENPFREKKFFCAIL
γ2-SENPFREKKFFCAIL
γ1-DKNPFKELKGGCVIS

FIG. 1. Alignment of various γ subunits. The protein sequences predicted from the cDNAs for γ_1, γ_2, γ_3, γ_5, and γ_7 were aligned. Because the cDNA for γ_4 has not been cloned, the protein sequence predicted from the polymerase chain reaction (PCR) fragment for γ_4 has not been included.[2] The cDNAs for γ_2 and γ_6 were cloned simultaneously in two different laboratories [N. Gautam, M. Baetscher, R. Aebersold, and M. I. Simon, *Science* **244**, 971 (1989); J. D. Robishaw, V. K. Kalman, C. R. Moomaw, and C. A. Slaughter, *J. Biol. Chem.* **264**, 15758 (1989)]. Since the sequences for γ_2 and γ_6 were subsequently found to be identical, we have included only the sequence of γ_2. Regions shown underlined correspond to peptides used for antibody production.

m-maleimidobenzoyl-*N*-hydroxysuccinimide ester (MBS; Sigma) is used to link the peptides to KLH through cysteine residues. A detailed description of the conjugation procedure has been included previously in this series.[4]

By way of comparison, one of the peptides is also conjugated to tuberculin (purified protein derivative of tuberculin; Statens Seruminstitute, Copenhagen, Denmark), which has been reported to boost the antibody response to the peptide while giving rise to virtually no antibody response against itself.[7] In our hands, however, peptide conjugation to tuberculin

---

[7] P. J. Lachman, L. Strangeways, A. Vyakarnam, and G. Evan, *Ciba Found. Symp.* **119**, 25 (1986).

TABLE I
ANTIBODIES WITH SPECIFICITY FOR VARIOUS γ SUBUNITS[a]

Desig-nation	Antigen			Antiserum		
	Region	Peptide	Carrier	Reactivity	Specificity	Code
γ_1	37–49	CEEFRDYVEERSG	KLH	3 of 3	γ_1	A-4
γ_2	2–14	CASNNTASIAQARK	KLH	1 of 4	γ_2	A-75
	2–14	ASNNTASIAQARK	Heptalysine	0 of 2		
	35–46	CDLMAYCEAHAK	KLH	2 of 4	γ_2	A-25
	35–49	DLMAYCEAHAKEDP	Heptalysine	1 of 2	γ_2	B-60
	48–61	CDPLLTPVPASENPF	KLH	4 of 4	$\gamma_2 \approx \gamma_3$	B-17
	48–62	DPLLTPVPASENPFR	Heptalysine	2 of 2	$\gamma_2 \approx \gamma_3$	B-63
	48–61	CDPLLTPVPASENPF	Tuberculin	1 of 2	$\gamma_2 \approx \gamma_3$	B-23
γ_3	2–17	CKGETPVNSTMSIGQAR	KLH	1 of 2	γ_3	B-53
γ_5	50–63	CTGVSSSTNPFRPQK	KLH	2 of 2	γ_5	D-9
γ_7	46–59	CDPLLVGVPASENPF	KLH	2 of 2	γ_7	A-67
	46–60	DPLLVGVPASENPFK	Heptalysine	1 of 2	γ_7	B-65

[a] The region numbers indicate the location of the peptide sequences in the various γ subunits. The reactivity refers to the number of rabbits that reacted with the protein of interest by immunoblotting out of the total number of rabbits that were immunized with the peptide. The code refers to the numbers assigned to the rabbits by the investigators (J.D.R. and E.A.B.).

results in a lower proportion of rabbits that produce antibodies to the protein of interest and no improvement in the titer of the antibodies produced (Table I). Also, by way of comparison, three of the peptides are attached to heptalysine backbones, which has been reported to enhance the antibody response by presenting multiple copies of the peptide on each heptalysine backbone.[5] Again, however, this procedure results in no improvement in the proportion of rabbits that produce antibodies to the proteins and no improvement in the titer of antibodies produced (Table I). Given the greater degree of difficulty in synthesizing peptides on the heptalysine backbone and the lack of improvement in the titers of the antibodies produced, we do not generally recommend the use of the latter procedure.

Immunization and Bleeding Schedule

Each of the conjugated or heptalysine peptides (equivalent to 200 μg of peptide) is solubilized with 500 μl of phosphate-buffered saline (PBS), emulsified with 500 μl of Freund's complete adjuvant (GIBCO Laboratories, Grand Island, NY), and administered to New Zealand White rabbits by subcutaneous injection at four sites on the back. Two weeks after the first injection, the same amount of conjugated peptide is emulsified with Freund's incomplete adjuvant and administered to rabbits by subcutaneous injection. One week after the second injection, the same amount of conjugated peptide is adsorbed to alum,[8] as described in detail elsewhere,[4,8] and administered to rabbits by intraperitoneal injection at two sites in the abdomen. In general, this immunization protocol results in the production of antibodies that recognize the peptides and, in most cases, the proteins of interest. However, the titer and specificity of the antibodies produced are found to vary widely between rabbits. Thus, whenever possible, we recommend that each of the conjugated peptides be injected into two or more rabbits.

Two conjugated peptides are emulsified with RIBI adjuvant in place of Freund's complete and incomplete adjuvants. The RIBI adjuvant (RIBI Immunochem Research, Inc., Hamilton, MT) has been reported to induce a strong antibody response, without causing the granuloma formation that is a frequent side effect of the Freund's adjuvant. The rabbits receiving the conjugated peptides with the Freund's adjuvant produce antibodies that react specifically with either the β_1 or the β_2 subunit (three out of four rabbits in each case). In contrast, the rabbits receiving the conjugated peptides with the RIBI adjuvant produce antibodies of similar titers that react indiscriminately with both the β_1 and β_2 subunits. Thus, the use of RIBI adjuvant adversely affects the selectivity of the antibody response to peptides based on the sequences of two closely related proteins. Although the applicability of this finding to other closely related proteins, such as the γ subunits, is not known, the use of RIBI adjuvant is not generally recommended for producing antibodies to closely related proteins in a limited number of rabbits.

Beginning 1 week after the final injection, rabbits are bled every 2 weeks to monitor antibody production. The blood (~20 ml) is allowed to clot for 1 hr at 37°, and the clot is allowed to retract for several hours at 4°. Following removal from the clot, the serum is centrifuged at 10,000 g for 10 min at 4° to remove residual blood cells, then divided into aliquots for storage at $-80°$. To monitor the presence of antibodies, the serum is tested for reactivity to the peptide by an enzyme-linked immunosorbent assay (ELISA) procedure.[9] If desired, antibodies to the carrier protein are effectively removed by passage through a peptide affinity column.[4]

Immune Response as Monitored by Immunoblotting

Because discrepancies between antibody titers for the peptide versus the protein of interest have been observed, the serum is routinely tested

---

[8] C. A. Williams and M. W. Chase, *Methods Immunol. Immunochem.* **1**, 201 (1967).
[9] E. Engvall, this series, Vol. 70, p. 419.

for reactivity to the purified protein by an immunoblotting procedure.[10] The transfer and immunoblotting procedures are presented in some detail, since both of these procedures have been modified to optimize the detection of the γ subunits.[11]

Purification of a mixture of the $\beta\gamma$ subunits is carried out from bovine brain, as described previously.[12] The different γ subunits are resolved on a 15% polyacrylamide separating gel (acrylamide to bisacrylamide ratio of 29:1) by sodium dodecyl sulfate–polyacrylamide gel electrophoresis (SDS–PAGE). In general, 2 μg of purified $\beta\gamma$ subunits (equivalent to ~300 ng of purified γ subunit) is loaded on each lane of the gel. After electrophoresis, the proteins are transferred to Nitroplus nitrocellulose (0.45 μm pore size; Micron Separations, Inc., Westboro, MA), using the Hoeffer Transphor system (TE series, Hoefer Scientific Instruments, San Francisco, CA). Transfer is carried out at 30 V overnight in transfer buffer [25 mM Tris, 190 mM glycine, and 20% (v/v) methanol, pH 8.2] that had been heated to 70°, using a circulating water bath (Forma Scientific, Model 2006).

As shown in Fig. 2, heating the transfer buffer results in a greater than 20-fold increase in the sensitivity of detection of the γ subunits by immunoblotting,[11] presumably by maintaining the proteins in a denatured state during transfer and/or binding to the nitrocellulose. Denaturation of the γ subunits would be expected to have a beneficial effect, since antipeptide antibodies generally recognize the epitopes more effectively in denatured proteins than in native proteins. In contrast, inclusion of the protein denaturant SDS in the transfer buffer or heating the transferred proteins on the nitrocellulose by baking or autoclaving are not nearly so effective in enhancing the sensitivity of detection of the γ subunits by immunoblotting (Fig. 2).

Another factor found to enhance the sensitivity of detection of the γ subunits by immunoblotting is the type of paper used for the transfer. Of the various types of paper tried, the Nitroplus nitrocellulose and the diazotized papers (aminobenzoyloxymethyl cellulose and aminophenylthioether cellulose; Schleicher and Schuell, Inc., Keene, NH) have been found to exhibit a 5-fold higher binding capacity than nitrocellulose (Schleicher and Schuell) and Immobilon (Millipore, Inc., Bedford, MA). Although Nitroplus nitrocellulose and diazotized papers are found to have similar binding capacities, the use of the Nitroplus nitrocellulose is favored by the comparatively lower cost and longer shelf life of this paper.

[10] H. Towbin, T. Staehelin, and J. Gordon, *Proc. Natl. Acad. Sci. U.S.A.* **76**, 4350 (1979).
[11] J. D. Robishaw and E. A. Balcueva, *Anal. Biochem.* **208**, 283 (1993).
[12] P. C. Sternweis and J. D. Robishaw, *J. Biol. Chem.* **259**, 13806 (1984).

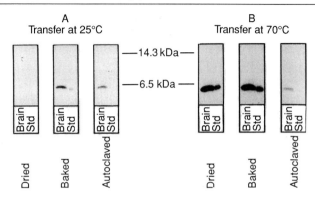

FIG. 2. Effect of high-temperature transfer to greatly enhance detection of G-protein γ subunits with antipeptide antibodies. Solubilized bovine brain membrane proteins (100 μg brain) and partially purified bovine brain G protein standard (10 μg Std) were resolved on a 15% polyacrylamide gel. After transfer of the resolved proteins to nitrocellulose, the blots were air-dried, baked *in vacuo* at 80°, or autoclaved at 120°. The nitrocellulose pieces were incubated with anti-γ_7 antibody (A-67), as the primary antibody, and ^{125}I-labeled goat anti-rabbit antibody, as the secondary antibody. To obtain autoradiographic exposures, the nitrocellulose pieces were exposed to film for 18 hr. Numbers between (A) and (B) indicate apparent molecular masses, based on the mobilities of aprotinin and lysozyme standards. (Reproduced by permission from Ref. 11.)

To determine the efficiency of transfer, the Nitroplus nitrocellulose blots are stained with Ponceau S or amido black (Sigma) prior to immunological processing. For processing, the nitrocellulose blots are incubated for 1 hr in high-detergent buffer A [50 mM Tris (pH 8), 2 mM CaCl$_2$, 80 mM NaCl, 5% nonfat dry milk, 2% Nonidet P-40 (NP-40) and 0.2% SDS] with the appropriate antipeptide antibodies.[3] After three 5-min washes in high-detergent buffer A the nitrocellulose blots are incubated for 1 hr in high-detergent buffer A with ^{125}I-labeled goat anti-rabbit F(ab')$_2$ fragment [1 × 10^6 disintegrations/min (dpm)/ml; New England Nuclear, Boston, MA]. After three 5-min washes in high-detergent buffer A followed by three 5-min washes in detergent-free buffer A, the nitrocellulose blots are air-dried at room temperature. The incubation and wash times are minimized to prevent loss of the low molecular weight γ subunits during immunological processing. To obtain autoradiographic images, the nitrocellulose blots are exposed with an intensifying screen to Kodak (Rochester, NY) XAR-5 film overnight. If desired, the intensities of the immunodetectable bands are quantitated by scanning the nitrocellulose blots with the AMBIS Radioanalytic Imaging System (AMBIS Inc., San Diego, CA).

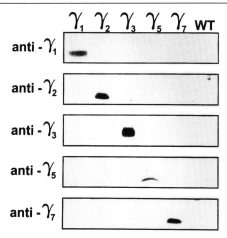

FIG. 3. Specificity of antibodies for recombinant γ subunits expressed in Sf9 cells. Cholate extracts of particulate fractions from Sf9 cells infected with recombinant viruses encoding the indicated γ subunits or the wild-type virus (100 μg) were subjected to SDS–PAGE followed by immunoblot analysis with the various antibodies. Antibodies used for blotting were the anti-γ_1 antibody A-4, the anti-γ_2 antibody A-75, the anti-γ_3 antibody B-53, the anti-γ_5 antibody D-9, and the anti-γ_7 antibody A-67.

Characterization of Antibody Specificity

The specificity of the antipeptide antibodies for the various γ subunits is best demonstrated by immunoblotting proteins of the appropriate sizes in cholate extracts from Sf9 cells (*Spodoptera frugiperda* ovary) infected with recombinant baculoviruses encoding the various γ subunits.[3,13] As shown in Fig. 3, the cholate-soluble membrane extracts from Sf9 cells infected with recombinant baculoviruses encoding the γ_1, γ_2, γ_3, γ_5, and γ_7 subunits were resolved on a 15% polyacrylamide–SDS gel, transferred to nitrocellulose, and then immunoblotted. As expected, each of the antibodies generated against a peptide based on sequence unique to the γ_1, γ_2, γ_3, γ_5, and γ_7 subunits recognized a protein of the appropriate size (5–7 kDa range) only in the cholate-soluble membrane extract from cells infected with the virus for the corresponding γ subunit. The specificity of each of these antibodies was further demonstrated by the failure of each of these antibodies to detect similarly sized proteins in extracts from cells infected with viruses for the noncorresponding γ subunits, or with the wild-type virus.

[13] J. D. Robishaw, V. K. Kalman, and K. L. Proulx, *Biochem. J.* **286,** 672 (1992).

Usefulness of Antibodies

With the production of antibodies to the different γ subtypes, it is now possible to determine the distribution of the γ subtypes within various tissues, to examine the localization of the γ subtypes within particular cell types, and to follow the synthesis, posttranslational processing, and assembly of particular α, β, and γ subunits of the G proteins. This information is ultimately needed to determine the role of specific combinations of αβγ subunits of the G proteins in a large number of different receptor-mediated signaling pathways.

Tissue Distribution and Localization of γ Subunits

The identification of the γ subunits in a tissue or cell type is greatly facilitated by the use of subtype-specific antibodies.[3] Such studies have lagged behind those with the α and β subunits owing to the inability of antibodies against the γ subunits to detect these proteins in whole cell or membrane extracts by immunoblotting. As discussed above, by incorporating a high-temperature transfer step to allow the antipeptide antibodies to gain access to the epitopes, it has become possible for the first time to utilize these antibodies to determine the tissue distribution of the γ subunits.

The preparation and solubilization of membranes from several bovine tissues with 0.9% cholate have been described previously.[3] In contrast to the β subtypes, the various γ subtypes showed a more selective pattern of expression. Thus, the γ_2 and γ_3 subunits were preferentially expressed in brain, whereas the γ_5 and γ_7 subunits were widely expressed in brain, heart, kidney, spleen, liver, and lung (Fig. 4). The γ_1 subunit was exclusively expressed in the retina (data not shown). This pattern of distribution for the γ subunits is reminiscent of that for the α subunits, that is, if not found largely in the brain or retina, the α subunits were found in a variety of tissues. Ascertainment of the tissue distribution of specific subunits will be of importance in determining which of the large number of possible αβγ combinatorial associations actually occur in a physiological setting.

The subtype-specific antibodies have also been successfully used for the cellular localization of the γ subunits in monkey retina,[14] which contains readily distinguishable cell types with highly specialized functions. The β_1 and γ_1 subunits were found to be localized in the rod outer segments (ROS), whereas the β_3 and γ_2 subunits were found to be localized in the cone outer segments. It is likely that the localization of particular α as

[14] Y. W. Peng, J. D. Robishaw, M. A. Levine, and K. W. Yau, *Proc. Natl. Acad. Sci. U.S.A.* **89**, 10882 (1992).

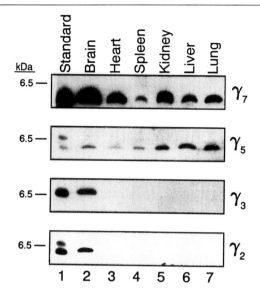

FIG. 4. Tissue distribution of γ subunits. Cholate extracts of particulate fractions from the indicated tissues (100 μg) were subjected to SDS-PAGE and immunoblot analysis, as described in the text. The standard shown in lane 1 represents 4 μg of purified bovine brain G proteins. The 6.5-kDa molecular mass marker was aprotinin. Exposure times were 2 and 4 days for the γ_2 and γ_3 blots, respectively, and 18 hr for the γ_5 and γ_7 blots.

well as βγ subunits within different cell types in the retina reflects the association of these subunits into different G proteins, with the phototransduction properties characteristic of the rod and cone outer segments. A similar approach to colocalize α, β, and γ subunits in specific cell types in other tissues may provide a useful way of defining the number of possible αβγ associations that can occur in a physiological setting.

Processing of γ Subunits

The subtype-specific antibodies have also been used to examine the posttranslational processing of the γ subunits, which first involves the attachment of a prenyl group to a cysteine residue near the carboxyl terminus of these proteins.[15,16] Insect (Sf9) cells expressing the γ_2 and γ_3 subunits were fractionated into cytosolic and particulate fractions. Following extraction with 0.9% cholate, the particulate fractions were then frac-

[15] W. A. Maltese and J. D. Robishaw, *J. Biol. Chem.* **265,** 18071 (1990).
[16] S. M. Mumby, P. J. Casey, A. G. Gilman, S. Gutowski, and P. C. Sternweis, *Proc. Natl. Acad. Sci. U.S.A.* **87,** 5873 (1990).

tionated into cholate-soluble and cholate-insoluble particulate fractions. As shown in Fig. 5, the nonprenylated forms of the γ_2 and γ_3 subunits, which migrated more quickly on the gel,[13] were expressed largely in the cytosolic and insoluble particulate fractions, respectively (compare upper and lower portions of Fig. 5). Although the reason for the difference in subcellular localization is not known, the localization of the γ_3 subunit in the insoluble particulate fraction may reflect an increased tendency of the γ_3 subunit to aggregate on overexpression. Conversely, the prenylated forms of the γ_2 and γ_3 subunits, which migrated more slowly,[13] were expressed exclusively in the soluble and insoluble particulate fractions. The origin of a third form of the γ_3 subunit in the insoluble particulate fraction has not been determined, but it appears to represent an aggregated form of the γ_3 subunit. Nevertheless, the localization of the prenylated

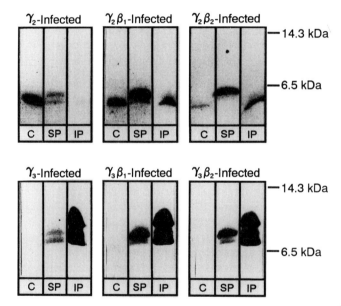

FIG. 5. Expression and subcellular distribution of various combinations of the β_1, β_2, γ_2, and γ_3 subunits of the G proteins in Sf9 cells. At 72 hr after infection with various combinations of viruses encoding the β_1, β_2, γ_2, and γ_3 subunits, cells were fractionated to yield the cytosolic (C), cholate-soluble particulate (SP), and cholate-insoluble particulate (IP) fractions. The same percentage (15%) of each fraction was resolved by SDS–PAGE on a 15% polyacrylamide gel, then transferred to nitrocellulose blots. The nitrocellulose blots were incubated with either the γ_2 antibody B-17 (*top*) or the γ_3 antibody B-53 (*bottom*), using ^{125}-labeled goat anti-rabbit secondary antibody for detection. The blots were exposed to film for 2 days with an intensifying screen.

forms of the γ_2 and γ_3 subunits in the soluble particulate fraction is consistent with the role of prenylation in promoting membrane interaction.[17,18]

Interestingly, the proportion of the prenylated forms of the γ_2 and γ_3 subunits was significantly increased by coexpression of the β_1 or β_2 subunits in insect cells,[13] as demonstrated by an increase in the amount of the slower migrating forms in the soluble particulate fractions from $\beta\gamma$-infected cells compared to γ-infected cells (Fig. 5). Although the mechanism of this effect is not known, association of the β and γ subunits into a functional heterodimer does not appear to be required for two reasons. First, the γ subunit is able to undergo prenylation when expressed alone. Second, both the β_1 and β_2 subunits are able to enhance prenylation of the γ_3 subunit even though a number of investigators have shown that the β_2 and γ_3 subunits do not associate to form a functional dimer.[19-21]

Acknowledgments

This work was supported by National Institutes of Health Grant GM 39867 and an American Heart Association Established Investigatorship to J.D.R.

[17] W. F. Simonds, J. E. Butrynski, N. Gautam, C. G. Unson, and A. M. Spiegel, *J. Biol. Chem.* **266,** 5363 (1991).
[18] K. H. Muntz, P. C. Sternweis, A. G. Gilman, and S. M. Mumby, *Mol. Biol. Cell* **3,** 49 (1992).
[19] J. Iniguez-Lluhi, M. Simon, J. D. Robishaw, and A. G. Gilman, *J. Biol. Chem.* **267,** 23409 (1992).
[20] C. J. Schmidt, T. C. Thomas, M. A. Levine, and E. J. Neer, *J. Biol. Chem.* **267,** 13807 (1992).
[21] A. N. Pronin and N. Gautam, *Proc. Natl. Acad. Sci. U.S.A.* **89,** 6220 (1992).

[40] Isoprenylation of γ Subunits and G-Protein Effectors

By Bernard K.-K. Fung, Janmeet S. Anant, Wun-Chen Lin, Olivia C. Ong, and Harvey K. Yamane

Introduction

Protein isoprenylation is a recently discovered posttranslational modification.[1] The covalent addition of a isoprenyl moiety to a polypeptide was first described in the yeast mating factors.[2] These short polypeptides,

[1] S. Clarke, *Annu. Rev. Biochem.* **61,** 355 (1992).
[2] Y. Kamiya, A. Sakurai, S. Tamura, and N. Takahashi, *Biochem. Biophys. Res. Commun.* **83,** 1077 (1973).

which mediate the mating response between opposite cell types in yeast, were found to contain a farnesylcysteine residue at their carboxyl termini.[3] Since then, a diverse group of mammalian proteins have also been shown to be modified by isoprenylation. Among these are nuclear lamins, the γ subunits of heterotrimeric G proteins, nearly all small G proteins, the α and β subunits of muscle phosphorylase kinase, and certain G-protein-related enzymes, such as the catalytic subunits of retinal cGMP phosphodiesterase and rhodopsin kinase. Despite the apparent diversity in structures and functions, these proteins all share the common characteristic of possessing a cysteine residue at their carboxyl termini in a -CXXX, -CC, or -CXC motif (where C is cysteine and X is any amino acid). However, it is important to note that not all proteins bearing these distinctive carboxyl-terminal sequences are modified by isoprenylation. Most notably, the α subunits of G_i, G_o, and transducin terminate in a -CXXX motif, but are not isoprenylated. Significantly, the same cysteine residue in these proteins is the site of pertussis toxin-catalyzed ADP-ribosylation (see [6] in this volume).

With respect to proteins terminating in a -CXXX motif, such as γ subunits of G proteins and retinal cGMP phosphodiesterase, the term isoprenylation is generally intended to encompass a closely coupled series of posttranslational modifications (Fig. 1). These modifications include attachment of either a 15-carbon farnesyl or a 20-carbon geranylgeranyl group to the cysteine through a thioether linkage, proteolytic removal of three residues downstream of the isoprenylated cysteine, and methyl esterification of the newly exposed carboxyl-terminal cysteine residue.[1]

The first evidence that G proteins are modified by isoprenylation came from labeling studies of purified brain γ subunits in the presence of ³H-labeled S-adenosylmethyl-L-methionine and a methyltransferase in brain membranes.[4] The methylation was found to occur on the α-carboxyl group of a carboxyl-terminal cysteine residue, indicating that the -CXXX terminal sequence of the γ subunits must have undergone multiple modification steps. Subsequently, the γ subunits of brain G proteins and transducin were found to be modified by geranylgeranylation[5,6] and farnesylation,[7,8]

---

[3] R. J. Anderegg, R. Betz, S. A. Carr, J. W. Crabb, and W. Duntze, *J. Biol. Chem.* **263,** 18236 (1988).

[4] B. K.-K. Fung, H. K. Yamane, I. M. Ota, and S. Clarke, *FEBS Lett.* **260,** 313 (1990).

[5] H. K. Yamane, C. C. Farnsworth, H. Xie, W. Howald, B. K.-K. Fung, S. Clarke, M. H. Gelb, and J. A. Glomset, *Proc. Natl. Acad. Sci. U.S.A.* **87,** 5868 (1990).

[6] S. M. Mumby, P. J. Casey, A. G. Gilman, S. Gutowski, and P. C. Sternweis, *Proc. Natl. Acad. Sci. U.S.A.* **87,** 5873 (1990).

[7] Y. Fukada, T. Takao, H. Ohguro, T. Yoshizawa, T. Akino, and Y. Shimonishi, *Nature (London)* **346,** 658 (1990).

[8] R. K. Lai, D. Pérez-Sala, F. J. Cañada, and R. R. Rando, *Proc. Natl. Acad. Sci. U.S.A.* **87,** 7673 (1990).

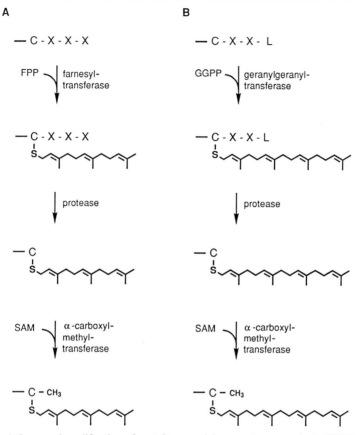

FIG. 1. Isoprenyl modification of proteins containing a carboxy-terminal -CXXX motif. The modifications include attachment of an isoprenoid to the cysteine through a thioether linkage, proteolytic removal of three residues downstream of the isoprenylated cysteine, and methyl esterification of the newly exposed carboxyl-terminal cysteine residue. The isoprenyl transfer reaction is catalyzed by two different protein:prenyltransfererases, a farnesyltransferase that uses farnesyl pyrophosphate (FPP) as the donor molecule (A) and a geranylgeranyltransferase that uses geranylgeranyl pyrophosphate (GGPP) (B). The specificity for the transfer reaction is known to reside in the -CXXX sequence of the acceptor protein. The two transferases have different specificities. The farnesyltransferase prefers moderately polar residue such as Ser, Met, or Gln at the carboxyl terminus, and the geranylgeranyltransferase prefers Leu at that position. α-Carboxylmethylation of the isoprenylated cysteine is mediated by a membrane-bound methyltransferase using S-adenosyl-L-methonine (SAM) as the substrate.

respectively. A number of G-protein-related enzymes such as retinal cGMP phosphodiesterase[9,10] and rhodopsin kinase[11,12] have also been shown to be isoprenylated.

Structural Analysis of Isoprenyl Groups

Several approaches have been developed for the identification of isoprenyl groups on modified proteins. The first, and most rigorous, involves chemical cleavage of the modifying isoprenyl group with Raney nickel, followed by gas chromatography-coupled mass spectrometric (GC-MS) analysis of the products. The advantage of GC-MS analysis is that definitive identification of the isoprenyl group, as well as its stereochemical configuration, can be determined. This technique was used to show that the G-protein γ subunits purified from bovine brain are modified by all-*trans*-geranylgeranyl groups.[5] A detailed protocol for the GC-MS analysis of isoprenylated proteins has been described.[13] There are also two variations of this approach. The first involves the release of the modifying isoprenoid with methyl iodide, followed by the separation of the resulting products by high-performance liquid chromatography (HPLC) and structural analysis by GC-MS.[8,14] This method, however, does not allow the rigorous assignment of the stereochemistry of the modifying isoprenyl group. The second variation employs atom bombardment mass spectrometry to determine the combined mass value of the isoprenylated polypeptide. This information is then used to deduce the identity of the isoprenyl group if the exact amino acid sequence of the polypeptide is known.[7]

Whereas GC-MS identification rigorously identifies the structure of the modifying isoprenoid, it does not reveal the amino acid to which the isoprenyl group is attached nor its location within the polypeptide. To alleviate these shortcomings, methods have been developed to radiolabel the isoprenyl group[9] or the α-carboxyl methyl group.[4,15,16] The radiolabeled

[9] J. S. Anant, O. C. Ong, H. Xie, S. Clarke, P. J. O'Brien, and B. K.-K. Fung, *J. Biol. Chem.* **267**, 687 (1992).
[10] N. Qin, S. J. Pittler, and W. Baehr, *J. Biol. Chem.* **267**, 8458 (1992).
[11] J. Inglese, J. F. Glickman, W. Lorenz, M. G. Caron, and R. J. Lefkowitz, *J. Biol. Chem.* **267**, 1422 (1992).
[12] J. S. Anant and B. K.-K. Fung, *Biochem. Biophys. Res. Commun.* **183**, 468 (1992).
[13] C. C. Farnsworth, S. L. Wolda, M. H. Gelb, and J. A. Glomset, *J. Biol. Chem.* **264**, 20422 (1989).
[14] P. J. Casey, P. A. Solski, C. J. Der, and J. E. Buss, *Proc. Natl. Acad. Sci. U.S.A.* **86**, 8323 (1989).
[15] O. C. Ong, I. M. Ota, S. Clarke, and B. K.-K. Fung, *Proc. Natl. Acad. Sci. U.S.A.* **86**, 9238 (1989).
[16] H. Xie, H. K. Yamane, R. C. Stephenson, O. C. Ong, B. K.-K. Fung, and S. Clarke, *Methods (San Diego)* **1**, 276 (1990).

polypeptide is then converted to its constituent amino acids by exhaustive proteolytic digestion, followed by reversed-phase HPLC separation of the digest and comparison to authentic isoprenylcysteine or isoprenylcysteine methyl ester standards. These methods can be used to confirm the modification of a cysteine residue and, under carefully controlled conditions, the presence of an α-carboxyl methyl ester. The procedure described in this chapter is based on this latter approach developed for the analyses of isoprenyl proteins in retinas.[9]

Procedures

Synthesis of S-(all-trans-Farnesyl)cysteine, S-(all-trans-Geranylgeranyl)cysteine, and Methyl Esters

Isoprenylcysteine methyl esters are synthesized by first converting the isoprenyl alcohols to the chlorides, which are then used to react with cysteine methyl ester. Farnesol and geranylgeraniol are commercially available from Aldrich (Milwaukee, WI) and USF Research Foundation (University of Southern Florida, Box 3004, Tampa, FL), respectively. The procedure for the synthesis of S-(all-*trans*-geranylgeranyl) chloride is as follows. Methanesulfonyl chloride (65 mg, 0.56 mmol) is added to 1 ml of *N,N*-dimethylformamide containing all-*trans*-geranylgeraniol (150 mg, 0.51 mmol), s-collidine (80 mg, 0.66 mmol), and LiCl (50 mg, 1.1 mmol). After stirring the mixture for 3 hr at room temperature under nitrogen, the reaction is terminated by the addition of 4 ml of ice-cold water. The product is then extracted two times with 7 ml of cold ether/hexane (1:1, v/v). The combined extract is washed two times with 2 ml of saturated copper sulfate solution, washed two times with 2 ml of saturated NaCl solution, and dried with anhydrous sodium sulfate. Approximately 180 mg of geranylgeranyl chloride is obtained after removing the solvent. This material is used in the subsequent reaction without further purification. If the geranylgeranyl chloride is not used immediately, it can be stored under anhydrous conditions at $-20°$ for several months.

S-(all-*trans*-Geranylgeranyl)cysteine methyl ester is synthesized by reacting geranylgeranyl chloride (150 mg, 0.5 mmol) with cysteine methyl ester (90 mg, 0.5 mmol) in 3 ml of 2.0 *M* ammonia in methanol (Aldrich) under nitrogen for 3 hr at 0° and then for an additional 6 hr at room temperature. After removing the solvent under reduced pressure, the residue is redissolved in ethyl acetate and purified by flash chromatography on a 1 × 26 cm silica gel column (Fisher Scientific, Pittsburgh, PA, Davilsil grade 643, 235–400 mesh) eluted with ethyl acetate/hexane (2:1, v/v).

The eluent is monitored by silica gel-60 thin-layer chromatography using ethyl acetate/hexane (2 : 1, v/v) as the developing solvent. Geranylgeranyl cysteine methyl ester migrates with an R_f 0.34 and can be easily detected by vanillin–sulfuric acid and ninhydrin spray reagents.[17] The overall yield of the synthesis is between 30 and 40%, and the purity of the products is better than 95% based on HPLC analysis. The identity of the product has been confirmed by nuclear magnetic resonance (NMR) and mass spectroscopies.

S-(all-*trans*-Geranylgeranyl)-L-cysteine methyl ester is readily converted to S-(all-*trans*-geranylgeranyl)-L-cysteine under alkaline conditions. The cleavage of the methoxyl group is achieved by adding 0.1 ml of ammonium hydroxide solution (7.4 M) to 0.1 ml of S-(all-*trans*-geranylgeranyl)-L-cysteine methyl ester (0.2 mg) in acetonitrile. The reaction is allowed to proceed at 60° for 5 hr. The mixture is then applied to a C_{18} reversed-phase HPLC column (Alltech, Deerfield, IL, Econosphere, 4.6 mm × 25 cm) equilibrated in solvent A (0.1% trifluoroacetic acid/99.9% water, v/v). Elution is performed at 1 ml/min with a gradient over 50 min of 0–100% solvent B (90% acetonitrile/0.1% trifluoroacetic acid/9.9% water, v/v/v). Under these chromatographic conditions, S-(all-*trans*-geranylgeranyl)-L-cysteine and S-(all-*trans*-geranylgeranyl)-L-cysteine methyl ester are normally eluted at 27 and 29 min, respectively. However, the elution time may vary slightly depending on the HPLC column and the exact conditions used. The yield of this reaction is approximately 90%.

S-(all-*trans*-Farnesyl)-L-cysteine methyl ester and S-(all-*trans*-farnesyl)-L-cysteine can be synthesized from S-all-*trans*-farnesol and purified using the same methods and reaction conditions. These compounds elute from the C_{18} reversed-phase HPLC column at 21 and 24 min, respectively.

Preparation of [³H]Mevalonate for Intravitreal Injection

[³H]Mevalonate, a precursor of isoprenyl pyrophosphates that serve as substrates for protein isoprenylation,[18] is used to label the isoprenyl group metabolically. To reduce the level of endogenous mevalonate, an inhibitor of hydroxymethylglutaryl (HMG)-CoA reductase, such as mevinolin or compactin, is usually included in the injection mixture. The radioactive mixture is prepared as follows. An aliquot of 5 μl of mevinolin (20 μg) in dimethyl sulfoxide is added to a microcentrifuge tube and the solvent removed by vacuum centrifugation. The solid mevinolin is then resuspended in 20 μl of phosphate-buffered saline solution (0.2 M NaCl, 0.1 M NaH$_2$PO$_4$, pH 11), followed immediately by the addition of 0.2 ml

---

[17] E. Stahl, "Thin-Layer Chromatography." Springer-Verlag, New York, 1969.
[18] J. L. Goldstein and M. S. Brown, *Nature (London)* **343**, 425 (1990).

of [^{3}H]mevalonolactone in ethanol (200 μCi, specific activity 35 Ci/mmol, New England Nuclear, Boston, MA) and 0.2 ml of distilled water. Under these alkaline conditions, [^{3}H]mevalonolactone is readily converted to [^{3}H]mevalonate. The solvent is again removed by vacuum centrifugation (2 hr) and the solid redissolved in 20 μl of distilled water.

In Vitro Radiolabeling of Isoprenyl Proteins in Rat Retinas

The isoprenyl proteins in rat retinas can be efficiently labeled by intravitreal injection of [^{3}H]mevalonate into rat eyes. Because the eye is a closed, fluid-filled sphere, injection volumes should be kept to a minimum. We routinely inject 2 μl of a saline solution containing 2 μg of mevinolin and 20 μCi of [^{3}H]mevalonate into each eye. To eliminate variation in biosynthesis among animals, we also use at least three rats for each experiment. The injection is routinely performed using a 10-μl Hamilton syringe equipped with a 1-cm, 30-gauge needle. The eye of the slightly anesthetized rat is held by a pair of small forceps, and the needle is inserted through the sclera about 2 mm posterior to the iris edge and at an angle that avoids penetrating the lens or the retina. The injected rats are then maintained for 6 hr to 3 days. Light cycles, light intensity, and temperature should be stringently controlled during this period. The incorporation of radioactivity into the retinal isoprenyl proteins occurs as early as 30 min postinjection and reaches a plateau between 12 and 24 hr, followed by a slow and steady decay of radioactivity over a period of 7 to 10 days.

Preparation of Rod Outer Segments Containing ^{3}H-Isoprenylated Proteins

Rats metabolically labeled with [^{3}H]mevalonic acid are dark-adapted overnight and sacrificed with CO_2. The eye is held by a pair of small forceps and bisected with a sharp scalpel along the cornea. The lens, which usually adheres to the blade of the scalpel, is removed and discarded. The retina is then squeezed out of the collapsed eye cup with the forceps holding the eye and quickly dropped into 1 ml of ice-cold sucrose (14%, w/v) in Ringer's buffer. With practice, the entire procedure usually takes less than 30 sec per eye and can easily be performed under dim red light. After the dissection is completed, the retina suspension is transferred to a Potter-Elvehjem homogenizer, diluted with the same sucrose solution to two retinas per milliliter, and homogenized for 6 full strokes. Three milliliters of the retinal suspension is then carefully overlayed on top of a linear 25–50% (w/v) sucrose density gradient (8 ml) and centrifuged at 25,000 g for 1 hr at 4°. After removing the soluble retinal proteins from the top of the gradient, the red rod outer segment (ROS) membranes banding at 34% (w/v) sucrose near the top of the gradient are recovered with a syringe equipped with a long 18-gauge needle. The ROS suspension is then diluted with an equal volume of Ringer's buffer and pelleted by centrifugation at 18,000 g for 30 min at 4°.

Figure 2 shows a typical Coomassie blue-stained gel of the retinal homogenate, the ROS membranes, and the dense membrane pellet at the bottom of the gradient prepared by this procedure. The corresponding fluorogram shows four groups of ^{3}H-isoprenylated polypeptides. The 90-, 65-, and 7-kDa polypeptides enriched in the ROS preparation (lane 2, Fig. 2) have been positively identified as cyclic GMP phosphodiesterase,[9] rhodopsin kinase,[12] and the γ subunit of transducin,[9] respectively. The radiolabeled polypeptides with molecular masses ranging between 20 and 30 kDa are largely associated with the dense membrane pellet at the bottom of the gradient. This group of ^{3}H-isoprenylated proteins is most likely comprised of small GTP-binding proteins. The amount of radioactivity associated with the protein bands can be determined by excision of the appropriate gel regions followed by digestion in 20% (w/v) hydrogen peroxide at 65° overnight and liquid scintillation counting. The specific activity of the labeled proteins is typically 0.1 Ci/mmol.

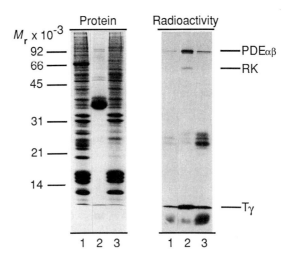

FIG. 2. Analysis by SDS–polyacrylamide gel electrophoresis of isoprenylated proteins in retinal preparations. *Left:* Coomassie blue-stained gel of rat retinal homogenate (lane 1), ROS membranes (lane 2), and cellular debris (lane 3) metabolically labeled with [^{3}H]mevalonic acid. The retinas were isolated 6 hr postinjection. *Right:* Fluorograph of the same gel. Tγ, γ subunit of transducin; PDE, cGMP phosphodiesterase; RK, rhodopsin kinase.

Isolation of ^{3}H-Isoprenylated Polypeptides

Sodium dodecyl sulfate (SDS)–polyacrylamide gel electrophoresis is a convenient method for obtaining small amounts of radiolabeled polypeptides suitable for subsequent chromatographic analysis. The choice of an appropriate gel system will depend on the size of the polypeptide under investigation. Resolution of the α and β catalytic subunits of phosphodiesterase (M_r 90,000 and 88,000, respectively) can be achieved by prolonged electrophoresis (16 hr at 30 mA) on a low cross-linked Tricine SDS–polyacrylamide gel (8.5% acrylamide/1.5% bisacrylamide with 20% glycerol).[9,19] For small polypeptides such as the γ subunit of G protein, a highly cross-linked Tricine SDS–polyacrylamide gel system[19] capable of separating the labeled polypeptide from the radioactive metabolites which migrate near the tracking dye position is recommended. There are many published protocols for eluting the polypeptides from the gel pieces, and commercial instruments designed for this purpose are also available. The following is a description of a simple elution procedure. Immediately following electrophoresis, a portion of the gel is stained to locate the positions of the polypeptides under study. The unstained gel band is excised, broken into small pieces by homogenization, and incubated in elution buffer (0.1 M Tris–acetate, pH 6.0, 0.2% SDS, 1% Triton X-100) for 8 to 12 hr. The volume of elution buffer should be large enough to cover all the gel pieces, and incubation should be carried out under constant agitation. After removing the gel pieces by filtration, SDS in the extraction buffer is precipitated by the addition of 0.1 ml of 2 M KCl per milliliter of filtrate and removed by centrifugation. The recovery ranges from 50% for a large polypeptide (e.g., phosphodiesterase) to over 80% for a small polypeptide (e.g., γ subunit).

Identification of ^{3}H-Labeled Isoprenyl Group by Chromatographic Analysis

The radiolabeled polypeptides eluted from the SDS–polyacrylamide gels are exhaustively digested for 5 hr at 37° with 2.5 mg/ml pronase (*Streptomyces griseus*, Calbiochem, La Jolla, CA). These digestion conditions have been shown to hydrolyze the carboxyl methyl esters and generate [^{3}H]isoprenylcysteine with high yield. The final digests [1000–2000 counts/min (cpm)] are then mixed with 0.15 nmol of isoprenylcysteine standards and applied to a C_{18} reversed-phase column (Alltech Econosphere; 4.6 mm × 25 cm) equilibrated in solvent A (0.1% trifluoroacetic

---

[19] H. Schagger and G. von Jagow, *Anal. Biochem.* **166**, 368 (1987).

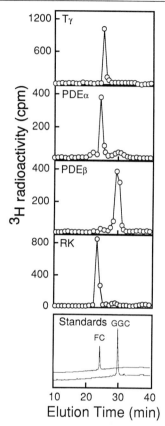

FIG. 3. Chromatographic identification of the isoprenyl groups of signal-transducing proteins in rod photoreceptors. The radiolabeled γ subunit of transducin (Tγ), α subunit of PDE (PDEα), β subunit of PDE (PDEβ), and rhodopsin kinase (RK) were separated by SDS–polyacrylamide gel electrophoresis, digested with pronase, and analyzed by reversed-phase HPLC using the method described in this chapter. Fractions (1 ml) were collected and assayed for radioactivity. *Bottom:* UV traces of authentic farnesylcysteine (FC) and geranylgeranylcysteine standards (GGC).

acid/99.9% water, v/v). Elution is performed at 1 ml/min with a gradient of 0–100% solvent B (90% acetonitrile/0.1% trifluoroacetic acid/9.9% water, v/v/v) over 50 min. The UV absorbance of the eluent is monitored at 214 nm to determine the exact elution time of the standards. Fractions (1 ml) are collected, and aliquots (500 μl) are counted in 10 ml of scintillation cocktail. The isoprenyl groups are then identified by comparing the radioactivity profiles with the elution profiles of authentic farnesylcysteine and geranylgeranylcysteine standards. It is important to note that the elution times of farnesylcysteine and geranylgeranylcysteine may vary slightly, depending on the condition of the HPLC column and the volume of the sample. To ensure accurate identification of the isoprenyl group, the proteolytic digests should cochromatograph with the farnesyl cysteine and geranylgeranyl cysteine standards. Figure 3 shows the analyses of four different isoprenylated polypeptides obtained from the radiolabeled ROS using this technique. Comparison with the standards indicates that the γ subunit of transducin and rhodopsin kinase are modified by farnesylation, whereas the α and β catalytic subunits of cyclic GMP phosphodiesterase are differentially modified by farnesylation and geranylgeranylation, respectively.

Concluding Remarks

The method described in this chapter is also applicable for the analysis of [^{3}H]mevalonate-labeled isoprenyl proteins derived from cell cultures and from *in vitro* translation of mRNAs, as well as proteins radiolabeled by carboxylmethylation using ^{3}H-labeled S-adenosylmethonine as the substrate. In the latter case, the conditions of proteolysis and the choice of proteases are much more critical since most proteases are capable of hydrolyzing the α-carboxyl methyl group. A combination of *Staphyloccoccus aureus* V8 protease and leucine aminopeptidase has been employed successfully for the analysis of the γ subunits radiolabeled by carboxylmethylation.[5] A detailed description of the analysis of isoprenylated carboxyl-terminal cysteine methyl esters in proteins has also been published.[16]

Author Index

Numbers in parentheses are footnote reference numbers and indicate that an author's work is referred to although the name is not cited in the text.

A

Abood, M. E., 24
Aboul-Ela, N., 73, 77(22), 78(22)
Abramovitz, J., 3
Abramowitz, J., 38, 213, 286(m), 287, 457
Abrams, E. S., 310, 314(19)
Adair, G. M., 367
Adamik, R., 46–47, 49, 49(21), 50(25), 53(45), 237
Adams, S. P., 254, 255(4), 257, 267(15)
Adra, C. N., 359
Aebersold, R., 474, 476, 483, 500
Aeschmann, S., 307
Ahn, T. G., 478, 482(9), 483
Ahn, Y. H., 349
Akino, T., 449, 451, 510, 512(7)
Aktories, K., 15, 17(25), 18, 18(17, 18, 25, 27, 29), 24(17), 25, 25(19, 30), 26(22)
Alessandra, V., 147
Alexander, A., 499
Alexander, T., 499
Alkhatib, G., 214
Alvarez-Gonzalez, R., 89
Amano, T., 25
Amatruda, T. T., 330, 334–335, 478, 482(7), 483, 486(1), 487, 491
Amatruda, T. T., III, 192, 195(16)
American, N., 317
Anant, J. S., 509, 512, 513(9), 516(9, 12), 517(9)
Anderegg, R. J., 510
Andersen, D., 329
Anderson, S. R., 37
Anjard, C., 388
Antonarakis, S. E., 319
Aragay, A. M., 327
Arai, M., 319
Arendt, A., 424, 427(5), 431(5), 436
Argiolas, A., 37
Arkinstall, S. J., 272, 275(23)
Armstrong, V. W., 76
Arnheim, N., 328, 471
Aronson, N. N., 476, 483

Arshavsky, V. Y., 14
Artemyev, N. O., 409, 424, 427(6)
Asakawa, T., 25
Asano, T., 296
Aslanidis, C., 53
Audigier, Y., 227, 238(27), 239–241, 241(9), 242(10), 243(10), 244, 245(9), 246, 246(11), 248
Aunis, D., 70
Ausubel, F. M., 324, 326(3)
Avger, K. R., 423
Axelrod, J., 457
Axton, J. M., 234, 235(21)
Ayres, M. D., 217

B

Baehr, W., 512
Baetscher, M., 474, 476, 483, 500
Baggiolini, M., 5, 14, 16(12), 17(12)
Balcueva, E. A., 476, 483, 498, 503, 504(3, 11), 505(3), 506(3)
Baraban, J. M., 296
Barber, A., 47
Barbetti, F., 319, 320(46)
Beames, B., 198
Becker, E. L., 16, 25(43)
Beiderman, B., 18, 240, 242, 270, 409, 427
Beiderman, D., 437
Beidler, J. L., 316
Beltz, G. A., 475
Berg, P., 239, 367
Berlot, C. H., 71, 409, 437
Bernard, H.-U., 51
Berns, A., 356, 358
Berridge, M. J., 182, 192
Berstein, G., 14, 35, 107, 181
Bertrand, P., 286, 286(i)
Betz, R., 510
Beverley, S. M., 356
Bigay, J., 139, 143, 145(3), 146, 146(3), 248, 449, 453
Biltonen, R. L., 113

Birnbaumer, L., 3, 14, 38–39, 41(3), 42(3), 45, 110–111, 114, 116, 124(8, 12), 131, 165, 177, 213, 227, 238(3, 5), 239(3), 240, 275, 286, 286(i, m), 287, 327, 366, 386, 451, 457, 459
Birnbaumer, M., 110
Birren, B. W., 478, 482(7), 483
Birren, S., 329
Bitensky, M. W., 110–111, 145
Bitonti, A. J., 15, 18(16)
Blackmore, P. F., 175
Blake, K. R., 351
Blake, R. D., 309
Blanchard, S. G., 499, 501(5)
Blank, J. L., 14, 35, 107, 174, 179, 180(13), 181, 181(13), 192, 276, 277(32)
Bliziotes, M. M., 46
Bobak, D. A., 46, 49
Bockaert, J., 70, 116, 124(12), 227, 240–241, 241(9), 242(10), 243(10), 245(9), 246, 246(11), 248, 252(25), 253
Boer, P. H., 359
Boiziau, C., 346
Bokoch, G. M., 457
Bominaar, A. A., 391, 397, 399, 406(32)
Bonora, E., 319, 320(46)
Borkovich, K., 327, 335
Borleis, J., 388
Borregard, N., 17
Botstein, D., 47, 59(41), 61(41)
Bouillon, T., 13
Boulay, F., 72, 81(13), 85(13), 86(13)
Boulikas, T., 115
Bourne, H. R., 14, 18, 24–26, 45, 64, 71, 74, 112, 191, 240, 242, 251, 270, 295–296, 300, 306–307, 308(20), 317, 319, 319(36), 327, 332(5), 409, 427, 436–437, 457
Bowden, D. W., 319
Bownds, M. D., 14, 424
Boyd, A. E. III, 286, 286(i)
Boyer, J. L., 182, 184(4, 11), 190(4, 11), 192, 276
Brabet, P., 386
Bradford, M., 76, 96(26), 150, 248
Bradley, A., 359, 366–367, 369, 369(10), 370–371, 371(2, 3, 11), 374(11), 386, 386(3)
Bradley, M. K., 218
Brann, M. R., 246, 249(19), 254, 255(3), 256, 256(3), 275, 286, 286(e)
Brass, L. F., 286, 286(d)
Braunagel, S., 198
Bray, P., 308
Brendel, S., 145
Brent, R., 324, 326(3)
Briedis, D., 214
Brier, M. E., 5
Brizuela, B. J., 47
Broecker, M., 307
Bronstein, I., 490
Brotherman, K. A., 367
Brown, A., 327
Brown, A. M., 3, 45, 213, 457
Brown, D. A., 269
Brown, J. H., 296
Brown, K. E., 346
Brown, K. O., 35, 172, 174, 182, 192, 194(9), 213, 286(q), 287, 423
Brown, M. S., 514
Brown, P. K., 429
Bryan, J., 110
Bujard, H., 148
Bukoch, G. M., 452
Burkart, W., 310
Bürki, E., 388
Burnett, W. N., 63, 66(8)
Burnier, J., 26, 27(1), 28(1), 29(1), 31(1), 33(1), 35(1), 36(1)
Burns, D. L., 15, 25, 25(33), 63–64, 65(4, 5, 10), 66(5), 81
Bushfield, M., 286, 286(g)
Buss, J. E., 147, 150, 154, 254–256, 266(14), 268, 268(1), 286, 286(c), 512
Butrynski, J. E., 482, 486(8), 487–488, 509

C

Cai, S.-P., 319
Caillou, B., 307
Cali, J. J., 476, 483, 498, 504(3), 505(3), 506(3)
Cama, A., 319, 320(46)
Camps, M., 16, 26, 452
Cañada, F. J., 510, 512(8)
Cantley, L. C., 423
Cantor, C., 106
Cao, Q., 309, 319(5)
Capecchi, M. R., 356, 358–359, 359(3, 10), 366–367, 367(1), 372(1)
Carlson, K. E., 16, 286, 286(d)

Caron, M. G., 213, 457, 512
Carpenter, C., 423
Carr, C., 283
Carr, S. A., 510
Carter, A., 308, 486(4), 487
Carter, H., 37
Cartwright, I. L., 76
Carty, D. J., 38–39, 41(3, 4), 42(3), 43(4), 63, 131, 165, 213, 451, 454
Caruso, D. R., 307
Caruthers, M. H., 349
Carvalho-Alves, P. C., 72
Casey, P. J., 45, 147, 153(5), 213, 240, 254–255, 258, 268, 268(1), 286, 286(c), 457, 507, 510, 512
Cassel, D., 3, 14, 15(8), 17(8), 18, 18(8, 10), 24(9, 10), 45, 243
Cavanaugh, E., 46
Cazenave, J.-P., 5
Celis, J. E., 347, 352(15), 353(15)
Cerione, A. G., 409
Cerione, R. A., 269, 409, 410(6, 7), 414, 416(7, 18), 418(8, 10), 419(9, 10), 420(18), 421(9), 422, 422(9), 423, 423(8), 457
Chabre, M., 139, 143, 145(3), 146, 146(3), 252(25), 253, 449, 453
Chabre, O., 26
Chae, H. Z., 174, 192, 423
Chan, J., 451
Chang, F.-H., 18
Chang, K. J., 499, 501(5)
Chang, L.-F.H., 71
Chang, P., 46, 49(21)
Chang, P. P., 47
Chao, S., 356, 364, 364(9)
Chase, M. W., 502
Chen, C., 246
Chen, H.-C., 46, 49(21), 50(25)
Chen, S., 72, 73(20), 75(20), 81
Chen, X. N., 330, 334
Cherbas, P. T., 475
Chi, M.-H., 18
Chiu, I.-M., 307
Cho, K. S., 176, 193
Choi, K. D., 182, 295
Christophe, J., 15, 18(13), 24
Christy, K. G., Jr., 228
Chuang, D. M., 45
Clapham, D. E., 449, 452
Clark, J., 47, 59(37), 297, 331

Clark, O. H., 300, 308(20), 317, 319(36)
Clark, R. A., 17
Clarke, A., 356
Clarke, S., 509–510, 510(1), 512, 512(4, 5), 513(9), 516(9), 517(9), 519(5, 16)
Clementi, E., 317
Coburn, C. M., 356
Coccuci, S. M., 389
Codina, J., 39, 41(3), 42(3), 45, 110–111, 131, 165, 177, 213, 227, 238(3, 5), 239(3), 240, 269, 275(8), 281(8), 283, 286, 286(i, m), 287, 443, 457, 459
Coffino, P., 112, 240
Cohen, J. S., 346–347, 347(6), 351(6)
Collier, R. J., 44
Collins, R. M., 256, 306, 309, 316(3), 317(3)
Collins, S. J., 16
Conklin, B. R., 26, 457
Conner, D. A., 356, 364, 364(9)
Copeland, N. G., 330, 334
Costa, E., 15
Costa, T., 25
Coulson, A. R., 61
Cox, D. R., 311
Crabb, J. W., 510
Crowe, J., 148
Crowl, R., 50, 51(47)
Cruz, A., 356
Czarnecki, J. J., 72, 81(12), 89, 90(45)

D

Dalbon, P., 72, 81(13), 85(13), 86(13)
Das, H. K., 51
Dash, P., 346
Daugherty, B. L., 349
Davies, S.-A., 274
Davis, R. L., 300
Dawson, R.M.C., 23
De Goede, J., 394
De Gunzburg, J., 388
De Jong, C.C.C., 406
de Jong, P., 330, 334
de Jong, P. J., 53
del C. Vila, M., 269
Delcourt, S. G., 309
de Mazancourt, P., 317
Deng, C., 358–359, 359(10)
Denker, B. M., 226–227, 229, 229(2), 234(2), 236, 457

Der, C. J., 512
Deretic, D., 71, 424, 427(2, 5), 431(5)
Derwahl, M., 307
Deterre, P., 14, 143, 145(3), 146, 146(3), 248, 252(25), 253
Devine, C. S., 55, 256
DeVit, M. J., 355
Devreotes, P. N., 335, 387–388, 388(6, 7), 397, 400(6, 7), 405–406, 406(43)
Dewald, B., 5, 14, 16(12), 17(12)
De Wit, R.J.W., 394, 400(33), 401
Dhanasekaran, N., 70–71, 87, 213
Diacumakos, E. G., 347
Dieterich, K., 5, 7(10), 10(10), 20
Dietzel, C., 452
Dingus, J., 457
Dinsart, C., 328, 329(14)
Disher, R. M., 427
Dobberstein, B., 148
Doetschman, T., 356
Doolittle, R. F., 427, 478, 482(6), 483
Dottin, R. P., 388
Doty, P., 310
Dougherty, W. G., 161
Downes, C. P., 182
Dratz, E. A., 429, 432, 433(20), 434(20)
Drayer, A. L., 388
Dreyer, W. J., 4, 476, 483
Duckworth, B., 423
Duh, Q.-Y., 300, 308(20), 317, 319(36)
Dumont, J., 295, 307(3), 328, 329(14)
Dunn, W. J. III, 100
Duntze, W., 510
Duronio, R. J., 55, 147, 149(7), 164, 173(3), 256–257, 257(9, 11, 13), 258(13), 261(11), 267, 267(15), 268(11, 13)
du Villard, J. A., 307

E

Eckstein, F., 18
Economou-Petersen, E., 319
Edie, B., 275
Edwards, B., 490
Ehrlich, H. A., 328
Eickbush, T. H., 475
Eil, C., 314
Elliot, D. C., 23
Elliot, W. H., 23

Emeis, D., 429
Engvall, E., 502
Enomoto, K., 25, 65
Entman, M. L., 114, 124(8), 275
Erdos, J. J., 145
Erickson, J. W., 409, 419(9), 421(9), 422(9)
Erlich, H. A., 328, 329(16), 471
Erneux, C., 387
Evan, G., 500
Evans, F. E., 93
Evans, M., 366, 371(2)
Evans, T., 182, 307, 319
Eversole-Cire, P., 64, 286(n), 287, 297
Ewald, D. A., 104, 147, 149(4), 152(4), 153(4), 213, 258, 260(18), 261(18), 262(18), 273
Ewel, C., 270, 327, 348, 354(17), 355(18)
Exton, J. H., 14, 35, 107, 174–176, 176(4), 177, 178(12), 179, 180(13), 181, 181(4, 13), 182, 192, 213, 276, 277(32), 284, 286(r), 287, 291(10), 423

F

Falls, K., 319
Faloona, F. A., 327–328, 328(11), 471
Farese, R. V., 269
Farfel, Z., 18
Farnsworth, C. C., 510, 512, 512(5), 519(5)
Faurobert, E., 145
Fawzi, A. B., 6, 8(14), 145
Fay, D. S., 145
Federman, A., 26, 307, 319, 457
Feichtinger, H., 300, 308(20), 317, 319(36)
Feinberg, A. P., 375
Feltner, D. E., 16
Ferguson, K. M., 27, 35, 260, 409–410, 410(4, 5)
Figler, R. A., 212, 220, 226(31), 273
Fink, G. R., 59, 60(55)
Firtel, R. A., 327, 335, 387–388, 388(6, 7), 400(6, 7)
Fischer, S. G., 309–311, 314(8)
Fisher, K. J., 476, 483
Fixman, M., 311
Fleming, J. W., 15, 25(37)
Fling, S. P., 76, 77(28), 91(28)
Florio, V. A., 6, 7(16), 291
Fong, H. K., 64, 286(n), 287

Fong, H.K.W., 147, 153(5), 258, 297, 476, 478, 482(6, 7), 483, 486(1), 487
Forquet, F., 146
Forte, M., 335, 436–438, 443, 443(6), 444(6)
Foster, C., 302
Francomano, C. A., 306, 318
Frank, R., 486(10), 487
Franke, C. A., 437
Franklin, N., 51
Franklin, P. H., 15
Fraser, M. J., 213–214
Freier, S. M., 349
Freissmuth, M., 25, 39, 147, 149(3), 153(3), 213, 269, 273, 281(10), 282, 457
Friedman, E., 301, 306, 306(22), 309, 314(4), 316(3), 317(3, 4), 318(4)
Friere, J. J., 311
Fritsch, E. F., 147, 149(10), 195, 204(25), 228, 331, 342(21), 347, 373
Fukada, Y., 254–255, 268(5), 449, 451, 510, 512(7)
Fukuda, Y., 61
Fung, B.K.-K., 24, 45, 75–76, 78(29), 145, 156, 229, 230(14), 242, 457, 509–510, 512, 512(4, 5), 513(9), 516(9, 12), 517(9), 519(5, 16)
Furstenau, J. E., 429, 432, 433(20), 434(20)

G

Gachet, C., 5
Gallagher, C., 471, 476, 488
Gallagher, R. E., 16
Gallego, C., 308
Gallo, R. C., 16
Galper, J., 452
Gao, B., 452, 482(8), 483, 486(6), 487
Garotta, G., 148, 149(12)
Garrison, J. C., 212, 220, 226(31), 273
Gaskin, F., 106
Gausepohl, H., 486(10), 487
Gautam, N., 25, 146, 182, 191, 212, 213(1), 284, 327, 330, 334(6), 345, 408, 449, 471, 474, 476–477, 482, 482(3), 483, 486(1–3, 8), 487–488, 494, 495(20), 498, 500, 500(2), 509
Gehrke, L., 218
Geisterfer, L. A., 356, 364, 364(9)
Gejman, P. V., 301, 306, 306(22), 308–309, 314(4), 316, 316(3), 317, 317(3, 4), 318(4), 319, 319(5), 320(46)
Gelb, M. H., 510, 512, 512(5), 519(5)
Gelfand, D. H., 321
Gelmann, E. P., 47, 59(37)
Gentz, R., 148
Gershon, E. S., 306, 309, 316(3), 317(3), 319(5)
Gerton, G. L., 64
Gething, M.-J., 239
Gierschik, P., 4–7, 7(10, 17), 10(10), 13–14, 16, 16(12), 17(12), 20, 21(44), 22, 22(44, 45), 24–25, 25(11, 44), 26, 269, 275, 275(4), 283
Giershick, P., 452
Gilbert, D. J., 330, 334
Gilbertson, J., 302
Gill, D. M., 63, 65, 68
Gill, R., 302
Gilman, A. G., 3, 14, 27, 34–35, 39, 45–46, 64, 78, 104, 111, 112(3), 113, 124, 124(3), 125(3), 135, 146–147, 149(3, 4), 152(4), 153, 153(3–5), 154, 164, 169, 173(3), 191, 193, 213, 218, 230, 239–240, 246, 249(18), 254–255, 255(2), 256, 256(2), 257(11), 258, 260, 260(18), 261(11, 18), 262(18), 266, 268(1, 11), 269, 271, 273, 286, 286(c), 326, 327, 391, 409, 410(4, 5), 437, 451–452, 457–458, 482(8), 483, 486(6, 9, 12), 487–488, 490, 497(2), 498, 500(4), 502(4), 507, 509–510
Gilman, G., 345, 351(2)
Gilmore, E., 307
Glaser, L., 254, 255(4)
Gless, C., 25
Glickman, J. F., 512
Glomset, J. A., 510, 512, 512(5), 519(5)
Glover, J. S., 413
Goddard, A., 47
Godfrey, P. P., 272, 275(23)
Goff, S. P., 356, 367
Goldman, D., 316
Goldsmith, P., 269, 275, 275(4, 8), 281(5, 8), 283, 286, 286(f), 295, 443, 486(4), 487
Goldstein, J. L., 514
Gollasch, M., 348
Gollet, P., 346
Goochee, C. F., 219
Gordon, J., 465, 503
Gordon, J. H., 107, 108(13)

Gordon, J. I., 55, 147, 149(7), 164, 173(3), 246, 249(18), 254–255, 255(2, 4), 256, 256(2), 257, 257(11, 13), 258(13), 261(11), 267, 267(15), 268(11, 13)
Gottikh, B. P., 72, 80(18)
Gould, G. W., 283
Gould, S. J., 473
Graber, S. G., 212, 220, 226(31), 273
Grace, T.D.C., 217
Graessmann, A., 347, 352(15, 16), 353(15, 16)
Graessmann, M., 347, 352(16), 353(16)
Graf, R., 213
Grandt, R., 15, 21(34), 26(22, 34)
Grant, G. A., 426
Graziani, A., 423
Graziano, M. P., 39, 147, 149(3), 153(3), 213, 240, 326
Green, K. C., 175
Green, N., 490
Green, N. H., 499
Greenwood, F. C., 413
Gregerson, D. S., 76, 77(28), 91(28)
Greiner, C., 15, 21(34), 26(34)
Grenet, D., 110
Griffin, B. E., 78
Griffiths, S. L., 274
Grinfeld, E., 309
Grünewald, K., 300, 308(20), 317, 319(36)
Gruppuso, P. A., 314
Guillon, G., 252(25), 253
Guillory, R. J., 72, 73(20), 75(20)
Gundersen, R., 387, 388(7), 400(7)
Guo, V., 308
Gupta, S. K., 71, 307–308, 319
Gutkind, S., 319
Gutowski, S., 193, 269, 507, 510
Guy, P. M., 409, 410(7), 416(7)

H

Hadwiger, J. A., 327, 335, 388
Haeuptle, M.-T., 148
Haga, K., 25, 423
Haga, T., 25, 286(k), 287, 423
Haley, B. E., 72, 86, 87(42), 95
Ham, R. G., 353, 354(27)
Hamm, H. E., 71, 108, 409, 423–424, 427, 427(2, 4–6), 429, 431(5), 432, 433(20), 434(20), 435
Hammes, G. G., 411
Hampton, A., 84
Hancock, R., 115
Happle, R., 306
Harden, T. K., 182, 184, 184(4, 11), 185(13), 190(4, 7, 11), 192, 276–277, 296
Hargrave, P. A., 424, 427(5), 431(5), 436
Haribabu, B., 388
Harris, B., 437
Harris, B. A., 240, 258, 427
Harsh, G., 300, 308(20), 317, 319(36)
Hasty, P., 359, 370–371, 386
Hatefi, Y., 81
Hatta, S., 107, 108(13), 434
Haun, R. S., 44, 46–47, 49, 49(38), 50(38), 53, 53(38, 45), 55, 61(38)
Hayflick, J. S., 438
Hazan, J., 319
Heasley, L. E., 71, 308
Heideman, H., 270
Heideman, W., 427
Heiple, J. M., 17
Hélène, C., 346
Helinski, D. R., 51
Helman, L. J., 478, 482(9), 483
Helmreich, E.J.M., 4
Henderson, J. T., 99
Henning, D., 214
Hepler, A.J.R., 345, 351(2)
Hepler, J. R., 14, 35, 146–147, 172, 174, 182, 191–192, 193(12, 13), 194(9, 12), 197(12), 201(12), 202(12), 203(12), 204(12), 205(12), 206(13), 207(13), 210(13), 211(12, 13), 213, 286(q), 287, 423, 451
Herbert, E., 437
Hermouet, S., 319
Herrmann, E., 4, 16, 22(45)
Hershfield, M. V., 51
Herson, D., 227
Herz, A., 25
Hescheler, J., 270, 345, 348, 354(17, 18), 355, 355(17, 18)
Hescheller, J., 327
Heuckeroth, R. O., 55, 254–255, 255(2), 256, 256(2), 257(9)
Heukeroth, R. O., 246, 249(18)

Hewlett, E. L., 15, 25, 25(33), 63–64, 65(5), 66(5), 81, 237
Hicks, J. B., 59, 60(55)
Higashijima, T., 26–27, 27(1), 28(1), 29, 29(1), 31(1), 33(1), 35, 35(1), 36(1), 37, 107, 181, 260, 409–410, 410(4, 5), 431
Higuchi, M., 319
Hildebrandt, J. D., 39, 41(3, 4), 42(3), 43(4), 110–111, 213, 454, 457–459, 462(15)
Hilf, G., 4–6, 12, 15
Hill, R. L., 235
Hilliard, P. R., 316
Hingorani, V. N., 71, 99, 431
Hinsch, K. D., 486(10), 487
Hirose, T., 482(5), 483
Ho, Y.-K., 71, 99, 431
Hoard, D. E., 87
Hoehe, M. R., 309, 319(5)
Hoffman, J. F., 72
Hofman, K. P., 71
Hofmann, K. P., 424, 427(2, 5), 429, 431(5), 436
Hokin-Neaverson, M., 72, 74(21)
Holbrook, S., 436
Holden, J., 47
Holland, S., 347
Homburger, V., 70, 248
Hooper, M., 356
Hopkins, R. S., 478, 482(6), 483
Hopp, T. P., 426
Horn, G. T., 328, 329(16), 471
Hoss, W., 15
Hotta, K., 349
Houslay, M. D., 274, 286, 286(g)
Hovens, C. M., 329
Howald, W., 510, 512(5), 519(5)
Howard, B. H., 239
Howard, E. D., 161
Hoyt, M. A., 47, 59(41), 61(41)
Hruby, D., 437
Hruby, V. J., 286, 286(g)
Hsai, J. A., 64
Hsia, J. A., 81
Hsu, W. H., 286, 286(i)
Hubbard, R., 429
Hudson, T. H., 231
Huff, R. M., 234, 235(21)
Hughes, J., 296
Hunt, J. B., 72, 81(14)

Hunt, T., 346
Hunter, T., 265
Hunter, W. M., 413
Hunzicker-Dunn, M., 116, 124(12)
Hurley, I., 309, 314(8)
Hurley, J. B., 24, 78, 254–255, 255(5), 268(5), 476, 478, 482(6), 483
Hurley, J. R., 230
Husimi, Y., 309–310
Hutchinson, D. W., 76
Hwang, S.-B., 15, 16(23), 25(23), 26(23)
Hwo, S., 106

I

Ibrahimi, I., 148
Ichiyama, A., 25, 286(k), 287
Iiri, T., 45
Ikehara, M., 93
Imayama, S., 482(5), 483
Inaba, H., 319
Inanobe, A., 131, 136(5, 6), 139(7), 453
Inglese, J., 512
Iñiguez-Lluhi, J., 193, 239, 509
Innis, M. A., 321
Insall, R., 406
Insel, P., 110
Ishi, S., 63
Ishizaka, Y., 307
Ito, H., 61, 131, 138(4), 213
Ito, K., 63
Itoh, H., 136, 138(12), 153, 191, 297, 308, 315(1)
Iyengar, R., 38–39, 41(3, 4), 42(3), 43(4), 110, 131, 165, 213, 451, 454

J

Jackson-Machelski, E., 55, 256, 257(9)
Jacobs, K. A., 475
Jacobson, M. K., 73, 77(22), 78(22)
Jaeger, J. A., 349
Jakobs, K. H., 3–7, 7(10, 17), 10(10), 12–16, 16(12), 17(12, 25), 18, 18(17, 18, 25, 27, 29), 19(39), 20, 21(34, 44), 22, 22(44, 45), 24, 24(17), 25, 25(11, 44), 25(19, 30), 26(22, 34), 283, 404

Jamieson, G. A., 72, 81(14)
Janssens, P.M.W., 387, 394, 405–406, 406(44)
Jardine, K., 359
Jarman, M., 78
Jasin, M., 367
Jasper, J., 110
Jauniaux, J. C., 295, 307(3)
Jefferson, J. R., 72, 81(14)
Jelsema, C. L., 457
Jenkins, N. A., 330, 334
Jhiang, S. M., 307
Jhon, D.-K., 452
Jhon, D.-Y., 14, 181
Jobson, E. L., 73, 77(22), 78(22)
Johns, D. R., 314
Johnson, G. L., 18, 71, 75, 87, 213, 231, 240, 257(13), 258(13), 307–308, 319, 321, 347
Johnson, J. E., 329
Johnson, M. S., 478, 482(6), 483
Johnson, R. A., 35
Johnson, R. L., 256, 268(13), 388
Johnson, R. S., 254–255, 255(5), 268(5)
Johnson, S. A., 355
Johnston, S. A., 161
Jones, D., 443
Jones, D. T., 257, 286, 286(a), 296, 321
Jones, K. M., 23
Jones, T. L., 286, 286(e)
Jones, T.L.Z., 246, 249(19), 254, 255(3), 256(3)
Journot, L., 227, 240–241, 241(9), 242(10), 243(10), 245(9), 246, 246(11)

K

Kadowaki, T., 306, 309, 316(3), 317(3)
Kafatos, F. C., 475
Kahler, K. R., 6
Kahlert, M., 436
Kahn, C. R., 286(l), 287
Kahn, R. A., 46–47, 47(26), 59(26, 37, 41), 61(41), 266, 269, 486(9), 487
Kalman, V. K., 483, 488, 500, 505, 508(13), 509(13)
Kamb, A., 328
Kamholz, J., 308
Kamiya, Y., 509
Kan, Y. W., 319

Kanaho, Y., 237
Kandel, E. R., 346
Kaneko, K., 310
Kangawa, K., 286(k), 287, 482(5), 483
Kapeller, R., 423
Kaplan, J., 386
Kaplan, R., 108
Karschin, A., 437
Kasch, L., 319
Kaslow, D., 251
Kaslow, H. R., 18, 63–64, 65(3, 5, 10), 66(5, 8), 240, 251
Kasper, C. K., 319
Katada, T., 16, 25, 25(40), 45, 63–64, 131, 134, 135(2, 3), 136, 136(2, 3, 5, 6), 138(4, 12), 139(7), 213, 452–453, 455, 457
Kato, K., 296
Kato, T., 85
Kaufman, M. H., 366, 371(2)
Kawasaki, E., 300, 308(20), 317, 319(36)
Kawasaki, H., 136
Kay, R. R., 387–388
Kazazian, H. H., Jr., 319
Kaziro, Y., 131, 136, 136(6), 138(12), 191, 297, 308, 315(1)
Keith, T. P., 319
Kelleher, D. J., 71
Kelly, E. C., 283
Kelly, M. T., 44
Kennison, J. A., 47
Kenyon, C., 328
Kern, F. G., 47, 59(37)
Kesbeke, F., 387, 394, 397(4), 399, 399(4), 400(4, 5)
Khorana, H. G., 483
Kierzek, R., 349
Kikkawa, S., 131, 135(3), 136(3)
Kim, H., 86, 87(42)
Kim, S.-H., 436
Kimmel, A. R., 388
Kimura, A., 61
Kimura, N., 15, 18(15)
Kindy, M. S., 346
Kingston, R. E., 324, 326(3)
Kipins, D. M., 391
Kirschner, M., 106
Kissinger, M., 47
Kitts, P. A., 217
Klee, W. A., 15, 18(31, 32), 21(32)
Klein, P. S., 388

Kleuss, C., 145, 193, 270, 283, 345, 348, 354(17, 18), 355, 355(17, 18)
Kluess, C., 327
Klump, H., 310
Knowler, J. T., 274
Kobayashi, I., 131, 135(3), 136(3, 5), 138(4), 213, 453
Koenig, B., 424, 427(5), 431(5), 436
Kohl, B., 71, 424, 427(2)
Kohnken, R. E., 457–458, 462(15)
Kokame, K., 254–255, 268(5)
Koland, J. G., 409, 410(7), 416(7)
Kontani, K., 131, 139(7), 453
Kopf, G. S., 64
Korenberg, J. R., 330, 334
Kosack, C., 110
Koski, G., 15, 18(31, 32), 21(32)
Kozak, M., 227
Kozasa, T., 131, 136(6), 191–192, 193(12, 13), 194(12), 197(12), 201(12), 202(12), 203(12), 204(12), 205(12), 206(13), 207(13), 210(13), 211(12, 13), 297, 308, 315(1)
Krantz, M. J., 77
Krapp, M., 346
Krawczak, M., 343
Krayevsky, A. A., 72, 80(18)
Kreiss, J., 6, 7(17)
Kroll, S., 269, 422
Krumlauf, R., 371
Kühn, H., 140, 146, 429
Kumagai, A., 387, 388(6, 7), 400(6, 7)
Kumar, C., 349
Kumar, R., 331
Kung, H.-F., 297
Kupper, R., 5, 14, 16(12), 17(12)
Kurachi, Y., 131, 138(4), 213, 452
Kurban, R. R., 329
Kurgan, J., 452
Kurose, H., 25
Kusakabe, K., 131, 135(2), 136(2)
Kyte, J., 427

L

Lachman, P. J., 500
Lacombe, M. L., 388
Laemmli, U. K., 70, 113, 114(7), 141, 218, 228, 248, 267, 287
Lai, R. K., 510, 512(8)
Lalumiere, M., 214
Lambert, C. G., 429, 432, 433(20), 434(20)
Lambert, M., 15, 18(13), 24
Landau, E. M., 39, 41(4), 43(4), 454
Landis, C. A., 18, 296, 300, 308(20), 317, 319(36)
Lanford, R. E., 198
Lang, J., 25
Lanzer, M., 148
Larsson, P.-O., 72
Lasch, P., 18
Laskey, R. A., 267
Latart, D. B., 228
Laugwitz, K.-L., 283–284, 286, 286(b, o, t), 287, 291(12)
Lavan, B. E., 283, 286, 286(g)
Lavu, S., 297
Law, S. F., 270, 271(15), 283, 286(j), 287, 486(7), 487
Lazarevic, M., 108
Lazarevic, M. B., 429, 435
Lee, C. H., 93, 295, 308(7)
Lee, C.-M., 46–47, 61(40)
Lee, C.-W., 452
Lee, E., 146
Lee, F.-J.S., 44, 47, 59(42), 61(40)
Lee, H.-H., 452
Lee, K.-H., 452
Lee, K.-Y., 176, 193
Lee, Y. C., 77
Lefkowitz, R. J., 15, 18(14), 457, 512
Lefort, A., 328, 329(14)
Lerman, L. S., 309–311, 311(20), 312(20), 314(8, 10), 318
Lerner, R., 499
Lester, H. A., 15, 18(20), 24(20), 437
Leuther, K. K., 161
Levi, S., 302
Levin, D., 214
Levine, M. A., 229, 306, 314, 317–318, 478, 482(9), 483, 506, 509
Levitzki, A., 15, 18(20), 24(20)
Levkovitz, H., 14, 18(10), 24(10)
Lewin, B., 352, 353(26)
Libert, F., 328, 329(14)
Licko, V., 251
Liebmann, C., 5
Lilly, P., 388
Lim, L. K., 63, 65(3)

Lin, G., 387
Lin, K. C., 388
Lin, W.-C., 509
Lindberg, M., 72
Linden, J., 283
Linder, M. E., 32, 33(5), 104, 146–147, 149, 149(4, 7), 152(4), 153(4), 164, 173(3), 213, 254–256, 257(11, 13), 258, 258(13), 260(18), 261(11, 18), 262(18), 268(11, 13), 273
Litman, B. J., 413
Liu, T., 64
Liu, T. Y., 64
Lo, W.W.Y., 296
Lochrie, M. A., 327, 335
Lockwood, A., 106
Loflin, P. T., 73, 77(22), 78(22)
Logothetis, D. E., 452
Lok, J. M., 64, 124, 235, 237(23)
Loke, S. L., 347
Lomedico, P., 50, 51(47)
Londos, C., 116, 455
Loomis, W. F., 387
Lopez, N., 437
Lopez, N. G., 18, 240, 409
Lorenz, W., 512
Lotan, I., 346
Lowe, M., 18
Lowndes, J. M., 72, 74(21), 307, 319
Loyter, A., 347, 352(15), 353(15)
Luckow, V. A., 196, 214–215
Lumelsky, N., 309–310, 314(8)
Lyons, J., 295, 300, 302, 308(20), 317, 319(36)

M

Ma, H., 327, 335
Maandag, E. R., 356, 358
MacKinnon, C., 367
Maeda, N., 356
Maehama, T., 134
Maenhaut, C., 328, 329(14)
Magee, A. I., 147, 263, 265(22)
Maguire, M. E., 113
Malbon, C. C., 347
Malech, H. L., 269, 275(4), 281(5)
Malencik, D. A., 37
Malgaretti, N., 317

Malinski, J. A., 424, 427(4)
Maltese, W. A., 227, 238(6), 507
Manclark, C. R., 25, 63–64, 65(4, 5), 66(5), 74, 110, 427
Manganiello, V. C., 45
Maniatis, T., 147, 149(10), 195, 204(25), 228, 309–311, 314(10), 331, 342(21), 347, 373
Mann, S.K.O., 388
Manning, D., 270, 271(15), 283, 286(j), 287, 486(7), 487
Manning, D. R., 16, 266, 269, 286, 286(d), 486(9), 487
Mansour, S. L., 356, 359(3), 367
Mar, V. L., 63, 66(8)
Marasco, W. A., 16
Marcus, A., 227
Marcus, F., 72, 95
Marcus, M. M., 107, 108(13), 434
Marcus-Sekura, J. C., 351
Margolskee, R. F., 334–335, 336(28)
Marmur, J., 310
Marshall, C. J., 302
Martin, M. W., 296
Martinez, M., 309, 319(5)
Masters, S. B., 18, 240, 242, 270, 296, 306, 317, 409, 427, 436–437
Masuda, M., 310
Matile, H., 148, 149(12)
Matsumoto, T., 16, 25(43)
Matsuo, M., 286(k), 287, 482(5), 483
Matsuoka, M., 297
Matsushima, K., 297
Mattera, R., 45, 213, 227, 238(5), 240, 286(m), 287
Matthews, R. G., 429
Maurice, D. H., 182, 190(7), 276–277
Mazarguil, H., 246
Mazzaferri, E. L., 307
Mazzoni, M. R., 424, 427(4)
McAndrew, S., 50, 51(47)
McBurney, M. W., 359
McClue, S. J., 269, 273, 280–281, 281(10), 282
McCormick, F., 191, 295, 300, 308(20), 327, 332(5)
McCusick, V. A., 306
McDermed, J., 499, 501(5)
McDowell, J. H., 436
McElligott, S. G., 355
McFadzean, I., 269

McKeehan, W. L., 353, 354(27)
McKenzie, F. R., 269, 275(9), 280(9), 281(9), 283
McKinnon, P. J., 334–335, 336(28)
McLaughlin, S. K., 334–335, 336(28)
McLeish, K., 5, 16, 22
McMahon, A. P., 367, 369(10)
Meisenhelder, J., 265
Mekalanos, J. J., 44
Meldolesi, J., 317
Mendel, J. E., 327, 335
Meren, R., 68
Merendino, J. J., 246, 249(19)
Merendino, J. J., Jr., 254, 255(3), 256(3), 286, 286(e), 319
Merino, M. J., 301, 306(22), 309, 314(4), 317(4), 318(4)
Merril, C. R., 316
Meyerowitz, E., 327, 335
Miake-Lye, R., 387, 388(7), 400(7), 478, 482(6), 483
Middleton, P., 147
Miller, L. K., 196, 214–215
Miller, P. S., 351
Miller, R., 437
Miller, R. J., 104, 147, 149(4), 152(4), 153(4), 213, 258, 260(18), 261(18), 262(18), 273
Miller, R. T., 18, 240, 242, 270, 409, 427
Milligan, G., 147, 182, 190(8), 268–270, 272–275, 275(4, 9, 23), 279(31), 280, 280(9), 281, 281(9, 10), 282–283, 286, 286(g)
Mills, J. S., 409, 424, 427(6)
Milona, N., 388
Minamino, N., 286(k), 287, 482(5), 483
Mingmuang, M., 73, 77(22), 78(22)
Minshull, J., 346
Mintz, P. W., 38
Minuth, M., 18
Miric, A., 317
Mishima, K., 47, 61(40)
Mitchell, F. M., 272, 274, 275(23)
Miyada, C. G., 472
Miyazawa, T., 29
Moen, P. T., 478, 482(9), 483
Moffatt, B. A., 150
Moghetti, P., 319, 320(46)
Moghtader, R., 5, 7(10), 10(10), 20
Molski, T.F.P., 16, 25(43)
Monaco, L., 46, 47(28)
Monastrinsky, B., 177

Monier, R., 307
Moomaw, C. R., 483, 488, 500
Mori, K., 347
Moriarty, T. M., 39, 41(4), 43(4), 454
Morishima, N., 45
Morris, A. J., 182, 184, 184(4), 185(13), 190(4, 7), 192, 276–277
Mortensen, R. M., 356, 359(7), 364, 364(9), 365(7)
Mosbach, K., 72
Moss, J., 15, 18(16), 25, 25(33), 44, 44(2), 45–47, 47(28, 29), 48–49, 49(21, 38), 50(25, 29, 38), 52–53, 53(38, 45), 55, 55(29), 59(42), 61(38, 40), 63–64, 65(5), 66(5), 73, 77(22), 78, 78(22), 81, 237, 297
Moss, L. G., 286, 286(i)
Mowatt, M. R., 47, 61(40)
Mueller, M., 148
Muggeo, M., 319, 320(46)
Mukai, H., 37
Mullaney, I., 269, 272, 275, 275(23), 279(31), 283
Mullhofer, G., 359
Mulligan, R. C., 239
Mullis, K. B., 327–328, 328(11), 471
Mumby, S., 437, 486(6, 9), 487
Mumby, S. B., 149–150
Mumby, S. M., 147, 153–154, 218, 240, 246, 249(18), 254–255, 255(2), 256, 256(2), 258, 266, 266(14), 268(1), 269, 271, 286, 286(c), 486(10), 487–488, 490, 497(2), 498, 500(4), 502(4), 507, 509–510
Munekata, E., 37
Munshi, R., 283
Munson, P. J., 25
Muntz, K. H., 486(11), 487–488, 497(2), 509
Murad, F., 391
Murai, S., 63
Murakami, A., 351
Murakami, T., 482(4), 483, 486(5), 487, 488
Murata, K., 61
Murhammer, D. W., 219
Murphy, E. A., 145
Murphy, P. M., 275, 286, 286(g)
Murtagh, J. J., 46, 47(28)
Murtagh, J. J., Jr., 47, 61(40)
Musci, T. S., 356
Mutzel, R., 388
Myers, G. A., 64, 81
Myers, R. M., 309–311, 314(10)

N

Nagao, M., 307
Nagata, K., 134, 136, 138(12)
Nahorski, S. R., 37
Nair, B. G., 269
Nairn, R. S., 367
Nakafuku, M., 191
Nakajima, T., 26
Nakanishi, M., 347
Nash, C. R., 76, 78(29), 229, 230(14), 242
Nash, T. E., 47, 61(40)
Navon, S. E., 24, 45, 457
Nawrath, M., 5
Neckers, L. M., 347
Neer, E. J., 64, 111, 124, 226–227, 229, 229(2), 230(15), 231(15), 233–234, 234(2), 235, 235(20, 21), 236(24, 35a), 237(23), 242, 296, 356, 359(7), 365(7), 449, 457, 509
Neff, N. H., 45
Nelson, C., 286, 286(i)
Nelson, T., 349
Neubert, T. A., 254–255, 255(5), 268(5)
Neuhoff, V., 115, 124(9)
Neumann, P. E., 356
Newell, P. C., 387
Newkirk, M., 47
Newman, C.M.H., 147
Newman, K. B., 46, 47(28)
Nicholas, R. A., 182, 190(7), 276–277
Nicklen, S., 61
Nightingale, M., 46
Nightingale, M. S., 46, 50(25)
Nirenberg, N., 308
Nishigaki, K., 310
Nishikawa, Y., 286(k), 287
Nishina, H., 131, 453
Nishizuka, Y., 192
Noda, L., 92
Noda, M., 46, 49(21), 286(k), 287, 482(5), 483
Nogimori, K., 63
Northrup, J. K., 39, 182
Northup, J. K., 6, 8(14), 34–35, 45, 111, 112(3), 124, 124(3), 125(3), 135, 145, 154, 169, 458, 476–477, 483, 486(1), 487–488
Nowak, L., 193, 269
Nozawa, Y., 134, 136, 138(12)
Nukada, T., 286(k), 287
Numa, S., 286(k), 287, 482(5), 483

O

O'Brien, P. J., 512, 513(9), 516(9), 517(9)
Offermanns, S., 283–284, 284(8), 285(8), 286, 286(b, o, t), 287, 291, 291(12)
Offermans, S., 103, 107, 107(4), 287
Ogasawara, N., 296
Ogino, Y., 25
Ohguro, H., 449, 451, 510, 512(7)
Ohlmann, P., 5
Ohoka, Y., 45, 131, 134, 453
Ohtsuka, T., 134
Ohya, M., 29
Oinuma, M., 131, 135(2), 136(2), 457
Okabe, K., 213
Okada, A., 29
Okajima, F., 16, 25(40)
Okayama, H., 246
Okuma, Y., 271, 286, 286(h)
Olate, J., 227, 238(5), 240, 286, 286(i)
Oldenburg, J., 319
Olek, K., 319
Olianas, M. C., 15
Olins, P. O., 55, 256, 257
Oliveira, C. R., 72
Olsen, A. S., 330, 334
Olson, S., 499
Onali, P., 15
Onaran, H. O., 25
Ong, O. C., 509, 512, 513(9), 516(9), 517(9), 519(16)
Ooi, B. G., 215
Oppenheimer, N. J., 45
Oprian, D. D., 483
O'Reilly, D. R., 196
Osawa, S., 71, 307, 319
Osterhoudt, H. W., 228
O'Sullivan, W.J.O., 23
Ota, I. M., 512, 512(4)
Ott, D. G., 87
Ozawa, H., 108

P

Pace, A. M., 18, 296, 307, 317, 319
Padrell, E., 39, 41(3, 4), 42(3), 43(4), 213, 454

AUTHOR INDEX

Pagès, F., 14
Pallat, M., 145
Pandya, B. V., 483
Pang, I., 213
Pang, I.-H., 39, 147, 149(7), 164–165, 170(1), 171(2), 172(1), 173(2, 3), 192–193, 193(18), 198(19), 256, 257(11), 261(11), 268(11), 283
Pang, I. H., 183, 291
Pantaloni, C., 227, 240–241, 242(10), 243(10), 246, 246(11), 248
Papermaster, D. S., 4
Pappone, M.-C., 24
Parenti, M., 147
Parikh, B., 269
Park, D., 452
Parker, C. W., 391
Parker, P. J., 286, 286(g)
Parks, T. D., 161
Parmentier, C., 307, 329(14)
Parmentier, M., 328
Patel, T. B., 269
Patten, J. L., 314
Paturu, K., 297
Paulson, J. C., 235
Paylian, S., 429, 432, 433(20), 434(20)
Pecora, P., 347
Pelegrini, O., 387
Pelz, C., 16, 25(43)
Peng, Y. W., 506
Pennel, R. B., 231, 232(18)
Pennington, S. L., 367
Pérez-Sala, D., 510, 512(8)
Perrin, O. D., 176
Pestka, S., 349
Peters, D.J.M., 399
Pfeuffer, T., 4, 100, 103
Pfister, C., 14, 143, 145(3), 146, 146(3), 252(25), 253
Phelan, S. A., 218
Phillips, W. J., 409, 410(6), 414, 416(18), 418(8, 10), 419(10), 420(18), 422–423, 423(8)
Phipps, P., 319
Pike, L. J., 15, 18(14)
Pinaud, S., 388
Pisano, J. J., 37
Pitcher, J. A., 423
Pitt, G. S., 387–388, 388(6), 400(6), 406
Pittler, S. J., 512

Pizzo, P., 319, 320(46)
Poehlig, H. M., 115, 124(9)
Poland, D., 311
Possee, R. D., 217
Potten, H., 359
Poul, M.-A., 246
Pouysségur, J., 307, 319
Pratt, M. E., 316
Price, S. R., 44, 46–47, 47(29), 49, 49(38), 50(25, 29, 38), 53(38), 55(29), 61(38)
Priedemuth, V., 351
Pronin, A., 482(3), 483, 486(2), 487–488
Pronin, A. N., 482, 486(3), 487–488, 494, 495(20), 509
Proulx, K. L., 505, 508(13), 509(13)
Prpic, V., 175
Puchwein, G., 4
Puckett, C., 308
Pugh, W., 499, 501(5)
Pulsifer, L., 457
Pupillo, M., 387–388, 388(6, 7), 400(5, 6, 7), 406
Purygin, P. P., 72, 80(18)

Q

Qin, N., 512
Quan, F., 335, 436–437, 443, 443(6), 444(6)
Quinnan, G. V., 351

R

Raben, N., 319, 320(46)
Rajaram, R., 269
Rall, T., 427
Ralph, S. J., 329
Ramachandran, J., 45, 240, 437
Ramanchandran, J., 409
Ramdas, L., 427
Ramírez-Solis, R., 369, 371, 371(11), 374(11)
Randerath, K., 90
Rando, R. R., 510, 512(8)
Rangwala, S. H., 257
Rankin, C., 215
Rarick, H. M., 409, 423–424, 427(6), 429, 432, 433(20), 434(20)
Rasenick, M. M., 100, 106–108, 108(7, 13, 14), 110, 429, 434–435

Rashidbaigi, A., 96
Rasnas, L., 110
Ravid, S., 388
Rearick, J. L., 235
Reed, R., 257, 286, 286(a), 296, 321, 330, 334, 388, 443, 457
Reed, S. I., 267
Reese, C. B., 78
Reichert, J., 429
Reichlin, M., 271
Reisine, T., 269–271, 271(15), 283, 286, 286(h, j), 287, 486(7), 487
Reiss, J., 343
Remaut, E., 51
Reymond, C. D., 388
Rhee, S. G., 14, 174, 176, 181–182, 192–193, 295, 308(7), 423
Ribeiro-Neto, F., 110
Richardson, C., 153, 214
Riedy, M., 355
Rill, R. L., 316
Riquelme, P. T., 72, 81(12)
Rivera, P. J., 359
Rivera-Pérez, J., 369–370, 371(11), 374(11)
Robert, M., 146
Robertson, E., 366, 371(2)
Robertson, E. J., 356, 361, 367, 371, 386(15)
Robishaw, J. D., 64, 78, 165, 193, 225, 227, 230, 238(6), 239–240, 258, 437, 458, 476, 482(8), 483, 488, 498, 500, 503, 504(3, 11), 505, 505(3), 506, 506(3), 507, 508(13), 509, 509(13)
Rodbard, D., 25
Rodbell, M., 116, 455
Roeber, J. F., 231
Roger, P. P., 295, 307(3)
Romig, W. R., 44
Rosenthal, W., 103, 107, 107(4), 110–111, 145, 270, 283, 284(8), 285(8), 291, 327, 348, 354(17), 355(17), 459, 486(10), 487
Rosler, U., 343
Ross, A. H., 179, 180(13), 181(13), 192, 276, 277(32)
Ross, E. M., 14, 26, 27(1), 28(1), 29(1), 31(1), 32, 33(1, 5), 35, 35(1), 36(1), 45, 107, 111, 112(3), 113, 124(3), 125(3), 181, 409, 410(4), 431
Ross, F. M., 387
Rossiter, K., 486(4), 487
Roth, J., 319, 320(46)
Rothenberg, P. L., 286(l), 287

Rouot, B., 70
Roychowdhury, S., 106
Rubenstein, R. C., 32, 33(5)
Rudnick, D. A., 256–257, 257(13), 258(13), 267(15), 268(11)
Rudolph, U., 286, 286(i), 366, 386
Rudy, B., 328
Rulka, C., 47, 59(37)
Runge, D., 181
Ruoho, A. E., 70–72, 74(21), 87, 96
Ruscetti, S. J., 16
Russell, M., 321
Rybalkin, I., 476, 483, 498, 504(3), 505(3), 506(3)
Ryu, S. H., 176, 193

S

Sadler, J. E., 235
Saggerson, E. D., 274
Saiki, R. K., 328, 471
Saito, T., 449
Sakurai, A., 509
Salomon, M. R., 251
Salomon, Y., 116, 455
Sambrook, J., 147, 149(10), 195, 204(25), 228, 239, 331, 342(21), 347, 373
Sanders, D. A., 191, 295, 327, 332(5)
Sanford, J., 227, 238(3), 239(3), 286, 286(i), 355
Sanger, F., 61
Sarma, R. H., 93
Satoh, T., 191
Saxe, C. L. III, 388
Sayce, I. G., 176
Schaap, P., 397, 399
Schacht, J., 176
Schaeffer, H. J., 82, 84(38), 95(38)
Schagger, H., 517
Scharf, S., 328, 471
Scharf, S. J., 328, 329(16)
Schatz, H., 307
Scheerer, J., 367
Scheffler, I. E., 473
Schenker, A., 295
Schepers, T., 5, 16, 429, 432, 433(20), 434(20)
Scherer, N. M., 114, 124(8), 275
Scherle, P. A., 335, 344
Scherly, P. A., 192

Scherübl, H., 270, 327, 348, 354(18), 355, 355(18)
Schessinger, J., 423
Schiller, D. L., 26
Schimerlik, M. I., 6
Schleicher, A., 71, 424, 427(2)
Schleifer, L. S., 35, 45, 111, 112(3), 124(3), 125(3)
Schlotterbeck, J. D., 63, 66(8)
Schlumberger, M., 307
Schmidt, A., 227, 234(2)
Schmidt, C. J., 226–227, 229, 229(2), 236(25a), 457, 509
Schmidtke, J., 343
Schnabel, P., 26
Schnittler, M., 5
Scholder, J. C., 388
Schöneberg, V., 351
Schrader, K. A., 457
Schreck, R., 26
Schultz, G., 15, 17(25), 18, 18(18, 25, 27, 29), 25, 25(19, 30), 103, 107, 107(4), 145, 270, 283–284, 284(8), 285(8), 286, 286(b, o, t), 287, 291, 291(12), 327, 345, 348, 354(17, 18), 355, 355(17, 18), 486(10), 487
Schütz, W., 25
Schwartz, J. P., 15
Schwartzberg, P. L., 356, 367
Schwindinger, W. F., 306, 318
Scott, M. P., 47
Seal, S., 227
Seamans, C., 50, 51(47)
Sedivy, J. M., 367
Sefton, B. M., 147, 254–255, 286, 286(c)
Seidman, J. G., 324, 326(3), 356, 359(7), 364, 364(9), 365(7)
Seidman, M. M., 367
Sekura, R. D., 25, 63–64, 65(3), 74, 110, 427, 459
Selinger, Z., 3, 14, 15(8), 17(8), 18, 18(8, 10), 24(9, 10), 45, 243
Selzer, E., 269, 273, 281(10), 282
Semba, R., 296
Senogles, S. E., 213
Serino, K., 319
Serventi, I. M., 46, 55
Sewell, J. L., 46, 47(26), 59(26)
Sha'afi, R. I., 16, 25(43)
Sharp, P. A., 367
Shaw, K., 175

Sheffield, V. C., 311
Shelanski, M., 106
Shenker, A., 15, 301, 306(22), 309, 314(4), 317(4), 318(4)
Sherman, F., 59, 60(55)
Sherrin, J. A., 397
Shibasaki, H., 131, 135(3), 136(3, 5, 6), 138(4), 213
Shieh, W.-T., 309, 319(5)
Shimada, N., 15, 18(15)
Shimonishi, Y., 254–255, 268(5), 451, 510, 512(7)
Shinnick, T., 499
Shinozuka, K., 351
Sidiropoulos, D., 4, 7, 14, 16, 21(44), 22(44, 45), 25(11, 44), 283
Silberras, O., 101
Silverstein, K., 309–311, 311(20), 312(20), 314(8), 318
Simon, M., 174, 269, 509
Simon, M. I., 25, 64, 78, 146–147, 153(5), 182, 191–193, 195(16, 17), 212, 213(1), 230, 239, 258, 284, 286(n, p, s), 287, 295, 297, 308(7), 327, 330, 333–334, 334(6, 23), 335, 344, 344(7, 23–25), 345, 408, 449, 474, 476–479, 479(7), 482(6, 7, 10), 483, 486(1), 487–488, 498, 500, 500(2)
Simon, M. N., 387–388
Simonds, M., 443, 482(4), 483
Simonds, W. F., 246, 249(19), 254, 255(3), 256, 256(3), 269, 275(8), 281(5, 8), 283, 286, 286(e, f), 482, 486(4, 5, 8), 487–488, 509
Simons, C., 308
Simons, E. R., 17
Simons, M.-J., 328, 329(14)
Sinigel, M. D., 409, 410(4)
Skiba, N. P., 409, 424, 427(6)
Sladek, T., 236
Slaughter, C., 181
Slaughter, C. A., 483, 488, 500
Slepak, V. Z., 192, 195(16), 335, 344
Slice, L. W., 55
Smallwood, P. M., 314, 478, 482(9), 483
Smart, F. J., 101
Smigel, M. D., 27, 34–35, 39, 45, 111, 112(3), 124, 124(3), 125(3), 135, 154, 169, 260, 409, 410(5), 452, 457–458
Smith, F., 236
Smith, G. E., 195, 196(26), 213, 215, 217(24), 218(24)

Smith, J. A., 175, 176(4), 181(4), 182, 213, 324, 326(3)
Smith, R. S., 16
Smithies, O., 356, 370
Smrcka, A., 107, 269, 291(10)
Smrcka, A. V., 35, 164, 172, 174, 181–182, 192–193, 193(12, 13), 194(9, 12), 197(12), 201(12), 202(12), 203(12), 204(12), 205(12), 206(13), 207(13), 210(13), 211(12, 13), 213, 284, 286(q, r), 287, 423
Snaar-Jagalska, B. E., 16, 19(39), 387, 397, 397(4), 399, 399(4), 400(4, 5), 401, 404–405, 406(44)
Sninsky, J. J., 321
Snyder, S. H., 296
Sobell, M., 93
Solski, P. A., 512
Soltoff, S., 423
Sommer, R., 329
Sonenshein, G. E., 346
Spada, A., 18, 296, 317
Spicher, K., 283–284, 286, 286(b, o, t), 287, 291(12)
Spiegel, A., 275, 281(5, 8), 301, 306, 306(22), 308, 443
Spiegel, A. M., 15, 213, 246, 249(19), 254, 255(3), 256, 256(3), 269, 275(4, 8), 283, 286, 286(e, f), 295, 309, 314(4), 316(3), 317, 317(3, 4), 318(4), 319, 319(5), 482, 482(4), 483, 486(4, 5, 8), 487–488, 509
Spudich, J. A., 388
Staehelin, T., 465, 503
Stahl, E., 514
Standaert, M. L., 269
Staniszewski, C., 457
Stanley, S. J., 45, 47, 49(38), 50(38), 53(38), 61(38), 63–64, 65(5), 66(5), 81
Stanton, V. P., Jr., 310, 314(19)
Starmer, W. T., 330, 334
Stearns, T., 47, 59(41), 61(41)
Steele, D. A., 192, 195(16), 335
Steele, G., 314
Steer, M. L., 15, 18(20), 24(20)
Stein, C. A., 346–347, 347(6), 351(6)
Stein, P. J., 110
Steinberg, F., 25, 64, 74, 427
Steiner, A. L., 390(30), 391
Steisslinger, M., 4, 7, 16, 21(44), 22(44, 45), 25(44)
Stephenson, R. C., 512, 519(16)
Sternberg, P. W., 327, 335
Sternweis, P., 39, 107
Sternweis, P. C., 6, 7(16), 35, 45, 111, 112(3), 124, 124(3), 125(3), 135, 147, 149(7), 164–165, 170(1), 171, 171(2), 172, 172(1), 173(2, 3), 174, 181–183, 192–193, 193(18), 194(9), 198(19), 213, 225, 256, 257(11), 261(11), 268(11), 269, 283–284, 286(q, r), 287, 291, 291(10), 409, 410(4), 423, 458, 486(11), 487–488, 497(2), 503, 507, 509–510
Sternweis, P. W., 64
Stevens, L., 46
Strangeways, L., 500
Strathman, M., 174, 335
Strathman, M. P., 335, 498, 500(2)
Strathmann, M., 25, 192, 195(17), 286(p, s), 287, 295, 335, 388, 478–479
Strathmann, M. P., 146, 182, 191–192, 212, 213(1), 284, 327, 330, 333–334, 334(6, 23), 335, 344, 344(7, 23–25), 345, 408, 449, 482(10), 483
Straub, C., 5, 7(10), 10(10), 20
Streaty, R. A., 15, 18(32), 21(32)
Stround, R., 436
Struhl, K., 324, 326(3)
Stryer, L., 14, 24–25, 74–75, 409, 411, 427
Stuber, D., 148, 149(12)
Studer, H., 307
Studier, F. W., 150
Stueber, D., 148
Sturgill, T. W., 113
Styer, L., 64
Suarez, G., 307
Subasinghe, C., 347
Subramani, S., 473
Sugimoto, K., 286(k), 287, 482(5), 483
Sugimoto, N., 349
Suh, P.-G., 193
Suh, S. G., 176
Suki, W. N., 286(m), 287
Sullivan, K., 240, 242, 270, 409, 427, 437
Sulston, J. E., 78
Summers, M. D., 195, 196(26), 198, 213–215, 217(24), 218(24)
Sun, T. J., 388
Sundralingam, M., 93
Sunyer, T., 177
Sussman, M., 389

Sutcliff, J., 499
Suzuki, H., 286(k), 287
Suzuki, K., 136
Svoboda, M., 15, 18(13), 24
Svoboda, P., 110
Swartz, T. L., 38
Swarup, R., 297
Symons, R. H., 86, 87(41), 89(41)

T

Tabor, S., 153
Tahira, T., 307
Takahashi, H., 286(k), 287, 482(5), 483
Takahashi, K., 131, 134, 135(3), 136(3, 5, 6), 138(4), 139(7), 213, 453
Takahashi, N., 509
Takao, T., 254–255, 268(5), 451, 510, 512(7)
Takemoto, D. J., 71
Takemoto, L. J., 71
Tallent, M., 269
Talluri, M., 100
Tam, J. P., 499
Tamir, H., 145, 476–477, 483, 488
Tamkun, J. W., 47
Tamura, M., 63
Tamura, S., 509
Tan, J. L., 388
Tanabe, T., 286(k), 287, 482(5), 483
Tanaka, T., 310
Tandon, N. N., 15, 18(16)
Tanford, C., 233
Tang, W., 457
Tang, W.-J., 452
Taramelli, R., 317
Tarussova, N. B., 72, 80(18)
Tatsumi, M., 145
Taussig, R., 147
Tautz, D., 329
Tavale, S. S., 93
Taylor, S. I., 319, 320(46)
Taylor, S. J., 174–176, 176(4), 177, 178(12), 181(4), 182, 192, 213, 423
Taylor, S. S., 55, 256, 388
Te, R. H., 356, 358
Teplow, D. B., 78, 230, 476, 483
Theibert, A., 405, 406(43)
Thierault, D. L., 429, 432, 433(20), 434(20)

Thomas, D. M., 302
Thomas, G., 437–438
Thomas, J., 82, 84(38), 95(38)
Thomas, K. R., 356, 359, 359(3), 366–367, 367(1), 372(1)
Thomas, L., 437, 443(6), 444(6)
Thomas, R., 103
Thomas, T. C., 226, 229, 236(25a), 509
Thorne, B. A., 437
Thuong, N. T., 346
Timir, H., 476
Tobias, D. T., 99
Tohkin, M., 45, 136, 138(12)
Tohyama, K., 131, 138(4), 213
Tomita, U., 131, 136(5)
Tomkins, G. M., 112, 240
Toro, M.-J., 114, 124(8), 275
Toshihide, N., 482(5), 483
Tota, M. R., 6
Toulmé, J.-J., 346
Toutant, M., 70
Towbin, H., 465, 503
Towler, D. A., 254, 255(4), 257, 267(15)
Toyama, R., 297
Trask, B. J., 330, 334
Traystman, M. D., 319
Treco, D. A., 61
Trenthan, D. R., 78
Trepel, J. B., 45
Trukawinski, S., 414, 416(18), 420(18)
Tsai, S.-C., 44, 46–47, 47(28, 29), 48–49, 49(21), 50(25, 29), 53(45), 55(29), 237
Tsilevich, T. L., 72, 80(18)
Tsubokawa, M., 45
Tsuchiya, M., 46–47, 47(29), 50(29), 55(29)
Tsukamoto, T., 297, 308, 315(1)
Turner, D. H., 349
Turner, G. H., 327
Tychowiecka, I., 486(10), 487

U

Ueno, K., 134
Uesugi, S., 93
Ui, M., 9, 16, 25, 25(40), 45, 63–64, 131, 134, 135(2, 3), 136, 136(2, 3, 5, 6), 138(4, 12), 139(7), 213, 452, 457
Ullrich, A., 423

Unson, C., 269, 443
Unson, C. G., 15, 269, 275(4, 8), 281(5, 8), 283, 286, 286(f), 295, 486(4), 487, 509
Unson, G., 482, 486(8), 487–488
Urbano-Ispizua, A., 302
Uzu, S., 26

V

Vahrson, W., 351
Vaillancourt, R. R., 70–71, 87
Valancius, V., 370
Vallar, L., 18, 296, 300, 308(20), 317, 319(36)
Valle, D., 314
Van Amsterdam, J. R., 229, 230(15), 231(15), 242
Van Dongen, A.M.J., 45
Van Dop, C., 25, 45, 64, 74, 296, 427
Van Haastert, P.J.M., 16, 19(39), 387–388, 391, 394, 397, 397(4), 399, 399(4), 400(4, 5, 33, 40), 401, 404–406, 406(32, 44)
Van Lookeren Campagne, M. M., 387
Van Sande, J., 328, 329(14)
van Vloten-Doting, L., 227
Varmus, H., 328
Vassart, G., 328, 329(14)
Vaughan, M., 15, 25, 25(33), 44, 44(2), 45–47, 47(28, 29), 48–49, 49(21, 38), 50(25, 29, 38), 52, 53(38, 45), 55(29), 59(42), 61(38, 40), 64, 78
Verjovski-Almeida, S., 72
Vernet, T., 214
Veron, M., 387–388
Vialard, J., 214
Vignais, P. V., 72, 81(13), 85(13), 86(13)
Villarejo, M. R., 151
Vinitsky, R., 269, 275(4), 281(5), 486(4), 487
Vink, A. A., 406
Vogelstein, B., 375
von Jagow, G., 517
Von Weizacker, E., 478–479, 482(10), 483
Voyta, J. C., 490
Vyakarnam, A., 500

W

Wada, A., 309
Wakamatsu, K., 29
Wakelam, M.J.O., 272, 274, 275(23)
Wald, G., 429
Waldo, G. L., 182, 184, 184(4, 11), 185(13), 190(4, 7, 11), 192, 276–277
Wallace, J. D., 369, 371(11), 374(11)
Wallace, M. A., 37
Wallace, R. B., 472
Walseth, T. F., 35
Walsh, K. A., 254–255, 255(5), 268(5)
Wang, M., 397
Wang, N., 106–107, 108(7)
Wange, R. L., 284, 291(10)
Watanabe, A. M., 15, 25(37)
Watanabe, M., 108, 429, 435
Watkins, D. C., 347
Watkins, P. A., 25, 45, 63, 65(5), 66(5)
Watson, A. J., 327
Webb, A. C., 218
Webb, N. R., 214
Weber, J. L., 319
Weingarten, M., 106
Weinstein, L. S., 301, 306, 306(22), 308–309, 314(4), 316, 316(3), 317, 317(3, 4), 318(4), 319(5)
Weir, M., 328
Weiss, O., 47
Welker, D. L., 388
Welsh, C. F., 44, 47, 49(38), 50(38), 52, 53(38), 61(38)
Wensel, T., 409, 427
Wessling-Resnick, M., 71, 75
West, R. E., 64, 78
Wheeler, G. L., 110
White, T. J., 321
Wickstrom, E., 349
Wieland, T., 3, 5–6, 7(17), 20
Wilcox, M. D., 457
Wilde, M. W., 16
Wilkie, T. M., 192, 295, 327, 330, 333–334, 334(23), 335, 344, 344(23, 24), 388, 477, 479(7)
Wilks, A. F., 328–329
Williams, C. A., 502
Williams, J. G., 388
Williams, R. S., 355
Williamson, K. C., 46, 50(25)
Wilson, J. H., 367
Wims, M., 369, 371(11), 374(11)
Winitz, S., 321
Winslow, J. W., 229, 230(15), 231(15), 242

Wittig, B., 270, 283, 327, 345, 348, 351, 354(17, 18), 355, 355(17, 18)
Woerner, A. M., 351
Wojcikiewicz, R.J.H., 37
Wolda, S. L., 512
Wolf, L. G., 64, 124, 235, 237(23)
Wolfgang, W., 335, 438, 443
Wong, S. C., 423
Wong, Y. H., 26, 307, 319
Woodard, C. J., 286, 286(f)
Woods, K. R., 426
Woodward, C. J., 269, 281(5)
Woolkalis, M. J., 64
Woon, C. W., 71
Worley, P. F., 296
Wray, G. W., 115
Wray, V. P., 115
Wu, D., 193, 269, 295, 308(7)
Wu, J. Q., 218
Wu, L., 335, 388

Yamane, H. K., 509–510, 512, 512(4, 5), 519(5, 16)
Yamazaki, A., 111, 145
Yan, K., 107, 110
Yanofsky, C., 51
Yanofsky, M., 327, 335
Yatani, A., 45, 213
Yathrinda, N., 93
Yatsunami, K., 483
Yau, K. W., 506
Yi, F., 236
Yokoyama, S., 330, 334
Yonemoto, W., 55, 256, 388
Yoo, D. J., 452
Yoshida, K., 93
Yoshikawa, M., 85
Yoshimoto, K. K., 64, 286(n), 287, 297
Yoshizawa, T., 254–255, 268(5), 449, 451, 510, 512(7)
Yost, D. A., 64, 81

X

Xie, H., 510, 512, 512(5), 513(9), 516(9), 517(9), 519(5, 16)

Y

Yabuki, S., 309
Yajima, M., 63
Yamaguchi, M., 81
Yamanaka, G., 25, 64, 74, 427

Z

Zabin, I., 151
Zachary, I., 306–307, 319
Zamecnik, P. C., 78
Zhang, X. H., 347
Zheng, H., 369, 371(11), 374(11)
Zhou, Z., 213
Zhu, J., 349
Zigmond, S. H., 16
Zon, G., 351
Zubiaur, M., 356, 359(7), 365(7)
Zubin, P., 15, 21(34), 26(34)

Subject Index

A

Acetic acid staining, 115
Acrylodan, 417
Adenosine receptor agonists, stimulation of GTP hydrolysis, 15
Adenylyl cyclase, 296, 408
　activation, 437
　assay, 35, 116, 451–456
　GTPase stimulation, 18
　GTP modulation, 108
　isoproterenol-stimulated, 434
　in reconstituted cyc^- membranes, 252–254
　regulation, 269–270, 452
　stimulation by GTPγS $in\ vitro$, 404–406
Adipocytes, hamster, GTP hydrolysis in, 15
ADP-ribose, hydrolysis, 45
ADP-ribosylagmatine, formation catalyzed by cholera toxin ADP-ribosyltransferase, 48
α-ADP-ribosylarginine, formation, 45
ADP-ribosylation
　cholera toxin-catalyzed, 24–26, 45, 71, 243–245, 268
　G_{sa} proteins, 48
　pertussis toxin-catalyzed, 24–26, 71, 77–79, 93–94, 117, 126–128, 132–133, 268
　　G-protein subunits, 236–238
　　(N)-[^{125}I]iodoazidophenylpropionyl-NAD$^+$ in, 77–78, 93
　transducin, 76–79, 91, 93–94
ADP-ribosylation factors
　bovine
　　expression, 50–52
　　purification, 49–50
　cDNA, ligation-independent cloning
　　PCR amplification, 53–54
　　reactions, 53–54
　　screening transformants, 54
　cholera toxin activation, 44–63
　　assays, 48–49

　class I, 46
　class II, 46–47
　class III, 47
　complementation, 61–63
　expression, 50–52
　　vector construction for, 55–57, 60–61
　genes, 46–47
　myristoylated
　　expression, 57–58
　　purification, 58–59
　myristoylation, 47, 55–59
　purification, 49–50, 58–59
　recombinant, synthesis and purification, 50–52
　　as fusion protein, 52–53
　$Saccharomyces\ cerevisiae$, genes, 47, 59–63
　synthesis, as glutathione S-transferase fusion protein, 53–55
ADP-ribosyltransferase, activity
　cholera toxin, 45, 48
　pertussis toxin, 63–64
$α_2$-Adrenoceptor agonists, stimulation of GTP hydrolysis, 15
$β$-Adrenoceptor agonists, stimulation of GTP hydrolysis, 15
　G_s interaction sites, 434–436
　G-protein interactions, 8–9, 14
Affinity chromatography, βγ-agarose
　applications, 171–174, 201–204
　Gα subunit purification technique, 171–174, 192–193, 201–204
　matrix synthesis, 164–174
Affinity purification, antibodies, 491–492
βγ-Agarose
　affinity chromatography, 171–174, 201–204
　characterization, 168–170
　Gα subunit purification, 171–174, 192–193, 201–204
　preparation, 164–168
Agarose gel, 325–326, 338–339
Agmatine, cholera toxin-catalyzed formation of ADP-ribosylagmatine, 48–49

Albright hereditary osteodystrophy, 309, 316–317
Albumin, bovine serum, molecular weight calibration standard, 92
Alcohol dehydrogenase, NAD^+ photoaffinity labeling, 72
Amino acids
  alignment
    at Gα subunit amino-terminal, 255
    in Gβγ subunits, 484–485
  analysis, 429
  mutations
    in $G_s\alpha$ gene, 296–297, 299
    in GTPase-deficient Gα subunits, 325–326
  sequences
    analysis by Gβ and Gγ subunit cloning, 471–474
    Gβγ subunit types, 476, 478
    interfacial peptides, 427
    oligonucleotide primers, 332–333, 473
    specification, 472–474
    for peptide synthesis, 483–490
Aminobenzoyloxymethyl cellulose, 503
ω-Aminobutyl-agarose, βγ subunit immobilized on
  characterization, 168–170
  preparation, 164–168
Aminophenylthioether cellulose, 503
Ammonium sulfate pellet assay, cAMP binding, 395
Ampicillin resistance
  expression vector, 148
  as selection factor
    in ADP-ribosylation factor synthesis, 54, 58
    in Gα subunit expression, 195, 258, 439
Antibodies
  affinity purification, 491–492
  to Gβγ subunits, characterization, 482–498
  to Gγ subunits, 498–509
    applications, 506–509
    preparation, 499–504
    specificity, characterization, 505
  to Gγ subunit subtypes, usefulness, 506–509
  monoclonal, 4A, 424, 427, 430–431
  testing, dot-blot procedure, 492
  uncoupling experiments, membrane preparations, 279–282
Antisera
  antipeptide, for identification of Gα subunits, 268–283
    agonist inhibition, 282
    cross-reactivity, 275–276
    IgG fraction preparation, 279–280
    immunoprecipitating, 286
    preparation, 284–285
    receptor–guanine nucleotide-binding protein interaction, 280–282
    specificity, 271–279
    synthetic sequence, 270–271
  AS7, 273, 275
  B4-1, 485
  BN1, 485
  CQ2, 278
  for Gα subunit purification, 193
  to Gβγ subunits, 486–488
    specificity, 494–495
  KT, 485
  MS, 485
  OC2, 279
  ON1, 279
  rabbit, 490
  polyclonal, for Gα subunit purification, 193
  SG1, 275
  testing, dot-blot procedure, 492
Arabidopsis, Gα subunits, PCR amplification, 327
Arginine, ADP-ribosylation, cholera toxin-catalyzed, 45
Atrium, porcine, GTPγ binding in, 6
Autographica californica nuclear polyhedrosis virus, 214
2-Azidoadenosine, synthesis, 84
2-Azido-ADP-ribose, ^{32}P-labeled, intra- and intermolecular transfer from Gα subunit carboxy terminus, 95–99
2-Azido-AMP
  nonradioactive, synthesis, 85, 93
  ^{32}P-labeled at α position
    chemical synthesis, 82–86
    in preparation of 2-azido-[^{32}P]NAD$^+$, 82
4-Azidoanilido-GTP
  G protein labeling, 100–110
  preparation, 101–103

structure, 101
tubulin labeling, 106–107
4-Azidoaniline
  GTP analog preparation from, 101, 102–103
  synthesis, 101–102
2-Azido-ATP, ^{32}P-labeled at α position
  enzymatic synthesis, 87–89, 92
  in preparation of 2-azido-[^{32}P]NAD$^+$, 82
2-Azido-NAD$^+$
  as G-protein structure probe, 71
  kinetic constants, determination, 91–92
  nonradioactive, synthesis, 86–87
  ^{32}P-labeled
    enzyme cleavage analysis, 91
    synthesis, 81–95
      chemical, 82–87

B

Baculovirus
  Gα subunit expression, 212–226, 273
  virus and vector preparation, 195–196
  G$\beta\gamma$ subunit expression, 193
  recombinant
    production and purification, 195–196, 217–218
    transfer vectors, construction, 195–196, 214–216
Biotinylation, G$\beta\gamma$ subunits, 458–466
Bisacrylamide, 274
Blastocysts, in chimera generation, 366, 386
Bordetella pertussis, 63, 268
Bradykinin, stimulation of GTP hydrolysis, 15, 21
Brain
  bovine
    ADP-ribosylation factors
      expression, 50–52
      purification, 49–50
    cytosol, ammonium sulfate precipitate from, 49
    G proteins, 171
      biotinylated G$\beta\gamma$ complexes, 457–471
      GTPγ binding assay, 39–44
      isolation, 165
      pertussis toxin-substrate, 131–139
      preparation, 133, 458

G$\beta\gamma$ subunits
  purification, 503
  stimulation of S49 cyc^- adenylyl cyclase, 453–456
  membrane extract, preparation, 133
G proteins, GTPγ binding activity, 38–39
rat
  G proteins
    pertussis toxin labeling, 66–69
    pertussis toxin-substrate, 131–139
    photolabeling, 107–108
  Gα subunits, 279
  GTP hydrolysis, 15
Brain tumor, Gα subunit mutation detection in, 300–301
Butylamineagarose, conjugation to G$\beta\gamma$ subunit, effect of sulfo-MBS concentration, 167

C

Caenorhabditis elegans, Gα subunits, PCR amplification, 327
Calcium channels, inhibition and stimulation in GH$_3$ cells by hormones, antisense oligonucleotide effects, 348, 354–355
Calcium phosphate
  –DNA complex, plasma membrane *in vivo* targeting, 248
  in transfection of COS cells, 246–249
Capillaries, for microinjection of antisense oligonucleotides, 353
Carbachol, 4, 13
  calcium channel inhibition, 345–355
Carbodiimidazole, 74–81, 83, 92
Carbonate dehydratase, 92, 115
Cations, monovalent, effects on receptor-mediated GTP hydrolysis, 21–22
cDNA
  for ADP-ribosylation factors
    expression, 50–52
    ligation-independent cloning
      PCR amplification, 53–54
      reactions, 54
      recombinant clones from, screening transformants for, 54
  cloning
    directly from library, 480–482

Gα subunit, 257–258
Gβγ subunit, 471–474
Gβγ subunit clones, 474–480
Gγ subunit clones, 477, 479
for G$_s$α subunits, *in vitro* translation, 240
isolation
  by PCR, 472–474
  by plaque hybridization, 472
related, cloning, low-stringency hybridization conditions for, 475
subcloning, 215
synthesis, 337–338
Cell division, suppression by antisense oligonucleotides, 348–349
Cell lines
  AB-1, gene targeting in, 367–369, 371, 373, 377–386
    clone selection, propagation, and identification, 377, 381
  BL21/DE3, Gα subunit expression, 149–150, 152
  Chinese hamster ovary
    Gα subunit mutation detection, 307
    leukemia inhibitory factor-producing, for gene targeting, 372–373
  GH$_3$, calcium channels, antisense oligonucleotides and, 345–355
  JM109, 258
  mutant
    applications, 364–366
    biochemical analysis, 356–366
    homozygous, production, 357–364
    permeable, G proteins, photolabeling, 108–110
  Sf9
    culture, 196, 216–217
    membrane extraction, 198–200
    Gα subunits, recombinant, purification, 191–212, 214–226
    Gγ subunits, antibody expression, 505, 508
    G$_q$α subunits, recombinant, 196–197
    G$_{16}$α subunits, recombinant, cellular location and activity, 205–206
    transfection, recombinant baculovirus production, 217–218
Centrifugation
  cesium chloride gradient, recombinant insertion vector purification, 439–440
  sucrose density gradient, solubilized membrane proteins, 10–11, 13
CHAPS, 4, 10–11, 13, 66
Chimeras
  embryonic stem cell, 366
  Gα subunits, construction, 321–326
Chloramphenicol acetyltransferase, gene, 148–149
3-[(3-Cholamidopropyl)dimethylammonio]-1-propane sulfonate, 4, 10–11, 13, 66
Cholecystokinin, stimulation of GTP hydrolysis, 15
Cholera toxin
  activation
    by ADP-ribosylation factors, 44–63
    assays, 48–49
    by GTP, 46
  ADP-ribosylation catalyzed by, 24–26, 45, 71
    G$_s$α subunits, 243–245, 268
  ADP-ribosyltransferase activity, 45, 48
  A1 protein, catalytically active, 44–46
  A2 protein, carboxyl-terminal, 44
  effect on GTP hydrolysis, 14
  hydrolysis, 45
  substrates, 45
Choline, effect on receptor-mediated GTP hydrolysis, 22
Chromatofocusing, in Gα subunit purification, 222–225
Chromatography
  activated G$_s$α and G$_i$α, 119, 126–129
  adenylyl cylase, assay, 454
  ADP-ribosylation factor, 49–50
  affinity, see Affinity chromatography
  azidonicotinamide adenine nucleotide, 89–90
  fast protein liquid, see Fast protein liquid chromatography
  G proteins
    pertussis-toxin substrate proteins, 133–139
    recombinant, 154–158
  G$_q$ proteins, 175–179, 181
  G$_{11}$ proteins, 187–190
  Gα subunits, 153–154, 200–202, 207–212, 220–223, 261
    recombinant, 197, 262
  G$_i$α$_2$ subunit, 260–261
  Gβγ subunits, 138–139, 165
    biotinylated, 461–466

high-performance liquid, *see* High-performance liquid chromatography
in ^{3}H-isoprenylated protein identification, 517–519
histidine-tagged proteins, 160–161
holotransducin, 75
ion-exchange, *see* Ion-exchange chromatography
liquid, *see* Fast protein liquid chromatography; High-performance liquid chromatography
transducin subunits, 143–145, 450–451
translation products generated *in vitro*, 228–231
Clones
G$\beta\gamma$ subunit, families, novel members, 474–480
homozygous, selection, 363–364
identification based on PCR dependent on gap repair, 370–371
recombination events, analysis, 383–385
screening, 341–342
Cloning, *see also* Subcloning
cDNA
directly from library, 480–482
related, low-stringency hybridization conditions for, 475
Gα subunits, 257–258
frequency analysis, 333
PCR in, 327–344
G$\beta\gamma$ subunits
amino acid sequence information for, 471–474
probe design, 471–482
ligation-independent
PCR amplification, 53–54
reactions, 54
screening transformants, 54
PCR products, 329–331, 340–341
TA, kits, 331
Complementation, ADP-ribosylation factors, 61–63
Complement C5a, stimulation of GTP hydrolysis, 16
Coomassie Brilliant Blue, staining
in analysis of Gα subunit expression, 160–161
human erythrocyte G proteins, 115, 129
pertussis toxin labeling, 70

Creatine kinase, in photoactivatable probe synthesis, 88–89, 93
Cross-linking
association analysis with, 235–237
2-azido-ADP-[^{32}P]-ribosylated transducin, 96–98
G-protein subunits, 236
Culture media
aqueous, solvent exchange for, 264–265
cellular uptake from, in generation of antisense oligonucleotides, 346–347
Cyclic AMP
binding
assays, 394–397
GTP inhibition, 399–400
extracellular, *Dictyostelium*, 408
isotope dilution assays, 390–394
second messenger responses, 398–399
stimulation
GTPase activity, 402–404
GTPγS binding, 401
trophic hormones, 296
Cyclic AMP receptor agonists, stimulation of GTP hydrolysis, 16
Cyclic GMP
isotope dilution assays, 390–394
second messenger response, 398
Cyclic GMP phosphodiesterase, 387, 410, 419

D

1-(Deoxy-2-fluoro-β-D-arabinofuranosyl)-5-iodouracil, 367
Detergents
effects on GTPγS binding assay, 40–41
effects on mastoparan-stimulated G-protein function, 28–30
in G$\beta\gamma$ subunit affinity matrix synthesis, 168
Diacylglycerol, 174, 192
Diazotized papers, immunoblotting, 503
Dictyostelium, 387
culture, 389–390
effector enzymes, 408
extracellular signal, 408
G-protein assays in, 387–408
Gα subunits, PCR amplification, 327
membranes, GTP hydrolysis, 16, 18
Dimethylformamide, 74, 78–80

Dimethyl sulfoxide, solvent exchange for, 264
2,5-Diphenyloxazole, 5
DNA
  complementary, see cDNA
  in degenerate PCR primers, sequence, 328–329
  denaturation, partial, 310
  electrophoretic mobility, 310
  isolated, Southern analysis, 362
  manipulation, 311–312
  melting
    maps, 311–312
    principles, 309–310
  minipreparations
    for plasmid isolation, 342–343
    sequencing, 343
  modeling, 311–312
  stability, 309–312
  targeting construct, for Gα subunit inactivation, 359–361
  viral, AcPR-lacZ, for Gα subunit expression, 195
DNA–RNA heteroduplexes, 349–350
Dot-blot procedure
  antisera/antibody testing, 492
  in Gα subunit point mutation detection, 297–301, 307
Drosophila, Gα subunits
  expression in S49 cyc^- cells, 437, 443
  PCR amplification, 327

E

Electron microscopy, DNA melting study, 309
Electroporation, gene targeting, 373–374, 377
Enzyme cleavage, 2-azido-[^{32}P]NAD$^+$ analysis by, 91
Enzyme-linked immunosorbent assays, anti-G protein antisera specificity, 271–272, 502
Enzymes
  effector, in Dictyostelium, 408
  GTP modulation, 108
Eosin isothiocyanate, in fluorescence assays for G-protein interactions, 420–422

Epitope mapping studies, interfacial peptides, 426–427
Erythrocyte membranes
  frog, GTP hydrolysis, 15
  human
    extraction, 118
    G proteins, purification, 110–130
  turkey
    G_{11} proteins, purification, 182–191
    Gα subunit antipeptide antisera, 276–277
    GTPγ binding, 8–9
    GTP hydrolysis, 14
    preparation, 184–186
Escherichia coli
  ADP-ribosylation factors, 47, 51–52
  culture, 151–152
  Gα subunit expression, 104, 146–164, 213, 258–259, 439
  anti-G protein antisera specificity, 273
  Gα subunit myristoylation, 256–258, 267
  lysis, 153
Ethanol, solvent exchange for, 264
Ethidium bromide, staining with, 316, 325
N-Ethylmaleimide, 4
  membrane treatment, 8–10

F

S-(all-trans-Farnesyl)cysteine, synthesis, 513–514
Fast protein liquid chromatography
  activated G_q proteins, 175–176
  biotinylated Gβγ subunits, 464–465
  myristoylated Gα subunit purification, 262
  transducin subunit purification, 143
Fibroblasts
  NIH 3T3, Gα subunit mutation detection, 307
  QT6, transfectants for immunochemical studies, 495
Fluorescamine, protein determination with, 113
Fluorescence
  in assays of G-protein interactions, 409–423
  IAF-PDE, 421–422

intrinsic, in studies of receptor-stimulated GTP binding and GTPase, 412–415
labeled protein studies, 417–419
reporter groups, 415–422
  extrinsic fluorescence, 411–412
  resonance energy transfer measurements, 420–422
  tryptophan, intrinsic, 410–411
  measurement, 414–416
Fluorography, in metabolic radiolabeling, 267
Fluorophores, environmentally sensitive, as G-protein interaction probes, 415–417
N-Formylmethionylleucylphenylalanine
  and GTPγ binding, 4, 7, 11
  and GTP hydrolysis, 13–14, 16, 20, 22

G

β-Galactosidase, as molecular weight calibration standard, 92
Gancyclovir, 367
Ganglioside G_{M1}, 44
Gas chromatography-mass spectrometry, isoprenyl groups, 512
GDP
  effect on Gα subunit photolabeling, 289–291
  effect on GTP hydrolysis, 22, 24
  equilibrium exchange, 32–34
  regulation of GTPγS binding, 6–8
  release, 13
Gel electrophoresis
  activated $G_2\alpha$ and $G_i\alpha$, 113–115
  antipeptide antisera specificity analysis, 273–277
  2-azido-[^{32}P]ADP-ribosylated transducin, 96–97
  biotinylated Gβγ subunits, 463–466
  denaturing gradient
    analysis strategies, 319–320
    application of DNA melting principles, 309–310
    $G_s\alpha$ subunit gene, 308–320
    Gα subunit mutation detection, 301–302

mutation pattern analysis, 312–314
  parallel, 313–314
  perpendicular, 313–315
gene targeting, 375
in G-protein labeling, 104–105
G proteins, 114–115, 126–128
  α subunits, 204–205, 211–212
    recombinant, 219–220, 222–224
  recombinant, 156–157
    histidine-tagged, 162–163
G_q proteins, 177–178
G-protein subunit translation studies, 228–229
$G_s\alpha$ subunit gene, 314–315
Gβγ subunit antibodies, 492–493
immunoprecipitated Gα subunits, 287–288
isoprenylated proteins, 516, 518–519
in metabolic radiolabeling, 265–267
in pertussis toxin labeling, 70
pertussis toxin-substrate G proteins, 138–139
proteins, 77
recombinant $G_i\alpha$ subunits, 262–263
transducin subunits, 141
urea–sodium dodecyl sulfate polyacrylamide, in G protein purification, 114–115, 126–128
Gel filtration
  G_q proteins, 176–177, 180–181
  Stokes radii determination, 234
Gel staining, Gα subunit genes, 316
Genes
  ADP-ribosylation factors, 46–47
    Saccharomyces cerevisiae, 47, 59–63
  arl, 47
  cat, 148–149
  EagI, in Gα subunit mutation detection, 302–306
  gip2, in Gα subunit mutation detection, 307
  G proteins, inactivation, 356–366
  G-protein subunits, 345
  gsp, in Gα subunit mutation detection, 302–306
Gα subunit, 147, 269
  expression in S49 cells, 436–445
$G_{i2}\alpha$ subunit
  homologous recombination, 362

inactivation, targeting construct for, 359–360
targeting, 366–386
$G_s\alpha$ subunit
DGGE analysis, 308–320
genetic mapping, 319
mutations, activation, 317–319
hyg, 358
neo, 358–359, 363
NMT1, 257, 267
polyhedrin, 214
ras, in Gα subunit mutation detection, 302
tk, 438
Gene targeting
AB-1 cell line, 367–369, 371, 373, 377–386
blastocyst injection in, 366, 386
cell lines for, 371–372
cell rescue and expansion for, 385–386
cell samples for, 96-well plates, 374–375
Chinese hamster ovary cells for, 372–373
electrophoresis in, 375
electroporation in, 373–374, 377
embryonic stem cells for, 367–369, 371–373, 377–386
hybridization in, 376
probes for, preparation, 375–376
mini-Southern blotting in, 374–376
mouse strains for, 372
mouse tail biopsies for, 375
plasmids for, 372
plating, 377
selection, 373–374
replacement or insertion and excision, 367–368
Southern analysis in, 384–385
tissue culture for, 372–373
Genetic mapping, $G_s\alpha$ subunit, 319
S-(all-*trans*-Geranylgeranyl)cysteine, synthesis, 513–514
Giardia, ADP-ribosylation factors, 47
complementation, 61–63
Glucagon, stimulation of GTP hydrolysis, 15
Glutathione-S-transferase
fusion proteins, ADP-ribosylation factors as, 53–55

in Gα subunit expression vector, 160–161
Glycine, N-terminal, 254
G proteins, *see also specific proteins*
activation/deactivation cycle, 13–14
activation by mastoparans and cationic peptides, 26–37
α subunits, 45
affinity for G$\beta\gamma$ subunits, 256
amplified, cloning, 340–341
antipeptide antisera, 268–283
binding assay, 467–468
carboxy terminus, 2-azido-[^{32}P]ADP-ribose transfer, 95–99
chimeric, construction, 321–326
classification, 191, 295, 345
cloning, PCR, 327–344
cloning frequency analysis, 333
coexpression in *Escherichia coli*, 257–258
cross-linking with $\beta\gamma$ subunits, 236
expression
baculovirus-mediated, 212–226
in *Escherichia coli*, 146–164, 213, 258–259
host strains, 150–151
in S49 cyc^- cells, 443–444
in Sf9 cells, 218–219
time course, 149–150
vector construction, 147–149, 158–160
genes, 147, 269
vaccinia virus-mediated expression in S49 cells, 436–445
GTPase-deficient, single amino acid mutations, 325–326
immunoprecipitated, incorporated [α-^{32}P]GTP azidoanilide, determination, 287–288
$\beta\gamma$ matrix binding
affinities, 169–170
sites, 168–169
membrane-associated
agonist-dependent photolabeling, 285
isolation from solubilized membrane, 172–173
mutations, 295–308
construction, 321–326
detection, 296–303

SUBJECT INDEX

myristoylated
  purification, 260
  recombinant, purification, 262
myristoylation, 254-268
PCR, 298, 338-340
pertussis toxin-substrate
  heterogeneous, separation, 138
  separation, 137
photolabeled, immunoprecipitation, 283-294
  efficiency, 294
  specificity, 292-293
photolabeling
  agonist-dependent, 285
  receptor-dependent stimulation, 289-292
purification, 219-225
recombinant
  accumulation, peak time, 151
  purification, 212-226
subcloning strategy, 215-216
transfer vectors, preparation, 195-196
translated
  native molecular weight, 233-235
  structure and function, 229-239
translation *in vitro*, 226-239
$\beta\gamma$ subunits, 45
  adenylyl cyclase assay, 451-456
  affinity matrix containing, synthesis, 164-174
  amino acid sequences, 476, 478
    alignment, 484-485
  antibodies, characterization, 482-498
  antisera, 486-488
  biotinylated, 457-471
    binding assays, 466-468
    properties and characterization, 468-469
  bovine brain, S49 cyc^- adenylyl cyclase, 453-456
  cloning, probe design, 471-482
  cross-linking with α subunits, 236
  dimerization, 229-230, 235-237
  effect on $G_s\alpha$ subunit sedimentation rate, 243
  effect on mastoparan-stimulated G-protein function, 30
  Gα subunit affinity, 256
  immobilized, binding of α subunits, efficacy, 170

novel members, cDNA cloning, 474-480
pertussis toxin-substrate
  homogeneous, separation, 139
  separation, 137
preparation, 165-166
  adenylyl cyclase assay, 453-455
translated
  native molecular weight, 233-235
  structure and function, 229-239
translation *in vitro*, 226-239
classification, 345
effectors, isoprenylation, 509-519
G_i
  activation by mastoparan, 27
  ADP-ribosylation
    cholera toxin-catalyzed, 45
    pertussis toxin-catalyzed, 64
  α subunits
    activated, preparation, 110-130
    ADP-ribosylation, 24-25, 45
    cDNA, cloning, 257-258
    chimeras, construction, 321-326
    recombinant, purification, 152-158
  GTPγ binding activity, 38-44
  purification, 117-123, 125-127
G_{i1}, 171
  α subunits
    purification, 171
    SDS-PAGE resolution, 274
    subcloning strategy, 215-216
G_{i2}, 14, 171
  α subunits
    antipeptide antisera, 281-282
    chimeras, construction, PCR in, 321-326
    gene inactivation, 359-360, 376-380
    gene targeting, 366-386
    homologous recombinants, screening, 362-363
    myristoylated recombinant, purification, 259-262
    purification, 171
    recombinant, purification, 222-223
    SDS-PAGE, 274
    subcloning strategy, 215-216
  antipeptide antisera, 275
G_{i3}, 14
  α subunits
    antipeptide antisera, 281-282

homologous recombination in embryonic stem cells, frequency, 363
subcloning strategy, 215–216
antipeptide antisera, 275
G_o, 171
activation by mastoparan, 27, 29
ADP-ribosylation, cholera toxin-catalyzed, 45
α subunits
ADP-ribosylation, 24–25
calcium channel inhibition in GH_3 cells, 354–355
cDNA, cloning, 257–258
immunoprecipitation, 279
intrinsic tryptophan fluorescence, 410
purification, 152–158, 171
removal, 171–172
subcloning strategy, 215–216
GTPγ binding activity, 38–44
G_q, 171, 182
activated, purification, 175–178
activation by mastoparan, 27
α subunits, 278
activation and solubilization, 175
affinity chromatography, 203
classification, 191–192
purification, 196–200
recombinant, purification, 191–212
GTPγ binding activity, 39
G_s
activation by mastoparan, 27
adenylyl cyclase stimulation, 437
β-adrenergic receptor interaction sites, 434–436
α subunits
activated, preparation, 110–130
ADP-ribosylation, cholera toxin-catalyzed, 24
$\beta\gamma$ subunit interactions, 243
chimeras, construction, 321–326
expression in S49 cyc^- cells, Western analysis, 444
function, direct assay, 238
gene sequences, PCR and DGGE conditions, 315
genetic mapping, 319
GTPγ binding activity, 39
melting maps, 311–312
membrane association, assays, 245–254
membrane-bound, properties, assays, 251–254
mutations, detection by denaturing gradient gel electrophoresis, 308–320
peptides, isoproterenol-stimulated adenylyl cyclase, 434
point mutations, detection, 296–297
recombinant, purification, 152–158
sedimentation rate, G$\beta\gamma$ subunit effects, 243
single-base change combinations in encoding gene, 299
soluble, properties, assays, 242–245
translated *in vitro*, properties, 239–254
translation *in vitro*, 240–241
dissociation rate for GDP, 8
GTPγ binding assay, 35
preactivation, 116–117
purification, 117–127
G_{sa}, ADP-ribosylation, assay, 48
G_T, *see* Transducin
G_z, activation by mastoparan, 27
G_{11}, 182
α subunits
activation and solubilization, 175
immunoprecipitation, antiserum CQ2, 278
purification, 196–200
recombinant, purification, 191–212
avian, purification, 182–191
GTPγ binding activity, 39
purification, membrane solubilization, 186
G_{16}, α subunits
biochemical properties, 205
membranes and extracts, preparation, 207
purification, 205–212
recombinant
purification, 191–212
stabilization, 205, 207
Sf9 cells, recombinant, cellular location and activity, 205–206
γ subunits
alignment, 499–500
antibodies, 498–509

applications, 506–509
preparation, 499–504
specificity, characterization, 505
antisera, 488
cDNA clones, 477, 479
isoprenylation, 509–519
processing, 507–509
tissue distribution and localization, 506–507
genes, see also Genes
inactivation, 356–366
heterotrimeric, 345
purification, 179–181
histidine-tagged, expression and purification, 160–163
vector construction, 158–160
interactions
fluorescence assays, 409–423
with receptors, 423–436
prevention, antipeptide antisera, 280–282
stoichiometry, 10
interfacial region, synthetic peptide studies, 426–427
labeled, fluorescence studies, 417–419
labeling
pertussis toxin, 63–70
photoaffinity analog, 100–110
in membranes
agonist-induced [^{35}S]GTPγS binding to, 4–10
pertussis toxin labeling, 66–69
photolabeling, 107–108
predenaturation, effect on immunoprecipitation, 293
preparation, 133
solubilization and sucrose density gradient centrifugation, 10–11, 13
mutant, 321–326
in permeable cells, photolabeling, 108–110
pertussis toxin–substrate, see Pertussis toxin-substrate G proteins
receptor-activated, identification, 283–294
recombinant, purification, 152–158
solubilized, GTPγS binding assay, 38–44, 132
soluble, photolabeling, 104–105

structure, photoactivatable probes, 70–99
subunits, 45, 191, 212–213
binding assays, 466–468
functional assessment by antisense oligonucleotide injection, 345–355
genes, 345
interactions, 173–174
lipid modification, 238–239
GTP
analogs
photolabeling, 100–110
preparation, 102–103
binding assay, 225
receptor-stimulated, intrinsic fluorescence assays, 412–415
cholera toxin activation, 46
effects on agonist–receptor interaction affinity, in cyc^- membranes, 250–251
hydrolysis, 14
in membrane preparations, determination, 17–21
receptor-stimulated, 13–26
parameters, 21–26
inhibition, cAMP binding, 399–400
GTPase
activity
cAMP stimulation, 402–404
in HL-60 cell membranes, 20
intrinsic fluorescence assays, 412–415
mastoparan stimulation, 29
in plasma membranes, 18–19
receptor-stimulated, in membranes, 403–404
steady-state, 35–36
assay, reaction mixture for, 403
Gα subunits deficient in, amino acid mutations resulting in, 325–326
reduction by pertussis toxin-catalyzed ADP-ribosylation, 25
GTP azidoanilide, ^{32}P-labeled at α position
G protein α-subunit photolabeling
product immunoprecipitation, 292–294
receptor-dependent stimulation, 289–292
G protein α-subunits photolabeled with
immunoprecipitated, quantification of photolabel, 287–288

selective immunoprecipitation, 283–294
preparation, 284–285
GTPγS, see Guanosine 5'-O-(γ-thio)triphosphate
Guanidino compounds, ADP-ribosylation, cholera toxin-catalyzed, 45
Guanine nucleotides
  binding of Gα subunits, 230–232
  binding proteins, see G proteins
  effects on receptor-mediated GTP hydrolysis, 22, 24
Guanosine 5'-diphosphate, see GDP
Guanosine 5'-O-(γ-thio)triphosphate
  adenylyl cyclase activation by, 252–254
    in vitro, 404–406
  binding, 34–35
    cAMP stimulation, 401
    to cells and membranes, 396–397
    receptor-stimulated, 3–13
    to solubilized G proteins, 38–44, 132
  equilibrium exchange, 32–34
  $G_{16}\alpha$ stabilization by, 205, 207
  $G_2\alpha$ subunit activation by, 242
  guanylyl cyclase stimulation by, 406–407
  membrane treatment, 4–10
  in pertussis toxin labeling, 68
  phospholipase C regulation by, 406
  preactivation, for heterotrimeric G protein purification, 177–178, 181
  ^{35}S-labeled, binding, 34–35, 38–44, 225
    agonist-induced, 4–10
    receptor-stimulated, 11–13
Guanosine 5'-triphosphate, see GTP
Guanylyl cyclase, 408
  stimulation by GTPγS, 406–407
Guanylyl imidodiphosphate, 68, 108

H

Herpes simplex virus thymidine kinase, 358–359
Heteroduplexes, 313
  DNA-RNA, 349–350
  mRNA–antisense oligonucleotide, 349–350
Hexahistidine, sequences, in histidine-tagged protein expression vectors, 158
High-performance liquid chromatography
  in [adenylate-^{32}P]-azidonicotinamide adenine nucleotide synthesis, 89–90
  isoprenyl groups, 512
  peptides, 32, 429
Histidine, proteins tagged with, expression and purification, 160–163
  vectors for, construction, 158–160
Holotransducin
  extraction from bovine retinal membranes, 140–141
  purification, 75–76, 141–146
  purified
    [^{125}I]AIPP-ADP-ribosylated, proteolysis, 76
    [^{125}I]AIPP-ADP-ribosylation, 76
Homoduplexes, 313
Hormones
  effects on calcium channels in GH$_3$ cells, antisense oligonucleotides and, 348
  mitogenic effects, 296, 307
H$_6$pQE-60 vector, histidine-tagged protein expression, 158–160
H$_6$TEVGα, purification, 160–161
H$_6$TEVpQE-60 vector, histidine-tagged protein expression, 158–160
Hybridization
  gene targeting method, 376
  low-stringency conditions, for cloning related cDNAs, 475
  plaque, cDNA isolation by, oligonucleotide probes for, 472
  probes, for gene targeting, preparation, 375–376
2-Hydrazinoadenosine, synthesis, 82–84
Hydrolysis
  cholera toxin, 45
  GTP
    in membrane preparations, determination, 17–21
    receptor-mediated, 13–26
      effects of guanine nucleotides, 22, 24
      effects of magnesium ions, 22–23
      effects of monovalent cations, 21–22
      parameters, 21–26
    phosphatidylinositol 4,5-bisphosphate, 174
  steady-state, 35–36
Hydrophilicity, analysis in interfacial peptides, 426–427

SUBJECT INDEX

Hygromycin B phosphoribosyltransferase, gene, 358

I

IAEDANS, 417
IANS, 417
Immune response, to Gγ subunit antibodies, monitoring, 502–504
Immunoblotting
  antipeptide antisera, 276–277
  Gα subunit, 204–205, 211–212
  Gβγ subunit antibodies, 485, 489, 493–494, 503
    immune response monitoring, 502–504
  visualization techniques, 494
Immunocytochemistry
  Gβγ subunit antibodies, 497–498
  transfectants, 495
Immunoglobulin G, preparation from anti-G protein antipeptide antisera, 279–280
Immunological probes, G-protein fractions, 269
Immunoprecipitation
  antipeptide antisera, 277–279, 286
  effects of preclearing and predenaturation, 293
  Gβγ subunit antibodies, 495–496
  in metabolic radiolabeling, 266–267
  photolabeled Gα subunits, 283–294
    efficiency, 294
    specificity, 292–293
Immunosorbent assays, enzyme-linked, anti-G protein antisera specificity, 271–272, 502
Inositol lipid signaling cascade, 182
Inositol 1,4,5-trisphosphate, 174, 192
  isotope dilution assays, 390–394
  as second messenger, 398
Insertion vectors
  IV-1, in $G_{i2}\alpha$ gene inactivation, 376–380, 384
  recombinant, purification, with cesium chloride gradient centrifugation, 439–440
Interleukin 8, effects on GTP hydrolysis, 14, 16, 19
Iodoacetamidoaminonaphthalene derivatives, 416–417

Iodoacetamidofluorescein, 420
3-Iodo-4-azidophenylpropionic acid, 72
  radioiodinated, ADP-ribosylation
    pertussis toxin-catalyzed, 71, 77–78, 93
  purified holotransducin, 76
Iodoazidophenylpropionyl-NAD$^+$, radioiodinated
  as G-protein structure probe, 71
  synthesis, 71–81
Ion-exchange chromatography
  activated $G_s\alpha$ and $G_i\alpha$, 119, 121–122
  biotinylated Gβγ subunits, 461, 464–466
  in transducin subunit purification, 143–145, 450–451
Islet-activating protein, see Pertussis toxin
Isoprenylation, Gγ subunits and G-protein effectors, 509–519
Isoprenylcysteine methyl esters, synthesis, 513–514
Isoprenyl proteins
  rat retinal, radiolabeling in vitro, 515
  structural analysis, 512–513
  tritiated
    chromatographic identification, 517–519
    rod outer segments containing, preparation, 515–516
2′,3′-O-Isopropylidene 2-azidoadenosine, synthesis, 84–85
Isopropyl-β-D-thiogalactopyranoside, 52, 54, 148–149
Isoproterenol, 4, 8–10
Isotope dilution assays
  cAMP, cGMP, and inositol 1,4,5-trisphosphate, 390–394
  preparation, 390–391

K

Kanamycin resistance, 58, 149, 151, 258
Keyhole limpet hemocyanin, 489, 499
Kidney cells, COS
  Gα subunits, 256
  transfection mediated by calcium phosphate, 246–249

L

α-Lactalbumin, 115
Laemmli's sample buffer, 115, 218, 278

Leukotriene B_4, stimulation of GTP hydrolysis, 16
Lineweaver–Burk analysis, transducin ADP-ribosylation, 77–79, 93–94
Lipids
　mastoparan-stimulated G-protein function and, 28–30
　modification of G protein-subunits, 238–239
Lipofection, homologous recombination with, 439–440
Lithium, effect on receptor-mediated GTP hydrolysis, 22
Lubrol
　effects on mastoparan-stimulated G-protein function, 30
　in GTPγS binding assay, 40–41

M

Magnesium
　concentration, effect on GTPγS binding assay, 41–44
　effect on mastoparan-stimulated G-protein function, 30–31
　effect on receptor-mediated GTP hydrolysis, 22–23, 30
m-Maleimidobenzoyl-N-hydroxysuccinimide ester, 500
2-(4′-Maleimidylanilino)naphthalene-6-sulfonic acid, fluorescence studies, 416–419
Maltose-binding protein, 52–53
Mas7, 30
Mass spectrometry
　fast atom bombardment, peptide purity testing with, 429
　gas chromatography-coupled, isoprenyl groups, 512
Mast cells, effects of mastoparan, 37
Mastoparans
　analogs, structure–activity relationships, 27–28
　cellular effects, G-protein-mediated, 36–37
　G protein activation by, 26–37
McCune–Albright syndrome, 306–307, 309, 317

Membranes
　cerebral
　　G proteins
　　　pertussis toxin labeling, 66–69
　　　pertussis toxin-substrate, 131–139
　　　photolabeling, 107–108
　　Gα subunits, immunoprecipitation, 279
　　GTP hydrolysis, 15
　erythrocyte
　　G_{11} proteins, purification, 182–191
　　Gα subunit antipeptide antisera, 276–277
　　GTPγ binding, 8–9
　　GTP hydrolysis, 14
　　preparation, 184–186
　mononuclear, GTP hydrolysis, 15
　plasma, see Plasma membranes
Mercuric acetate, 74, 77–80, 98
Metarhodopsin II, 423–424
　assay, rhodopsin–G_T interaction, 430–431
　stabilization
　　by peptide analogs, 433, 435
　　by transducin, 428–432
Methionine, $G_s\alpha$ mRNA translation with, 244
Mevalonate, tritiated, preparation, 514–515
Microinjection, nuclear, antisense oligonucleotides, 347, 352–353
Mini-Southern blotting, in gene targeting, 374–376
Mitogenesis, and mutant proteins, 296
Molecular weight
　G-protein subunits synthesized in vitro, 233–235
　standards, 92, 115
Monoclonal antibody 4A, 424, 427, 430–431
Mononuclear cells, membranes, GTP hydrolysis, 15
Mouse
　cDNA library, direct cloning, 480–482
　$G_{i2}\alpha$ gene inactivation, 376–380
　lymphoma cells, see Tumor cells T-lymphoma S49
　strains for gene targeting, 372
　tail, biopsies for gene targeting, 375
　tissues, cDNA, amplification, 477, 479
Muscarinic cholinoceptor agonists, stimulation of GTP hydrolysis, 15

SUBJECT INDEX

Mutations, point, Gα subunits, detection, 296–297
Myokinase, in photoactivatable probe synthesis, 88–89, 93
Myometrium, rat, GTPγ binding in, 5–6
Myosin, as molecular weight calibration standard, 92
Myristic acid, 254
  metabolic radiolabeling, 263–268
Myristoylation
  ADP-ribosylation factors, 47
  Gα subunits, 254–268
Myristoyl-CoA, yeast, construction, 55–56
N-Myristoyltransferase, 47, 256
  coexpression with G-protein α-subunits
    in *Escherichia coli*, 257–258
    in *Saccharomyces cerevisiae*, 149
  expression factor, 55–56

N

NAD⁺
  derivatives, as G-protein structure probes, 70–99
  hydrolysis, 45
  kinetic constants, determination, 91–92
  photoaffinity label, preparation, 72
  radiolabeled, ADP-ribosylagmatine formation, measurement, 48–49
β-NAD⁺, α-ADP-ribosylarginine formation from, 45
NAD⁺ glycohydrolase, 63–64
Neomycin phosphoribosyltransferase, gene, 358–359, 363
Neomycin resistance, encoding gene, and $G_{i2}α$ gene inactivation, 376–380
Neoplasia, human, mutant G-protein α subunits in, identification, 295–308
Neurospora, Gα subunits, PCR amplification, 327
Neutrophils, human
  GTPγ binding, 5–6
  GTP hydrolysis, 14, 16, 19, 25
  membranes
    GTPase activity, 19
    preparation, 16–17
Nicotinamide, hydrolysis, 45
Nitrocellulose
  filters, for GTPγ binding assay, 13–14, 34
  sheets
    Nitroplus, 503
    for plasma membrane targeting, 248–249
Northern blot analysis, mutant cell lines, 364
Nuclear magnetic resonance, transferred nuclear Overhauser effect, 425, 432
Nucleic acids
  antisense
    application, 346–349
    suppression, 346–355
  transfer, 347
Nucleotide exchange
  assays *in vitro*, 32–36
  mastoparan and, 27–28
  detergent and lipid effects, 28–30
Nucleotides
  free, removal, transducin subunits, 141–142
  guanine-containing
    binding, Gα subunits, 230–232
    receptor-mediated GTP hydrolysis, 22, 24

O

Octylglucoside, Mono Q chromatography, in Gα subunit purification, 209–210
Oligonucleotide primers, 325, 473
  amino acid sequence specification, 472–474
  degenerate, 334
  design, 328–330
  Gα subunit cloning, 327–344
  sequence, 330
  for Gβγ subunit isolation, 475–477
Oligonucleotides
  antisense
    application, 346–349
    concentration, 351–352
    design, 349–352
    nuclear microinjection, 347, 352–353
    preparation, 351
    suppression, 345–355
  as cDNA plaque hybridization probes, 472
  for Gα subunit chimera construction, 323–325
  as Gβγ subunit cloning probes, 471–482

SG09, thyroid sample amplification, 303, 305
Oncogenes
  G-protein, 295–308
  p21^{ras}, activation by mastoparans, 27, 30
Opioid receptor agonists, stimulation of GTP hydrolysis, 15, 21
Orthophosphate, 35
Osteodystrophy, Albright hereditary, 309, 316–317
Ovalbumin, 92, 115

P

Partial specific volume, determination, 234–235
Peptide receptor agonists, stimulation of GTP hydrolysis, 16
Peptides
  analogs, metarhodopsin II stabilization, 433, 435
  antisera, specificity, 494–495
  conjugation, in Gγ subunit antibody preparation, 499–501
  G-protein-activating, 26–37
    purification, 31–32
  as G-protein–receptor interaction probes, 423–436
  G$_s\alpha$ subunits, effects on isoproterenol-stimulated adenylyl cyclase, 434
  purification and evaluation, 427–429
  sequences
    selection for Gγ subunit antibody preparation, 499–501
    synthetic, 270–271
  synthesis, 427–429
    amino acid sequences for, 483–490
  synthetic
    in protein interfacial region studies, 426–427
    in protein–protein interaction mapping, 424–425
Pertussis toxin, 63
  A component, 63–64
  activation, 64–66
  ADP-ribosylation, 24–26, 71, 117, 126–128, 132–133, 268
    (N)-[^{125}I]AIPP, transducin, 77–79, 93–94

G-protein subunits, 236–238
NEM, 9
ADP-ribosyltransferase activity, 63–64
B component, 63–64
labeling of G proteins, 63–70
substrates, 64
Pertussis toxin-substrate G proteins, 131–139
  α subunits
    heterogeneous, separation, 138
    separation, 137
  $\beta\gamma$ subunits
    homogeneous, separation, 139
    separation, 137
  purification, 133–134
2,2'-p-Phenylenebis(4-methyl-5-phenyloxazole), 5
Phosphate buffer pellet assay, cAMP binding, 395
Phosphatidylethanolamine, 183, 194
Phosphatidylinositol 4,5-biphosphate, 183, 194
  hydrolysis, 174
Phosphatidylserine, 183
Phosphodiesterase, snake venom, 74
Phosphoglycerate kinase, 359
Phospholipase C, 25, 182, 192, 408
  assay, 176, 194–195
  in Gα subunit purification, 193
  in GTPγ binding assay, 35
  GTPγS regulation, 406
  stimulation by G$\beta\gamma$ subunits, 452
Phospholipids
  reconstitution assay, 183
  vesicles
    phospholipase C in, measurement, 194
    rhodopsin incorporation, 412–414
Phosphorylase b, 92, 115
Photoaffinity analogs, G protein labeling, 100–110
Photocross-linking, 2-azido-ADP-[^{32}P]-ribosylated transducin, 96–98
Photolabeling
  G proteins
    in membranes, 107–108
    in permeable cells, 108–110
    soluble, 104–105
  Gα subunits
    agonist-dependent, 285

receptor-dependent stimulation, 289–292
GTP analogs, 100–110
Photolysis, 2-azido-ADP-[^{32}P]-ribosylated transducin, 96
Pituitary cells, calcium channel inhibition, antisense oligonucleotide effects, 345–355
Plaques
  hybridization, cDNA isolation by, oligonucleotide probes for, 472
  purification
    recombinant baculoviruses, 218
    recombinant vaccinia virus, 440–441
Plasma membranes
  GDP hydrolysis, receptor-stimulated, 13–26
  G$_s\alpha$ subunit association with
    in vitro reconstitution assays, 245–246
    in vivo targeting, 246–249
  GTPase activity, 18–21
  GTP hydrolysis, 15
  hepatic
    G proteins
      activation, 175–178
      GTPγ binding activity, 38
      heterotrimeric, 179–181
      purification, 174–181
    GTP hydrolysis, 15
  pancreatic, rat, GTP hydrolysis, 15
  preparation, 16–17, 117–118
Plasmids
  ADP-ribosylation factor expression, construction, 60–61
  Bluescript, 227
  gene targeting, 372
  G-protein subunits
    for coexpression in Escherichia coli, 257–258
    construction, 147–149, 158–160
    for in vitro translation, 227–228
  isolation, for sequencing, 342–343
  npT7-5, 152, 258
  pBB131, 257–258
  pQE-6, 147–150, 257–258
    cell culture, 151–152
    construction, 149
    host strains, 150
  pREP4, 149, 151–152

pSP64t, 227
pVL1392/pVL1393 transfer vector, 195
Qiagen vector pQE-60, 158
replacement vector RV-6, 376–380, 384
Platelet-activating factor, stimulation of GTP hydrolysis, 16
Platelets, human
  Gα subunits, immunoprecipitation, 290–292
  GTPγ binding, 5–6, 10
  GTP hydrolysis, 15
Plating, gene targeting, 377
  selection, 373–374
Point mutations, Gα subunits, detection, 296–297
Polyisoprenylation, G-protein subunits, 238–239
Polymerase chain reaction
  amplification
    ADP-ribosylation factor gene
      ligation-independent cloning, 53–54
      Saccharomyces cerevisiae, 60
    Gα subunits, 338–340
    products, cloning, 340–341
  chimera construction, Gα subunits, 321–326
  clones, screening, 341–342
  cloning
    cDNA, directly from library, 480–482
    Gα subunits, 327–344
  conditions for G$_s\alpha$ subunit gene sequences, 314–315
  contamination, 339
  gap repair-dependent, in clone identification, 370–371
  G-protein oncogenes, 295–308
  Gα subunits, 298
  isolation
    cDNA, 472–474
    G$\beta\gamma$ subunit family novel members, 475–480
  primers, 325, 473
    degenerate, design, 328–330
    targeting Gα subunits, 332–336
  products, cloning, 329–331
  template, 473–474
    crossover, 333–334
Polyoxyethylene 10 lauryl ether, 168

Polypeptides, ³H-isoprenylated, isolation, 517
Potassium, in receptor-mediated GTP hydrolysis, 22
Prostaglandins, stimulation of GTP hydrolysis, 15
Protein kinase A, 387
Proteins
  electrophoresis, 77
  fluorometric determination, 113
  fusion, ADP-ribosylation factor
    with glutathione S-transferase, 53–55
    with maltose-binding protein, 52–53
    purification, 54–55
  guanine nucleotide-binding, see G proteins
  histidine-tagged, expression and purification, 160–163
  vectors for, construction, 158–160
  islet-activating, see Pertussis toxin
  isoprenylated, tritiated
    identification, 517–519
    rod outer segments containing, preparation, 515–516
  N-myristoyltransferase expression factor, 55–56
Proteolysis
  [¹²⁵I]AIPP-ADP-ribosylated purified holotransducin, 76
  trypsin
    $G_2\alpha$ subunits translation products, 242
    patterns, analysis, 229–233
Pyrophosphate, 35

R

Rabbit
  antisera, 490
    polyclonal, Gα subunit purification, 193
  Gγ subunit antibody preparation in
    bleeding schedule for, 501–502
    immunization for, 501–502
Radiolabeling
  metabolic, myristic acid, 263–268
  rat retinal isoprenyl proteins, in vitro, 515
Receptor coupling, membrane-bound $G_s\alpha$ subunits, 251–253
Reconstitution assay, G_{11} proteins, 183–184
Replacement targeting construct, creation, 358–361, 367–369
Resonance energy transfer, 411–412
  between donor–acceptor pairs, as protein–protein interaction readout, 419–420
  fluorescence measurements, 420–422
Restriction enzyme-mediated integration, 408
Reticulocyte lysates
  $G_s\alpha$ subunits, in vitro translation, 240
  rabbit, 228
Retina
  bovine, rod outer segments, 139
  Gγ subunit antibodies, 506–507
  membranes
    holotransducin extraction, 140–141
    purification, 140
    Tβγ subunits, preferential extraction, 141, 449
  rat
    isoprenyl proteins, in vitro radiolabeling, 515
    rod outer segments, ³H-isoprenylated protein-containing, preparation, 515–516
  rod photoreceptors, signal-transducing proteins, isoprenyl groups, 518–519
Rhodopsin, 139, 410
  α_1-βγ systems, reconstitution, 414–415
  α_T interaction sites, 429–432
  –G_T interaction
    block by transducin α subunit amino-terminal peptides, 431–432
    metarhodopsin II assay, 430–431
    structural basis, 435–436
  incorporation into phospholipid vesicles, 412–414
  in protein–protein interaction studies, 423–424
RIBI adjuvant, 502
RNA, messenger, G-protein subunit, in vitro translation, 228, 241, 244

S

Saccharomyces cerevisiae
  ADP-ribosylation factors, 47, 59–63
  complementation, 61–63

expression, vector construction for, 55–57
alcohol dehydrogenase, in NAD$^+$ photoaffinity labeling, 72
myristoyl-CoA, construction, 55–56
N-myristoyltransferase, coexpression, 149
NMT1 gene, 257, 267
sporulation, 61–63
*Sac*I digests, targeting vectors, structure, 382–384
Scintillation cocktail, 5
Second messenger responses, cAMP-induced, 398–399
Secretin, stimulation of GTP hydrolysis, 15
Sedimentation coefficient, determination, 234–235
Silicone oil assay, cAMP binding, 395–396
Silicophosphate, 35
Silver staining, 115
Gα subunit genes, 316
Sodium, in receptor-mediated GTP hydrolysis, 21–22
Sodium chloride, regulation of receptor-stimulated GTPγS binding by G proteins, 6–8
Sodium cholate
GTPγS binding assay, 40–41
Mono Q chromatography, in Gα subunit purification, 211–212
Sodium dodecyl sulfate, in pertussis toxin activation, 65–66
Solubilization, membrane proteins, 10–11
Solubilized systems, agonist-induced [^{35}S]GTPγS binding in, measurement, 10–13
Somatostatin
calcium channel inhibition, antisense oligonucleotide effects, 345–355
stimulation of GTP hydrolysis, 15, 21–22
Southern analysis, *see also* Mini-Southern blotting
gene targeting, 384–385
isolated DNA, 362
Soybean trypsin inhibitor, 115
Spectrophotometry, DNA melting study, 309
Spodoptera frugiperda, see Cell lines, Sf9
Sporulation, yeast, 61–63
SP6 RNA polymerase, 227

Staphylococcus aureus, 267
Stem cells, embryonic
chimeras, 366
gene targeting, 367–369, 371–373, 377–386
heterozygosity, loss, 364
heterozygous, transfection and selection, 361
homologous recombinants, 362–363
Stokes radius, determination, 233–234
Streptavidin-agarose beads
Gβγ binding, 467–469
preparation, 466–467
Subcloning, Gα subunits, 215–216
Sulfo-MBS, in βγ affinity matrix synthesis, 167–168

T

TA cloning kits, 331
Taq polymerase, 331, 338
Temperature regulation, receptor-stimulated GTPγS binding by G proteins, 6–8
TEV protease, in Gα subunit expression vector, 160–161
Thioglycosidic bonds, chemical cleavage, 77, 98
Thrombin, cleavage, 54–55
Thromboxane, stimulation of GTP hydrolysis, 15
Thymidine kinase, 358–359, 366
gene, 438
Thyroid samples, amplification with oligonucleotides, 303, 305
Thyroid-stimulating hormone, mitogenic effects, 296, 307
Tissue culture, for gene targeting, 372–373
T_m, 310–311
Toluene, 5
Tosylphenylalanyl chloromethyl ketone, 232
Transducin, 139, 410
activation by mastoparan, 27
ADP-ribosylated, trypsin cleavage, 78, 80
ADP-ribosylation
2-azido-[^{32}P], time course, 91
cholera toxin-catalyzed, 45

pertussis toxin-catalyzed, 77–79, 93–94
α subunits, 254
  ADP-ribosylation
    cholera toxin-catalyzed, 24
    pertussis toxin-catalyzed, 25
  amino-terminal peptides, rhodopsin-G_T interaction, 431–432
  purification, 145–146
  rhodopsin interaction sites, 429–432
  2-azido-[^{32}P]ADP-ribosylated
    photolysis, 96
    trypsin treatment, 96–97
  2-azido-[^{32}P]ADP-ribosylation
    photocross-linking, 96–98
    time course, 91
βγ subunits, 449
  fluorescence studies, 417–419
  modified, separation, 451
  preferential extraction, rod outer segment membranes, 141, 449
  purification, 449–451
dissociation rate for GDP, 8
extracts, protein content, 141–142
Meta II stabilization by, 428–432
protein–protein interaction studies, 423
purification, 75–76, 139–146
purified
  [^{125}I]AIPP-ADP-ribosylated, proteolysis, 76
  [^{125}I]AIPP-ADP-ribosylation, 76
subunits
  extraction, 140–142
  purification, 141–146
Transfectants, for immunochemical experiments, 495
Transfer vectors
  baculovirus, construction, 214–216
  Gα subunits, preparation, 195–196
Translation, G-protein subunits, *in vitro*, 226–241
  mRNA transcription step, 228
  plasmids for, 227–228
  products
    native molecular weight, 233–235
    structure–function analysis, 229–239
    trypsin digestion, 229–232
    Ultrogel AcA 34 column chromatography, 229–231
  SDS–PAGE, 228–229

Tritium, detection in gels, 267
Trypsin
  cleavage
    ADP-ribosylated transducin, 78, 80
    G proteins, 231–233
  digestion, *in vitro* translation products, 229–232
  proteolysis
    $G_2α$ subunits translation products, 242
    patterns, analysis, 229–233
    treatment of 2-azido-[^{32}P]ADP-ribosylated transducin, 96–97
Tryptophan fluorescence, intrinsic, 410–411
  measurement, 414–416
Tubulin, labeling, 104–107
Tumor cells
  glioma, C_6, Gα subunit radiolabeling, 108–109
  human leukemia, GTPγ binding, 5–6
  neuroblastoma, SH-SY5Y, Gα subunits, immunoprecipitation, 291
  neuroblastoma × glioma, NG108-15, GTP hydrolysis, 15, 21, 25
  promyelocytic leukemia, HL-60
    G-protein–mediated mastoparan effects, 37
  membranes
    GTPase activity, 20
    GTPγ binding, 5–7, 10
    GTP hydrolysis, 14, 16, 22–24
    preparation, 16
  T-lymphoma, S49
    cyc^-
      adenylyl cyclase activity in reconstituted membranes, 252–254
      adenylyl cyclase assay in, 452–453
      agonist-receptor interactions, 249–250
      Gα subunit expression, 443–444
      $G_sα$ subunits, 240, 437
      GTP hydrolysis, 15, 22
      membranes, reconstituted, 251–254
      membranes, reconstitution *in vitro*, 245–247
      preparation, 112–113, 452–453
      recombinant vaccinia virus infection, 442–443
      vaccinia virus Gα gene expression, 436–445

unc, Gα subunit antipeptide antisera, 270

V

Vaccinia virus
  Gα gene expression in S49 cells, 436–445
  infection protocol, 439
  recombinant
    construction, 438–443
    detection, 440–441
    infection
      membrane preparation for, 443
      in S49 cyc^- cells, 442–443
    purification, 441
    stock, partially purified, preparation, 441–442
Vent polymerase, 331
Vibrio cholerae, 44, 268

W

Wasp venom, 26
Waters 650E Advanced Protein Purification System, 75, 219
Western blot analysis
  anti-G protein antisera specificity, 272–277
  in G-protein purification, 104–105
  $G_s\alpha$ subunit expression in S49 cyc^- cells, 444
  heterotrimeric G_q proteins, 180–181
  mutant cell lines, 364

X

XAR-5 film, fluorographic, 267
XRS Omni-Media 6cx scanner, 469–470

Y

Yeast, *see Saccharomyces cerevisiae*

ISBN 0-12-182138-2

283570